Singularities, Bifurcations and Catastrophes

Suitable for advanced undergraduates, postgraduates and researchers, this self-contained textbook provides an introduction to the mathematics lying at the foundations of bifurcation theory. The theory is built up gradually, beginning with the well-developed approach to singularity theory through right-equivalence. The text proceeds with contact equivalence of map-germs and finally presents the path formulation of bifurcation theory. This formulation, developed partly by the author, is more general and more flexible than the original one dating from the 1980s. A series of appendices discuss standard background material, such as calculus of several variables, existence and uniqueness theorems for ODEs, and some basic material on rings and modules. Based on the author's own teaching experience, the book contains numerous examples and illustrations. The wealth of end-of-chapter problems develop and reinforce understanding of the key ideas and techniques: solutions to a selection are provided.

James Montaldi is Reader in Mathematics at University of Manchester. He has worked both in theoretical aspects of singularity theory as well as applications to dynamical systems, and co-edited the books: *Geometric Mechanics and Symmetry: The Peyresq Lectures* (Cambridge, 2005), *Peyresq Lectures in Nonlinear Systems* (2000), and *Singularity Theory and its Applications Part 1* (1991).

Singularities, Bifurcations and Catastrophes

James Montaldi
University of Manchester

CAMBRIDGE
UNIVERSITY PRESS

University Printing House, Cambridge CB2 8BS, United Kingdom

One Liberty Plaza, 20th Floor, New York, NY 10006, USA

477 Williamstown Road, Port Melbourne, VIC 3207, Australia

314–321, 3rd Floor, Plot 3, Splendor Forum, Jasola District Centre, New Delhi – 110025, India

79 Anson Road, #06–04/06, Singapore 079906

Cambridge University Press is part of the University of Cambridge.

It furthers the University's mission by disseminating knowledge in the pursuit of education, learning, and research at the highest international levels of excellence.

www.cambridge.org
Information on this title: www.cambridge.org/9781107151642
DOI: 10.1017/9781316585085

First published 2021

Printed in the United Kingdom by TJ Books Limited, Padstow Cornwall

A catalogue record for this publication is available from the British Library.

ISBN 978-1-107-15164-2 Hardback
ISBN 978-1-316-60621-6 Paperback

Contents

Preface

HOW DO WE CREATE MATHEMATICAL MODELS for systems that have discontinuous jumps? This was the fundamentally new question addressed by the topologist René Thom in his book on morphogenesis [111]. Although the beginnings of the mathematics of singularities had been around for a decade or so before, that book gave an important impetus to the later development of the subject.

What are singularities? In as brief a description as is possible, they are points where the hypotheses of the inverse or implicit function theorems fail. Singularity theory is of course the study of such singularities, and *bifurcation theory* and *catastrophe theory* form part of singularity theory, and deal with how aspects of functions and solutions of systems of nonlinear equations, change in families. Over the decades, singularity theory has become a very broad subject, and of necessity this text cannot cover it all. The principal aim of this book is to introduce readers to the techniques of singularity theory with an eye on applications to bifurcation theory, although specific applications to the natural sciences or other areas are not covered in any detail.

The book is divided into four parts. The first is dedicated to catastrophe theory, which we take to mean the study of (degenerate) critical points of smooth real-valued functions and their deformations. The second part of the book is on the singularity theory of nonlinear equations, studied using contact equivalence, while the third considers its applications to bifurcation theory. We begin with catastrophe theory because, while more restricted in application, it is technically more straightforward, and contains all the main ideas of the more general theories. The final part, the appendices, contains necessary background material as well as solutions to a selection of the problems that are found at the end of each chapter.

A very brief history To quote Poston and Stewart from the start of their book [98], '*in the beginning there was Thom*'. In his fundamental book on the subject [111], René Thom described how a large number of phenomena could be understood in terms of a few models of families of functions and their critical points, which he called ***catastrophes***. This is the subject that Zeeman and others dubbed *catastrophe theory*. The work of Thom contains the essence of all the ideas contained in this text.

As part of his programme, Thom provided a classification of the *elementary catastrophes* – those of codimension at most 4. In the 1970s, Vladimir Arnold [3, 6] produced a classification up to codimension 10 and proved some important results relating this classification to other areas of mathematics and especially to Dynkin diagrams and hence to Lie algebras.

Prior to Thom's work, Hassler Whitney in the 1940s [116, 117] found what can be seen as the first results of singularity theory, where he studies the 'essential' singularities for a map of an n–dimensional manifold into $\mathbb{R}^{2n-1}$. In the 1950s, Whitney classified the local structure of mappings of the plane to itself [118]. More or less concurrently, Thom [110] proved some general results about singularities of smooth maps between manifolds of general dimensions, and more particularly when the target dimension is 2.

In the late 1960s and early 1970s, following the work of Thom, John Mather wrote a series of six fundamental and technical papers [76] which moved the fledgling subject of singularities of mappings to a different level. Since then, there have been many advances and refinements, most based on the ideas of Mather, enhanced by methods of commutative algebra and algebraic geometry.

Probably the single most important application of singularity theory is to the foundations of bifurcation theory, an approach developed principally by Martin Golubitsky, starting with two seminal papers he wrote with David Schaeffer at the end of the 1970s [48], which then developed into two monographs [49, 50], the second on symmetric bifurcations written together with Ian Stewart.

Thom's idea Many physical, biological and other systems can be modelled by equations (often differential equations) depending on a number of external parameters, and it is important to understand how the solutions (or equilibrium points) of the equations depend on these external parameters. Thom considered systems governed by a potential function V to be of particular importance, where the stable states of such a system are determined by local minima of V.

As parameters are varied, the number of local minima of V can change, and in particular, if there is more than one local minimum, then the system must somehow decide which one it occupies, possibly dependent on its history. Generally speaking, a small change in the parameter values will result in a small change in the state of the system, and the system has a continuous response to the parameters. However, at some parameter values there can be abrupt changes, where the state of the system is forced to jump from one local minimum to another, and these Thom called *catastrophes*. We describe these in more detail at the end of Chapter 2.

Thom's profound insight is that it is possible to explain and model these abrupt changes, and moreover there are only a few possible models (fold, cusp, swallowtail, umbilic, etc.), and, perhaps surprisingly, which of these models are relevant to a particular system depends more on the number of parameters and less on the number of variables modelling the system. Finding the form of these models is the principal aim of catastrophe theory, and the theory predicts that these give a description of how systems 'generically' respond to changes in parameters. 'Generic' here means that these models are likely to appear in any such system, and they are robust to small changes in the model, or, from the opposite perspective, other catastrophes are unlikely to occur, and if they do, they will not be robust.

Thom himself [111], and particularly Christopher Zeeman [120, 121], developed many applications of catastrophe theory in the 1970s. Some of these were well–founded scientifically, while others (such as using the cusp catastrophe to model when dogs bark or when prisoners riot) have more dubious foundations. These less well–founded applications provoked a certain amount of criticism, and led to catastrophe theory being branded in some quarters as pseudo-science. However, these critics often missed the point Thom was making, in particular that the possible responses of a system are independent of whether the equations represent the system precisely, or even whether one has modelled all the ingredients of the system. The language of folds and cusps is now part of everyday science, vindicating, I believe, Thom's initial vision.

The text As mentioned above, the text is divided into four parts.

The first part presents catastrophe theory, and the content is not very different from that of the book by Bröcker [12], although the presentation is different, as befits a textbook for undergraduates. In this part, we prove the fundamental theorems on finite determinacy and state the versal unfolding theorem in the context of right equivalence of functions (germs). The proof of the versal unfolding theorem is given in Part II.

I should point out that my use of the term 'catastrophe theory' is not universal. I use it for the study of critical points of functions, rather than solutions of more general equations. In a single variable the distinction is not essential: if one is interested in solving an equation $f(x) = 0$, then defining V to be any integral of f (a solution of $V'(x) = f(x)$), solving the given equation is the same as finding critical points of V. In more than one variable, the two notions are of course not equivalent (such a V exists only if, as a vector field, $f(x)$ is conservative).

Part II concerns the singularity theory of solutions of systems of (nonlinear) equations, studied using the technique of contact equivalence of map germs. Again, we discuss the relevant theorems on finite determinacy and versal unfoldings. We also discuss some of the known classification of map germs up to contact equivalence. Much of this material can be found in the literature, though not in text book form. This part also includes a discussion of the Malgrange preparation theorem (in a form due to Mather), and a proof of the versality theorems. It concludes with a short digression on left–right equivalence of maps.

Part III is more novel and describes in some detail the so-called *path approach* to bifurcation theory. This approach was first considered by Golubitsky and Schaeffer in the 1970s, but the underlying singularity theory was not available at the time and so they used their distinguished parameter approach, which does work well for 1–parameter bifurcations. In the 1980s, Damon introduced a new equivalence relation, called $\mathcal{K}_V$-equivalence, and it was pointed out by your current author [85] that this can be used to define a path approach to bifurcation theory. This was followed up in particular by Furter, Sitta and Stewart in a series of papers on symmetric bifurcations; see for example [37, 38, 39] and references therein.

Part III begins with a description of this path approach to bifurcation theory in the context of a few well-known bifurcations such as the pitchfork and hysteresis bifurcations. We then develop Damon's $\mathcal{K}_V$-equivalence for map germs relative to a variety V in the target. We illustrate the method by classifying all low–codimension bifurcation problems with up to three parameters, by classifying the paths relative to the discriminant of a versal unfolding. This part finishes with a discussion of some bifurcation problems having additional structure, or constraints, which is included to illustrate the flexibility and general methodology of this path approach.

The text concludes with a series of five appendices. The first four of these contain standard background material, such as calculus of several variables, existence and uniqueness theorems for ODEs, and some basic material on rings and modules. Appendix E contains solutions to a selection of the problems given at the ends of the chapters.

What's not included While I have provided proofs of almost all results, the Malgrange preparation theorem is not proved. This central theorem is used to prove the versal unfolding theorems, and is discussed in Chapter 16; the interested reader can find proofs of it in the literature cited in that chapter.

It is customary to present singularity theory equivalence relations as the result of group actions. This is used firstly to develop techniques for classifying singularities, and secondly to provide converse results to the finite determinacy theorems. In this text, we avoid any discussion of group actions, and leave the equivalence relations as equivalence relations. For the classifications, particularly in Part III, we use an approach based on 'constant tangent spaces'. The converse results to finite determinacy (e.g. Theorem 5.17) are stated without proof.

It might be thought natural to include a discussion of manifolds as the natural spaces on which smooth maps are defined. We simplify matters by considering only submanifolds of Euclidean space (of course, with no loss of generality thanks to Whitney's embedding theorem). The material on submanifolds, along with the implicit function theorem and its friends, is presented in the second Appendix. A proof of the inverse function theorem using the homotopy method is given in Chapter 5 as a precursor for the proof, by the homotopy method, of the finite determinacy theorem for right equivalence.

Although Part III is dedicated to the study of bifurcations, we do not study dynamical properties of bifurcations in dynamical systems. Many excellent texts cover this, for example the books of Guckenheimer and Holmes [55] and Kuznetsov [67].

Throughout the text, attention is restricted to the singularity theory of smooth real maps and their germs. However, all the finite determinacy and unfolding theorems carry through for the complex analytic setting – one simply replaces the ring $\mathcal{E}_n$ of smooth germs with the ring $\mathcal{O}_n$ of holomorphic germs in both statements and proofs. The only specifically (complex) analytic results included are a summary of some of the beautiful geometric criteria for finite determinacy. I make no apology for not giving the proofs, which would require a development of sheaf theory and Hilbert's Nullstellensatz, and would be a book of its own; the reader can find some details in the survey article of Wall [115] and in the books of Ebeling [34], of Greuel, Lossen and Shustin [53] and of Mond and Nuño Ballesteros [81].

We assume throughout that all maps are C^∞. The literature contains numerous results about the cases where the maps in question are only C^r for some $r < \infty$. An important question in this direction is topological equivalence of smooth maps. It was originally hoped that stable maps would be dense (in an appropriate topology) in the space of C^r maps ($r > 0$) between any pair of manifolds, but Mather showed that this was not the case: it depends on the dimensions of the manifolds in question (Mather's 'nice dimensions' are where the stable maps are dense; see [76, VI]). Instead, it was necessary to introduce the idea of topological stability of smooth maps, namely that nearby maps were topologically equivalent to the given one. We say nothing about topological equivalence, and the interested reader can find much information in the monograph of du Plessis and Wall [94].

One of the important properties (invariants) of a singularity is its local topology, particularly in the complex setting, including for example the relation between the codimension of the singularity and the homology of its smooth fibres. This text has no discussion of this, but the interested reader will find it a central theme in the recent book by Mond and Nuño Ballesteros [81].

Another omission is symmetry. Except for brief treatments of catastrophe theory for even functions (Chapter 9) and bifurcations of odd maps (Section 23.1), this book does not deal with the interesting issue of symmetry, which is particularly important in bifurcation theory. Discussion of symmetric systems can be found in the book of Golubitsky, Stewart and Schaeffer [50] and the more recent book of Golubitsky and Stewart [51].

There are presumably other omissions omitted from this list of omissions.

Teaching the course I have taught most of Part I of this text (on critical points of functions) as a 20-hour lecture course, to students in the third year of a UK undergraduate mathematics degree programme. As a more advanced 30-hour master's-level course, I have taught the majority of Parts I and II, but never thus far, Part III. Each chapter ends with a collection of problems, and a selection of solutions are given in the final appendix. The reader will notice that the chapters for the earlier parts have many more problems (and more solutions), corresponding to the teaching experience of the author.

Technology The text was typeset with LaTeX (of course). The diagrams were produced using two pieces of software. The line drawings (e.g. Figure 7.1) were made using the LaTeX package *pstricks* developed principally by Herbert Vöss, and part of the standard LaTeX distributions. The surfaces similar to those in Figure 7.3 were produced using the software Surfer [107] for visualizing algebraic surfaces and developed under the auspices of the Mathematisches Forschungsinstitut Oberwolfach. Finally, two pieces of software were used to help with or check calculations: Macaulay2 (which is free) and Maple (which is not).

Notation The index begins with a list of notation, with the page where each symbol is introduced. The symbol := (as in $A := x$) means 'is defined to be' (A is defined to be x). Proofs end with the symbol ✔, remarks with the symbol ❞, definitions with ✯ and examples with ✐. Those problems that end with (†) have solutions, some in greater detail than others, in Appendix E.

Thanks I thank the many people whose collaborations, discussions and seminars influenced my approach to singularity theory and its applications. I also thank students of my lecture courses for comments on the original lecture notes that helped improve this final version.

From a personal historical perspective, I would like to acknowledge in particular David Mond and Mark Roberts, with whom I learned singularity theory as a PhD student at the University of Liverpool in the early 1980s. Also influential were others, staff and visitors, at Liverpool at the time, including Bill Bruce, Jim Damon, Terry Gaffney, Peter Giblin, Chris Gibson, Ian Porteous and Terry Wall. Others from whom I have learned much include Marty Golubitsky, Andrew du Plessis, Ian Stewart, Duco van Straten and Christopher Zeeman.

There are many others whose seminars and occasional conversations have also been influential and should be mentioned, but I would be afraid of missing someone out from the long list. Thanks are also due to them.

I thank in particular Terry Gaffney for discussions on Martinet's theorems which contributed to Chapter 17.

Finally, since, as I write, no–one else has seen this text, all and any errors are indeed my own, and I cannot pretend to share the blame with anyone. I just hope there are not too many.

Glossop
August, 2020

1

What's it all about?

MATHEMATICAL MODELS OF NATURE almost always involve solving equations (often differential equations) and these models, and their equations, frequently depend on external parameters. Examples of parameters might be the temperature of a chemical reaction, or the load on a bridge, or the tension in a rope (will it snap?), or the temperature of the ocean for plankton populations. In such models, it is important to understand how phenomena associated to that model can change as the parameters are varied. Usually one finds that a small change in the value of the parameters produces a corresponding small change in the (set of) solutions of the equation. But occasionally, for particular values of the parameter there is a more radical change, and such changes are called *bifurcations*. Often these bifurcations involve simply a change in the number of solutions. This chapter illustrates these ideas with a few examples.

Bifurcation theory is the (mathematical) study of such qualitative changes arising as parameters are varied. In this book, we consider a subset of this very general theory, namely *local bifurcations*, which excludes for example, routes to chaos in dynamical systems and other global bifurcations: everything we study can be described by local questions and local changes.

The majority of applications of mathematics involve differential equations (ordinary or partial), and the theory of bifurcations can be applied to these in a straightforward manner, as we will see in the first example below. However, the ideas are more general, and can be applied to other systems that depend on parameters, not just differential equations.

The general approach is to consider an equation $g(x) = 0$, where g may have several components,

$$g(x) = (g_1(x),\ g_2(x), \dots,\ g_p(x)),$$

and indeed so may x, that is $x = (x_1, x_2, \dots, x_n)$. Then introduce a parameter λ (also possibly multi–dimensional), writing

$$g_\lambda(x) = 0,$$

or $G(x; \lambda) = 0$. This is called a ***family*** of equations, depending on the parameter λ. We shall always assume our families are smooth as functions of (x, λ) (i.e. of class C^∞). The basic question of bifurcation theory is, how do solutions in x of these equations change as λ varies? And a bifurcation occurs when the change is in some sense qualitative.

There are many applications where the equation is a so-called *variational problem*, which means that the equation $g(x) = 0$ is in fact of the form $\nabla V(x) = 0$ for some scalar function V, usually called the *potential*. Then solutions of the equations $g(x) = 0$ are critical points of the function V. Zeeman's catastrophe machine described in Section 1.3 is one such physical example. A more geometric example is described in Section 1.4.

1.1 The fold or saddle-node bifurcation

The simplest mathematical example exhibiting a bifurcation is provided by the ordinary differential equation (ODE),

$$\dot{x} = x^2 + \lambda. \tag{1.1}$$

Here $\lambda \in \mathbb{R}$ is the parameter, and one often refers to $x \in \mathbb{R}$ as the ***state variable***. The dot over the x denotes the time-derivative, and a solution to the equation would be a function $x(t)$. Since this is a first order ODE, an equilibrium point occurs wherever the right-hand side vanishes. The equilibria therefore occur where

$$x^2 + \lambda = 0.$$

Define $g_\lambda(x) = x^2 + \lambda$. Then we are interested in solutions of $g_\lambda(x) = 0$, that is in the zeros of g. We call this set Z. Thus,

$$Z = \left\{(x, \lambda) \in \mathbb{R}^2 \mid x^2 + \lambda = 0\right\}.$$

The question we address is how the number of points in Z depends on λ. In this example, the curve Z is a parabola in the left half of the plane, as illustrated in Figure 1.1A. For $\lambda < 0$ there are two solutions (two equilibrium points), at $x = \pm\sqrt{-\lambda}$, and as λ increases to 0 these coalesce and then for $\lambda > 0$ there are no solutions (or they become complex, but we are just interested in real solutions). The transition, or *bifurcation*, occurs when $\lambda = 0$ (marked in red). The set of parameter values where such a bifurcation occurs is called the ***bifurcation set*** or ***discriminant***. The map π shown in the diagram is simply the projection taking $(x, \lambda) \in Z$ to the parameter value λ.

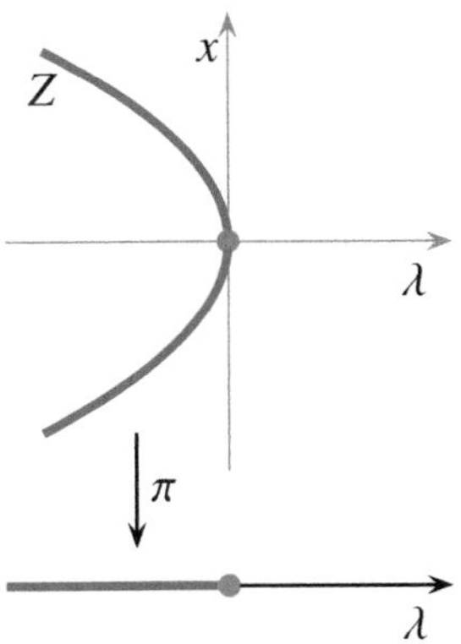

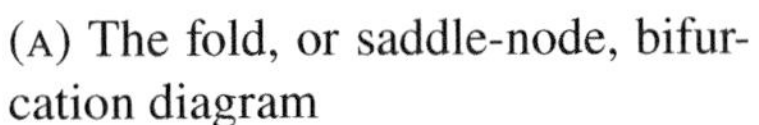

(A) The fold, or saddle-node, bifurcation diagram

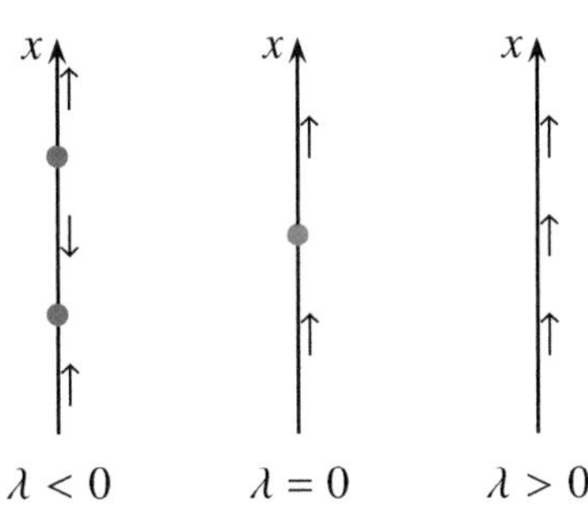

(B) Phase diagram for the saddle-node bifurcation (1.1)

FIGURE 1.1 (A) shows the equilibrium points, forming a smooth curve in the (x, λ)-plane. (B) shows whether x is increasing or decreasing (the sign of $\dot{x}$) for different values of λ; the dots represent the equilibrium points and correspond to points on the curve in (A).

The behaviour of the differential equation is illustrated in Figure 1.1B. There are two equilibrium points when $\lambda < 0$ and none for $\lambda > 0$. In differential equations, this transition is often called a ***saddle-node bifurcation*** because in two dimensions, when $\lambda < 0$, one of the equilibria would be a saddle and the other a node. In singularity theory, where the specific application is not of concern, it is more generally called a ***fold bifurcation***, because of the shape of the curve Z folding over with respect to the parameter space (the λ-axis).

Remark In this simple example, the differential equation is a standard one and can be solved explicitly (by separation of variables, the type of solution depends on the sign of λ). However, more generally, bifurcation theory can be used to study equilibria (and neighbouring dynamics) of systems of ODEs where this is not the case, such as for example the ODE $\dot{x} = x^2 e^x + \lambda$, which does not have a closed form solution but still exhibits a saddle-node bifurcation. ❞

The beauty of these ideas is that while the example above is so simple (g is quadratic), it contains essentially all that is expected to occur if there is just one parameter and no other restrictions. Imagine a small perturbation of the curve Z shown in Figure 1.1A; it seems reasonable to think that there will still be a single point where the curve 'folds over', with two solutions on the left and none on the

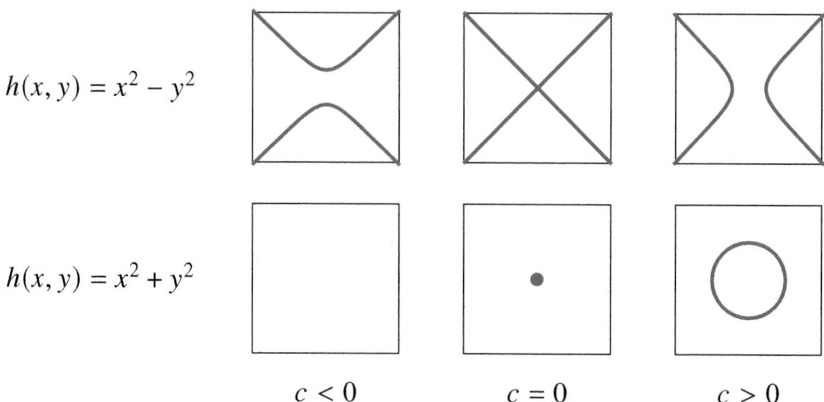

FIGURE 1.2 Two examples of bifurcation of contours $h(x, y) = c$ as c varies across the critical value of the function h.

right of this fold point (and one can prove this using the implicit function theorem; see Problem 1.6). This illustrates the *robustness* of the saddle-node bifurcation. In Section 1.5 below, we look briefly at an important bifurcation with one parameter, the pitchfork bifurcation, but where small perturbations do change its form. But first we look at two places bifurcations occur, the contours of a function as a parametrized set of equations, and a mechanical example with two parameters.

1.2 Bifurcations of contours

Landscape is determined in part by the height above sea level of each point of some region of the Earth. A ***contour*** is a curve on the landscape along which the height is constant; that is, for a given height the associated contour is the set of all points with that particular height. Let x, y be coordinates in the region in question, and $h(x, y)$ the height function. Then a contour at height c is the set of solutions of the equation

$$h(x, y) - c = 0.$$

Here we have a fixed function h and we can consider c as a parameter. Of course, height is only one example; another is the atmospheric pressure as a function on the surface, in which case the 'contours' are the familiar isobars from weather maps (although atmospheric pressure is best expressed as a function of three variables $P(x, y, z)$ as it varies with altitude z).

Consider a function $h(x, y)$ and the resulting equation $h(x, y) = c$. Most of the contours are curves, and a natural question to ask is, as c is varied, how can these curves change? The contours of a function are also called its ***level sets***.

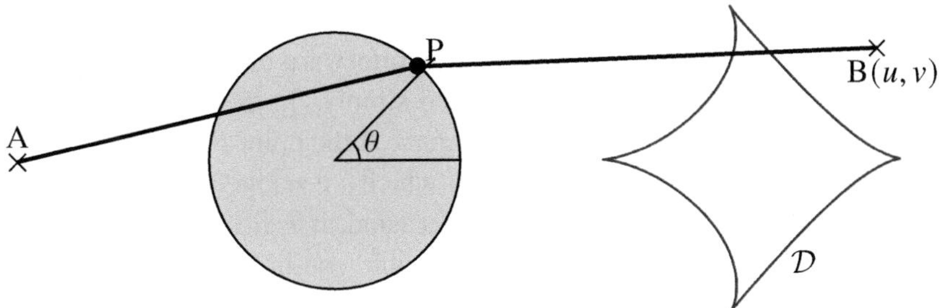

FIGURE 1.3 Schematic diagram of the Zeeman catastrophe machine. The red curve marked $\mathcal{D}$ is the discriminant or bifurcation set; notice the four cusps. Its precise size and position depend on the physical characteristics of the elastics and the position of the point A.

For example, suppose $h(x, y) = x^2 - y^2$. Then the contours are either hyperbolae or a pair of lines and the transition is depicted in the top row of Figure 1.2. For $h(x, y) = x^2 + y^2$ the contour is a circle for $c > 0$, a single point for $c = 0$ and is empty for $c < 0$. See the lower figures in Figure 1.2. In both cases a change occurs as one crosses the level $c = 0$, and one can show in general that qualitative changes only occur at *critical values* of the function; that is, the value the function takes at a critical point. We will study this in greater depth in later chapters.

A similar example in more variables is provided by $h(x, y, z) = x^2 + y^2 - z^2$. The zero level of this function is a circular cone in $\mathbb{R}^3$, while $h = 1$ is a one-sheeted hyperboloid and $h = -1$ is a two-sheeted hyperboloid.

1.3 Zeeman catastrophe machine

Conceived by Christopher Zeeman to illustrate the ideas of catastrophe theory, the Zeeman catastrophe machine consists of a wheel free to rotate about its centre, with a peg P attached at a point of its circumference. To the peg are attached two elastics: the other end of the first is pinned at a fixed point A in the plane of the wheel, while the other end of the second elastic is held by hand at a second point $B(u, v)$ in that plane. See Figure 1.3. The question is, how many equilibrium states are there of the wheel?

The answer will depend on where the end B is held; that is on the values of u and v, so these are the parameters. For each choice of point (u, v), the total elastic potential $V_{(u,v)}(\theta)$ is a function of θ, the position of the wheel (see Figure 1.4), and the equilibrium points are the points θ where V has a critical point: $\frac{\mathrm{d}}{\mathrm{d}\theta}V_{(u,v)}(\theta) = 0$.

The computation of the potential is straightforward but lengthy (and not relevant here), but the conclusion can be described simply. In the (u, v)-plane, there is a curve with four cusps, marked $\mathcal{D}$ in the figure. If the point B is within the curve, the wheel has four equilibrium points, two of which are stable (where $V'' > 0$) and two are unstable (where $V'' < 0$). On the other hand, if B lies outside this curve, then the wheel has only one stable and one unstable equilibrium point. The transition from four to two critical points happens when B approaches the curve $\mathcal{D}$ from the inside, and two of the critical points get closer and coalesce becoming degenerate in the process, and then disappearing; this curve $\mathcal{D}$ is therefore the discriminant of this family. This transition is the same as that in the fold bifurcation described above, although something more involved happens at the cusp points of the discriminant.

1.4 An example from geometry: the evolute

Consider a smooth simple closed curve C in the plane (e.g. an ellipse: a curve is said to be *smooth* if it has a parametrization whose derivative is nowhere zero). Let $P(u, v)$ be a point in the plane (possibly on C) with coordinates (u, v). The geometric question is, can you draw a perpendicular to the curve from the point P, and if so how many? (If P lies on the curve then we allow that the 'segment' (of zero length) from P to P is perpendicular to the curve.)

For example, if C is an ellipse, and P is at its centre, then it is not hard to see that there are 4 such perpendiculars – one to each of the points on the axes of the ellipse. What happens to those 4 points if P is perturbed? The feet of the perpendiculars will move, but can there be a different number of them? Imagine instead a point P' on the major axis of the ellipse, but outside the ellipse. It is easy to see that there are now only 2 perpendiculars from P' to C. See Figure 1.5. The bifurcation question is, how does 4 change to 2 as P is moved? And more completely, what is this number for all possible points P?

One observation is that for any P there are at least two such perpendiculars, and these arise at the nearest and furthest points of the curve to P as some thought should convince you (and which we prove below). This suggests defining the function on C which is the distance of each point of C to P. In fact we use the square of the distance which leads to simpler expressions after differentiating.

Let $\mathbf{r}(t)$ be a regular parametrization of the plane curve C, where 'regular' means that its derivative $\dot{\mathbf{r}}(t)$ is never zero, and for each point $\mathbf{c} = (u, v)$ in the

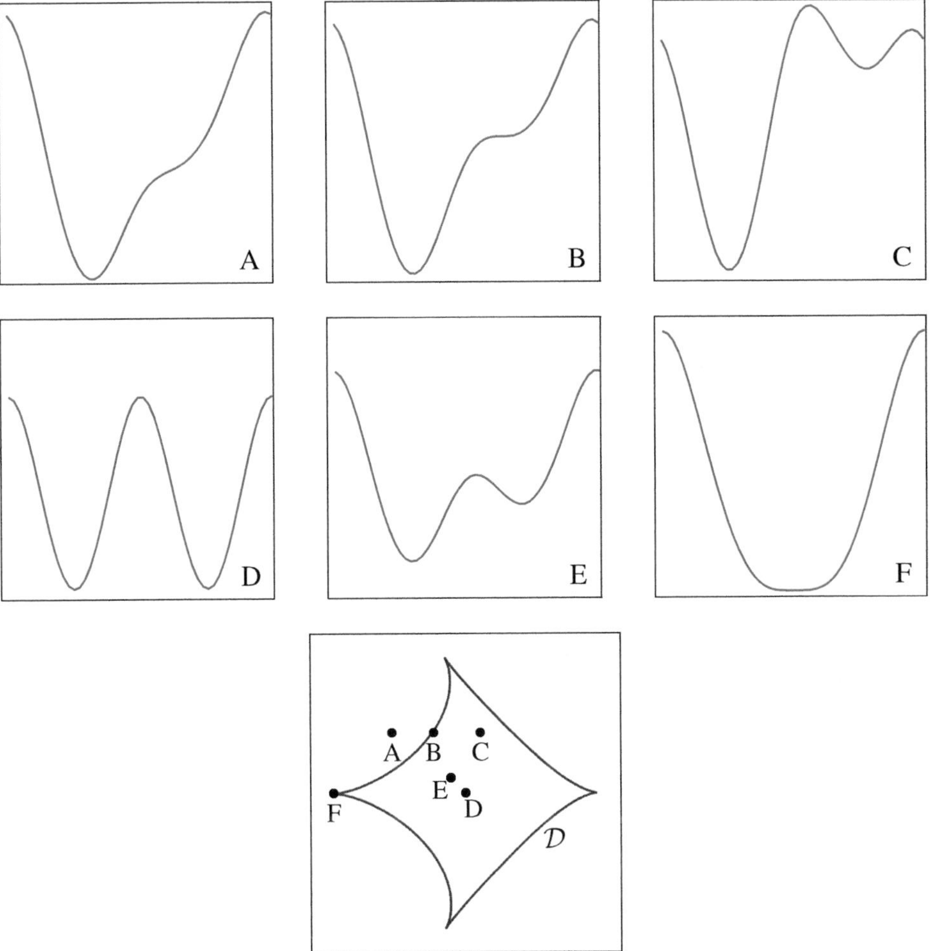

FIGURE 1.4 Graphs of the potential V in Zeeman's catastrophe machine for the six different parameter values shown in the bottom figure. Note that Figures B and F have degenerate critical points, and the corresponding points in the bottom diagram lie on the discriminant $\mathcal{D}$. The horizontal axis in diagrams A–F is $\theta \in [0, 2\pi]$.

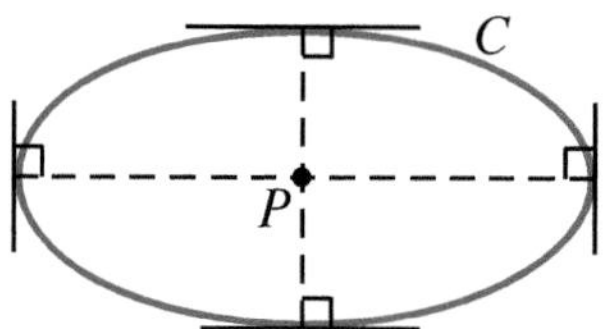

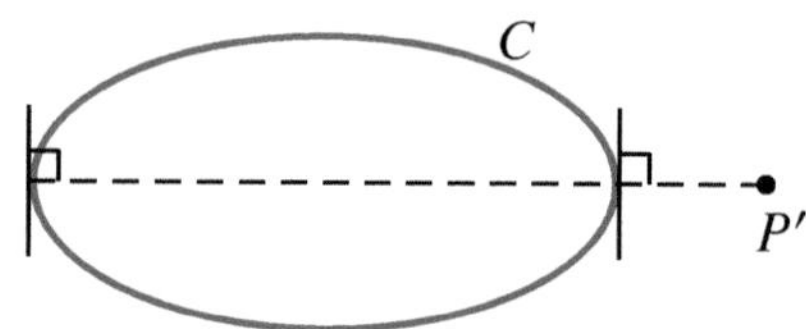

FIGURE 1.5 The dashed lines are perpendiculars from P and P' to the ellipse C.

plane, define the function

$$f_{\mathbf{c}}(t) = -\tfrac{1}{2}\,\|\mathbf{c} - \mathbf{r}(t)\|^2.$$

($\mathbf{c}$ is the position vector of the point P from the discussion above, and the factor $-\frac{1}{2}$ is for convenience.) This is a family of functions of t, with two parameters u and v. It measures the square of the distance from the point $\mathbf{c}$ to the point $\mathbf{r}(t)$; it's called the ***distance squared function***, or distance squared family.

Question: where does $f_{\mathbf{c}}$ have a critical point, and when is it degenerate?

First differentiate $f_{\mathbf{c}}$ (with respect to t),

$$f_{\mathbf{c}}'(t) = (\mathbf{c} - \mathbf{r}(t)) \cdot \dot{\mathbf{r}}(t). \tag{1.2}$$

Since $\dot{\mathbf{r}}(t)$ is the derivative of $\mathbf{r}(t)$, it represents a non-zero tangent vector to the curve. It follows that $f_{\mathbf{c}}'(t) = 0$, if $\mathbf{c}$ lies on the normal line to the curve at $\mathbf{r}(t)$. The set of critical points is therefore a very geometric object:

$$C = \{(t, u, v) \in \mathbb{R}^3 \mid (u, v) \text{ lies on the normal to the curve at } \mathbf{r}(t)\}. \tag{1.3}$$

Thus, for given P the original question is now, how many critical points does $f_{\mathbf{c}}$ have? In particular, the question of how many normals there are for a given point P is now cast as a variational problem.

Local changes in the number of critical points can only occur when a critical point is degenerate (as follows from the implicit function theorem). To see if the critical point is degenerate, we find the second derivative:

$$f_{\mathbf{c}}''(t) = (\mathbf{c} - \mathbf{r}(t)) \cdot \ddot{\mathbf{r}}(t) - \|\dot{\mathbf{r}}(t)\|^2. \tag{1.4}$$

Thus $f_{\mathbf{c}}$ has a degenerate critical point at t if both (1.2) and (1.4) are equal to zero. We can rewrite the two equations as,

$$\begin{cases} \dot{\mathbf{r}}(t) \cdot \mathbf{c} &= \mathbf{r}(t) \cdot \dot{\mathbf{r}}(t) \\ \ddot{\mathbf{r}}(t) \cdot \mathbf{c} &= \mathbf{r}(t) \cdot \ddot{\mathbf{r}}(t) + \|\dot{\mathbf{r}}(t)\|^2 \end{cases}$$

This is simply a pair of linear equations for $\mathbf{c}$, and if the coefficients $\ddot{\mathbf{r}}(t)$ and $\dot{\mathbf{r}}(t)$ are not parallel (they are both vectors), there is a unique solution $\mathbf{c}$, so giving a unique point[1] on that normal line. Call this point $\mathbf{e}(t)$: the resulting curve is called the ***evolute*** of the original curve. We have shown that the point $\mathbf{c} = \mathbf{e}(t)$ if and only if the function $f_{\mathbf{c}}$ has a degenerate critical point at t; the set of $\mathbf{e}(t)$ as t varies is therefore the discriminant of this family $f_{\mathbf{c}}$.

Example 1.1. As a specific example, consider the ellipse

$$\mathbf{r}(t) = (3\cos t,\ 2\sin t).$$

Then, with $\mathbf{c} = (u, v)$,

$$f_{\mathbf{c}}(t) = -\tfrac{1}{2}(u - 3\cos t)^2 - \tfrac{1}{2}(v - 2\sin t)^2. \tag{1.5}$$

The first two derivatives are $f_{\mathbf{c}}'(t) = -3u\sin t + 2v\cos t + 5\sin t\cos t$, and

$$f_{\mathbf{c}}''(t) = -3u\cos t - 2v\sin t + 10\cos^2 t - 5.$$

Solving $f'(t) = f''(t) = 0$ gives

$$u = \frac{5}{3}\cos^3 t, \quad v = -\frac{5}{2}\sin^3 t. \tag{1.6}$$

That is, $\mathbf{e}(t) = \left(\frac{5}{3}\cos^3 t,\ -\frac{5}{2}\sin^3 t\right)$; this curve is shown in Figure 1.6, together with the ellipse (notice that the ellipse is traversed anticlockwise, while the resulting parametrization of the evolute is clockwise). Note that this evolute or discriminant also has 4 cusps, like the ZCM above. We will see in later chapters that cusps occur very often on discriminants for 2 parameter families of functions, and using the theory of unfoldings we will explain why.

If $\mathbf{c}$ lies inside the evolute, the function $f_{\mathbf{c}}$ has 4 critical points, all nondegenerate, and if outside it has just 2. Indeed, using the symmetry of the ellipse if you take $\mathbf{c} = (0, 0)$ it is easy to see the 4 points of the curve for which the normal line passes through $\mathbf{c}$. If, on the other hand, $\mathbf{c}$ lies on the evolute but not at one of the cusps, then $f_{\mathbf{c}}$ has precisely 3 critical points, of which one is degenerate. Finally, if $\mathbf{c}$ lies at a cusp, $f_{\mathbf{c}}$ has a 'doubly' degenerate critical point and a nondegenerate one. As $\mathbf{c}$ varies from the interior of the evolute to the exterior, crossing at a regular point (ie, not at a cusp) then two of the critical points will coalesce and then disappear,

[1] it is in fact the *centre of curvature* of the curve at $\mathbf{r}(t)$; the evolute was originally defined by Huygens in the seventeenth century in his study of the pendulum, and it was later realised to be the locus of centres of curvature.

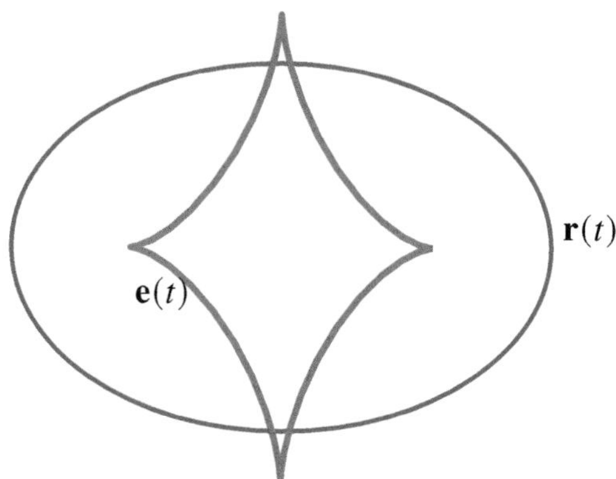

FIGURE 1.6 An ellipse with its evolute

just as they do for the fold family in Section 1.1 and the Zeeman Catastrophe Machine in Section 1.3. The 4 cusps are interesting geometrically: they are points on the evolute (centres of curvature) corresponding to points on the curve where the curvature has a local maximum or minimum.

If the major and minor axes of the ellipse were closer in value (here they are equal to 3 and 2 respectively), the evolute would be smaller, and in the limit as the ellipse tends to a circle, so the evolute tends to a single point: the centre of the circle. ✐

Applications of these ideas to the study of the geometry of curves and surfaces can be found in two books [18] and [61]; there is also a brief discussion in Chapter 15 in this book.

One question arising from the two very different examples, the evolute and Zeeman's catastrophe machine, is why do the bifurcation curves or discriminants have cusps? We will show in later chapters that this is very natural, given that we are studying a 2–parameter family of functions. The fact that in both cases there is only one variable θ or t turns out not to be important: it's the number of parameters that is central.

These two examples are both variational problems (arising from looking for critical points of functions), and such problems will be the study of the first part of this book. Later we will study more general (non-variational) bifurcation problems, but it will turn out that for two parameters, folds and cusps are still all that are to be expected.

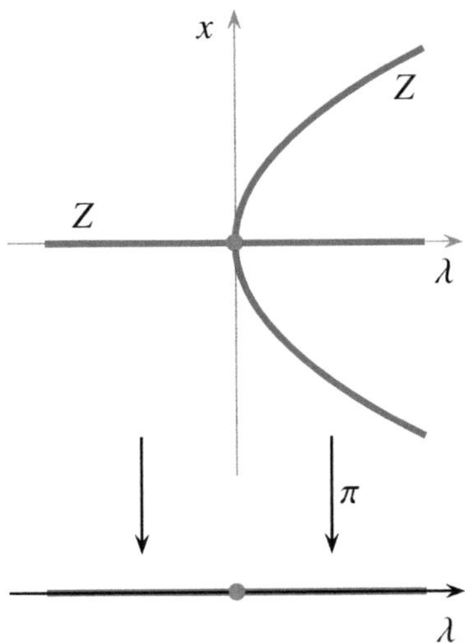

FIGURE 1.7 The pitchfork bifurcation: the zero-set Z consists of two intersecting curves, one of which is folded over relative to the projection.

1.5 Pitchfork bifurcation

Let us now look at a different 1–parameter example, namely the *pitchfork bifurcation*. Consider the family of ODES,

$$\dot{x} = x^3 - \lambda x.$$

Again λ is the parameter.

This bifurcation often arises in problems where there is an assumed symmetry in the problem: notice that both sides of the equation are odd functions of x.

In this example $g_\lambda(x) = x^3 - \lambda x$, and the zero-set is

$$Z = \left\{(x, \lambda) \in \mathbb{R}^2 \mid x(x^2 - \lambda) = 0\right\}.$$

This is the set shown in the top diagram in Figure 1.7. For $\lambda < 0$ there is just one solution (namely $x = 0$), while for $\lambda > 0$ there are three. Again, the bifurcation point is at $\lambda = 0$ (marked in red), and the map $\pi\colon Z \to \mathbb{R}$ is simply the projection $\pi(x, \lambda) = \lambda$.

In contrast to the fold bifurcation, this pitchfork bifurcation is not robust, or 'structurally stable', in the following sense. Consider the small perturbation given by $\dot{x} = x^3 - \lambda x + \varepsilon$ (where $\varepsilon \in \mathbb{R}$ is a small constant). The new zero-set is depicted in Figure 1.8, and looks very different (structurally different).

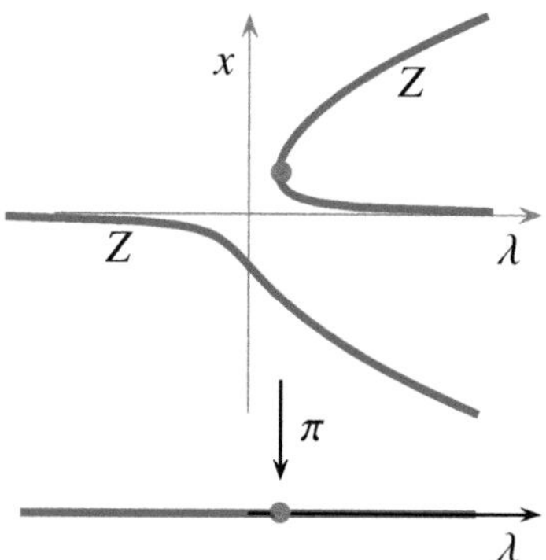

FIGURE 1.8 A perturbation of the pitchfork bifurcation

The symmetry of the pitchfork bifurcation was mentioned above, and the perturbation just given does not preserve this symmetry. It turns out that if only symmetric perturbations are allowed, then the pitchfork does not change qualitatively, and one says it is robust, or structurally stable, with respect to symmetric perturbations. This idea of symmetry breaking is important in applications: see for example the book [60] on *imperfect bifurcations*. These imperfect bifurcations arise by perturbing a 'perfect bifurcation': in this case the pitchfork would be the perfect bifurcation (with perfect symmetry), but in reality one expects systems not to have perfect symmetry and for the symmetry to be broken, leading to the 'imperfect bifurcation' shown in Figure 1.8.

An example of this scenario of pitchfork and perturbation can be seen within the Zeeman catastrophe machine described earlier. While the ZCM is a 2-parameter system, consider just the 1–parameter system where the point (u, v) lies on the line of symmetry $v = 0$, and let $(u_1, 0)$ and $(u_2, 0)$ be the two points where this line meets the bifurcation set $\mathcal{D}$ (the first is marked F in Figure 1.4). As u increases from $u = u_1$, a single solution splits into 3, giving a diagram similar to the one in Figure 1.7. A similar figure can be drawn for u varying through u_2, but with 3 solutions coalescing into 1, so the diagram should be reversed.

Now perturb this path a little, say by raising it (the line $v = 0.1$ say). The new bifurcation diagram near $u = u_1$ would be similar to the one in Figure 1.8, and near $u = u_2$ would be a reversed version, both with a single saddle-node bifurcation.

This idea of considering bifurcations as paths in some parameter space and their perturbations, was introduced in [48] and is central to the so-called *path approach* to bifurcation theory that we study in depth in Part III.

1.6 Conclusions

- A bifurcation problem is a family of equations, which we write $g_\lambda(x) = 0$ where λ is the parameter (in later chapters we will often use $u, v, \ldots$ as parameters). This might arise as the equation for equilibria of the differential equation $\dot{x} = g_\lambda(x)$, but it may arise as an equation in its own right (such as for contours). One is interested in how the solutions of $g_\lambda(x) = 0$ change as the parameter is varied.
- A bifurcation is a *qualitative* change in the set of solutions to the bifurcation problem, as the parameter is varied. For example, if there are finitely many solutions, then a bifurcation occurs where the number of solutions changes, while if the solution set is a curve or surface then a bifurcation would occur when the topology of the set changes. The set of parameter values where a bifurcation occurs is called the *bifurcation set* or the *discriminant.*
- In many applications, a bifurcation problem arises as an equation for critical points of a function (of one or several variables). Such equations are called *variational problems*. So our $g_\lambda(x)$ would be equal to $\frac{\partial}{\partial x}V_\lambda(x)$ for some 'potential' function $V_\lambda(x)$. These variational problems form the part of bifurcation theory we call ***catastrophe theory***, and are the subject of the first part of this book. This is for both historical and technical reasons. Catastrophe theory was introduced by René Thom in the 1960s and the use of similar techniques in more general bifurcation theory was developed a decade or so later, and uses many of the ideas Thom introduced. Moreover, the techniques for variational problems are more straightforward.

The principal aim of this text is to show that there are only a few basic models for bifurcations, provided there are not too many parameters. For example, if there is just a single parameter, and no other restrictions, then the only bifurcation arising 'generically' is the saddle-node, or fold bifurcation described in Section 1.1, while if there are two parameters with no further restriction then there are two possibilities: the fold and the cusp (a further restriction might be something like the problem having some symmetry as described above). These ideas were first developed in the 1960s by the French mathematician and Fields medal winner René Thom.

A short account of early bifurcation theory can be found in the book by Drazin [33], who discusses how the first bifurcation problem to be studied was by Euler in the eighteenth century, who addressed what is now called the Euler beam problem, where a beam buckles under a load and as the load increases there is a (pitchfork) bifurcation.

Problems

1.1 Consider the following perturbation of the ODE given in Section 1.1, namely $\dot{x} = x^2 + \lambda + \varepsilon x$ where $\varepsilon \in \mathbb{R}$ is a small consant. Show that the bifurcation diagram is similar to the one in Figure 1.1A, but with discriminant equal to $\{\varepsilon^2/4\}$.

1.2 Investigate the bifurcations in the set of equilibria occurring as λ varies, in the family $\dot{x} = x^2 + \lambda^2 - 1$. In particular, show the bifurcation set consists of two points on the λ-axis.

1.3 Investigate the bifurcations in the set of equilibria occurring as λ varies, in the two pitchfork families of ODEs,

$$\dot{x} = \pm(x^3 - \lambda x),$$

showing the phase diagram analogous to Figure 1.1B. Notice that in both of these, the origin is attracting (stable) for λ on one side of the origin and and repelling (unstable) on the other. Correspondingly, the bifurcating equilibria are stable in one of these and unstable in the other. When they are stable it is called a *supercritical* pitchfork bifurcation, and when they are unstable it is a *subcritical* pitchfork bifurcation. (This is the only place in the book we consider dynamical properties such as stability and instability.) (†)

1.4 Investigate the contours of the height function $h(x, y) = 2x^4 + 4y^4 - x^2 + y^2$, and relate the transitions (bifurcations) to what was seen in Section 1.2, specifically in Figure 1.2. [*Hint*: *find the critical points, and hence critical values, and then use a graphing calculator, or Wolfram Alpha on the internet. For the latter you can enter an instruction such as,* `plot 2*x^4+4*y^4-x^2+y^2=0` *or* `contour plot 2*x^4+4*y^4-x^2+y^2`].

1.5 Investigate the level sets (contours) of the function $h_u(x) = x^3 - 3ux$ for different values of u. (Here the level sets are finite sets of points, so the question is, how many points in each level set? The answer depends on the value of c as well as u.)
[*Hint*: *begin by sketching the curve* $y = h_u(x)$ *for* $u > 0, u = 0$ *and* $u < 0$. *You will see that bifurcations occur at critical points of* h_u (*i.e.*, $\frac{\mathrm{d}}{\mathrm{d}x}(h_u) = 0$). *Find the locus of points in the* (u, c)*-plane where these occur – that is the discriminant.*] (†)

1.6 Here we show the *robustness* of the saddle-node bifurcation. Consider the saddle-node bifurcation $g_\lambda(x) = x^2 + \lambda = 0$ shown in Figure 1.1A. Let $G(x, \lambda, u)$ be any smooth ‘perturbation’ of g; that is suppose G is a smooth function (with $u \in \mathbb{R}$) and that $G(x, \lambda, 0) = g_\lambda(x)$. Note that one cannot solve $g(x, \lambda) = 0$ for x as a function of λ, but one can instead solve for λ as a function of x, and this function has a nodegenerate extremum at the bifurcation point.

Use the implicit function theorem to show that for sufficiently small values of u, the perturbed bifurcation problem g_u given by $g_u(x, \lambda) = G(x, \lambda, u)$ (for u fixed) also has a saddle-node bifurcation, in the sense that one can solve (locally) for λ as a function of x (and u) and for each u this function λ has a nondegenerate extremum. (†)

1.7 In contrast to the saddle-node bifurcation the so-called *hysteresis bifurcation* is not robust. Consider $\dot{x} = x^3 + \lambda$, and its perturbation $\dot{x} = x^3 + \lambda + ux$. Sketch three phase portraits for this system similar to Figure 1.1B, one with $u > 0$, one with $u = 0$ and with with $u < 0$ (u fixed in each case). [See Figure 18.3C on p. 226 for the curves of equilibria.]

1.8 Investigate the change in contours for the family of surfaces $h(x, y, z) = x^2 + y^2 - z^2 = c$, distinguishing $c = 0$, $c > 0$ and $c < 0$. (†)

1.9 Find the evolute of the parabola $y = x^2$, and show it has a single cusp. On a sketch, show that from a point inside (or above) the evolute there are three lines perpendicular to the parabola, while from a point outside there is only one.

1.10 The diagram below shows a curve (in blue) and its evolute (in red). For a point P in each of the five components of the complement of the evolute find the number of perpendiculars to the curve from P.

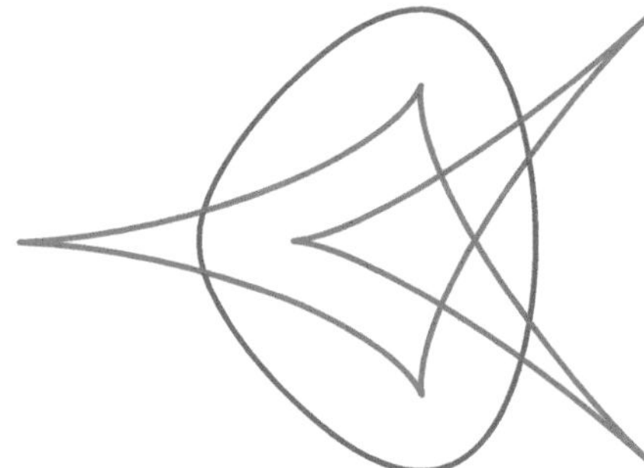

[*Hint*: (1) *If P is far from the curve there are only two perpendiculars, and* (2) *as P crosses a smooth point of the evolute, two solutions are either created or destroyed (depending on the direction of crossing), like in saddle-node bifurcations.*] (†)

Part I

Catastrophe theory

2

Families of functions

CATASTROPHE THEORY is the study of families of functions, and in particular of their critical points, the principal motivation being to study bifurcations in variational problems.[1] For example the function considered in Problem 1.5, namely $f_\lambda(x) = x^3 - 3\lambda x$ has a single (degenerate) critical point when $\lambda = 0$, it has two critical points when $\lambda > 0$ and none at all when $\lambda < 0$ (or two complex ones if one prefers to include those). On the other hand if a family has only nondegenerate critical points when $\lambda = 0$, then for nearby values of λ it will still have nondegenerate critical points and they will be close to the original ones. We will prove this later in the book.

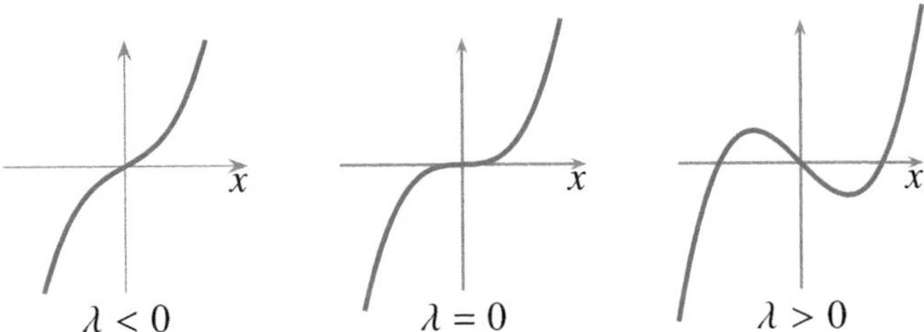

In most areas of mathematics the words 'function' and 'map' are more or less interchangeable; however it is traditional in this area to reserve the word 'function' to refer to scalar-valued functions (so $f\colon \mathbb{R}^n \to \mathbb{R}$ or $f\colon \mathbb{C}^n \to \mathbb{C}$), while 'map' refers to an $f\colon \mathbb{R}^n \to \mathbb{R}^p$ with $p > 1$, or the complex analogue.

2.1 Critical points

We consider smooth functions $f\colon \mathbb{R}^n \to \mathbb{R}$: here smooth means infinitely differentiable, although in practice, smooth just means 'as many times differentiable as is needed in the current context', so for example if we are talking about the second

[1] A *variational problem* is one involving finding critical points of functions. In applications these are often, but not always, required to be local maxima or minima.

derivative of f, we will need to assume it is twice differentiable, and the second derivatives are continuous (in short, f is of class C^2). However, we will make the blanket assumption that all functions we consider are infinitely differentiable.

In general, our function may not be defined on all of $\mathbb{R}^n$, and we adopt the notation $f\colon \mathbb{R}^n \rightarrowtail \mathbb{R}$ (notice the different arrow) to mean a function of n variables whose domain is some open subset of $\mathbb{R}^n$. This saves us writing things like, 'let $f\colon U \to \mathbb{R}$ be a smooth function, where U is an open subset of $\mathbb{R}^n$', and instead we write 'let $f\colon \mathbb{R}^n \rightarrowtail \mathbb{R}$ be a smooth function'. If we need to refer to the domain of such an f we will denote it $\mathrm{dom}(f)$.

Definition 2.1. A smooth function $f\colon \mathbb{R}^n \rightarrowtail \mathbb{R}$ has a ***critical point*** at x_0 if its differential there vanishes: $\mathsf{d}f_{x_0} = 0$. ☆

Here $\mathsf{d}f_{x_0}$ is the ***differential*** of f at x_0, so

$$\mathsf{d}f_{x_0} = \left(\frac{\partial f}{\partial x_1}(x_0),\ \frac{\partial f}{\partial x_2}(x_0),\ \dots,\ \frac{\partial f}{\partial x_n}(x_0)\right).$$

Thus x_0 is a critical point of f if all n partial derivatives of f vanish at x_0:

$$\frac{\partial f}{\partial x_1}(x_0) = \frac{\partial f}{\partial x_2}(x_0) = \cdots = \frac{\partial f}{\partial x_n}(x_0) = 0.$$

Critical points are also known as *singular points* of the function.

Definition 2.2. If x_0 is a critical point of $f\colon \mathbb{R}^n \rightarrowtail \mathbb{R}$, then the ***Hessian*** matrix of f at x_0 is the symmetric $n \times n$ matrix of second partial derivatives,

$$H_f(x_0) = \mathsf{d}^2 f_{x_0} = (h_{ij})$$

where

$$h_{ij} = \frac{\partial^2 f}{\partial x_i\, \partial x_j}(x_0).$$

A critical point x_0 of f is ***nondegenerate*** if $\det H_f(x_0) \neq 0$: otherwise it is ***degenerate***. ☆

Example 2.3. Find the critical points of $f(x, y, z) = x^3 - xy^2 + 3x^2 + y^2 + z^2$, and determine whether each is nondegenerate.

Solution: Differentiating f with respect to each of the three variables gives the equations

$$3x^2 - y^2 + 6x = 0, \quad -2xy + 2y = 0, \quad 2z = 0.$$

The solutions, 4 in all, are found to be

$$(0,0,0), \quad (-2,0,0), \quad (1,3,0), \quad (1,-3,0).$$

The second differential (Hessian matrix) of f is

$$\mathrm{d}^2 f(x,y,z) = \begin{pmatrix} 6x+6 & -2y & 0 \\ -2y & -2x+2 & 0 \\ 0 & 0 & 2 \end{pmatrix}.$$

The Hessian at each of the critical points is,

$$\begin{pmatrix} 6 & 0 & 0 \\ 0 & 2 & 0 \\ 0 & 0 & 2 \end{pmatrix}, \quad \begin{pmatrix} -6 & 0 & 0 \\ 0 & 6 & 0 \\ 0 & 0 & 2 \end{pmatrix}, \quad \begin{pmatrix} 12 & -6 & 0 \\ -6 & 0 & 0 \\ 0 & 0 & 2 \end{pmatrix}, \quad \begin{pmatrix} 12 & 6 & 0 \\ 6 & 0 & 0 \\ 0 & 0 & 2 \end{pmatrix}.$$

A quick inspection shows that all four matrices are invertible, and hence all four critical points are nondegenerate. ✐

Recall that the eigenvalues of a symmetric matrix are all real. Moreover, at a critical point, the Hessian matrix is 'intrinsic' in the following sense. Suppose f has a critical point at the origin and let ϕ be a change of coordinates about the origin (i.e. such that $\phi(0) = 0$). Write H_f for the Hessian matrix of f at the origin in the original coordinates, and H'_f for the matrix in the transformed coordinates. Then by the chain rule for second derivatives (see Proposition A.10) $\mathrm{d}^2(f \circ \phi)_0(\mathbf{u}^2) = \mathrm{d}^2 f_0(\mathrm{d}\phi_0 \mathbf{u})^2$ (using $\mathrm{d}f_0 = 0$), whence

$$\Phi^T H_f \Phi = H'_f, \tag{2.1}$$

where $\Phi = \mathrm{d}\phi_0$ (the Jacobian matrix of ϕ).

Definition 2.4. If $x_0 \in \mathbb{R}^n$ is a nondegenerate critical point of the smooth function f, then its ***index*** is the number of negative eigenvalues of the Hessian matrix, counting multiplicity. ✯

It is important to realize that while the eigenvalues of the Hessian do depend on the chosen basis (or coordinates), their signs do not; in particular the index of a critical point does not depend on the chosen coordinates, which follows from (2.1). For a symmetric $n \times n$ matrix, there can be anywhere between 0 and n negative eigenvalues, so the index of a nondegenerate critical point in n variables lies between 0 and n inclusive. The term 'counting multiplicity' means that for example if an eigenvalue is a double root of the characteristic polynomial of the matrix, then it should count twice; that way one always has exactly n eigenvalues 'counting multiplicity'. In particular, the identity matrix has eigenvalue 1 with multiplicity n, and the index of the critical point of the function $-x_1^2 - \cdots - x_n^2$ is n.

In Example 2.3 above, the origin is of index 0, while the other critical points are all of index 1. The index of a nondegenerate critical point is an important invariant and determines the geometry of the level sets of f near the critical point. For example, a point of index 0 is a local minimum of the function, and one of index n is a local maximum (if n is the number of variables). This follows from the following important result which we will return to later (Section 4.4).

Morse Lemma *Let $p \in \mathbb{R}^n$ be a nondegenerate critical point of f of index k. Then there is a change of coordinates $x = \phi(y)$ near p such that in these new coordinates y_i the function has the form*

$$f(y) = f(p) - y_1^2 - y_2^2 - \cdots - y_k^2 + y_{k+1}^2 + \cdots + y_n^2.$$

This is a particularly simple form of Taylor series for the function: since p is a critical point the linear terms in the Taylor series at p must all vanish, so this lemma is saying that if the critical point is nondegenerate then in some local coordinate system the Taylor series is purely quadratic.

Complex functions A complex analytic function $f\colon \mathbb{C}^n \to \mathbb{C}$ also has critical points, at points where all partial derivatives vanish, and one can form the Hessian matrix, but whose eigenvalues may now be complex. However, the notion of index is meaningless: firstly because the eigenvalues of the Hessian may not be real, and secondly if they are real, then a change of coordinates can change their sign: for example the function x^2 becomes $-y^2$ upon substituting $x = iy$.

2.2 Degeneracy in one variable

The story is fairly simple in one variable. Suppose $f\colon \mathbb{R} \rightarrowtail \mathbb{R}$ is a smooth function of a single variable. Then a point $x_0 \in \mathrm{dom}(f)$ is a critical point of f if $f'(x_0) = 0$. This critical point is nondegenerate if $f''(x_0) \neq 0$, otherwise it is degenerate. And one can continue looking at higher and higher derivatives as follows.

Definition 2.5. A critical point of a smooth function f of a single variable is of ***type*** A_k if the first k derivatives of f all vanish at that point, but the $(k+1)^{\text{th}}$ does not. ☆

The same definition is made for critical points of complex analytic functions $f\colon \mathbb{C} \to \mathbb{C}$. Thus in particular, a critical point (real or complex) of type A_1 is a nondegenerate critical point, and one sometimes says that a degenerate critical point is of type $\mathsf{A}_{\geq 2}$, or of type *at least* A_2.

Example 2.6. Consider the simple monomial $f(x) = x^{k+1}$, with $k \geq 1$ and $x \in \mathbb{R}$ or $\mathbb{C}$. Then

$$f'(0) = f''(0) = \cdots = f^{(k)}(0) = 0, \quad \text{but} \quad f^{(k+1)}(0) = (k+1)! \neq 0.$$

It follows that the function $x \mapsto x^{k+1}$ (whether real or complex) has a critical point of type A_k at $x = 0$.

We will see in Chapter 4 that any function of one variable with a critical point of type A_k is *equivalent* to x^{k+1} (in a neighbourhood of the critical point), in the sense that there is a local change of coordinates that turns the given function into the appropriate monomial.

We will see later the corresponding definitions for functions of several variables; note for now that any nondegenerate critical point is said to be of type A_1, for any number of variables.

2.3 Families of functions

We begin with a simple example, demonstrating the issues of interest.

Example 2.7. To set the scene, consider the 1–parameter family of functions

$$F(x; u) \;=\; f_u(x) \;=\; x^3 + ux,$$

parametrized by $u \in \mathbb{R}$. This is called the ***fold family***, or ***fold catastrophe***. Note that when $u = 0$, the function $f_0(x) = x^3$ which has a critical point of type A_2 at the origin. The 3 figures below show graphs of the function f_u for $u < 0$, $u = 0$ and $u > 0$ respectively. Notice that there are 2, 1 and 0 critical points in the three figures, and as u increases from negative values to 0 so the 2 critical points coalesce and then disappear for $u > 0$. At the point of coalescing the critical point is degenerate.

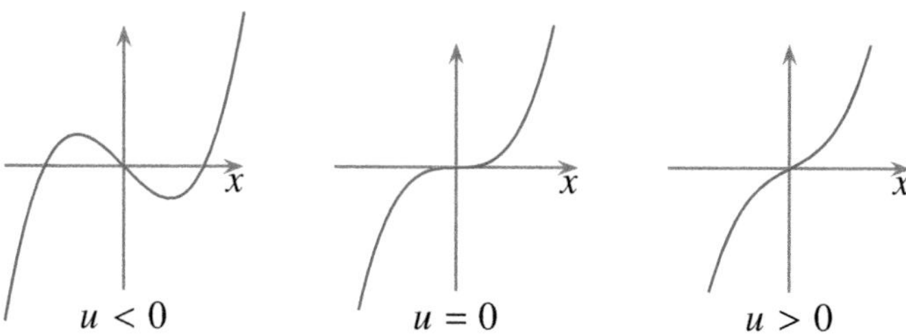

To visualize this better, we plot below the set of critical points in (x, u)-space:

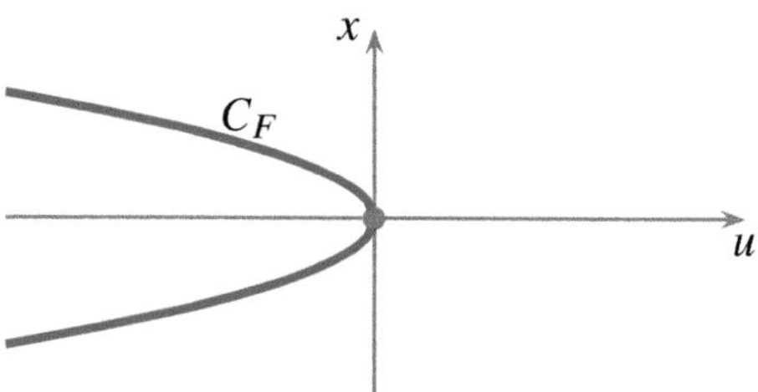

The curve $C_F = \{(x, u) \mid 3x^2 + u = 0\}$ is the set of critical points of f_u (since $(f_u)'(x) = 3x^2 + u$), and for each $u < 0$ there are two corresponding values of x (equal to $\pm\sqrt{-u/3}$), which coalesce when $u = 0$ and then for $u > 0$ there are no points on C_F (or at least, no real ones). The red dot is the point where the two critical points coalesce and in general is called the *singular set* of the family of functions, while C_F is called the *catastrophe set*. We will define these more generally below.

The reason this family is called the fold family is simply that C_F is folded over compared to the parameter space. This is the same as the saddle-node bifurcation described in Section 1.1. ✐

In general we consider a family of functions, $f_u(x)$, with $x \in \mathbb{R}^n$ and $u \in \mathbb{R}^a$. This needs to be smooth in u as well as x, so altogether we require that the map (function),

$$f\colon \mathbb{R}^n \times \mathbb{R}^a \rightarrowtail \mathbb{R}, \quad F(x, u) = f_u(x)$$

be smooth as a function of $n + a$ variables. We call these ***smooth families*** of functions.

Convention: As a rule, we will use x, y, z or $x_1, x_2, x_3, \ldots$ as variables, and u, v, w or $u_1, u_2, u_3, \ldots$ as parameters. A useful convention is to use a semicolon and write $F(x; u)$ instead of $F(x, u)$ to distinguish variables from parameters.

Definition 2.8. A smooth a–parameter ***family of functions*** on $\mathbb{R}^n$ is a smooth map

$$\begin{aligned} f\colon \mathbb{R}^n \times \mathbb{R}^a &\rightarrowtail \mathbb{R} \\ (x,\, u) &\longmapsto f_u(x). \end{aligned}$$

The notation $f_u(x)$ reflects the interpretation that u is a parameter and x the variable, so we are interested in the behaviour (especially critical points) of the function f_u for each u and how this varies as u changes. As already mentioned, we often write $F(x; u)$ rather than $F(x, u)$ to emphasize this distinction.

In applications x is often called the ***state variable***.

Since we are interested in the critical points of functions, it is natural to make the following definition. Let $f\colon \mathbb{R}^n \times \mathbb{R}^a \rightarrowtail \mathbb{R}$ be a smooth family of functions, as above, and define the ***catastrophe set*** of F to be

$$C_F = \{(x, u) \in \mathbb{R}^n \times \mathbb{R}^a \,|\, \mathrm{d}(f_u)_x = 0\}\,.$$

In other words, it is the set of points $(x, u) \in \mathbb{R}^{n+a}$ such that f_u has a critical point at x:

$$\frac{\partial f_u}{\partial x_1}(x) = \cdots = \frac{\partial f_u}{\partial x_n}(x) = 0.$$

In many important cases, this will be a subset of dimension equal to a (the number of parameters), as we shall see.

An important subset of the catastrophe set is the ***singular set***, denoted Σ_F, which is the set of points in C_F where the critical point is *degenerate*:

$$\Sigma_F := \{(x, u) \in C_F \mid \det(H_{f_u}(x)) = 0\}.$$

Finally, we define the ***discriminant*** or ***bifurcation set*** Δ_F. This is a subset of the set of parameters, equal to the set of parameter values u for which f_u has a degenerate critical point. In other words, if we let π_F be the map projecting C_F onto the parameter space,

$$\begin{aligned} \pi_F\colon C_F &\longrightarrow \mathbb{R}^a \\ (x, u) &\longmapsto u, \end{aligned}$$

then

$$\Delta_F = \pi_F(\Sigma_F) = \{u \in \mathbb{R}^a \mid \exists x \in \mathbb{R}^n,\ (x, u) \in \Sigma_F\}.$$

☆

The reason Δ_F is called the *bifurcation set* is that it is where transitions (or bifurcations) take place. The idea is that if $u_0 \notin \Delta_F$ then there is a neighbourhood U of u_0 such that $\forall v \in U$, the number of critical points of f_v is the same as for f_{u_0}. Thus Δ_F divides $\mathbb{R}^a$ into a number of connected components, and in each of these the number of critical points is constant. We will see this in examples.

In Example 2.7 we saw that the catastrophe set is the parabola

$$C_F = \{(x, u) \in \mathbb{R}^2 \mid 3x^2 + u = 0\}$$

and $\Sigma_F = \{(0, 0)\}$ so $\Delta_F = \{0\} \subset \mathbb{R}$ (ie, the point $u = 0$). The complement of Δ_F in $\mathbb{R}$ consists of two components $\{u < 0\}$ and $\{u > 0\}$, and the number of critical points is 2 and 0 respectively. Notice that C_F can be parametrized by x, with $u = -3x^2$; we will find that catastrophe sets often have good parametrizations.

2.4 Cusp catastrophe

The ***cusp family*** is the 2–parameter family given by

$$F(x; u, v) = \tfrac{1}{4}x^4 + \tfrac{1}{2}ux^2 + vx$$

(the coefficients of $1/4$ and $1/2$ are for convenience). With $u = v = 0$ the function $f_{(0,0)}$ has a critical point at the origin of type A_3 (see Definition 2.5). The catastrophe set C_F is given by the equation $f'_{(u,v)}(x) = 0$, so is given by

$$C_F = \left\{(x, u, v) \in \mathbb{R}^3 \,\middle|\, x^3 + ux + v = 0\right\}.$$

This can be parametrized by (x, u) with $v = -x^3 - ux$ and is the curved sheet in Figure 2.1. In this parametrization, the projection $\pi_f : C_F \to B = \mathbb{R}^2$ is given by

$$\pi_F(x, u) = (u, v) = (u, -x^3 - ux). \tag{2.2}$$

Now consider the Hessian matrix at the critical points (this is why it is useful to be able to parametrize C_F):

$$H_f = f''_{(u,v)}(x) = 3x^2 + u.$$

This means that the singular set Σ_F is

$$\Sigma_F = \left\{(x, u, v) \in C_F \,\middle|\, 3x^2 + u = 0\right\}.$$

So on Σ_F we have $u = -3x^2$. Given that $v = -x^3 - ux$ on C_F we have that

$$\Sigma_F = \left\{(x, u, v) \,\middle|\, u = -3x^2,\ v = 2x^3\right\}.$$

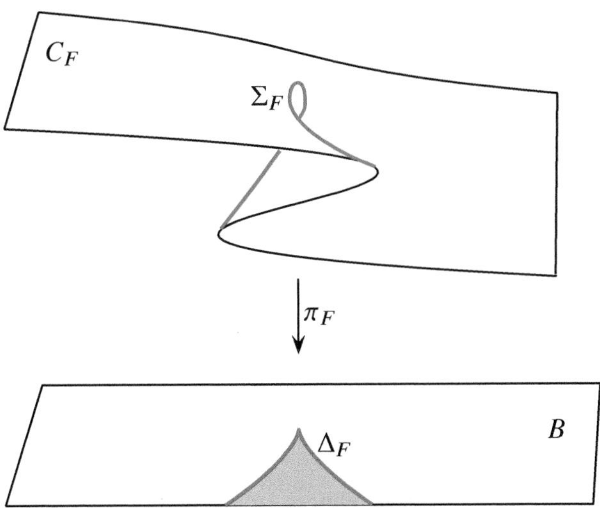

FIGURE 2.1 The cusp catastrophe: x is vertical.

This is a smooth curve in $\mathbb{R}^3$ (parametrized by x) – see the red curve in the top sheet of Figure 2.1.

The image of this singular set in $B = \mathbb{R}^2$ is,

$$\Delta_F = \left\{(u, v) \in \mathbb{R}^2 \,\middle|\, \exists x,\ u = -3x^2,\ v = 2x^3\right\}$$

which is the *discriminant* of F. Eliminating x gives a curve with equation

$$\left(\frac{u}{3}\right)^3 + \left(\frac{v}{2}\right)^2 = 0,$$

or $27v^2 + 4u^3 = 0$, which is the classical ***semicubical parabola*** or ***cusp***, and is why this is called the cusp family. See the red curve on the bottom sheet B in Figure 2.1.

Figure 2.2 shows the graph of $f_{(u,v)}$ for a few different points $(u, v) \in B$. Over points in the grey region in that and Figure 2.1, π_F has 3 preimage points, and so $f_{(u,v)}$ has 3 critical points, and over the other points it has 1, except along the discriminant, where it has precisely two (one of which is degenerate).

In both the Zeeman Catastrophe Machine and the evolute of the ellipse, described in Chapter 1, the catastrophe set has 4 cusp catastrophes occurring in different places, and this is a challenge to visualize globally.

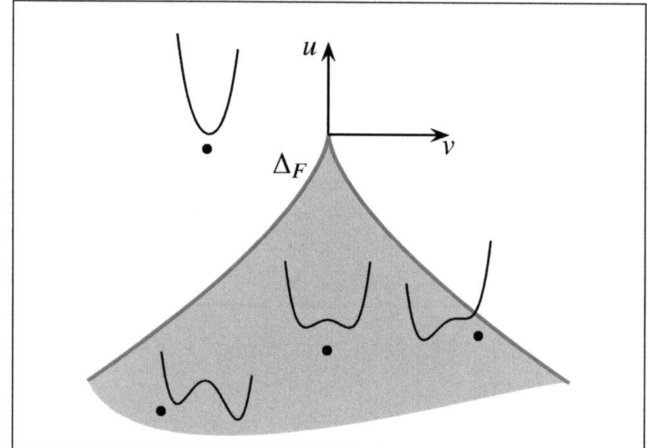

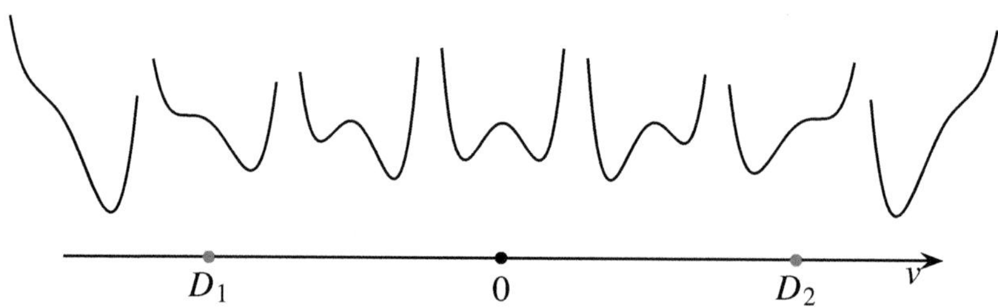

FIGURE 2.2 (TOP) The shape of the graph of $f_{(u,v)}$ in the cusp family, for different values of (u, v). (BOTTOM) The potential function $f_{u,v}(x)$ for $u = -1$ and different values of v. The points D_1 and D_2 represent the points where the line $u = -1$ crosses the discriminant.

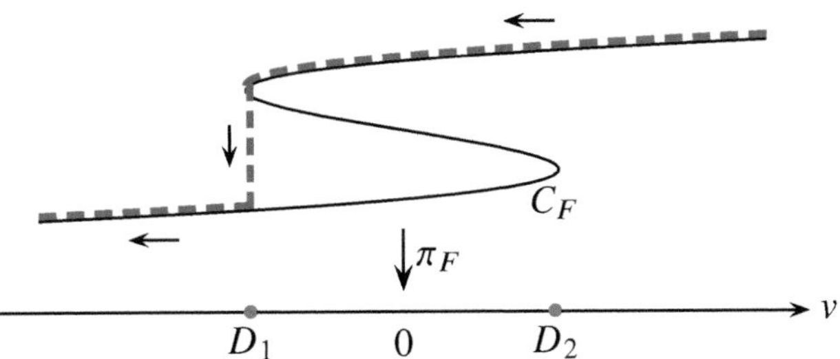

FIGURE 2.3 The effect of moving around the cusp point in the cusp catastrophe. See the text for discussion.

2.5 Why 'catastrophes'

In his fundamental work [111], René Thom describes his approach to 'qualitative' modelling, which we discuss in the preface. He is particularly concerned with the origin of and changes in form or shape, or *morphogenesis*. He covers examples of forms that are physical, biological (at many different scales), or linguistic. A particular assumption in Thom's work is that a form is selected by minimizing some potential function, which justifies the particular interest in critical points of functions.

Thom writes about *regimes of conflict*, which occur where there are coexisting local minima, and a 'catastrophe' occurs when the state of the system jumps from one local minimum to another. There are two types of conflict point, firstly where there are more than one critical point with the same critical value (the importance of these is attributed to Maxwell) and secondly where a local minimum disappears after collision with a saddle point and local maximum, as shown in the figure on p. 19, which is a (local) bifurcation point (a saddle-node or fold bifurcation). See also the changes that occur in Figure 2.2 as v crosses D_1 or D_2; the origin on that axis $v = 0$ corresponds to the Maxwell point.

The catastrophic jump resulting from the local bifurcation is illustrated in Figure 2.3 which represents a tour across (or around) the cusp catastrophe. Compare also with Figs 2.1 and 2.2. The curve in Figure 2.3 is a section through the catastrophe set C_F in Figure 2.1 with $u < 0$ constant (a similar picture is obtained from a path around the cusp point). If the parameters (u, v) are outside the shaded region within the discriminant (say on the side with $v > 0$) then there is only one possible state for the system (the unique minimum in the bottom right graph in Figure 2.2). Call this the top sheet. Now as the parameters vary to cross the discriminant (e.g.

v is decreased), then the state changes slowly as the parameter is varied, until the parameter passes the point D_1, where there is no more 'top sheet', and the system is forced to jump to the other equilibrium, the one on the 'bottom' sheet: this is a discontinuous, or 'catastrophic', change. Reversing the path in parameter space, the system would remain on the lower sheet until the parameter reaches the point D_2, when it would again jump to the upper sheet, giving a hysteresis effect.

Thom's remarkable work shows that for a small number of parameters (he considered up to four), there are only a few local models describing how these bifurcations occur. Without further constraints (such as symmetries, as in the pitchfork bifurcation discussed in Chapter 1), the only structurally stable bifurcations are named as follows:

- for 1 parameter, only the fold (saddle-node),
- for 2 parameters, the cusp,
- for 3 parameters, the swallowtail and the elliptic and hyperbolic umbilics,
- for 4 parameters, the butterfly and the parabolic umbilic.

These seven local models are called the *elementary catastrophes* by Thom, and we discuss these in more detail in Chapter 7 on 'unfoldings'.

Remark 2.9. The two types of conflict set described above give rise to two notions of bifurcation set. The one we have descibed is sometimes called the *local* bifurcation set, to distinguish from the *full* bifurcation set. The latter includes points in parameter space (the Maxwell set) where two or more critical *values* coincide, where a critical value is a value $f(x)$ where x is a critical point. Which type of bifurcation set is relevant to a particular problem depends on the context. Except for a brief metion in the final chapter of the book, we only consider the local bifurcation set as defined above. ❞

Problems

2.1 Let $f\colon \mathbb{R}^2 \to \mathbb{R}$ be the function $f(x, y) = x^4 - 2x^2 + y^2 + 11$. Find all the critical points of f, and determine whether each is nondegenerate, and if it is find its index.

2.2 Repeat the previous question for the function $f\colon \mathbb{R}^3 \to \mathbb{R}$ given by

$$f(x, y, z) = \tfrac{1}{4}x^4 + x^3 - x^2 + 2xy - y^2 - z^2.$$

2.3 The functions $f(x) = x^2 \sin^5(x)$ and $g(x) = x^2 \cos^5(x)$ have critical points at the origin. What are their types (ie A_k for which k)? (†)

2.4 Sketch the catastrophe set of the family $F(x;u) = x^3 - 3u^2x$. What is the singular set for this family? (†)

2.5 Let $F(x, y; u, v) = \frac{1}{3}x^3 - xy^2 - u(x^2 + y^2) - vx$. Find C_F, Σ_F and Δ_F. For each component of the complement of the discriminant, find the number of critical points of $f_{(u,v)}$. What are their indices?

2.6 Let $F(x;u) = \frac{1}{4}x^4 - \frac{3}{2}u^2x^2 + 2u^3x$. Sketch the catastrophe set C_F [*Hint: the expression for C_F factorizes*]. Find Σ_F, and show that $\Delta_F = \mathbb{R}$.

3

The ring of germs of smooth functions

T THE HEART OF ANY BIFURCATION is a degeneracy and for now we leave the world of families of functions and begin a study of degenerate critical points. We will return to families and 'unfoldings' later.

Many of the calculations required in this subject are algebraic, involving rings of smooth functions and their ideals. Since all the results we encounter are local, it is natural to use the language of 'germs', which we now introduce. The set of germs of functions at a point forms a ring, and after defining germs we discuss this ring and the most important ideals and how to represent them and calculate with them.

3.1 Germs: making everything local

Theorems such as the inverse function theorem and implicit function theorem (see Appendix B), as well as many others we will meet later, are of a *local* nature. That is, they state that, given a property of a map at a single point, one can deduce a property that holds on some unspecified neighbourhood of that point. It is not possible in general to do better: if all one knows (as in the inverse function theorem) about f is its differential $\mathsf{d}f_x$, then one cannot possibly estimate how large the open set U might be on which f is a diffeomorphism. For an explicit example, see Problem B.4.

This ignorance of the size of the neighbourhood in question is formalized by the notion of a germ of a function, as follows.

Definition 3.1. Two maps $f, g\colon \mathbb{R}^n \rightarrowtail \mathbb{R}^p$ are said to be ***germ equivalent*** at a point $q \in \mathbb{R}^n$ if q is in the domain of both and there is a neighbourhood U of q such that the restrictions coincide: $f|_U = g|_U$; that is, if,

$$\forall x \in U, \; f(x) = g(x). \qquad \star$$

It is not hard to show this relation is an equivalence relation (see Problem 3.1). A ***map germ*** or ***function germ*** at a point q is an equivalence class of germ equivalent maps or functions.[1] If η is such an equivalence class, then any $f \in \eta$ is called a ***representative*** of η.

Notation Given a map $f\colon \mathbb{R}^n \rightarrowtail \mathbb{R}^p$ with $q \in \operatorname{dom} f$, we denote the germ of f at q by $[f]_q$. In other words, $[f]_q$ is the set of all maps $g\colon \mathbb{R}^n \rightarrowtail \mathbb{R}^p$ with $q \in \operatorname{dom} g$ and for which there is a neighbourhood U of q such that $f\big|_U = g\big|_U$.

To denote the source and target of a germ, one should not write $[f]_q\colon \mathbb{R}^n \rightarrowtail \mathbb{R}^p$ since $[f]_q$ is not a map; indeed as the neighbourhood is unspecified the only point in its 'domain' is the point q itself. Instead we write

$$[f]_q\colon (\mathbb{R}^n, q) \to \mathbb{R}^p, \qquad \text{or} \qquad [f]_q\colon (\mathbb{R}^n, q) \to (\mathbb{R}^p, y)$$

if $y = f(q)$. If $\eta = [f]_q$ we write of course $\eta\colon (\mathbb{R}^n, q) \to \mathbb{R}^p$ etc.

If two maps are germ equivalent at q, then they have the same value there, so if η is a germ, then it makes sense to speak of $\eta(q)$ but not of $\eta(x)$ for $x \neq q$. Similarly, they have the same derivative at q, so $\mathrm{d}\eta_q$ makes sense: it is just $\mathrm{d}f_q$ for any representative f of the equivalence class η. See Problem 3.2.

Germs of other objects We have defined germs of smooth functions, but one can define germs of many other objects – indeed of anything that can be defined locally. For example, if v is a vector field on $\mathbb{R}^n$, then its germ at $q \in \mathbb{R}^n$ is $[v]_q$ which is the equivalence class consisting of all vector fields that coincide with v in some neighbourhood of the point q.

Similarly, if $A, B \subset \mathbb{R}^n$ are two subsets of $\mathbb{R}^n$, one says they are ***germ equivalent*** at q if there is a neighbourhood U of q such that $A \cap U = B \cap U$. In essence the notation $(\mathbb{R}^n, q)$ used above means the germ of $\mathbb{R}^n$ at q.

For example, the sets

$$A = \{(x, y) \in \mathbb{R}^2 \mid xy = 0\} \quad \text{and} \quad B = \{(x, y) \in \mathbb{R}^2 \mid y = 0\}$$

are germ equivalent at $q = (1, 0)$. They are not germ equivalent at the origin, nor at any point where $x = 0$. They are trivially germ equivalent at points such as $(1, 1)$, as this point has a neighbourhood where $A \cap U = B \cap U = \emptyset$. See Problem 3.3.

Given a subset $A \subset \mathbb{R}^n$ we denote the germ of A at q by $[A]_q$ or (A, q). One can therefore write that $[A]_q = [B]_q$, where A, B are the sets given in the paragraph above and $q = (1, 0)$. But $[A]_p \neq [B]_p$ for $p = (0, 0)$.

[1] Recall that we say f is a function if its values are real or complex numbers, otherwise it is a map (see p. 19.)

Notions of germs are often combined: let $f\colon \mathbb{R}^n \rightarrowtail \mathbb{R}^p$ (say) and $A \subset \mathbb{R}^n$, and let $q \in A \cap \mathrm{dom}(f)$. Then it makes sense to define the restriction of the function-germ $[f]_q$ to the set-germ (A, q). See Problem 3.4.

3.2 The ring of germs

Recall that two functions $f, g\colon \mathbb{R}^n \rightarrowtail \mathbb{R}$ are germ equivalent at 0 if there is a neighbourhood U of 0 in $\mathbb{R}^n$ such that $f|_U = g|_U$.

Denote by $\mathcal{E}_n$ the set of all germs at the origin of smooth functions $f\colon \mathbb{R}^n \rightarrowtail \mathbb{R}$ (only those for which $0 \in \mathrm{dom}\, f$ play a role here). It has a natural ring structure, given by addition and multiplication of functions as follows.

Let $\xi, \eta \in \mathcal{E}_n$ be two germs. To add and multiply them, we need to consider representatives. So let $f \in \xi$ and $g \in \eta$, be functions, and let $U = \mathrm{dom}\, f \cap \mathrm{dom}\, g$ (so that f and g are both defined on the open set U which necessarily contains the origin) then the operations are

$$\xi + \eta = \left[f|_U + g|_U\right]_0, \quad \text{and} \quad \xi\eta := \left[f|_U g|_U\right]_0.$$

In other words, to add or multiply two germs, one adds or multiplies representative functions. One has to check that these operations are well-defined (independent of the choice of representative). One also has to check that these do indeed give $\mathcal{E}_n$ the structure of a ring, but this follows easily from the corresponding properties of functions. The reader is encouraged to verify these details; see Problem 3.5.

The ring $\mathcal{E}_n$ has an important ideal $\mathfrak{m}_n$ consisting of all smooth germs vanishing at the origin:

$$\mathfrak{m}_n = \{\eta \in \mathcal{E}_n \mid \eta(0) = 0\}.$$

In fact this is a maximal ideal, and indeed the only maximal ideal in $\mathcal{E}_n$. The important point here is that if $\eta \notin \mathfrak{m}_n$ then η is a unit (that is, it has a multiplicative inverse in the ring, namely $1/\eta$). The reader is also encouraged to establish these statements for themselves. (The algebraic definitions we are taking for granted here are discussed in Appendix D.)

Recall (see Section D.2) that given any collection $\eta_1, \ldots, \eta_r$ of elements of $\mathcal{E}_n$, then the ideal they generate, which we write as $\langle \eta_1, \eta_2, \ldots, \eta_r \rangle$, is the set of all linear combinations of these generators, with coefficients taken from the ring. That is,

$$\langle \eta_1, \eta_2, \ldots, \eta_r \rangle = \left\{ \sum_{j=1}^{r} a_j \eta_j \mid a_j \in \mathcal{E}_n \right\}.$$

As so often with equivalence classes, we work with representatives rather than the whole equivalence class. In practice therefore, we will often just write $f\colon (\mathbb{R}^n, 0) \to \mathbb{R}$ for a germ of a function, rather than the $[f]_0$ or η used above. Thus x^2 in the following example means the germ at 0 of the function $(x, y) \mapsto x^2$. That is, we (lazily!) write $\eta = x^2$ in place of $\eta = \left[(x, y) \mapsto x^2\right]_0$.

It is important to become familiar with calculations with ideals. Almost all the ideals we meet are given in terms of generators (one notable exception is $\mathfrak{m}_n$, and hence the importance of Hadamard's lemma below), and for example if we want to show that one ideal I is contained in another ideal J, it is sufficient to show that each generator of I is contained in J (see Problem D.4).

Examples 3.2. (i). Let $I = \langle x^2, y^2\rangle \lhd \mathcal{E}_2$. Then, for example, $x^3 - xy^2 \in I$ because $x^3 - xy^2 = x(x^2) + (-x)(y^2)$. On the other hand, one can show $x + y \notin I$ (e.g. by showing that every element of I has a critical point at the origin while $x + y$ does not).

(ii). Let J be the ideal $J = \langle x^2 + x^3\rangle \lhd \mathcal{E}_1$, and $I = \langle x^2\rangle$. A short calculation shows that $I = J$. Indeed, $x^2 + x^3 = (1 + x)x^2 \in I$ so that $J \subset I$, and moreover $(1 + x)$ is invertible in $\mathcal{E}_1$ (because it is not in $\mathfrak{m}_1$) and so $x^2 = (1 + x)^{-1}(x^2 + x^3) \in J$, and $I \subset J$. Thus $J = I$.

(iii). Let $J = \langle x^2, x^2y + y^3\rangle$ and $I = \langle x^2, y^3\rangle$. Again we see $I = J$ because firstly x^2 is a generator of both I and J, secondly $x^2y + y^3 = (y)x^2 + y^3 \in I$ showing that $J \subset I$, and conversely, $y^3 = (x^2y + y^3) - y(x^2) \in J$ so $I \subset J$. ✐

The following simple but important result and its corollaries gives us a handle on calculating many ideals we will meet. Obviously, the coordinate functions (and their germs) $x_1, x_2, \dots, x_n$ all vanish at the origin so belong to the maximal ideal $\mathfrak{m}_n$. However, Hadamard's lemma (see below) tells us they generate that ideal, meaning that if $\eta \in \mathfrak{m}_n$ then there are smooth germs $\xi_1 \dots, \xi_n \in \mathcal{E}_n$ for which (as germs)

$$\eta = \sum_{j=1}^{n} x_j \xi_j.$$

The usual statement of Hadamard's lemma is that

$$\mathfrak{m}_n = \langle x_1, x_2, \dots, x_n\rangle .$$

It will be useful to have the following extension of this important fact,

Proposition 3.3 (Hadamard's lemma). *Write $\mathbb{R}^n = \mathbb{R}^a \times \mathbb{R}^b$ (with $a + b = n$), and use coordinates $(x_1, \ldots, x_a, y_1, \ldots, y_b)$. The ideal*

$$I_a = \{f(x, y) \in \mathcal{E}_n \mid f(0, y) \equiv 0\},$$

consisting of germs of functions vanishing on the subspace $\{0\} \times \mathbb{R}^b$, is generated by the x_i coordinates:

$$I_a = \langle x_1, x_2, \ldots, x_a \rangle .$$

The original version of Hadamard's lemma mentioned above is obtained from this by putting $b = 0$, in which case $a = n$ and $I_n = \mathfrak{m}_n$.

PROOF: Since each of the coordinate functions $x_1, \ldots, x_a$ is zero on the subspace $\{0\} \times \mathbb{R}^b$, it is clear that $\langle x_1, x_2, \ldots, x_a \rangle \subset I_a$, so we need only prove the reverse inclusion. Let $\eta \in I_a$, and let f be a representative of η defined on $U = B(0, \varepsilon)$ (for some $\varepsilon > 0$). For $i = 1, \ldots, a$, and $(x, y) \in U$, let

$$g_i(x, y) = \int_0^1 \frac{\partial f}{\partial x_i}(tx, y)\, \mathrm{d}t.$$

Note that $g_i(x, y)$ is a smooth function (because derivatives and integrals of smooth functions are smooth). Then we claim that $f(x, y) = \sum_{i=1}^a g_i(x, y) x_i$, which implies that $f \in \langle x_1, x_2, \ldots, x_a \rangle$, as required.

There remains to prove the claim. By the fundamental theorem of calculus, for each $(x, y) \in U$,

$$f(x, y) = \int_0^1 \tfrac{\mathrm{d}}{\mathrm{d}t} f(tx, y)\, \mathrm{d}t.$$

(Note that if $(x, y) \in U$ then $(tx, y) \in U$ for $t \in [0, 1]$ so this integral is defined.) Now, by the chain rule, the integrand satisfies

$$\frac{\mathrm{d}}{\mathrm{d}t} f(tx_1, tx_2, \ldots, tx_a, y_1, \ldots, y_b) = \sum_{i=1}^n x_i \frac{\partial f}{\partial x_i}(tx, y)$$

so that

$$\begin{aligned} f(x, y) &= \sum_{i=1}^n \int_0^1 x_i \frac{\partial f}{\partial x_i}(tx, y)\, \mathrm{d}t \\ &= \sum_{i=1}^n x_i \int_0^1 \frac{\partial f}{\partial x_i}(tx, y)\, \mathrm{d}t \\ &= \sum_{i=1}^n x_i\, g_i(x, y), \end{aligned}$$

as claimed. ✔

Remark 3.4. In almost all examples, we work with polynomial functions. Since derivatives and integrals of polynomials are polynomials, the proof of Hadamard's lemma carries over to the polynomial setting. Thus, with the notation used above, if a polynomial $p(x, y)$ vanishes on $x = 0$ then there are polynomials $q_1, \ldots, q_a$ such that $p(x, y) = \sum_{j=1}^{a} x_j q_j(x, y)$. However, it should be pointed out that there is a much more direct proof in the polynomial case: see Problem 3.9. ❞

Powers of the maximal ideal are defined inductively: for $r \geq 2$,

$$\mathfrak{m}_n^r := \mathfrak{m}_n . \mathfrak{m}_n^{r-1}.$$

They are also ideals, and by the corollary below are finitely generated.

Example 3.5. For $n = 2$, with coordinates x, y, successive powers of $\mathfrak{m}_2$ are as follows:

$$\begin{aligned}
\mathfrak{m}_2 &= \langle x,\, y \rangle \\
\mathfrak{m}_2^2 &= \left\langle x^2,\, xy,\, y^2 \right\rangle \\
\mathfrak{m}_2^3 &= \left\langle x^3,\, x^2y,\, xy^2,\, y^3 \right\rangle \\
\mathfrak{m}_2^4 &= \left\langle x^4,\, x^3y,\, x^2y^2,\, xy^3,\, y^4 \right\rangle
\end{aligned}$$

etc., and for $n = 3$

$$\mathfrak{m}_3^2 = \left\langle x^2,\, y^2,\, z^2,\, xy,\, xz,\, yz \right\rangle.$$

For example, $f \in \mathfrak{m}_2^3$ if and only if it can be written as

$$f(x, y) = x^3\, g_1(x, y) + x^2y\, g_2(x, y) + xy^2\, g_3(x, y) + y^3\, g_4(x, y),$$

for some smooth function germs $g_1, g_2, g_3, g_4 \in \mathcal{E}_2$.

The reader is invited to write down the 10 monomials of degree 3 in 3 variables (the generators of $\mathfrak{m}_3^3$). ✎

Corollary 3.6. *The ideal $\mathfrak{m}_n^r$ is generated by the monomials of degree r in $x_1, \ldots, x_n$.*

Proof: We proceed by induction on r. For $r = 1$ this is the content of Hadamard's lemma (with $b = 0$). Now suppose the statement is true for $r \leq k$, and let $f \in \mathfrak{m}_n^{k+1}$. Then by definition of this ideal, $f = \sum_j x_j g_j$, where $g_j \in \mathfrak{m}_n^k$, so each g_j is in the ideal generated by the monomials of degree k. It follows that f is in the ideal generated by monomials of degree $k + 1$ as required. ✔

Corollary 3.7. *The germ $f \in \mathcal{E}_n$ is in $\mathfrak{m}_n^r$ if and only if f and all of its partial derivatives of order less than r vanish at the origin.*

PROOF: This is also proved by induction (exercise for the reader). ✔

In particular, suppose $f(0) = 0$. Then f has a critical point at the origin if and only if $f \in \mathfrak{m}_n^2$.

Recall that the r-jet of a function f at the origin (i.e., its Taylor series to degree r) is the sum,

$$\mathrm{j}^r f(0)(x_1, \dots, x_n) = \sum_{|\alpha| \leq r} \frac{1}{\alpha!} \frac{\partial^{|\alpha|} f}{\partial x_1^{\alpha_1} \dots \partial x_n^{\alpha_n}}(0) x_1^{\alpha_1} \dots x_n^{\alpha_n},$$

where $\alpha = (\alpha_1, \dots, \alpha_n)$ is a multi-index (each $\alpha_i \in \mathbb{N}$), and $|\alpha| = \alpha_1 + \alpha_2 + \cdots + \alpha_n$, and finally $\alpha! = (\alpha_1!) \dots (\alpha_n!)$. See Section A.2.

Example 3.8. Let $f(x, y) = \mathrm{e}^{x+y}$. The 3-jet of f at the origin is

$$\mathrm{j}^3 f(0)(x, y) = 1 + x + y + \tfrac{1}{2}x^2 + xy + \tfrac{1}{2}y^2 + \tfrac{1}{6}x^3 + \tfrac{1}{2}x^2y + \tfrac{1}{2}xy^2 + \tfrac{1}{6}y^3.$$

For example, the term x^2 corresponds to $\alpha = (2, 0)$ while $x^2 y$ corresponds to $\alpha = (2, 1)$. (For this function, every partial derivative of any order at the origin is equal to 1.) ✏

The corollary above is therefore saying that $f \in \mathfrak{m}_n^r$ if and only if its Taylor series to degree $r - 1$ at the origin is identically zero.

Rings and jets The k-jet of a smooth function (germ) at the origin is obtained by keeping the terms up to order k in the Taylor series and ignoring the rest. It follows that one can identify the k-jet of a germ $f \in \mathcal{E}_n$ with the image of f under the projection π_k, where

$$\pi_k : \mathcal{E}_n \to \mathcal{E}_n / \mathfrak{m}_n^{k+1}$$

is the ring homomorphism, with kernel $\mathfrak{m}_n^{k+1}$.

If $f \in \mathfrak{m}_n^k$ then the $(k-1)$-jet is zero, and we say f is of ***order*** k, often written $f = O(k)$ (*big-oh* notation).

3.3 Newton diagram

There is a convenient way to represent ideals in $\mathcal{E}_n$, which is particularly useful for $n = 2$, but in principle can be used in any dimension though is obviously harder to visualize.

The skeleton of the Newton diagram for $\mathcal{E}_2$ is the lattice $\mathbb{N}\times\mathbb{N}$ (where $\mathbb{N}$ includes 0) of points in $\mathbb{R}^2$ with non-negative integer coefficients. Each point of the lattice represents a monomial, with (a, b) representing $x^a y^b \in \mathcal{E}_2$. See Figure 3.1A. More generally, for $\mathcal{E}_n$ it is the lattice $\mathbb{N}^n$, with the point $(a_1, a_2, \ldots, a_n) \in \mathbb{N}^n$ representing the monomial $x_1^{a_1} x_2^{a_2} \cdots x_n^{a_n}$.

Suppose now that I is an ideal generated by a single monomial $I = \langle x^a y^b \rangle$. Then a monomial $x^c y^d$ belongs to I if and only if $c \geq a$ and $d \geq b$, so I contains all the monomials corresponding to (c, d) above and to the right of (a, b). So we shade that region of the lattice. See Figure 3.1B for the ideal $\langle x^2 y^3 \rangle$. Figure 3.1C shows the Newton diagram for the ideal $\mathfrak{m}_2^4$.

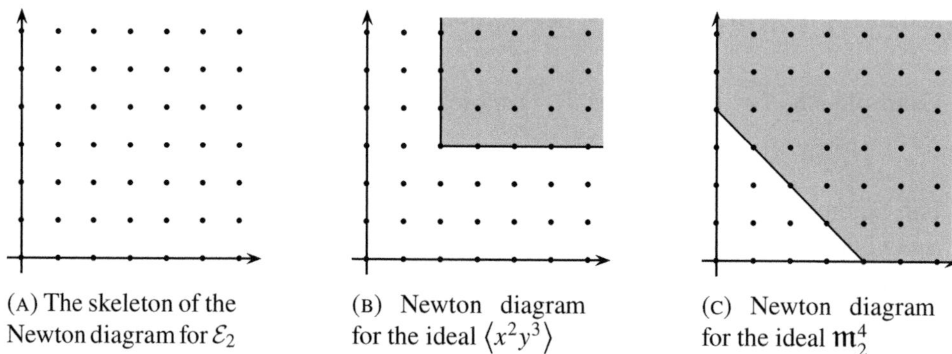

(A) The skeleton of the Newton diagram for $\mathcal{E}_2$

(B) Newton diagram for the ideal $\langle x^2 y^3 \rangle$

(C) Newton diagram for the ideal $\mathfrak{m}_2^4$

FIGURE 3.1 Newton diagrams.

Now suppose I has several generators. If they are all monomials, then one shades the regions from each generator, as in Figure 3.2A. If that is not the case, then there are two steps: firstly find as many generators as possible and shade in the corresponding regions, and then secondly illustrate in some way the fact that some of the unshaded monomials are related (the second part is a help in calculations that follow, but not strictly necessary). In the example shown in Figure 3.2B, the monomials in the ideal are generated by $x^2 y$, y^3 and x^4. There are then two monomials marked with a small square (x^2 and y^2), and two marked with a triangle (x^3 and xy^2) illustrating the fact that $x^2 + y^2$ and $x^3 + xy^2$ are in the ideal but not in the part generated by monomials.

An alternative way of depicting Newton diagrams is shown in Figure 3.3. Notice that in the last figure, the terms x^2 and y^2 are in squares and x^3 and xy^2 are underlined – these latter are alternative to the triangles in Figure 3.2B.

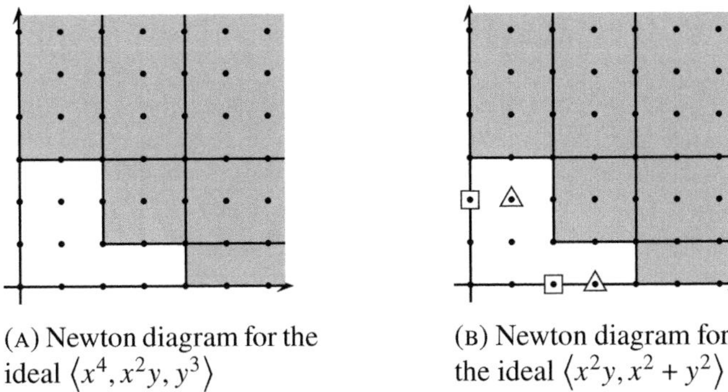

(A) Newton diagram for the ideal $\langle x^4, x^2y, y^3\rangle$

(B) Newton diagram for the ideal $\langle x^2y, x^2+y^2\rangle$

FIGURE 3.2 Newton diagram for more general ideals

3.4 Nakayama's lemma

It is often necessary to show that one ideal is contained in another, for example to show that an ideal is of finite codimension (see next section). For this one frequently uses Nakayama's lemma, which in the context of general rings is as follows:

Lemma 3.9 (Nakayama's lemma). *Let R be a ring, and $\mathcal{M}$ an ideal with the property that $a \in \mathcal{M} \Rightarrow 1+a$ is a unit (i.e., has an inverse in R). Let I, J be ideals in R with I finitely generated. Then*

$$I \subset J + \mathcal{M}I \implies I \subset J.$$

We wil need this general version in Chapter 5, but in the context of the ring of germs, this lemma is applied by setting $R = \mathcal{E}_n$ and $\mathcal{M} = \mathfrak{m}_n$ (which clearly has the required property). Thus, for $I \lhd \mathcal{E}_n$ finitely generated, and any ideal J, Nakayama's lemma states that

$$I \subset J + \mathfrak{m}_n I \implies I \subset J. \tag{3.1}$$

The lemma is stated and proved in a more general context (with modules not just ideals) as Theorem D.20 on p. 379. The following example is a typical application.

Example 3.10. Show $\mathfrak{m}_2^5 \subset \langle x^3, y^3 + x^2y^2\rangle$.

Solution: Let $I = \mathfrak{m}_2^5$ and $J = \langle x^3, y^3 + x^2y^2\rangle$. We first show that $I \subset J + \mathfrak{m}_2 I$, which is $J + \mathfrak{m}_2^6$, and then apply Nakayama's lemma. Denote the two generators of J by α, β (so $\alpha = x^3$, $\beta = y^3 + x^2y^2$), and check each generator of I in turn:

$$x^5 = x^2\alpha, \quad x^4y = xy\alpha, \quad x^3y^2 = y^2\alpha,$$

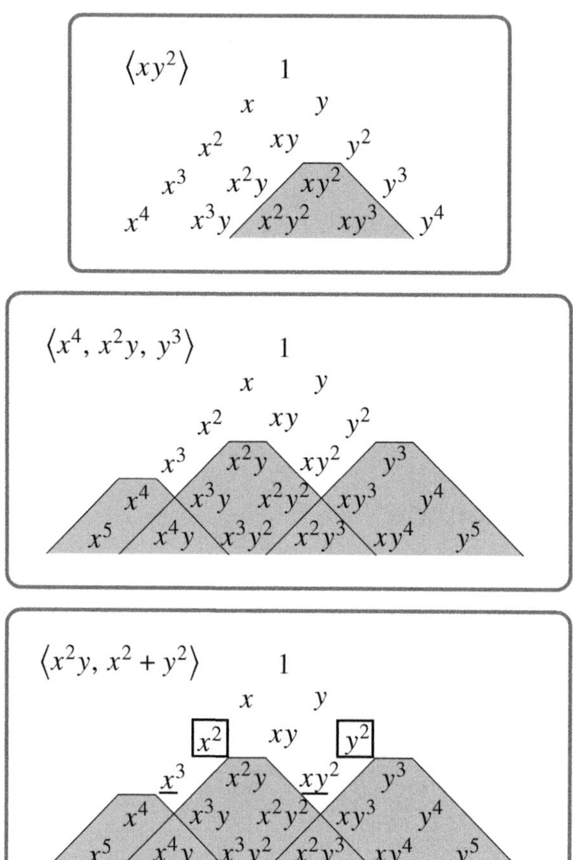

FIGURE 3.3 An equivalent representation of Newton diagrams

$$x^2y^3 = x^2\beta - x^4y^2, \quad xy^4 = xy\beta - x^3y^3,$$

$$\text{and} \quad y^5 = y^2\beta - x^2y^4.$$

The first three show directly that the monomials belong to J, while the others are all written in the form $p = q\beta + r$ with $r \in \mathfrak{m}_2^6$, so showing that $p \in J + \mathfrak{m}_2^6$. Thus $I \subset J + \mathfrak{m}_2 I$ so by Nakayama's lemma $I \subset J$ as required. ✐

It should be emphasized that the hypothesis of I being finitely generated is essential and in fact Nakayama's lemma can be used to prove that certain ideals are not finitely generated. This is in contrast to the ring of polynomials or analytic functions in which every ideal is finitely generated (the Noetherian property – see Section D.6.

3.5 Ideals of finite codimension

The ring $\mathcal{E}_n$ is a vector space, but it is of infinite dimension, which means that there is no finite basis spanning all of $\mathcal{E}_n$. And the same is true of every ideal in $\mathcal{E}_n$ (except for the zero ideal, which consists just of 0), so ideals are not finite–dimensional vector spaces. On the other hand many interesting ideals are of finite *codimension*:

Definition 3.11. An ideal $I \lhd \mathcal{E}_n$ is of ***finite codimension*** if $\mathcal{E}_n/_I$ is a finite–dimensional vector space. This is equivalent to the condition that there is a finite–dimensional vector subspace V of $\mathcal{E}_n$ such that $\mathcal{E}_n = V + I$, so any germ $f \in \mathcal{E}_n$ can be written as $f = g + h$ with $g \in V$ and $h \in I$. A ***cobasis*** for an ideal I of finite codimension in $\mathcal{E}_n$ is a linearly independent set of elements $\{h_1, \ldots, h_r\} \subset \mathcal{E}_n$ such that

$$\mathcal{E}_n = \mathbb{R}\{h_1, \ldots, h_r\} \oplus I. \qquad ✯$$

See also the discussion of cobasis in Section A.1c.

We will frequently use the following notation: if $v_1, \ldots, v_r$ are elements of a vector space V over a field $\mathbb{F}$, then

$$\mathbb{F}\{v_1, v_2, \ldots, v_r\} = \left\{ \sum_{j=1}^{r} \lambda_j v_j \mid \lambda_j \in \mathbb{F} \right\},$$

called the ***span*** over $\mathbb{F}$ of $v_1, \ldots, v_r$ in V.

The direct sum $\oplus$ means in particular that for any $f \in \mathcal{E}_n$ there is a *unique* choice of $a_1, \ldots, a_r \in \mathbb{R}$ and $g \in I$ for which

$$f = \sum_j a_j h_j + g.$$

(See Problem 3.15.)

Example 3.12. The maximal ideal $\mathfrak{m}_n$ is of finite codimension in $\mathcal{E}_n$ because

$$\mathcal{E}_n = \mathbb{R} + \mathfrak{m}_n.$$

So here $V = \mathbb{R}$. This is because for any $f \in \mathcal{E}_n$ one has $f(0) \in \mathbb{R}$ and we let $\bar{f}$ be the germ $\bar{f} = f - f(0)$ which is in $\mathfrak{m}_n$. Then f can be written as

$$f = f(0) + \bar{f} \in \mathbb{R} \oplus \mathfrak{m}_n.$$

Moreover, $\mathcal{E}_n = \mathbb{R} \oplus \mathfrak{m}_n$ (because this decomposition is unique) so $\{1\}$ is a cobasis for $\mathfrak{m}_n$ in $\mathcal{E}_n$. Similarly,

$$\mathcal{E}_n = \mathbb{R}\{1, x_1, x_2, \dots, x_n\} \oplus \mathfrak{m}_n^2,$$

so $\mathfrak{m}_n^2$ is also of finite codimension. More generally, write V_r for the vector space of all polynomials of degree less than r. Then it follows from Taylor's theorem that $\mathcal{E}_n = V_r \oplus \mathfrak{m}_n^r$, so each $\mathfrak{m}_n^r$ is of finite codimension. In fact, V_r has a basis consisting of all monomials of degree less than r, so that this set of monomials forms a cobasis for $\mathfrak{m}_n^r$. ✐

In practice, the easiest way to determine whether a given ideal is of finite codimension is to use the following result.

Proposition 3.13. *An ideal $I \lhd \mathcal{E}_n$ is of finite codimension if and only if there is $r \in \mathbb{N}$ such that $\mathfrak{m}_n^r \subset I$.*

For example, one can see from Figure 3.2B that $\mathfrak{m}_2^4 \subset \langle x^2y,\, x^2 + y^2\rangle$ so that this latter ideal is of finite codimension.

PROOF: Suppose first that $\mathfrak{m}_n^r \subset I$. Let W be the vector space spanned by all monomials of degree *less* than r, which is finite–dimensional. Using Taylor series, let f_r be the Taylor series of f to degree $r - 1$, and write $\bar{f} = f - f_r$. Then the Taylor series of $\bar{f}$ to degree $r - 1$ is zero, which is to say that $\bar{f} \in \mathfrak{m}_n^r$ (by Corollary 3.7). Consequently, every $f \in \mathcal{E}_n$ can be written as a sum,

$$f = f_r + \bar{f} \in W + I,$$

and so $\mathcal{E}_n = W + I$. Since $\dim W < \infty$, this shows I has finite codimension.

The converse is a little more involved, and relies on Nakayama's lemma. For each $r > 0$ define $I_r = I + \mathfrak{m}_n^r$ (which is of finite codimension). Then as r increases, the I_r get smaller:

$$I_1 \supset I_2 \supset \cdots \supset I_{r-1} \supset I_r \supset I_{r+1} \supset \cdots \supset I.$$

Write $c_r = \dim\left(\mathcal{E}/I_r\right)$. Since I is of finite codimension, the c_r are all finite, and it follows that

$$0 \le c_1 \le c_2 \le \cdots \le c_{r-1} \le c_r \le c_{r+1} \le \cdots \le c.$$

Since c is finite, it must happen that two successive c_k are equal say $c_k = c_{k+1}$. Since $I_k \supset I_{k+1}$ and they have the same codimension they must be equal:

$$I + \mathfrak{m}_n^k = I + \mathfrak{m}_n^{k+1}.$$

This implies in particular that $\mathfrak{m}_n^k \subset I + \mathfrak{m}_n^{k+1}$ so that by Nakayama's Lemma $\mathfrak{m}_n^k \subset I$ as required. ✔

Remark 3.14. Consider the ring $R = \mathbb{R}[x_1, \ldots, x_n]$ of polynomials in n (real) variables, and let $J \lhd R$ be an ideal of finite codimension. By general theory, J is finitely generated (R is a Noetherian ring; see Appendix D). Now let $I \lhd \mathcal{E}_n$ be the ideal in $\mathcal{E}_n$ generated by the generators of J. Then it follows from the arguments above that I is also of finite codimension in $\mathcal{E}_n$, and that a cobasis for J in R is also a cobasis for I in $\mathcal{E}_n$. ❞

3.6 Geometric criterion for finite codimension

If I is an ideal generated by polynomials, or more generally by analytic functions, then there is a geometric criterion for finite codimension which can be very useful. If $I \subset \mathfrak{m}_n$ then every element of I vanishes at the origin. The geometric criterion says that I is of finite codimension if, when considering the generators as complex functions, the origin is the only common zero in a neighbourhood of the origin for the set of generators of I. More precisely, we have the following theorem.

Theorem 3.15 (Geometric criterion). *Let $I = \langle h_1, \ldots, h_k \rangle \subset \mathfrak{m}_n$ be an ideal generated by finitely many analytic germs. Then I has finite codimension in $\mathcal{E}_n$ if and only if the origin is isolated in the set of complex zeros of I; that is, there is a neighbourhood U of the origin in $\mathbb{C}^n$ such that $U \cap V_{\mathbb{C}}(I) = \{0\}$, where*

$$V_{\mathbb{C}}(I) := \{z \in \mathbb{C}^n \mid h_1(z) = \cdots = h_k(z) = 0\}.$$

It is important that $V(I)$ is taken in $\mathbb{C}^n$ and not just in $\mathbb{R}^n$. A simple example illustrating this is to take $I = \langle x^2 + y^2 \rangle \subset \mathfrak{m}_2$. This does not have finite codimension, although the real version of $V(I)$ consists only of the origin. The set $V_{\mathbb{C}}(I) \subset \mathbb{C}^2$ consists of two complex 'lines', given by $x = \pm \mathrm{i}y$, so the origin is not isolated.

This is a consequence of the *Nullstellensatz*, and is discussed in Section D.10, and in particular Corollary D.28.

Problems

3.1 Show that germ equivalence of maps is indeed an equivalence relation.

3.2 If f and g are germ equivalent functions at 0, show that they have the same derivative at x, and more generally the same Taylor series, and hence the same k-jets at 0, for all $k \in \mathbb{N}$.

3.3 Let $A = \{(x, y) \in \mathbb{R}^2 \mid xy = 0\}$ and $B = \{(x, y) \in \mathbb{R}^2 \mid y = 0\}$. Show A and B are germ equivalent at any point $q = (u, v)$ if and only if $u \neq 0$. [You will have to treat several cases.] (†)

3.4 Let $\eta\colon (\mathbb{R}^n, x) \to \mathbb{R}^p$ be a map germ, and let $A \subset \mathbb{R}^n$ with $x \in A$. Show that there is a well-defined map germ $\eta\big|_A\colon [A]_x \to \mathbb{R}^p$. That is, show that if $f, g \in \eta$ and $B \subset \mathbb{R}^n$ is such that $[A]_x = [B]_x$ then $f\big|_A$ and $g\big|_B$ are germ equivalent at x. [Here $f\big|_A$ denotes the restriction of f to A, and is a map with domain equal to $A \cap \operatorname{dom}(f)$.]

3.5 Show that addition and multiplication of functions determines a well-defined ring structure on the set of germs $\mathcal{E}_n$. In particular, show that if $[f_1]_0 = [f_2]_0$ (so f_1 and f_2 are germ equivalent), and $[g_1]_0 = [g_2]_0$, then $[f_1 + g_1]_0 = [f_2 + g_2]_0$ and $[f_1 g_1]_0 = [f_2 g_2]_0$. (†)

3.6 Let $A \subset \mathbb{R}^n$ be any subset, and consider the set of functions defined on neighbourhoods of A. By replacing q by A in Definition 3.1 write down a definition of two functions being ***germ equivalent along*** A. Show that the ring structure in Problem 3.5 continues to hold for these germs. (†)

3.7 Show that $\mathfrak{m}_n$ is the *only* maximal ideal in $\mathcal{E}_n$, making $\mathcal{E}_n$ into a *local ring* (which is a ring with a unique maximal ideal).

3.8 Let $V \subset (\mathbb{R}^n, 0)$ be any subset germ. Show that the set $I(V) := \{\eta \in \mathcal{E}_n \mid \eta\big|_V = 0\}$ is an ideal in $\mathcal{E}_n$ (here as in Problem 3.4 above, this notation means that there is a representative function $h \in \eta$ for which $x \in V \cap \operatorname{dom}(h) \Rightarrow h(x) = 0$).

3.9 Let $p(x, y)$ be a polynomial, for $x \in \mathbb{R}^a$, $y \in \mathbb{R}^b$, and suppose $p(0, y) = 0$. Show directly (without using Hadamard's lemma) that there are polynomials $q_1, \dots q_a$ such that $p(x, y) = \sum_i x_i q_i(x, y)$. [*Hint*: *A polynomial is a (finite) sum of monomials*; *separate out those that are independent of the* x_i.]

3.10 Let $\phi\colon (\mathbb{R}^n, 0) \to (\mathbb{R}^p, 0)$ be a map germ and define the *pullback* map $\phi^*\colon \mathcal{E}_p \to \mathcal{E}_n$ by $\phi^*(g) = g \circ \phi$ (i.e. $\phi^*(g)$ is the function germ defined on any neighbourhood by $(\phi^* g)(x) := g(\phi(x))$). Show
(i) that ϕ^* is well-defined at the level of germs,
(ii) that ϕ^* is a ring homomorphism, and
(iii) if $n = p$ and ϕ is a diffeomorphism germ then ϕ^* is an isomorphism. (†)

3.11 Let $I = \langle ax^r + bx^s \rangle \lhd \mathcal{E}_1$, with $a \neq 0$ and $s > r$. Show that $I = \langle x^r \rangle$.

3.12 By induction on k (or otherwise), show that if $f \in \mathfrak{m}_n^k$, and $\phi\colon (\mathbb{R}^n, 0) \to (\mathbb{R}^n, 0)$ is a diffeomorphism germ then $f \circ \phi \in \mathfrak{m}_n^k$.

3.13 Let $I = \langle x^2, y^2 \rangle$. Show that if $f \in I$ then $f(0) = f_x(0) = f_y(0) = f_{xy}(0) = 0$ (where $f_x = \frac{\partial f}{\partial x}$ etc). Deduce that $xy \notin I$.

3.14 Show that $y^5 \in \langle x^2, xy + y^3 \rangle$. (†)

3.15 Some linear algebra: let W be a real vector space and $U, V \subset W$ be two subspaces such that $W = U + V$ (see Appendix A.1c for details).

(i). Suppose that V is finite–dimensional. Show that W/U is finite-dimensional. [*Hint*: *either use one of the fundamental isomorphism theorems, or show that any finite set of vectors which spans V provides a finite set of vectors spanning* W/U.]

(ii). Show moreover that if V is finite–dimensional and $W = U \oplus V$ then any basis of V descends to a basis of W/U, and that given such a basis $\{\mathbf{e}_1, \ldots, \mathbf{e}_r\}$ of V then every element $w \in W$ can be written *uniquely* as

$$v = \sum_j \lambda_j \mathbf{e}_j + u$$

with $u \in U$ and $\lambda_1, \ldots, \lambda_r \in \mathbb{R}$.

3.16 Use the previous problem to justify the statements in Example 3.12.

3.17 (i). Show $\mathfrak{m}_2^3 \subset \langle x^2, y^2 \rangle$.

(ii). Show $\mathfrak{m}_2^4 \subset \langle x^3 + y^3, xy \rangle$.

(iii). Find the smallest value of k such that $\mathfrak{m}_2^k \subset \langle x^3, y^3 \rangle$.

(iv). Find the smallest value of k such that $\mathfrak{m}_3^k \subset \langle x^3, y^3, z^2 \rangle$.

For each of the ideals in this question, find a cobasis for I in $\mathcal{E}_2$ or $\mathcal{E}_3$.

3.18 Use Nakayama's lemma to show each of the following inclusions:

(i). $\mathfrak{m}_2^3 \subset \langle x^2, y^2 + x^2 y \rangle$.

(ii). $\mathfrak{m}_2^4 \subset \langle x^3 + y^3, xy + y^3 \rangle$.

3.19 Let $I = \langle x^a, y^b \rangle \lhd \mathcal{E}_2$. By considering the Newton diagram (or otherwise), show that $\dim \left(\mathcal{E}_2/I\right) = ab$.

3.20 Let $I \lhd \mathcal{E}_2$ be an ideal. Show that I is of finite codimension if and only if $\exists k \geq 0$ such that $x^k, y^k \in I$.

3.21 Extend the previous two exercises to n variables.

3.22 Show by induction on n (or otherwise) that there are $\binom{n+r-1}{r}$ monomials of degree r in n variables. (†)

4

Right equivalence

In singularity theory, there are several notions of equivalence of germs, all based on changes of coordinates. The idea is that any 'intrinsic' property of a germ should be independent of the particular coordinates used. The basic equivalence, which suffices for germs of functions $f\colon (\mathbb{R}^n, 0) \to \mathbb{R}$, is *right equivalence*, and involves just changing coordinates in the source $\mathbb{R}^n$ (but not the target $\mathbb{R}$), and is the subject of this chapter. For map germs $f\colon (\mathbb{R}^n, 0) \to (\mathbb{R}^p, 0)$ with $p > 1$ right equivalence is not a broad enough equivalence relation to be useful, and one needs to extend it to *left–right equivalence* or the even broader notion of *contact equivalence*; these involve changes of coordinates in both source ($\mathbb{R}^n$) and target ($\mathbb{R}^p$), and are discussed in Part II. If the reader is not completely comfortable with the notion of diffeomorphism and the inverse function theorem, they should consult Appendix B, and especially Secs B.1 and B.2.

4.1 Right equivalence

From now, and for the following few chapters, we consider just germs of functions ($p = 1$, suitable for variational problems). The definition applies more generally but turns out to be less useful if $p > 1$.

Definition 4.1. Two map germs $f, g\colon (\mathbb{R}^n, 0) \to \mathbb{R}$ are ***right equivalent***, or $\mathcal{R}$-***equivalent***, if there is a diffeomorphism germ ϕ of $(\mathbb{R}^n, 0)$ such that

$$f = g \circ \phi.$$

In other words, f is transformed into g by a change of coordinates. One often writes $f \sim_{\mathcal{R}} g$ in this case.

Notice that in particular $f(0) = g(0)$ is necessary for $\mathcal{R}$-equivalence. This is not always a desirable restriction, and it is useful to extend the equivalence to one called $\mathcal{R}^+$-***equivalence*** by allowing f and g to be $\mathcal{R}^+$-equivalent if

$$f = g \circ \phi + a$$

for some constant $a \in \mathbb{R}$. ★

Example 4.2. The germs $f, g\colon (\mathbb{R}, 0) \to \mathbb{R}$ given by $f(x) = x^2 - x^4$ and $g(x) = x^2$ are $\mathcal{R}$-equivalent. To see this, let $y = \phi(x)$ be the change of coordinate. We require

$$f(x) = g \circ \phi(x) = g(y),$$

which means $x^2 - x^4 = y^2$. Taking the positive square root gives $y = x\sqrt{1 - x^2}$. By the inverse function theorem $\phi(x) = x\sqrt{1 - x^2}$ is a diffeomorphism in a neighbourhood of 0 (as $\phi'(0) = 1 \neq 0$) and so gives the desired change of coordinate.

We can understand this example geometrically by looking at the graphs of f and g: ϕ is providing the necessary shrinkage of a neighbourhood of 0 so that one graph is taken to the other.

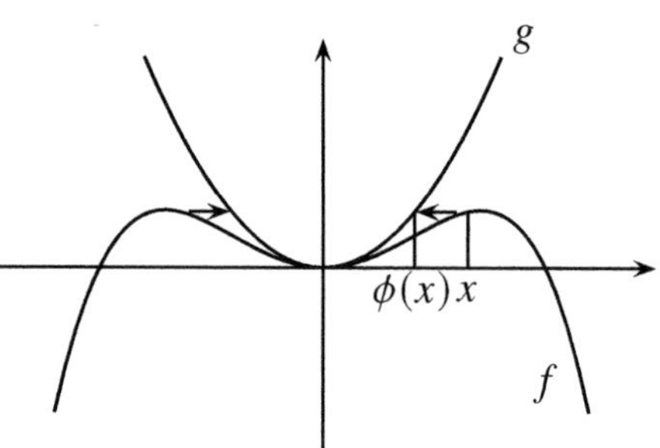

Example 4.3. The germs at 0 of $f(x) = x^3$ and $g(x) = x^5$ are not $\mathcal{R}$-equivalent. Suppose to the contrary that they are; so there is a diffeomorphism (germ) ϕ such that $g \circ \phi = f$. That is, $\phi(x)^5 = x^3$, so $\phi(x) = x^{3/5}$. This map is not differentiable at 0, so is not a diffeomorphism.

A similar argument shows that x^a and x^b are not equivalent unless $a = b$. If we allowed homeomorphisms (ϕ and ϕ^{-1} continuous, not necessarily differentiable) then x^3 and x^5 become equivalent, as the map $\phi\colon \mathbb{R} \to \mathbb{R}$, $\phi(x) = x^{3/5}$ is a homeomorphism. This marks the difference between the study of *topology* (using homeomorphisms) and *differential topology* (using diffeomorphisms).

The reader may have realized that it is straightforward to determine whether or not two germs of 1 variable with non-zero Taylor series are $\mathcal{R}$-equivalent. For more than 1 variable, this is not easy, and much of the following few chapters is dedicated to this end.

Remark 4.4. We have been discussing germs of smooth functions of real variables, $f\colon (\mathbb{R}^n, 0) \to \mathbb{R}$. One could equally well study right equivalence for germs of complex functions, $f\colon (\mathbb{C}^n, 0) \to \mathbb{C}$. Recall that to be differentiable, a complex function is necessarily analytic, or holomorphic, and this applies to the diffeomorphism ϕ of the definition of $\mathcal{R}$-equivalence as well. The ring of germs of holomorphic functions in n (complex) variables is denoted $\mathcal{O}_n$. The theory and examples we encounter below hold equally well in $\mathcal{O}_n$ as in $\mathcal{E}_n$, as well as in the ring of germs of real-valued analytic functions. For example, in the definition below, the Jacobian ideal becomes an ideal in $\mathcal{O}_n$ instead of $\mathcal{E}_n$, but it has the same expression. The main difference is that while, for example, the germs x^2 and $-x^2$ are not $\mathcal{R}$-equivalent for real functions, they are for complex functions (using the change of coordinates $\phi(x) = ix$). ❞

4.2 Jacobian ideal

Critical points are determined by the partial derivatives of the function, and this suggests forming the ideal they generate, as follows.

Definition 4.5. Let $f\colon (\mathbb{R}^n, 0) \to \mathbb{R}$ be a smooth function-germ. The ***Jacobian ideal*** Jf of f is the ideal in $\mathcal{E}_n$ generated by the partial derivatives of f. ✯

So for a germ $f \in \mathcal{E}_n$, we have

$$Jf \;:=\; \left\langle \frac{\partial f}{\partial x_1}, \frac{\partial f}{\partial x_2}, \ldots, \frac{\partial f}{\partial x_n} \right\rangle \lhd \mathcal{E}_n.$$

Example 4.6. (i) Let $f(x, y) = x^2 - y^2 \in \mathcal{E}_2$ then $Jf = \langle 2x, -2y\rangle = \mathfrak{m}_2$.
(iii) Let $f(x, y) = x^2y + y^4$. Then $Jf = \langle 2xy, x^2 + 4y^3\rangle = \langle xy, x^2 + 4y^3\rangle$.
(iii) Let $f(x, y, z) = x^2 + y^2 + z^2 \in \mathcal{E}_3$. Then

$$Jf = \langle 2x, 2y, 2z\rangle = \langle x, y, z\rangle = \mathfrak{m}_3.$$

(iv) More generally, if $f = \sum_{j=1}^n x_j^2 \in \mathcal{E}_n$ then $Jf = \mathfrak{m}_n$. ✐

Recall from Chapter 2 that a function $f\colon \mathbb{R}^n \to \mathbb{R}$ has a ***critical point*** at 0 if $\mathsf{d}f_0 = 0$; that is, if

$$\frac{\partial f}{\partial x_1}(0) = \frac{\partial f}{\partial x_2}(0) = \cdots = \frac{\partial f}{\partial x_n}(0) = 0.$$

Essential observation: f has a critical point at the origin if and only if each of the generators of Jf belongs to $\mathfrak{m}_n$, and hence if and only if $Jf \subset \mathfrak{m}_n$.

4.3 Codimension

Since we are not actually interested in the value of f at the critical point, we will assume for the moment that $f(0) = 0$, so $f \in \mathfrak{m}_n$. In this case, f has a critical point at 0 if and only if $f \in \mathfrak{m}_n^2$. Since we are studying critical points in this chapter, from now on we assume that $f \in \mathfrak{m}_n^2$, and (as pointed out above) $Jf \subset \mathfrak{m}_n$.

Recall that an ideal $I \lhd \mathcal{E}_n$ is of finite codimension if the vector space ${\mathcal{E}_n}/{I}$ is finite–dimensional (see Section 3.5).

Definition 4.7. A germ $f \in \mathfrak{m}_n^2$ is said to be of ***finite codimension*** if the Jacobian ideal is of finite codimension in $\mathfrak{m}_n$, and in this case we say the ***codimension*** of f is

$$\operatorname{codim}(f) := \dim \left({\mathfrak{m}_n}/{Jf} \right).$$

This number is finite if and only if f is of finite codimension. ☆

The codimension of a critical point is a measure of its complexity, as will be understood later.

Examples 4.8. (i). If $f = x_1^2 + x_2^2 + \cdots + x_n^2 \in \mathcal{E}_n$ then $Jf = \mathfrak{m}_n$, so $\operatorname{codim}(f) = 0$.

(ii). For $f = x^3 + y^3 \in \mathcal{E}_2$ one has $Jf = \langle x^2, y^2 \rangle$. Using the Newton diagrams discussed in Chapter 3, one sees that $\{x, y, xy\}$ forms a cobasis for Jf in $\mathfrak{m}_2$ so f has codimension 3.

(iii). Let $f(x, y) = x^2 y$. Then $Jf = \langle xy, x^2 \rangle$ and f is of infinite codimension, as for all $k > 0$, $y^k \notin Jf$ (cf. Proposition 3.13). ✐

We show in Problem 4.9 that $\mathcal{R}^+$-equivalent germs have the same codimension.

4.4 Nondegenerate critical points

We continue with the study of germs of functions, so $f \colon (\mathbb{R}^n, 0) \to \mathbb{R}$. And we assume $f(0) = 0$ for simplicity, so $f \in \mathfrak{m}_n$. And f has a critical point (at 0) if and only if $f \in \mathfrak{m}_n^2$.

A very important case is that of *nondegenerate* critical points. Recall that a function germ $f \in \mathfrak{m}_n^2$ has a ***nondegenerate*** critical point if the Hessian matrix $\mathsf{d}^2 f(0)$ is nondegenerate. An example is the function (germ) $f = x^2 + y^2 + z^2$ in the example above, as $\mathsf{d}^2 f(0) = 2I$.

Proposition 4.9. *A germ $f \in \mathcal{E}_n$ has a nondegenerate critical point if and only if $Jf = \mathfrak{m}_n$.*

Notice that by definition, $Jf = \mathfrak{m}_n$ if and only if $\operatorname{codim}(f) = 0$: nondegenerate critical points are precisely those of codimension 0.

PROOF: The proof of the 'only if' part is based on Nakayama's lemma as follows. Assume f has a nondegenerate critical point at 0 and let $Q = \mathsf{d}^2 f_0$, which is therefore a nondegenerate symmetric matrix. Then by Taylor's theorem with remainder, we can write $f(x) = \frac{1}{2}x^T Qx + h(x)$ with $h \in \mathfrak{m}_n^3$ so that for each i, $\frac{\partial f}{\partial x_i} = \sum_j q_{ij}x_j + \frac{\partial h}{\partial x_i}$. Clearly $\frac{\partial h}{\partial x_i} \in \mathfrak{m}_n^2$ so it follows that

$$Jf + \mathfrak{m}_n^2 = \left\langle \sum_j q_{ij}x_j \mid i = 1, \dots, n \right\rangle + \mathfrak{m}_n^2.$$

Write $I = \langle \sum_j q_{ij}x_j \mid i = 1, \dots, n \rangle$. We claim that $I = \mathfrak{m}_n$.

It is clear that $I \subset \mathfrak{m}_n$. To prove the reverse inclusion let $A = Q^{-1}$ (the nondegeneracy implies Q is invertible). Then

$$x_k = \sum_j \left(\sum_i a_{ki}q_{ij} \right) x_j = \sum_i a_{ki} \left(\sum_j q_{ij}x_j \right) \in I,$$

and this is true for each k so that indeed $\mathfrak{m}_n \subset I$. Therefore we have that

$$\mathfrak{m}_n = Jf + \mathfrak{m}_n^2,$$

and it follows from Nakayama's lemma that $\mathfrak{m}_n = Jf$.

Now we prove the 'if' statement, which proceeds by contradiction. Firstly, if f does not have a critical point at 0 then $Jf = \mathcal{E}_n \neq \mathfrak{m}_n$. Now suppose f does have a critical point but that it is degenerate. Then, using Taylor's theorem again, we write $f(x) = \frac{1}{2}x^T Qx + h(x)$, with now $Q = \mathsf{d}^2 f_0$ being degenerate, and as before $h \in \mathfrak{m}_n^3$. Suppose Q has rank $m < n$ (as Q is degenerate). Choose a basis in $\mathbb{R}^n$ so that

$$Q = \begin{pmatrix} Q' & 0 \\ 0 & 0 \end{pmatrix},$$

where Q' is an invertible $m \times m$ symmetric matrix. Write the corresponding coordinates as $(x_1', \dots, x_m', y_1, \dots, y_{n-m})$. Then

$$f(x', y) = \tfrac{1}{2}(\mathbf{x}')^T Q'\mathbf{x}' + h(x', y)$$

with $h \in \mathfrak{m}_n^3$. In these coordinates,

$$\begin{aligned}\frac{\partial f}{\partial x_i'} &= \sum_{j=1}^m q_{ij}' x_j' + \frac{\partial h}{\partial x_i'} \quad (i = 1, \ldots, m)\\ \frac{\partial f}{\partial y_j} &= \frac{\partial h}{\partial y_j} \quad (j = 1, \ldots, n-m).\end{aligned}$$

Thus $Jf + \mathfrak{m}_n^2 = \left\langle \sum_{j=1}^m q_{ij}' x_j' \mid i = 1, \ldots, m \right\rangle + \mathfrak{m}_n^2$, so in particular $y_j \notin Jf$, so $\mathfrak{m}_n \not\subset Jf$ as required. ✔

Proposition 4.10. *Suppose $f\colon (\mathbb{R}^n, 0) \to (\mathbb{R}, 0)$ has a nondegenerate critical point at 0, then there is a change of coordinates ϕ such that*

$$f \circ \phi(x) = \tfrac{1}{2} x^T \, \mathrm{d}^2 f_0 \, x.$$

In short, f is right equivalent to its quadratic part (Taylor series to degree 2). If one does not assume $f(0) = 0$ then the expression should be modified to

$$f \circ \phi(x) = f(0) + \tfrac{1}{2} x^T \, \mathrm{d}^2 f_0 \, x. \tag{4.1}$$

We will see in Example 5.12 that this result is a consequence of the main 'finite determinacy theorem' (Theorem 5.10), but for now we outline a direct proof.

Outline proof: Since $f \in \mathfrak{m}_n^2$ we can use the corollary to Hadamard's lemma (Corollary 3.6) to write

$$f(x) = \sum_{i,j} \psi_{ij}(x) x_i x_j = x^T \Psi(x) x.$$

Here $\psi_{ij}(x)$ are smooth functions and the square matrix $\Psi(x) = (\psi_{ij}(x))$ can be chosen to be symmetric. Then $\mathrm{d}^2 f(0) = \Psi(0)$ which is therefore nondegenerate. Now, nearby nondegenerate quadratic forms are similar, so for each x near 0 there is an invertible matrix P_x, with $P_0 = I$, such that

$$\Psi(x) = P_x^T \, \Psi(0) \, P_x.$$

It follows from the inverse function theorem that the map $x \mapsto P_x x$ is a diffeomorphism germ at the origin, so we can define new coordinates $y = P_x x$. Then

$$f(x) = x^T \Psi(x) x = x^T P_x^T \Psi(0) P_x x = y^T \Psi(0) y.$$

This is the required statement, where ϕ is defined by $x = \phi(y)$ (the inverse of $y = P_x x$). ✔

Recall from linear algebra that any quadratic form can be diagonalized by a change of basis, and if it is nondegenerate the diagonal terms can be made equal to ± 1. This leads to the following important result.

Corollary 4.11 (Morse lemma). *If $f\colon (\mathbb{R}^n, 0) \to (\mathbb{R}, 0)$ has a nondegenerate critical point at 0 then there is a change of coordinates ϕ for which,*

$$f \circ \phi(x) = \pm x_1^2 \pm x_2^2 \pm \cdots \pm x_n^2.$$

Remark 4.12. When a nondegenerate critical point is put in this 'normal form', the number of negative squares is the index of the critical point (Definition 2.4). The Morse lemma and the index were introduced by Marston Morse in the 1930s when he showed that the Morse indices of the critical points of a smooth function on a manifold give information about the global nature (topology) of that manifold. See for example, J. Milnor [77]. The index is consequently often known as the *Morse index* of the critical point. ❞

4.5 Splitting Lemma

The splitting lemma is important as a first step in classifying degenerate critical points in that it separates out the degenerate part from the nondegenerate part. It is very similar to choosing linearly adapted coordinates for a map which has non-maximal rank (see Appendix B).

Consider the simple example, $f(x,u) = x^2 + 2xu^2$. The origin is a critical point of f, but the Hessian there is degenerate. Now complete the square to write f as $f(x,u) = (x+u^2)^2 - u^4$. If we change coordinates, and put $X = x + u^2$, then $f(x,u) = F(X,u) = X^2 - u^4$. And in this new expression for the function the nondegenerate part X^2 is split from the degenerate part $-u^4$. This is the idea of the splitting lemma – though in general it is not possible to find the change of coordinates explicitly.

Theorem 4.13 (Splitting lemma). *Let $f \in \mathcal{E}_{m+k}$, which we write as $f(x, u)$ with $x \in \mathbb{R}^m$, $u \in \mathbb{R}^k$. Suppose the restriction of f to $\mathbb{R}^m \times \{0\}$ has a nondegenerate critical point at $x = 0$. Then there is a change of coordinates in a neighbourhood of the origin $(x,u) = (x(X,u), u)$ such that*

$$f(x(X,u),\, u) = Q(X) + h(u),$$

where Q is the restriction of $\frac{1}{2}\mathrm{d}^2 f$ to $\mathbb{R}^m \times \{0\}$, and h is a smooth function.

Furthermore, the 'remainder function' h can be found implicitly as follows: for each u near 0 there is a unique point $x = \chi(u)$ such that

$$\mathsf{d}_x f(\chi(u), u) = 0.$$

Then

$$h(u) = f(\chi(u), u). \tag{4.2}$$

Here $\mathsf{d}_x f(x,u)$ is the differential of f with respect to the x variables alone (that is, partial differentiation holding u constant).

Remarks 4.14. (1) This result is sometimes called the parametrized Morse lemma, where the variables u take the place of parameters.
(2) Notice that it is only the x-coordinate that is changed, while the u-coordinate remains unchanged.
(3) One can further simplify Q by diagonalizing it so that $Q(X)$ takes the form $\sum \pm X_i^2$.
(4) The expression $\chi(u)$ can be found in principle by using the implicit function theorem (as one sees from the proof). In practice, this means that the Taylor series of h can be found to any given order. ❞

The application of the splitting lemma in singularity theory usually involves the following set-up. Recall that the ***corank*** of a square matrix is the dimension of its kernel.

Corollary 4.15. *If $f \in \mathcal{E}_n$ has a critical point at the origin with Hessian matrix of corank k then there are coordinates $X \in \mathbb{R}^{n-k}$ and $u \in \mathbb{R}^k$ such that*

$$f = Q(X) + h(u),$$

where Q is a nondegenerate quadratic form and $h \in \mathfrak{m}_k^3$.

Proof of corollary: Since the Hessian matrix of f has corank k, one can choose a basis so that it takes the form

$$\begin{pmatrix} A & 0 \\ 0 & 0 \end{pmatrix}$$

where A is an invertible symmetric $(n-k) \times (n-k)$ matrix. Call the coordinates relative to this basis $(x_1, \ldots, x_{n-k}, u_1, \ldots, u_k)$, so that the form of the matrix means that $\mathsf{d}_x^2 f = A$, which is invertible, $\mathsf{d}_u^2 f(0) = 0$ and $\mathsf{d}_u \mathsf{d}_x f(0) = 0$. Then the hypotheses of the splitting lemma are satisfied, and after a further change of coordinates of the form $(x, u) \mapsto (X(x, u), u)$ we can write f in the form $f(x, u) = Q(X) + h(u)$.

There remains to show that $h \in \mathfrak{m}_k^3$. Clearly, $h \in \mathfrak{m}_n^2$ as otherwise f would not have a critical point at 0. Furthermore, if we represent the quadratic from Q by a symmetric matrix $\widehat{Q}$, so that $Q(X) = X^T \widehat{Q} X$, then

$$\mathsf{d}^2 f_0 = \begin{pmatrix} 2\widehat{Q} & 0 \\ 0 & \mathsf{d}^2 h_0 \end{pmatrix}.$$

Since $\widehat{Q}$ is invertible of rank m it follows that $m = \mathrm{rk}(\mathsf{d}^2 f_0) = m + \mathrm{rk}(\mathsf{d}^2 h_0)$ so that $\mathsf{d}^2 h_0 = 0$ as required. ✔

Outline Proof of Theorem 4.13: The proof is a 'parametrized version' of the proof of the Morse lemma. We begin by finding the map $u \mapsto \chi(u)$ referred to at the end of the statement.

The map $(x, u) \mapsto \mathsf{d}_x f(x, u)$ is of rank m at the origin, because its differential there is $[\mathsf{d}_x^2 f,\ \mathsf{d}_u \mathsf{d}_x f]$, and the first $m \times m$ block of this $m \times (m+k)$ matrix is $\mathsf{d}_x^2 f$ which is invertible, by hypothesis. It follows from the implicit function theorem that the equation $\mathsf{d}_x f(x, u) = 0$ can be solved uniquely for x as a function of u, so defining the map $u \mapsto \chi(u)$. That is, in a neighbourhood of the origin,

$$\mathsf{d}_x f(x, u) = 0 \iff x = \chi(u).$$

Now change coordinates by defining $y = x - \chi(u)$; the map

$$(x, u) \longmapsto (x - \chi(u),\ u)$$

is a diffeomorphism germ at 0, as its inverse is simply $(y,\ u) \mapsto (y + \chi(u),\ u)$. Let g be the function f expressed in the coordinates (y, u), so $g(y, u) = f(x, u) = f(y + \chi(u), u)$. Then $\mathsf{d}_y g = 0$ if and only if $y = 0$, as is readily checked.

Now for each fixed value of u (near 0) there is a function of y, call it g_u, which has a critical point at the origin. Moreover, it is a nondegenerate critical point because the Hessian of g_0 is invertible and so too is the Hessian of g_u for sufficiently small values of u (by continuity).

We now mimic the proof of the Morse lemma, but with the extra variables u as parameters. Write $\mathcal{M}_x$ to be the ideal of functions $f \in \mathcal{E}_{m+k}$ such that $f(0, u) \equiv 0$. Then

$$\mathcal{M}_x = \langle x_1, \ldots, x_m \rangle \subset \mathcal{E}_{m+k}.$$

This follows from Hadamard's lemma, Proposition 3.3. Define

$$F(y,u) = g(y,u) - g(0,u).$$

Then $F \in \mathcal{M}_x$, so we can write

$$F(y,u) = \sum_{i,j} y_i y_j \psi_{ij}(y,u) = y^T \Psi(y,u) y.$$

Now nearby symmetric and nondegenerate matrices are all similar, so for each (y,u) near 0 there is an invertible matrix $P_{(y,u)}$ such that

$$\Psi(y,u) = P^T_{(y,u)} \Psi(0,0) P_{(y,u)}.$$

It follows from the inverse function theorem that the map $(y,u) \mapsto (P_{(y,u)}y,\ u)$ is a diffeomorphism germ. Let us write $(X,u) = (P_{(y,u)}y,\ u)$. Then

$$F(y,u) = y^T \Psi(y,u) y = y^T P^T_{(y,u)} \Psi(0,0) P_{(y,u)} y = X^T \Psi(0,0) X.$$

Write $Q(X) = X^T \Psi(0,0) X$, so that

$$g(y,u) = g(0,u) + Q(X).$$

But $y = 0$ corresponds to $x = \chi(u)$ so that $g(0,u) = f(\chi(u),u) = h(u)$, so that finally

$$f(x,u) = h(u) + Q(X)$$

as required. ✔

The following proposition will be useful when we come to classify critical points in Chapter 6.

Proposition 4.16. *Let* $f \in \mathcal{E}_{m+k}$ *with* $f(x,u) = \sum_{i=1}^m \pm x_i^2 + h(u)$ *(where* $x \in \mathbb{R}^m$, $u \in \mathbb{R}^k$*), then* $\operatorname{codim}(f) = \operatorname{codim}(h)$.

The proof of this is based on the following algebraic lemma.

Lemma 4.17. *Let* R, S *be two rings and let* $J \lhd S$*. Suppose* $\phi\colon R \to S$ *is a surjective homomorphism and let* $I = \phi^{-1}(J)$*. Then* $I \lhd R$ *and* ϕ *induces an isomorphism* $\bar{\phi}\colon R/I \to S/J$*, defined by* $\bar{\phi}(r + I) = \phi(r) + J$.

That I is an ideal is easily checked (and doesn't need ϕ to be surjective). The rest follows from the first isomorphism theorem for rings (Theorem D.10), applied to the homomorphism $R \to S/_J$ defined by composing ϕ with the natural homomorphism $S \to S/_J$. Details are left to the reader.

PROOF OF THE PROPOSITION: First note that $Jf = \langle x_1, \dots, x_m, h_1, \dots, h_k \rangle$, where $h_i = \frac{\partial h}{\partial y_i}$. Let $\phi\colon \mathcal{E}_{m+k} \to \mathcal{E}_k$ be the homomorphism obtained by putting $x = 0$, so if $g \in \mathcal{E}_{m+k}$ we have $\phi(g) \in \mathcal{E}_k$ is the function (germ) $\phi(g)(u) = g(0, u)$. This homomorphism is clearly surjective, as given any $h \in \mathcal{E}_k$ then the germ $H(x, u) = h(u)$ (independent of x) satisfies $\phi(H) = h$.

Let $g \in \mathcal{E}_{m+k}$, then $g \in Jf$ if and only if $\phi(g) \in Jh$ (this follows from Hadamard's lemma). The result now follows from the lemma above. ✔

Problems

4.1 Let $f\colon U \to \mathbb{R}$ be a smooth function with domain $U \subset \mathbb{R}^n$, and let $\phi\colon V \to U$ be a diffeomorphism, and put $g = f \circ \phi$. Show that $x \in V$ is a critical point of g if and only if $\phi(x)$ is a critical point of f, and moreover that if the first is nondegenerate so is the second. [*Hint*: *Use the chain rule for second derivatives given in Proposition A.10.*]

4.2 By following Example 4.2, show that $f(x) = x^r + bx^s$ is $\mathcal{R}$-equivalent to $g(x) = x^r$ for any $s > r$ and any $b \in \mathbb{R}$.

4.3 Extending the previous problem, let $k \geq 1$ and let $f = ax^k + h(x)$ with $a \neq 0$ and $h \in \mathfrak{m}_1^{k+1}$. Prove that f is $\mathcal{R}$-equivalent to $g(x) = \varepsilon x^k$, where $\varepsilon = 1$ unless k is even and $a < 0$ in which case $\varepsilon = -1$.

4.4 Show in Example 4.2 that ϕ is a diffeomorphism on the interval $(-a, a)$ where a is such that $f'(a) = 0$, and, referring to the diagram, explain why the domain of the diffeomorphism could not be larger.

4.5 Show that $f \in \mathfrak{m}_n^r \Rightarrow Jf \subset \mathfrak{m}_n^{r-1}$.

4.6 Let $f = x^3 + y^3$ and $g = x^3 + y^3 + x^2y^2$. Use Nakayama's lemma to show that $Jf = Jg$. (†)

4.7 Extend the previous problem to show that if $\mathfrak{m}_n^k \subset \mathfrak{m}_n\, Jf$ and $g - f \in \mathfrak{m}_n^{k+1}$ then $Jf = Jg$. [*Hint*: *each partial derivative of f differs from the corresponding one of g by an element of* $\mathfrak{m}_n^k$.]

4.8 Determine the codimension of each of the following germs, and where the germ is of finite codimension, write down a cobasis for Jf in $\mathfrak{m}_n$.

(i) $x^3 + y^3 \in \mathcal{E}_2$ (ii) $x^3 + y^3 + z^2 \in \mathcal{E}_3$

(iii) $x^3 + y^3 + z^3 \in \mathcal{E}_3$ (iv) $x^{a+1} + y^{b+1} \in \mathcal{E}_2$, for $a, b \geq 1$ (†)

(v) $x^2y^2 \in \mathcal{E}_2$ (vi) $x^3 + xy^a \in \mathcal{E}_2$ for $a \geq 2$

4.9 Suppose $f, g \in \mathcal{E}_n$ are right equivalent, and in particular $g = f \circ \phi$, for a diffeomorphism germ ϕ. Recall the natural pull-back homomorphism $\phi^*: \mathcal{E}_n \to \mathcal{E}_n$ (see Problem 3.10).
(i) Show that $Jg = \phi^* Jf$.
(ii) Show that if f (and hence also g) are singular and $\{h_i\}$ is a cobasis for Jf in $\mathfrak{m}_n$ then $\{h_i \circ \phi\}$ is a cobasis for Jg in $\mathfrak{m}_n$.
(iii) Deduce that $\mathcal{R}^+$-equivalent germs have the same codimension. (†)

4.10 Show that the germs $f = x^2y + y^4$ and $g = x^2y + ay^4$ for fixed $a > 0$ are $\mathcal{R}$-equivalent. [*Hint*: *first rescale y, then rescale x to compensate for the first term. This type of argument is used often and will usually be taken for granted.*]

4.11 A function germ $f \in \mathcal{E}_n$ is said to be *homogeneous of degree* $d > 0$ if for each $\lambda \in \mathbb{R}$ and $(x_1, \ldots, x_n) \in \mathbb{R}^n$,

$$f(\lambda x_1, \ldots, \lambda x_n) = \lambda^d f(x_1, \ldots, x_n).$$

By differentiating this identity with respect to λ show that $f \in \mathfrak{m}_n Jf$. [In fact the converse almost holds: K. Saito [100] proves that if $f \in \mathfrak{m}_n Jf$ (or indeed in Jf) then f is right equivalent to a weighted homogeneous germ. See also [31].]

4.12 (i) Determine the corank of the germ $xy + y^2z + z^3 \in \mathcal{E}_3$. Show that the restriction of f to the plane $z = 0$ has a nondegenerate critical point at the origin. Apply the second part of the splitting lemma (Theorem 4.13) to find the 'remainder function' $h(z)$ in the splitting lemma.
(ii) Repeat this question with the germ $f(x, y, z) = xy + y^2z - xz^2 + z^3$. (†)

4.13 Let $f(x, y, z) = x^2 + 2xy^2 + yz^2 \in \mathcal{E}_3$.
(i) Determine the corank of the germ f. Show that if we write $\mathbb{R}^3 = \mathbb{R} \times \mathbb{R}^2$ then the restriction of f to $\mathbb{R} \times \{0\}$ (i.e. to the line $y = z = 0$) has a nondegenerate critical point at the origin. Find the remainder function $h(y, z)$ in the splitting lemma.
(ii) Now write $f(x, y, z) = X^2 + h(y, z)$ (where h is the expression you found

in part (i)), and solve for X as a function of x, y, z. This is the change of coordinates from the splitting lemma. [*Note*: *this can be done by hand here as the* x*-variable is just* 1*–dimensional, and further there are no higher–order terms in* x^3 *etc.*] (†)

4.14 In this problem we show that nondegenerate critical points are robust, or 'structurally stable' in the following sense. Let $f_t\colon \mathbb{R}^n \rightarrowtail \mathbb{R}$ be a smooth 1–parameter family of functions, for $t \in I$, an interval containing 0. That is, the function $F(t,x) := f_t(x)$ is a smooth function of $n+1$ variables. Suppose that f_0 has a nondegenerate critical point at $x = 0$. Show that there is a smooth function $\psi\colon \mathbb{R} \rightarrowtail \mathbb{R}^n$, with $\psi(0) = 0$, such that for $t \in \operatorname{dom}(\psi)$ the function f_t has a nondegenerate critical point at $x = \psi(t)$. [*Hint*: *use the splitting lemma.*] (†)

5

Finite determinacy

HITHERTO, ALL THE COMPUTATIONS and examples have involved polynomial functions rather than more general smooth ones, as the reader may have noticed. We show in this chapter that in fact all germs of finite codimension are equivalent to polynomials after a change of coordinates. This is a consequence of *finite determinacy*, a central notion in singularity theory.

In order to prove two maps are right equivalent, one needs to construct a diffeomorphism (or at least, to show one exists), and the only general technique for constructing diffeomorphisms is by integrating vector fields (first order differential equations) as described in Appendix C. The proofs of this chapter assume the reader is familiar with the ideas of that appendix. In the first theorem (the Thom–Levine principle for right equivalence), the vector field is given, while in the later ones part of the problem is to construct the relevant vector field, and that is done using the so-called homotopy method.

Throughout this subject, as we have seen in the first chapters, we consider families of functions or more general maps. As we often need to differentiate with respect to the parameter, we require the family to be smooth in the following sense.

Definition 5.1. A family of maps $f_u\colon \mathbb{R}^n \rightarrowtail \mathbb{R}^p$ ($u \in \mathbb{R}^a$) is a ***smooth family*** if the map

$$\begin{aligned} F\colon \mathbb{R}^n \times \mathbb{R}^a &\rightarrowtail \mathbb{R}^p \\ (x, u) &\longmapsto f_u(x) \end{aligned}$$

is a smooth map. One can similarly define smooth families of germs, or germs of smooth families of germs. ☆

As indicated above, we make considerable use of vector fields in this chapter. We denote the set of germs at 0 of smooth vector fields on $\mathbb{R}^n$ by θ_n. One of the important constructions with vector fields is the so-called ***tangent map*** of f.

Definition 5.2. Let $f\colon \mathbb{R}^n \rightarrowtail \mathbb{R}^p$ be a smooth map and $\mathbf{v}$ be a smooth vector field on the domain of f. Then $\mathrm{t}f(\mathbf{v})$ is the ***vector field along*** f given by

$$\mathrm{t}f(\mathbf{v})(x) = \mathrm{d}f_x(\mathbf{v}(x)) = \sum_j \frac{\partial f}{\partial x_j}(x)\, v_j(x). \tag{5.1}$$

The map $\mathrm{t}f$ is the ***tangent map*** of f. Note that a vector field $\mathbf{w}$ along a map $f\colon \mathbb{R}^n \rightarrowtail \mathbb{R}^p$ is not a vector field in the traditional sense. It is rather a vector $\mathbf{w}(x) \in \mathbb{R}^p$ for each $x \in \mathrm{dom}(f) \subset \mathbb{R}^n$ – in particular x and $\mathbf{w}(x)$ are in different spaces. ✯

To illustrate this with a simple example let $f(x,y) = x^2 + y^3 \in \mathcal{E}_2$, and $\mathbf{v}(x,y) = (y^2, x^2)^T \in \theta_2$, then

$$\mathrm{t}f(\mathbf{v})(x,y) = (2x,\ 3y^2)\begin{pmatrix} y^2 \\ x^2 \end{pmatrix} = 2xy^2 + 3x^2y^2.$$

This is just a function (germ) of (x,y). In this chapter, $f \in \mathcal{E}_n$ so $p = 1$ and $\mathrm{t}f(\mathbf{v}) \in \mathcal{E}_n$ (a function germ); that is, $\mathrm{t}f\colon \theta_n \to \mathcal{E}_n$. The more general case with $p > 1$ is not used until Part II.

By letting $\mathbf{v} \in \theta_n$ be an arbitrary vector field, we see from (5.1) that the image of $\mathrm{t}f$ is the Jacobian ideal. That is, $\mathrm{t}f(\theta_n) = Jf$.

5.1 Trivial families of germs

When we are given a family of germs, one of the first questions is whether all the germs in the family are right equivalent or not. The approach taken is based on ideas introduced in some famous lecture notes [70] by Harold Levine based on lectures of René Thom. We begin therefore with the following definition.

Definition 5.3. A 1–parameter family of germs $f_s\colon (\mathbb{R}^n, 0) \to \mathbb{R}$ (with $s \in I$ an interval containing 0) is said to be a ***trivial family*** if there is a smooth family of diffeomorphism germs ϕ_s with $s \in I$ and $\phi_0 = \mathrm{Id}$ such that $f_s \circ \phi_s = f_0$. ✯

In particular, all the germs in the family are right equivalent. To be specific, one says this is an $\mathcal{R}$-trivial family. Note that all the germs are at 0, so necessarily $\phi_s(0) = 0$ for all s. The assumption that $\phi_0 = \mathrm{Id}$ is convenient (Id denotes the identity map), but not in fact necessary as one can always modify the family of diffeomorphisms so that indeed $\phi_0 = \mathrm{Id}$ (see Problem 5.3). For the remainder of this section, I will denote an interval in $\mathbb{R}$ containing the origin.

Theorem 5.4 (Thom–Levine principle). *Let $f_s \in \mathcal{E}_n$ be a smooth family of germs ($s \in I$). Then f_s is a trivial family if and only if $\dot{f}_s \in \mathfrak{m}_n Jf_s$ smoothly in s.*

Here $\dot{f}_s$ denotes the function germ $\frac{\mathrm{d}}{\mathrm{d}s} f_s$. The 'smoothly in s' condition requires that there are smooth functions $u_j(x, s)$, with $u_j(0, s) = 0$ for which

$$\dot{f}_s(x) = \sum_{j=1}^{n} u_j(x, s) \frac{\partial f_s}{\partial x_j}(x). \tag{5.2}$$

The smooth functions $u_j(x, s)$ are germs in x, but defined for all $s \in I$. This smoothness assumption is necessary: a simple example where the conclusion fails is provided in Problem 5.4.

When we make use of this theorem to prove a family is trivial, the difficulty will often be to establish this smoothness condition.

Proof: First suppose the family f_s is trivial: $f_s \circ \phi_s = f_0$, where $\{\phi_s\}$ is a smooth family of diffeomorphisms. Differentiating this with respect to s (see Lemma A.14) gives

$$\dot{f}_s(y) + \mathrm{t}f_s(\mathbf{v}_s)(y) = 0 \tag{5.3}$$

where $y = \phi_s(x)$ and $\mathbf{v}_s(\phi_s(x)) = \frac{\mathrm{d}}{\mathrm{d}s}(\phi_s(x))$. Since $\phi_s(0) = 0$ (for all s) it follows by differentiating that $\mathbf{v}_s(0) = 0$. Write

$$\mathbf{v}_s(x) = ((v_s)_1(x), (v_s)_2(x), \ldots, (v_s)_n(x));$$

then $\mathbf{v}_s(0) = 0$ implies each component $(v_s)_j \in \mathfrak{m}_n$ (for all s). Equation (5.3) then reads

$$\dot{f}_s = -\sum (v_s)_j \frac{\partial f_s}{\partial x_j} \in \mathfrak{m}_n Jf_s,$$

as required.

For the converse, we rely on the notion of the flow of a vector field, which is described in some detail in Appendix C. Suppose then that $\dot{f}_s \in \mathfrak{m}_n Jf_s$ smoothly in s. This means there are smooth functions $u_j(x, s)$ with each component $(u_s)_j \in \mathfrak{m}_n$ (for each s), satisfying (5.2). Now let ϕ_t be the flow corresponding to the time-dependent vector field $-\mathbf{u}(x, t)$. Taking representatives of the germs, the flow exists for all x in a neighbourhood U of 0 in $\mathbb{R}^n$ by Theorem C.7 (the theorem requires I to be compact, but every $s \in I$ is contained in a compact interval containing 0). Let $x \in U$, and consider $f_s(\phi_s(x))$ for $s \in I$. Differentiating this with respect to s gives

$$\frac{\mathrm{d}}{\mathrm{d}s}(f_s \circ \phi_s)(x) = \dot{f}_s(y) + \mathrm{d}(f_s)_y(-\mathbf{u}_s(y)) = 0.$$

where $y = \phi_s(x)$. Consequently, for each $x \in U$, $f_s \circ \phi_s(x)$ is constant (independent of s), so that $f_s \circ \phi_s = f_0$, as required. ✔

In applications of this theorem, one often considers families of the form $f_s = f + sh$ for some germ $h \in \mathcal{E}_n$. In this case $\dot{f}_s = h$ (independently of $s \in \mathbb{R}$) and the condition for triviality is $h \in \mathfrak{m}_n Jf_s$ (for all s and smoothly in s). The theorem can also be used for showing that a given family is *not* trivial, as follows.

Example 5.5. Let $f = x^5 + y^5$, and $h = x^3y^3$. Then $f_s(x, y) = x^5 + y^5 + sx^3y^3$. This is not a trivial family because $Jf = \langle x^4, y^4 \rangle$, and therefore $h \notin Jf$, so certainly $h \notin \mathfrak{m}_2 Jf$. ✎

By analogy with the tangent space to a curve, this theorem provides a motivation for defining the 'tangent space' to the set of germs that are right equivalent to a given germ f.

Definition 5.6. The ***right tangent space*** to f is the ideal

$$T\mathcal{R}{\cdot}f = \mathrm{t}f(\mathfrak{m}_n\theta_n) = \mathfrak{m}_n Jf.$$ ☆

For example, for $f = x^3 + y^3$ one has $T\mathcal{R}{\cdot}f = \mathfrak{m}_2^3$. On the other hand for $f = x^4 + y^4$, $T\mathcal{R}{\cdot}f = \langle x^4,\ x^3y,\ xy^3,\ y^4 \rangle \neq \mathfrak{m}_2^4$, but $\mathfrak{m}_2^5 \subset T\mathcal{R}{\cdot}f$.

Thus the Thom–Levine principle says that a smooth family f_s is trivial if and only if $\dot{f}_s \in T\mathcal{R}{\cdot}f_s$, smoothly in s. Comparing with the definition of tangent space to a submanifold given in Section B.4, this justifies calling $T\mathcal{R}{\cdot}f$ the tangent space to the right equivalence class (Chapter 12 contains further discussions of tangent spaces).

We now show, under a hypothesis on $T\mathcal{R}{\cdot}f$ we will see again, that the tangent space does not depend on certain higher–order terms of f. The result is not directly used again, though it is useful in classification questions.

Proposition 5.7. *Let $f \in \mathcal{E}_n$ and suppose $\mathfrak{m}_n^k \subset T\mathcal{R}{\cdot}f$ (i.e. $\mathfrak{m}_n^k \subset \mathfrak{m}_n Jf$). Now let $h \in \mathfrak{m}_n^{k+1}$ and put $g = f + h$. Then*

$$T\mathcal{R}{\cdot}f = T\mathcal{R}{\cdot}g.$$

In short, if $\mathfrak{m}_n^k \subset T\mathcal{R}{\cdot}f$, then $T\mathcal{R}{\cdot}f$ depends only on the k-jet of f.

Proof: This follows from Nakayama's lemma. Write $I = \mathfrak{m}_n^k$. Since $h \in \mathfrak{m}_n I$ it follows that $\frac{\partial h}{\partial x_i} \in I$ (e.g. by Corollary 3.6), and hence $\frac{\partial f}{\partial x_i} - \frac{\partial g}{\partial x_i} \in I$. Therefore $\mathfrak{m}_n Jg \subset \mathfrak{m}_n Jf + \mathfrak{m}_n I$ and $\mathfrak{m}_n Jf \subset \mathfrak{m}_n Jg + \mathfrak{m}_n I$.

The first of these, $\mathfrak{m}_n Jg \subset \mathfrak{m}_n Jf + \mathfrak{m}_n I$ shows that $\mathfrak{m}_n Jg \subset \mathfrak{m}_n Jf + \mathfrak{m}_n^2 Jf = \mathfrak{m}_n Jf$ so that $T\mathcal{R}{\cdot}g \subset T\mathcal{R}{\cdot}f$. The second gives

$$\mathfrak{m}_n Jf \subset \mathfrak{m}_n Jg + \mathfrak{m}_n I \subset \mathfrak{m}_n Jg + \mathfrak{m}_n^2 Jf.$$

Hence by Nakayama's lemma $\mathfrak{m}_n Jf \subset \mathfrak{m}_n Jg$, or $T\mathcal{R}{\cdot}f \subset T\mathcal{R}{\cdot}g$, whence the two tangent spaces coincide. ✔

In Lemma 5.19 we prove a 'smoothly in s' version of this, which is important for applying the Thom–Levine principle.

5.2 Finite determinacy

Finite determinacy is a very powerful and practical idea. It allows us to study a smooth function (germ) by replacing it with a polynomial which is right equivalent to it. In one variable, the criterion for finite determinacy is straightforward, but in more variables it is not so obvious.

Example 5.8. (One variable) Let $f = ax^k + h(x)$ for some $k \geq 1$ with $a \neq 0$ and $h \in \mathfrak{m}_1^{k+1}$; in other words, $f(x) = ax^k + O(x^{k+1})$. Then f is right equivalent to $g(x) = ax^k$. (See Problem 4.3.) ✎

This example can be restated in terms of jets as follows: if $\mathrm{j}^k f(0) = ax^k$ with $a \neq 0$ then f is $\mathcal{R}$-equivalent to ax^k. On the other hand, if $f \neq 0$ but $\mathrm{j}^k f(0) = 0$ for all k then f cannot be $\mathcal{R}$-equivalent to a polynomial, because only the zero polynomial has all its derivatives equal to zero.

Definition 5.9. A function germ $f \in \mathcal{E}_n$ is said to be ***k-determined*** if, whenever $g \in \mathcal{E}_n$ satisfies $\mathrm{j}^k g(0) = \mathrm{j}^k f(0)$ then g is $\mathcal{R}$-equivalent to f. We say f is ***finitely determined*** if there is a $k \in \mathbb{N}$ for which f is k-determined. ☆

In other words, f is k-determined if every germ with the same k-jet (the same kth-order Taylor series) as f is $\mathcal{R}$-equivalent to f. In practice this means that the geometry of f can be studied just by looking at its k-jet.

The conclusion of Example 5.8 can now be stated even more succinctly; namely, if $a \neq 0$ then the germ at 0 of ax^k is k-determined.

A well-known case is the implicit function theorem (see Appendix B) where $k = 1$. If $f \in \mathcal{E}_n$ is such that $\mathrm{d}f(0) \neq 0$ then f is right equivalent to its linear part, so f is 1-determined.

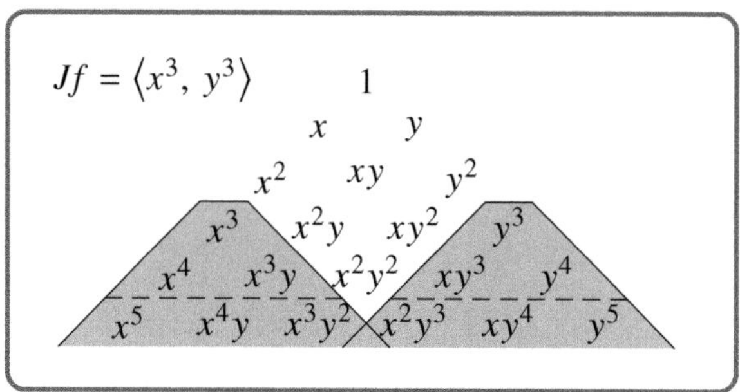

FIGURE 5.1 Newton diagram for Example 5.11(ii). Monomials in the shadow of the dashed lines are in $\mathfrak{m}_2^2\, Jf$.

In more variables, one cannot determine the degree of determinacy merely by looking at the degree of the lowest order terms. For example, it follows from Example 5.5 that $f = x^5 + y^5$ is not 5-determined. On the other hand, it follows from the theorem below that it is 6-determined.

Theorem 5.10 (Finite determinacy theorem). *If $f \in \mathcal{E}_n$ satisfies*

$$\mathfrak{m}_n^{k+1} \subset \mathfrak{m}_n^2\, Jf$$

then f is k-determined.

We defer the proof of this central theorem until Section 5.6, after we see the theorem in action and some variations on the statement. The proof uses the so-called homotopy method, which we illustrate in the simpler context of a proof of the inverse function theorem in Section 5.5.

First note that by multiplying both sides successively by $\mathfrak{m}_n$,

$$\mathfrak{m}_n^{k-1} \subset Jf \implies \mathfrak{m}_n^{k} \subset \mathfrak{m}_n\, Jf \implies \mathfrak{m}_n^{k+1} \subset \mathfrak{m}_n^2\, Jf. \tag{5.4}$$

So $\mathfrak{m}_n^{k-1} \subset Jf$ is a sufficient condition for k-determinacy, as is $\mathfrak{m}_n^k \subset \mathfrak{m}_n\, Jf$, but they give weaker statements (ie, less likely to hold) than the one in the theorem (see part (ii) in the examples below).

Examples 5.11. (i). Let $f = x^3 + y^3$: then $Jf = \langle x^2, y^2 \rangle$ so $\mathfrak{m}_2 Jf = \mathfrak{m}_2^3$, and f is 3-determined.

(ii). Let $f = x^4 + y^4$: then $Jf = \langle x^3, y^3 \rangle$ and $\mathfrak{m}_2^2\, Jf = \mathfrak{m}_2^5$ so f is 4-determined (note that $\mathfrak{m}_2^4 \not\subset \mathfrak{m}_2\, Jf$, so it is necessary to use the full statement of the theorem) – see Figure 5.1.

(iii). Let $f = x^5 + y^5$: then $\mathfrak{m}_2^2 Jf \supset \mathfrak{m}_2^7$ so f is 6-determined as claimed above. Note that $x^3y^3 \notin \mathfrak{m}_2 Jf$, so $\mathfrak{m}_2^6 \not\subset \mathfrak{m}_2 Jf$, and indeed by Theorem 5.4, $f + sx^3y^3$ is not a trivial family and f is not 5-determined (see Example 5.5).

(iv). Let $f = x^3 + 3xy^3$. Then $Jf = \langle x^2 + y^3, xy^2 \rangle$. Calculations show that $y^5 \notin \mathfrak{m}_2^2 Jf$ so we cannot deduce that f is 4-determined. However, $\mathfrak{m}_2^6 \subset \mathfrak{m}_2^2 Jf$ so it follows that f is 5-determined. (In fact this function *is* 4-determined, but more detailed methods are needed; see also Section 5.4 below.) ✏

Example 5.12. [Proof of Morse lemma, Proposition 4.10 (p. 54)] We want to show that if $f \in \mathcal{E}_n$ has a nondegenerate critical point at the origin, then f is 2-determined. In fact we showed in Proposition 4.9 that under this assumption $Jf = \mathfrak{m}_n$, and then the 2-determinacy follows using Equation (5.4). ✏

Proposition 5.13. *Suppose $f, g \in \mathcal{E}_n$ are right equivalent. Then f is k-determined if and only if g is.*

Proof: The underlying point here is that if $h \in \mathfrak{m}_n^{k+1}$ then so is $h \circ \phi$, for any diffeomorphism germ ϕ (see Problem 3.12). Suppose then that f is k-determined; we wish to deduce that g is too (the converse argument being the same, with the roles of f and g reversed). Suppose $f = g \circ \phi$. We have

$$g + h \sim (g + h) \circ \phi = f + h \circ \phi \sim f \sim g.$$

The $\sim$ here means $\mathcal{R}$-equivalence; the first $\sim$ is obtained by applying the diffeomorphism ϕ, and the second holds because f is k-determined and $h \circ \phi \in \mathfrak{m}_n^{k+1}$. ✔

Example 5.14. Let $f = x^2 + 2xy^2$. What is the degree of determinacy of f? Clearly f cannot be 2-determined as the 2-jet is just x^2, which is not finitely determined. It would seem reasonable to expect it to be 3-determined. However, it is only 4-determined as we now show.

First we see it is 4-determined. Now, $Jf = \langle 2x + 2y^2, 4xy \rangle = \langle x + y^2, xy \rangle$. Let us call these generators α and β respectively (i.e. $\alpha = x + y^2$ and $\beta = xy$). Then we find

$$\begin{aligned} x^2 &= x\alpha - xy^2 \\ &= x\alpha - y\beta \in \mathfrak{m}_2 Jf. \end{aligned}$$

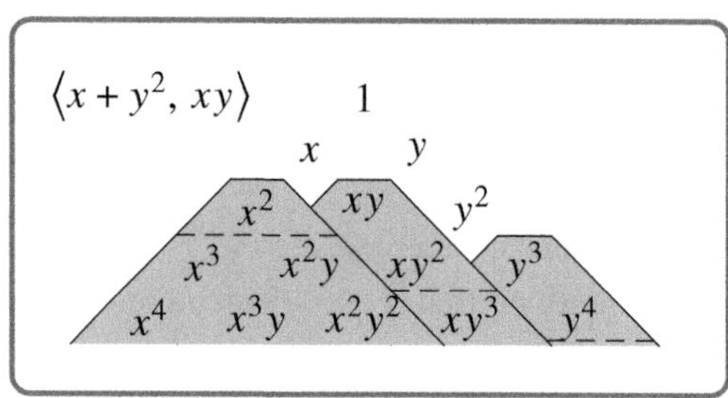

FIGURE 5.2 Newton diagram for Example 5.14. Monomials in the shadow of the dashed lines are in $\mathfrak{m}_2^2 Jf$.

It follows that $x^3, x^2y \in \mathfrak{m}_2^2 Jf$. Next, $xy^2 = x\alpha - x^2 \in \mathfrak{m}_2 Jf$, and so $xy^3 \in \mathfrak{m}_2^2 Jf$. Finally, $y^2 \notin Jf$, but

$$\begin{aligned} y^3 &= y\alpha - xy \\ &= y\alpha - \beta \in Jf. \end{aligned}$$

Then $y^4 \in \mathfrak{m}_2 Jf$ and $y^5 \in \mathfrak{m}_2^2 Jf$. So in all,

$$\mathfrak{m}_2^5 \subset \mathfrak{m}_2^2 Jf$$

which implies f is 4-determined. See Figure 5.2.

To see it is not 3-determined, consider the change of coordinates (see the beginning of Section 4.5 with this germ) putting $X = x + y^2$, and the function becomes $X^2 - y^4$ which is clearly not 3-determined. It now follows from Proposition 5.13 that f itself is not 3-determined. ✐

5.3 A partial converse

There is a partial converse of the finite determinacy theorem, but we do not give a proof. The argument uses the Thom–Levine principle, but the smoothness in s part relies on results from the theory of group actions on vector spaces which we do not cover. A proof can be found for example in the book of Dimca [31, Lemma 6.3].

Proposition 5.15. *Let $f \in \mathcal{E}_n$ be a k-determined germ (for $\mathcal{R}$-equivalence). Then*

$$\mathfrak{m}_n^{k+1} \subset T\mathcal{R}{\cdot}f.$$

Combining this with the finite determinacy theorem, we conclude

$$\mathfrak{m}_n^{k+1} \subset \mathfrak{m}_n^2 Jf \implies f \text{ is } k\text{-determined} \implies \mathfrak{m}_n^{k+1} \subset \mathfrak{m}_n Jf.$$

Finite codimension Recall from Section 4.3 that a function $f \in \mathfrak{m}_n^2$ is of finite codimension if the vector space $\mathfrak{m}_n/Jf$ is finite–dimensional.

Corollary 5.16. *A germ $f \in \mathcal{E}_n$ is finitely determined if and only if it is of finite codimension.*

One implication follows from Proposition 3.13 about finite codimension ideals together with the finite determinacy theorem above. The converse follows from Proposition 5.15.

5.4 A refinement of the finite determinacy theorem

As we have seen, the finite determinacy theorem is only a sufficient condition for k-determinacy, and in general it is not straightforward to find the exact value of k. However, one can define a restricted form of $\mathcal{R}$-equivalence, which we call $\mathcal{R}_1$-equivalence, where one can obtain a single necessary and sufficient condition giving the value of k for $\mathcal{R}_1$-finite determinacy. For $f, g \in \mathcal{E}_n$ we say f is $\mathcal{R}_1$-***equivalent*** to g if they are right equivalent via a diffeomorphism germ ϕ whose linear part is the identity; that is, the Taylor series at 0 of ϕ is

$$\phi(y) = y + O(y^2).$$

It turns out now that the partial converse we saw in the previous section becomes a full converse; that is, as already said, we have a condition for $\mathcal{R}_1$-finite determinacy that is both necessary and sufficient.

Theorem 5.17 ($\mathcal{R}_1$-finite determinacy). *A germ $f \in \mathcal{E}_n$ is k-$\mathcal{R}_1$-determined if and only if*

$$\mathfrak{m}_n^{k+1} \subset \mathfrak{m}_n^2\, Jf.$$

The important ‘if’ part of this theorem is proved in Section 5.6 below. We do not prove the ‘only if’ part; see the references given for the partial converse in Section 5.3. Note that if two germs are $\mathcal{R}_1$-equivalent, then they are a fortiori $\mathcal{R}$-equivalent. Consequently, Theorem 5.10 is a consequence of Theorem 5.17.

It is not straightforward to capture the difference between $\mathcal{R}_1$-equivalence and $\mathcal{R}$-equivalence. There is no theorem which gives a necessary and sufficient condition for k-$\mathcal{R}$-determinacy, though it is not hard to show a germ is finitely $\mathcal{R}$-determined if and only if it is finitely $\mathcal{R}_1$-determined – it is just the value of k that is not straightforward (see for example Problem 5.8). There are theorems that improve on Theorem 5.10 with equivalence relations situated between $\mathcal{R}$ and $\mathcal{R}_1$, but they go beyond the scope of this text (the key phrase is ‘unipotent group actions’ – see [21]).

5.5 The homotopy method

Before proceeding with the proof of the finite determinacy theorem, we illustrate the method of proof, the so-called homotopy method, with a proof of the inverse function theorem: first recall the statement of the theorem (see p. 339):

Let $f\colon \mathbb{R}^n \rightarrowtail \mathbb{R}^n$ be a smooth map with $x_0 \in \mathrm{dom}(f)$ and $y_0 = f(x_0)$, and suppose f has rank n at x_0. Then there is a neighbourhood U of x_0 such that the restriction $f_{|U}\colon U \to f(U)$ is a diffeomorphism.

Proof: We only prove the ‘only if’ part here: the ’if’ part is proved in the appendix (before the statement of the theorem). There are three steps to the proof. The first translates the question about existence of a diffeomorphism to an equation for a vector field; the second shows that this equation can be solved, and the third step integrates the vector field to obtain the required diffeomorphism. That is, differentiate, solve, integrate.

Without loss of generality, let $x = 0$ and $f(x) = 0$.

Using Taylor's theorem, we can write f as the sum of its linear part (i.e. its differential at 0) and the remainder: $f = L + h$, with $L = \mathsf{d}f_0$ and $\mathsf{d}h_0 = 0$. Note that by hypothesis L is invertible and thus itself a diffeomorphism. For each $s \in [0, 1]$ define the map

$$f_s = L + sh.$$

Then $f_0 = L$ and $f_1 = f$. Think of this as a straight line in the set of all smooth maps defined on $\mathrm{dom}(f)$, joining L and f: see the figure below.

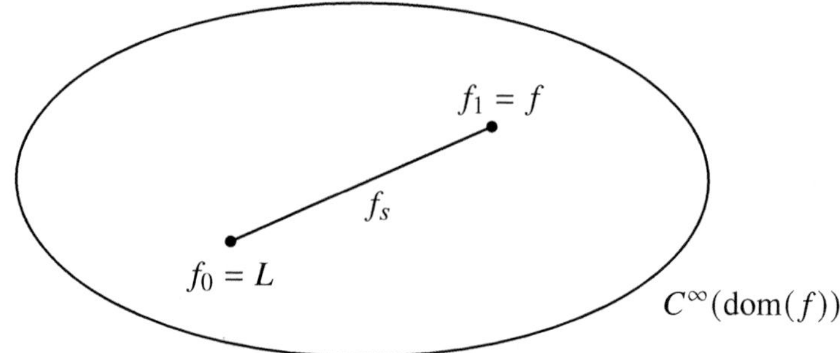

We want to construct (or rather, show there exists) a diffeomorphism ϕ of a neighbourhood U_1 of 0 in $\mathbb{R}^n$ satisfying $f \circ \phi = L$ on U_1. It will then follow that f, being the composite of two diffeomorphisms ($f = L \circ \phi^{-1}$), is itself a diffeomorphism, as required. In fact we will 'construct' a smooth family of diffeomorphisms ϕ_s of a neighbourhood of 0 for $s \in [0, 1]$ satisfying the so-called *homotopy equation*

$$f_s \circ \phi_s = L, \tag{5.5}$$

defined on some domain U_1 containing 0. As $f = f_1$, the required ϕ will then be ϕ_1.

The problem then is to show such a family of diffeomorphisms exists. To do this, we differentiate (5.5) with respect to s, so obtaining the *infinitesimal homotopy equation*, which we will see has a solution, and then integrate back to get the ϕ_s which by the general theory of ODEs will be a diffeomorphism. To differentiate (5.5) with respect to s, we use Lemma A.14.

Step 1 Thus, putting $y = \phi_s(x)$ and defining the vector fields $\mathbf{v}_s$ by,

$$\frac{\mathrm{d}}{\mathrm{d}s}\phi_s(u) = \mathbf{v}_s(\phi_s(u)) \tag{5.6}$$

(since ϕ_s is a diffeomorphism, $\mathbf{v}_s$ is defined everywhere by this equation). Lemma A.14 implies $h(y) + \mathrm{d}(f_s)_y(\mathbf{v}_s(y)) = 0$, or, writing in term of the tangent map,

$$\mathrm{t}f_s(\mathbf{v}_s)(y) = -h(y). \tag{5.7}$$

This so-called *infinitesimal homotopy equation* is the crucial equation and we need to solve it for $\mathbf{v}_s$.

Step 2 To do this, we use the fact that L is invertible: since $\mathrm{d}(f_s)_0 = L$ for all $s \in [0, 1]$, and $[0, 1]$ is compact, it follows that there is a neighbourhood U' of 0 in $\mathbb{R}^n$ such that for all $y \in U'$ the linear map $\mathrm{d}(f_s)_y$ is invertible (apply Lemma 5.18 to $u(y, s) = \det(\mathrm{d}(f_s)_y)$. Therefore we can define $\mathbf{v}_s(y)$ for all $(y, s) \in U' \times [0, 1]$ by

$$\mathbf{v}_s(y) = -[\mathrm{d}(f_s)_y]^{-1} h(y).$$

Notice that $\mathbf{v}_s(0) = 0$ (for all s) since $h(0) = 0$. (And furthermore $\mathrm{d}(\mathbf{v}_s)(0) = 0$ since $\mathrm{d}h_0(0) = 0$; that is, the components of $\mathbf{v}_s$ belong to $\mathfrak{m}_n^2$.)

Step 3 Now we come to the final step of the proof, which involves integrating the vector field $\mathbf{v}_s(y)$ and showing that the resulting diffeomorphisms ϕ_s do indeed satisfy (5.5).

By Theorem C.7 of Appendix C, there is a family of diffeomorphisms ϕ_s defined on some neighbourhood U'' of 0 satisfying (5.7) with $\phi_s(0) = 0$ for

$s \in [0, 1]$. (And furthermore, by Theorem C.8, $(\mathsf{d}\phi_s)_0 = \mathsf{Id}_n$.) To see that this family satisfies (5.5), differentiate $f_s \circ \phi_s(x)$ with respect to s. By construction of $\mathbf{v}_s$ this is zero, so that for each x, $f_s \circ \phi_s(x)$ is independent of s, meaning that $f_s \circ \phi_s = f_0 \circ \phi_0 = L$. Putting $s = 1$ concludes the proof. ✔

The principal difference between this proof and other applications of the homotopy method we encounter (in the finite determinacy theorems), is how we solve the infinitesimal homotopy equation (5.7). Problem 5.9 suggests a concrete example to elucidate the proof, and particularly the infinitesimal homotopy equation.

Lemma 5.18. *Let $u\colon \mathbb{R}^n \times [0, 1] \rightarrowtail \mathbb{R}$ be a continuous function defined on a neighbourhood of $\{0\} \times [0, 1]$. Suppose that $\forall s \in [0, 1]$, one has $u(0, s) \neq 0$. Then there is a neighbourhood U of $\{0\}$ in $\mathbb{R}^n$ such that $u(x, s) \neq 0$ for all $(x, s) \in U \times [0, 1]$.*

Proof: The proof relies on the familiar fact that every open set A in $\mathbb{R}^2$ contains a product of open sets $B \times C \subset A$, with B, C open in $\mathbb{R}$. A similar property holds in $\mathbb{R}^n$ for $n > 2$. Let $u_0\colon [0, 1] \to \mathbb{R}$ be defined by $u_0(s) = u(0, s)$. Since, by hypothesis, u_0 is nowhere zero, we may suppose $u_0(s) > 0$ (for all s). The argument if $u_0(s) < 0$ is similar. Now, since the domain of u_0 is compact, the function achieves its minimum, say $\varepsilon := \min_{s\in[0,1]}\{u_0(s)\}$. Thus $\varepsilon > 0$.

For each $s \in [0, 1]$ let V_s be a neighbourhood of $(0, s)$ in $\mathbb{R}^n \times [0, 1]$ on which $u(x, s) > \frac{1}{2}\varepsilon$. This contains an open set of the form $U_s \times S_s$, with U_s open in $\mathbb{R}^n$ and S_s open in $[0, 1]$. Now the collection of sets S_s form an open over of $[0, 1]$, which, by compactness, must have a finite subcover, say $[0, 1] = S_{s_1} \cup S_{s_2} \cup \cdots \cup S_{s_r}$.

Finally, one can take U to be the intersection $U = U_{s_1} \cap \cdots \cap U_{s_r}$, as is readily checked. ✔

5.6 Proof of finite determinacy theorems

In this section we provide proofs of the two main finite determinacy theorems (Theorems 5.10 and 5.17), using the homotopy method along the lines of the proof of the inverse function theorem given above – it is recommended to read that proof before this one as some details are skipped. The difference in the three proofs lies only in the part where the vector field is integrated.

The ‘smoothness in s’ condition of the Thom–Levine principle requires us to consider vector fields $\mathbf{v}_s(x)$ on $\mathbb{R}^n$ that depend smoothly on s. To this end we introduce a new ring of germs as follows.

Consider the space $\mathbb{R}^n \times I$ where I is the interval $[0, 1]$, with coordinates $x_1, \ldots, x_n, s$, and let $\mathcal{E}_{n,I}$ be the ring of germs of smooth functions *along* $\{0\} \times I$ where $I = [0, 1]$. That is, a representative of a germ in $\mathcal{E}_{n,I}$ is a smooth function defined on a neighbourhood of $\{0\} \times I$ in $\mathbb{R}^n \times I$ (in particular they are defined for all $s \in I = [0, 1]$).

Let $\mathfrak{m}_{n,I}$ denote the ideal in $\mathcal{E}_{n,I}$ of function germs which are identically zero on $\{0\} \times I$; that is,

$$\begin{aligned}\mathfrak{m}_{n,I} &= \left\{g \in \mathcal{E}_{n,I} \mid g(0, 0, \ldots, 0, s) = 0 \ \forall s \in [0, 1]\right\} \\ &= \langle x_1, \ldots, x_n \rangle .\end{aligned}$$

The second equality follows from Hadamard's lemma – Proposition 3.3 (here s takes the place of y in that proposition, and note that while that result was only proved for germs at the origin, the same proof works globally for $s \in [0, 1]$).

Proof of finite determinacy theorems: Suppose $f \in \mathcal{E}_n$ satisfies $\mathfrak{m}_n^{k+1} \subset \mathfrak{m}_n^2 Jf$. We wish to show that f is k-determined; in fact we show it is k-determined with respect to $\mathcal{R}_1$. In other words, if $h \in \mathfrak{m}_n^{k+1}$, we show $f + h$ is $\mathcal{R}_1$-equivalent to f. To do this we wish to construct a diffeomorphism ϕ of a neighbourhood of 0 and of the required form for $\mathcal{R}_1$, so that $(f + h) \circ \phi = f$.

We use the homotopy method introduced in the previous section. Form the path joining f to $f + h$: for $s \in [0, 1]$, define

$$f_s = f + s\,h.$$

For each s we will 'construct' a diffeomorphism-germ $\phi_s\colon (\mathbb{R}^n, 0) \to (\mathbb{R}^n, 0)$ satisfying the 'homotopy equation'

$$f_s \circ \phi_s = f. \tag{5.8}$$

In particular then, $\phi = \phi_1$ gives an $\mathcal{R}$-equivalence (indeed an $\mathcal{R}_1$-equivalence) between f and $f_1 = f + h$ as required.

As all the details except one are the same as in the proof of the inverse function theorem above, we refer to that for the details (the one difference is how the infinitesimal homotopy equation is solved). Differentiating (5.8) with respect to s and putting $\mathbf{v}_s(y) = \frac{\mathsf{d}}{\mathsf{d}s}\phi_s(x)$ with $y = \phi_s(x)$ gives the *infinitesimal homotopy equation*,

$$\mathsf{d}(f_s)_y(\mathbf{v}_s(y)) = -h(y). \tag{5.9}$$

For the proof of the inverse function theorem, $\mathsf{d}(f_s)_y$ is an invertible $n \times n$ matrix, which was inverted in order to find $\mathbf{v}_s$. However, here $\mathsf{d}(f_s)_y$ is a (row) vector, and written in components (5.9) becomes

$$v_{1,s}(y)\,\frac{\partial f_s}{\partial y_1} + \cdots + v_{n,s}(y)\,\frac{\partial f_s}{\partial y_n} = -h(y). \tag{5.10}$$

where $\mathbf{v}_s = (v_{1,s}, v_{2,s}, \ldots, v_{n,s})$.

Note that if there were no s-dependence in this equation, then it would have a solution $\mathbf{v}(y)$ with each component $v_i \in \mathfrak{m}_n^2$, precisely because $-h \in \mathfrak{m}_n^2\, Jf$. To show there is a solution $\mathbf{v}_s(y)$ (smooth also in $s \in [0,1]$) we apply Lemma 5.19 below which relies on Nakayama's lemma.

Since the hypotheses on f and h in the lemma below are satisfied, it follows that there are functions $v_{i,s} \in \mathfrak{m}_{n,I}^2$ satisfying equation (5.10). Let $\mathbf{v}_s(y)$ be the corresponding vector field germ for $s \in [0,1]$. By Theorems C.7 and C.8 of Appendix C, this vector field (or a representative of the germ) can be integrated to give a family of diffeomorphisms ϕ_s for $s \in [0,1]$ on some neighbourhood of 0 satisfying $\phi_s(0) = 0$ and satisfying the homotopy equation (5.8), which is what was required. Moreover, since the $v_{i,s} \in \mathfrak{m}_{n,I}^2$ it follows that the Taylor series of ϕ_s is $\phi_s(y) = y + O(y^2)$, so ϕ_s is of the form required for $\mathcal{R}_1$-equivalence. ✔

Observe that if $g \in \mathfrak{m}_{n,I}$ then $1+g$ is invertible in $\mathcal{E}_{n,I}$, as required in order to apply Nakayama's lemma using the ideal $\mathfrak{m}_{n,I}$.

For $f_s(x) \in \mathcal{E}_{n,I}$, let Jf_s denote the 'relative' Jacobian ideal in $\mathcal{E}_{n,I}$ generated by the partial derivatives of $f_s(x)$ with respect to just the x variables (not with respect to s):

$$Jf_s = \left\langle (f_s)_{x_1}, \ldots, (f_s)_{x_n} \right\rangle.$$

The final statement we need to complete the proof of finite determinacy is the following.

Lemma 5.19. *Suppose $f, h \in \mathcal{E}_n$ are such that $\mathfrak{m}_n^{k+1} \subset \mathfrak{m}_n^2\, Jf$ and $h \in \mathfrak{m}_n^{k+1}$. Then, for $f_s = f + sh$,*

$$\mathfrak{m}_{n,I}^{k+1} \subset \mathfrak{m}_{n,I}^2\, Jf_s.$$

In particular there are smooth germs $v_{i,s} \in \mathfrak{m}_{n,I}^2$ satisfying the infinitesimal homotopy equation (5.10).

Proof: Referring to the notation of Nakayama's lemma in Lemma 3.9, let $R = \mathcal{E}_{n,I}$ with $\mathcal{M} = \mathfrak{m}_{n,I}$ and

$$J = \mathfrak{m}_{n,I}^{k+1}, \quad K = \mathfrak{m}_{n,I}^2\, Jf_s,$$

where Jf_s is the relative Jacobian ideal defined above. We wish to show $J \subset K$, which is done as follows:

$$\begin{aligned} J &\subset \mathfrak{m}_{n,I}^2 \, Jf \\ &\subset \mathfrak{m}_{n,I}^2 \, (Jf_s + Jh) \\ &\subset K + \mathfrak{m}_{n,I}^{k+2} \quad (\text{as } Jh \subset \mathfrak{m}_{n,I}^k) \\ &= K + \mathfrak{m}_{n,I} \, J. \end{aligned}$$

Since J is a finitely generated ideal in $\mathcal{E}_{n,I}$, it follows from Nakayama's lemma that $J \subset K$. For the final part, we have $h \in J$ by hypothesis (considered as a function of x and s which happens to be independent of s), and hence h, and therefore also $-h$, belongs to K, which is precisely the statement required. ✔

5.7 Geometric criterion

A deep theorem from algebraic/analytic geometry (the Nullstellensatz) allows one to prove a useful geometric criterion for finite determinacy, rather than the algebraic criterion we have seen. This is appropriate for analytic germs rather than general smooth ones.

Before we state the theorem, we need a definition. If $f \in \mathcal{E}_n$ is the germ of an *analytic* function, then one can uniquely extend it to a function (germ) on $\mathbb{C}^n$, which is complex analytic, simply by replacing the x_i occurring in the Taylor series of f with complex variables z_i. We denote this complex version of f by $f^{\mathbb{C}}$; it is called the ***complexification*** of f. If f is merely smooth and not analytic, the function $f^{\mathbb{C}}$ is not defined, since the Taylor series does not converge to the function.

Theorem 5.20. *An analytic germ $f \in \mathcal{E}_n$ is finitely determined if and only if the origin is an isolated critical point for $f^{\mathbb{C}}$.*

Here, *isolated* means that if g is any representative of the germ $f^{\mathbb{C}}$, then there is a neighbourhood V of the origin contained in the domain of g in which g has only the one critical point (namely, the origin). The germ at 0 of $f(x, y) = (x^2 + y^2)^2$ provides an example of an analytic function for which 0 is an isolated critical point in $\mathbb{R}^2$ but not in $\mathbb{C}^2$, and it is not finitely determined.

The theorem follows immediately from the Nullstellensatz using the argument given in Corollary D.28. See [115] for general details.

Remark 5.21. In this complex analytic setting, there are two important geometric interpretations of the codimension of a critical point, which are usually stated in terms of the Milnor number $\mu(f) = 1 + \mathrm{codim}(f)$. These are as follows:

(i). given any complex analytic germ f of finite codimension, one can find an arbitrarily small analytic perturbation f_ε of f which has only nondegenerate critical points. It turns out that the number of critical points of such an f_ε coincides with the Milnor number $\mu(f)$;

(ii). for all $\delta \in \mathbb{C}$ sufficiently small, the *Milnor fibre* $f^{-1}(\delta) \cap U$ (where U is a neighbouhood of the origin in $\mathbb{C}^n$) is a submanifold of real dimension $2n - 2$, and Milnor [78] proves that this submanifold is homotopic to a 'wedge' of $\mu(f)$ spheres of dimension $n - 1$ (see Milnor's book for details).

❞

Problems

5.1 Show that $f_s(x, y) = x^4 + sx^2y^2 + y^4$ is not a trivial family.

5.2 Use the Thom–Levine principle to show that the family

$$f_s(x, y) = x^3 + 3sx^2y + y^3$$

is trivial for s in a neighbourhood of 0 (you will need to prove smoothness in s by direct calculation). (†)

5.3 In the definition of trivial family, it is written that $\phi_0 = \mathrm{Id}$. Show that without this assumption, replacing ϕ_s by $\psi_s = \phi_s \circ \phi_0^{-1}$ does satisfy $f_s \circ \psi_s = f$ with $\psi_0 = \mathrm{Id}$.

5.4 Let $f_s \in \mathcal{E}_2$ be given by $f_s(x, y) = (1 + s)x^2 + s^2y^2$. Show that $\dot{f}_s \in \mathfrak{m}_2 Jf_s$ for each $s \in (-1, 1)$, but that the family f_s is not trivial in any neighbourhood of $s = 0$. Contrast with the Thom–Levine principle. (†)

5.5 Prove the following variation of Proposition 5.7. Suppose $f \in \mathcal{E}_n$ and $k, p > 0$ satisfy $\mathfrak{m}_n^{k+p-1} \subset \mathfrak{m}_n^p Jf$, and suppose $g \in \mathcal{E}_n$ is such that $g - f \in \mathfrak{m}_n^{k+1}$. Prove that

$$\mathfrak{m}_n^p Jf = \mathfrak{m}_n^p Jg.$$

Let $f = x^5 + y^5$ and $g = f + x^3y^3$, and deduce that $\mathfrak{m}_2^3 Jf = \mathfrak{m}_2^3 Jg$.

5.6 In each of the following, f is to be understood as a germ at the origin.

(i). Show that $f(x,y) = \frac{1}{3}x^3 + xy^2$ is 3-determined.

(ii). Is $f(x,y) = x^3 + x^2y$ finitely determined?

(iii). Show that $f(x,y) = x^2 + y^k$ is k-determined, for $k \geq 2$.

(iv). Let $f(x,y) = x^2y + y^k$ where $k \geq 3$. Show f is k-determined.

(v). Use Nakayama's lemma to show that $f(x,y) = \frac{1}{3}x^3 + xy^3 + y^4$ is 4-determined.

(vi). Find the degree of determinacy given by the finite determinacy theorem for $f(x,y) = \frac{1}{2}x^2 + \frac{1}{3}x^3 - xy^2$.

(vii). Find the degree of determinacy given by the finite determinacy theorem for $f(x,y) = x^6 + y^6$. (†)

5.7 Let $f_a \in \mathcal{E}_2$ be defined by $f_a = x^2 + axy + y^2$ for $a \in \mathbb{R}$.
(i) Find the values of a for which f_a is not finitely determined, and
(ii) for those values of a, find the set of critical points.
(iii) For more of a challenge, repeat parts (i) and (ii) with the function $g_a = x^2y + axy^2 + y^3$.

5.8 In Example 5.11(iv) it is claimed that $f_0 = x^3 + xy^3$ is 4-determined. Prove this as follows. Let $f = f_0 + h$ with $h \in \mathfrak{m}_2^5$.

(i). Let $I = \langle x^4, x^3y, x^2y^2, xy^3, y^5 \rangle = x\mathfrak{m}_2^3 + \langle y^5 \rangle$. Use Nakayama's lemma to show that $I \subset T\mathcal{R}{\cdot}f$, for all $h \in \mathfrak{m}_2^5$.

(ii). Find two germs k_1, k_2, such that $T\mathcal{R}{\cdot}f = I \oplus \mathbb{R}\{k_1, k_2\}$, with k_1, k_2 independent of h.

(iii). Use the Thom–Levine principle to deduce that f_0 and f are right equivalent. (Note that they are not $\mathcal{R}_1$-equivalent.) (†)

5.9 Consider the map $f\colon \mathbb{R}^2 \to \mathbb{R}^2$, $f(x,y) = (x+y^2,\ y+x^2)$. Follow the proof of the inverse function theorem and write down L and h and f_s for this map. Write down explicitly the infinitesimal homotopy equation (5.7). [*Note*: *don't confuse* y *in that equation with the coordinate* y *on* $\mathbb{R}^2$. *You could use* $y = (u,v)$ *as coordinates for* y *in* (5.7).] Solve (5.7) to find an explicit expression for the vector field $\mathbf{v}_s(y)$.

5.10 Solve the infinitesimal homotopy equation (5.10) for the vector field $\mathbf{v}$ in the case that $f = x^3 \in \mathcal{E}_1$ and $h = x^4$, and show it satisfies the required properties that ensure it can be integrated to obtain a family of diffeomorphisms $\phi_s(x)$ such that $(f+h) \circ \phi_1 = f$. [*Note*: *in* 1 *variable,* $\mathbf{v}$ *is uniquely defined by* (5.10); *in more variables it's not.*]

5.11 Let $f \in \mathcal{E}_n$ and let $\mathbf{v}$ be a vector field germ for which $\mathbf{v}(0) = 0$ and $\mathrm{t}f(\mathbf{v}) = 0$. Show that $f \circ \phi_t = f$ where ϕ_t is the germ at 0 of the flow associated to $\mathbf{v}$.

5.12 Consider $f(x, y) = x^3 + y^3 \in \mathcal{E}_2$. Show that $\mathfrak{m}_2^3 \subset \mathfrak{m}_2 Jf$. Referring to the proof of the finite determinacy theorem, show there is a vector field $\mathbf{v}$ satisfying (5.9) with $h = -f$. Why can one not deduce that f is right equivalent to $f + h = f + (-f) = 0$? (†)

5.13 In the proof of the finite determinacy theorem, show that if $h \in \mathfrak{m}_n^{k+r}$ then each $v_{s,i}$ can be chosen in $\mathfrak{m}_n^{r+1}$ and hence that $\mathrm{j}^r\phi = \mathrm{id}$ (i.e. $\phi(x) = x + O(r + 1)$). [*Hint*: *for the last part, you will need to refer to Theorem* C.8.]

5.14 Let $f \in \mathfrak{m}_n^2$ and ϕ be a diffeomorphism germ at the origin. Show that f is k-$\mathcal{R}_1$-determined if and only if $f \circ \phi$ is. [*Hint*: *see Proposition* 5.13.]

5.15 Use the geometric criterion to show

(i). that the germ in Problem 5.6(ii) is not finitely determined, and

(ii). that $f(x, y) = (x^2 + y^2)^2$ has an isolated critical point in $\mathbb{R}^2$ but not in $\mathbb{C}^2$, and deduce that it is not finitely determined.

6

Classification of the elementary catastrophes

IN MANY BRANCHES OF mathematics there is a classification problem, where one aims to list all the objects of a certain type, up to some equivalence relation. For example, the classification of finite simple groups up to isomorphism. Proofs of classification theorems can be a rewarding mixture of calculations and theory, and the classification of critical points is no exception. In this chapter, we are going to classify critical points up to right equivalence, which means that we want to write down a list containing one germ from each equivalence class. As stated, however, this is not possible because the list would be not only infinitely long but infinitely complicated, in that the list cannot be organized into finitely many families. We therefore limit our level of complexity by considering germs of codimension at most 4, which is ample for most applications.

Before we begin the classification, we need to recall the splitting lemma from Section 4.5, and point out its relevance to the classification problem. Suppose $f\colon (\mathbb{R}^n, 0) \to \mathbb{R}$ is a function germ, with a critical point at the origin. We suppose $f(0) = 0$ so that $f \in \mathfrak{m}_n^2$. Let $H_f = \mathrm{d}^2 f_0$ be the Hessian matrix of f at the origin. If H_f is nondegenerate then f is equivalent to $\sum \pm x_i^2$, by the Morse lemma, so we already have the classification of nondegenerate critical points.

Now suppose H_f is degenerate. The ***corank*** of f at 0 is defined to be the dimension of $\ker H_f$. Let $\ell = \operatorname{corank}(f)$, then by the splitting lemma we can change coordinates so that

$$f(x, y) = h(x) + \sum_{i=1}^{n-\ell} \pm y_i^2. \qquad (6.1)$$

Here $x \in \mathbb{R}^\ell$ and $y \in \mathbb{R}^{n-\ell}$. The y-part is already as simple as possible, so there just remains to classify the x-part. Thus, *to classify germs on $\mathbb{R}^n$ of corank ℓ we need only classify functions of ℓ variables.*

Furthermore, if f is the function in (6.1) above, then

$$Jf = \left\langle \frac{\partial h}{\partial x_1}, \dots, \frac{\partial h}{\partial x_\ell}, y_1, \dots, y_{n-\ell} \right\rangle.$$

It follows that

$$\frac{\mathfrak{m}_n}{Jf} \simeq \frac{\mathfrak{m}_\ell}{Jh},$$

and consequently $\operatorname{codim}(f) = \operatorname{codim}(h)$, as stated in Proposition 4.16. Moreover, a cobasis for Jh in $\mathfrak{m}_\ell$ is also a cobasis for Jf in $\mathfrak{m}_n$ (in spite of the fact that the elements of the cobasis are only functions of the x variables). See also Problem 4.9.

Lemma 6.1. *Suppose* $\operatorname{corank}(f) = \ell$, *then* $\operatorname{codim}(f) \geq \frac{1}{2}\ell(\ell+1)$.

In particular if $\operatorname{corank}(f) > 2$ then $\operatorname{codim}(f) \geq 6$. The importance of this is that we are going to classify all germs up to codimension 4, so that we need only consider germs of corank 1 and 2.

PROOF: We show in fact that $\mathfrak{m}_\ell/(Jh + \mathfrak{m}_\ell^3)$ has dimension at least $\frac{1}{2}\ell(\ell+1)$. Now Jh has ℓ generators, all of which lie in $\mathfrak{m}_\ell^2$, so $Jh + \mathfrak{m}_\ell^3 \simeq \mathbb{R}^{\ell} \oplus \mathfrak{m}_\ell^3$ for some $\ell \leq \ell$. Now $\mathfrak{m}_\ell/\mathfrak{m}_\ell^3$ has dimension $\ell + \frac{1}{2}\ell(\ell+1)$ and the quotient by Jh reduces dimension by at most ℓ, so that

$$\dim\left(\mathfrak{m}_\ell/(Jh + \mathfrak{m}_\ell^3)\right) = \ell + \tfrac{1}{2}\ell(\ell+1) - \ell \geq \tfrac{1}{2}\ell(\ell+1)$$

as required. ✔

6.1 Classification of corank 1 singularities

For functions of one variable, one can give a complete classification of germs of finite codimension, which is not possible for 2 or more variables, so for this section we do not limit ourselves to codimension at most 5. And if the singularity is of corank 1 then as described above, the splitting lemma reduces the problem to one variable, the other variables appearing as squares as in (6.1) with $\ell = 1$.

Theorem 6.2. *Let* $f \in \mathfrak{m}_1^2$ *be of finite codimension. Then there is a* $k > 1$ *such that* f *is right equivalent to* $\pm x^k$.

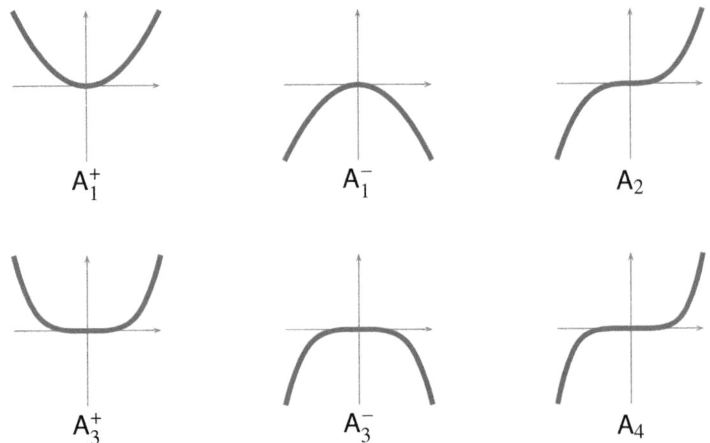

FIGURE 6.1 Graphs of A_k critical points in 1 variable, $k = 1, \dots, 4$.

If k is odd, then the two germs $+x^k$ and $-x^k$ are right equivalent, thanks to the change of coordinate $x \mapsto -x$. However, when k is even, the two functions are *not* equivalent. Thus, any finite codimension germ in one variable is right equivalent to one of

$$\begin{array}{ccccccc} & x^2 & & x^4 & & x^6 & \\ x & & x^3 & & x^5 & & \cdots \\ & -x^2 & & -x^4 & & -x^6 & \end{array}$$

As stated in Chapter 2, a germ equivalent to $\pm x^{k+1}$ is known as a *critical point of type* $\mathsf{A}_k^\pm$ (for reasons related to reflection groups and not of concern here). If k is even, $\mathsf{A}_k^+ = \mathsf{A}_k^-$, which is just called A_k, while if k is odd $\mathsf{A}_k^+ \neq \mathsf{A}_k^-$. Note that a critical point of type $\mathsf{A}_k^\pm$ is of codimension $(k-1)$.

Thus we have the following table:

type	$\mathsf{A}_1^\pm$	A_2	$\mathsf{A}_3^\pm$	A_4	$\mathsf{A}_5^\pm$
normal form	$\pm x^2$	x^3	$\pm x^4$	x^5	$\pm x^6$
codimension	0	1	2	3	4

Graphs of these functions are shown in Figure 6.1.

PROOF: Firstly, the fact that f is of finite codimension implies that the Taylor series of f starts at some order, say,

$$f(x) = ax^k + x^{k+1}h(x),$$

where $h(x)$ is a smooth function and $a \neq 0$. We know from Example 4.2 (or by the finite determinacy theorem) that such an f is equivalent to ax^k. Now by rescaling x by $\phi(x) = |a|^{-1/k}x$ reduces f (or rather $f \circ \phi$) to $\mathrm{sign}(a)x^k$ as required. ✔

6.2 Classification of corank 2 critical points

This is considerably more involved that the one variable case. Again we apply the splitting lemma to reduce the problem to just two variables, so we consider the classification of critical points of the form $f(x, y)$ satisfying $\mathrm{d}^2 f(0) = 0$; that is, $f \in \mathfrak{m}_2^3$.

The 3-jet Since the 2-jet of $f \in \mathfrak{m}_2^3$ is zero, the 3-jet is of the form,

$$\mathrm{j}^3 f = ax^3 + bx^2y + cxy^2 + dy^3,$$

for some $a, b, c, d \in \mathbb{R}$. This is a ***binary cubic form*** (i.e. a homogeneous cubic polynomial in two variables), and to proceed further we need to use the classification of such binary cubic forms, for which see the box on p. 85. Apart from the zero form, there are 4 types, called elliptic, hyperbolic, parabolic and symbolic cubics. We treat each of these possibilities in turn.

Elliptic umbilic After a linear change of coordinates, we can write

$$f(x, y) = x^2y - y^3 + h(x, y)$$

where $h \in \mathfrak{m}_2^4$. We claim that f is equivalent to its 3-jet. Indeed, write $f_0 = x^2y - y^3$, then

$$Jf_0 = \langle xy,\, x^2 - 3y^2 \rangle,$$

so that $\mathfrak{m}_2 Jf_0 = \langle x^2y,\, xy^2,\, x^3 - 3xy^2,\, x^2y - 3y^3 \rangle = \mathfrak{m}_2^3$. This shows that f_0 is 3-determined, and so f is equivalent to f_0. This ***elliptic umbilic*** is denoted D_4^-.

Hyperbolic umbilic After a linear change of coordinates as described in the box on p. 85, we can write

$$f(x, y) = x^2y + y^3 + h(x, y)$$

The same calculation as before shows that f is equivalent to its 3-jet, which is

$$f_0 = x^2y + y^3.$$

This ***hyperbolic umbilic*** is denoted D_4^+.

Binary Cubic Forms These are homogeneous polynomials of degree 3 in two variables which we show can be classified into 5 different types, including 0. Any binary cubic form can be written as

$$C = ax^3 + bx^2y + cxy^2 + dy^3,$$

with $a, b, c, d \in \mathbb{R}$. Assume first $a \neq 0$. Then we can factorize C over the complex numbers as

$$C = a(x - \alpha y)(x - \beta y)(x - \gamma y),$$

where for example $d/a = -\alpha\beta\gamma$. There are several cases to consider:

(i). The roots α, β, γ are real and distinct. The cubic is then said to be an ***elliptic cubic***, and after a linear change of coordinates, one can make

$$C = x^3 - xy^2 = x(x + y)(x - y).$$

(ii). The roots are distinct but not all real. Then one of them, say α is real and the other two are complex conjugates, $\gamma = \bar{\beta}$, and in this case C is said to be a ***hyperbolic cubic*** and a linear change of coordinates can reduce it to the form,

$$C = x^3 + xy^2 = x(x^2 + y^2) = x(x + \mathrm{i}y)(x - \mathrm{i}y).$$

(iii). A double root: $\alpha = \beta \neq \gamma$. The cubic is said to be a ***parabolic cubic***, and $C = a(x - \alpha y)^2(x - \gamma y)$. After replacing $(x - \alpha y)$ with x and $(x - \gamma y)$ with y, and then rescaling by $|a|^{1/3}$, one obtains $C = x^2y$.

(iv). Three equal roots: $\alpha = \beta = \gamma$. This is sometimes called a ***symbolic cubic***, or a perfect cube. So $C = a(x - \alpha y)^3$, and if we replace (x, y) by $(x - \alpha y, y)$ and then rescale by $|a|^{1/3}$ we get $C = x^3$.

If instead $a = 0$, then next suppose $d \neq 0$. Then we can repeat the above argument with the roles of x and y exchanged.

If $a = d = 0$ then $C = bx^2y + cxy^2 = xy(bx + cy)$. If $b = c = 0$ then we have the zero form; if one of b, c is zero then we have a parabolic form, either x^2y or xy^2; finally if both b and c are nonzero, then C has three distinct real factors, so it's an elliptic cubic. This concludes all possible cases, so we have shown that by linear changes of coordinates any nonzero binary cubic form can be reduced to one of the four types listed above. See Figure 6.2.

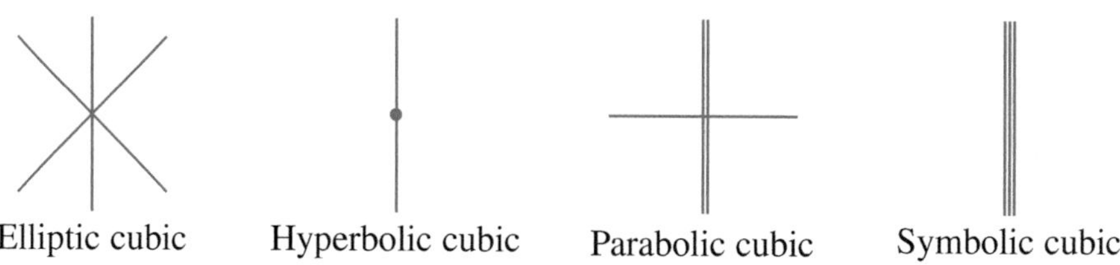

FIGURE 6.2 The zero-sets of the 4 types of binary cubic.

Parabolic umbilic After a linear change of coordinates, we can write

$$f(x, y) = x^2y + h(x, y)$$

In this case, the 3-jet is x^2y which is *not* 3-determined (it's not even finitely determined). We therefore turn to the 4-jet, and write $\hat{f} = \mathrm{j}^4 f$, whence

$$\hat{f} = x^2y + ax^4 + bx^3y + cx^2y^2 + dxy^3 + ey^4. \tag{6.2}$$

We assume $e \neq 0$ (we want to reduce to $x^2y \pm y^4$).

Proposition 6.3. *The germ $\hat{f}$ in (6.2) with $e \neq 0$ is right equivalent to*

$$f_0 := x^2y \pm y^4.$$

For the proof see Problem 6.4.

Symbolic cubic In this case one can show that the codimension is at least 5. Indeed the first singularity with this 3-jet is $x^3 + y^4$, the so-called E_6 singularity.

The diagrams in Figure 6.2 show the zero-set of the cubic form C in each of the 4 cases where $C \neq 0$. For the hyperbolic cubic, the dot at the origin represents the factor $(x^2 + y^2)$. In the two diagrams on the right, the double and triple lines represent the double and triple roots of C.

6.3 Thom's 7 elementary singularities

Here we show the table summarizing what we have found. These are at the heart of the so-called *elementary catastrophes* (see the next chapter), and were first classified by René Thom in the 1960s. In Table 6.1 we give them all as functions of two variables.

TABLE 6.1 Thom's seven elementary catastrophes.

Label	Normal form	Codim.
$\mathsf{A}_1^\pm$	$\pm x^2 \pm y^2$	0
A_2	$x^3 \pm y^2$	1
$\mathsf{A}_3^\pm$	$\pm x^4 \pm y^2$	2
A_4	$x^5 \pm y^2$	3
$\mathsf{A}_5^\pm$	$\pm x^6 \pm y^2$	4
$\mathsf{D}_4^\pm$	$x^2y \pm y^3$	3
D_5	$x^2y + y^4$	4

We could have given the A_k as functions of a single variable, which would involve ignoring the y^2 term. More generally, if there are n variables, then the $\mathsf{A}_k^\pm$ has the normal form (using the splitting lemma) $\pm x^{k+1} + Q(y)$, where Q is a nondegenerate quadratic form on $\mathbb{R}^{n-1}$, which can of course be diagonalized to give

$$\mathsf{A}_k^\pm : \quad \pm x^{k+1} + \varepsilon_1 y_1^2 + \varepsilon_2 y_2^2 + \cdots + \varepsilon_{n-1} y_{n-1}^2,$$

where $\varepsilon_i = \pm 1$.

6.4 Further classification

We list for interest the next singularities that occur in the classification. Table 6.2 gives the complete list of the so-called *simple singularities*, a notion we explain briefly in the next chapter (see Section 7.6). There are two infinite families, A_k and D_k, and three individual types, called the *exceptional simple singularities*, $\mathsf{E}_6, \mathsf{E}_7$ and E_8. The naming is due to V. I. Arnold, and is related to the geometry of reflection groups; for an explanation see [3].

The first corank 3 critical point is

$$\mathsf{T}_{3,3,3} : \quad x^3 + y^3 + z^3 + \lambda xyz$$

where $\lambda \in \mathbb{R}$, $\lambda \neq -3$ (or for complex functions one allows $\lambda \in \mathbb{C}$ with $\lambda^3 \neq -27$). Each of these has codimension 7, and is 3-determined. For different real values of λ these singularities are inequivalent. The appearance of the parameter λ giving a

TABLE 6.2 The simple singularities

Label	Normal form	Codimension
A_k	$x^2 \pm y^{k+1}$	$k-1 \quad (k \geq 1)$
D_k	$x^2y \pm y^{k-1}$	$k-1 \quad (k \geq 4)$
E_6	$x^3 + y^4$	5
E_7	$x^3 + xy^3$	6
E_8	$x^3 + y^5$	7

continuum of similar but inequivalent critical points is what makes these and others below into *non-simple* germs.

These are one of a family of critical points,

$$\mathsf{T}_{p,q,r}: \quad x^p + y^q + z^r + \lambda xyz. \tag{6.3}$$

If $\frac{1}{p}+\frac{1}{q}+\frac{1}{r} = 1$ these are called parabolic singularities: there are only 3 possibilities: $(p,q,r) = (3,3,3)$, $(2,3,6)$ and $(2,4,4)$. The $\mathsf{T}_{3,3,3}$ singularity is also variously denoted $\tilde{\mathsf{E}}_6$ and P_8, while $\mathsf{T}_{2,4,4}$ is called $\tilde{\mathsf{E}}_7$ and X_9, and $\mathsf{T}_{2,3,6}$ is called $\tilde{\mathsf{E}}_8$ or J_{10} ($\mathsf{T}_{2,4,4}$ and $T_{2,3,6}$ are of corank 2). If $\frac{1}{p}+\frac{1}{q}+\frac{1}{r} < 1$, $\mathsf{T}_{p,q,r}$ is called a hyperbolic singularity. It is easy to see that the parabolic singularities are weighted homogeneous, while the hyperbolic ones are not.

The classification of all degenerate critical points of codimension up to 10 was carried out by Arnold in the mid-1970s; many details can be found in books by Arnold and coauthors [5, 6]. Some further classification beyond ours is described in the book of Dimca [31].

Remark 6.4. While it is very useful to have a classification of all possible critical points (of low codimension), this is only half the story. The missing part is the recognition problem: given a critical point, how do we know to which one on the list it is equivalent? In one variable, this is simple enough: as we saw in Section 2.2, one just takes successive derivatives until one is non-zero.

If one has a specific expression for a critical point, then a combination of finite determinacy and codimension calculations can often be enough to identify a singularity type, but Ian Porteous [95, 97] developed a more geometric method, which he called *probing a singularity*. Very briefly, for $f \in \mathcal{E}_n$, a *probe* is a smooth *non-singular* curve (germ), $\gamma\colon (\mathbb{R},0) \twoheadrightarrow (\mathbb{R}^n,0)$ and the process is to optimize γ

to make $f \circ \gamma$, $\mathrm{d}f \circ \gamma$, $\mathrm{d}^2 f \circ \gamma$, etc. vanish to as high an order as possible. For example for an A_k critical point, one can find a probe so that $df \circ \gamma$ vanishes to order k. Details can be found in the references of Porteous cited above. ❞

Problems

6.1 Check the codimensions of all the corank 2 critical points described in Secs 6.3 and 6.4.

6.2 The hyperbolic umbilic is often given as $g(x, y) = x^3 + y^3$. By referring to the classification of binary cubics on p. 85, show that g is indeed of type D_4^+.

6.3 Consider the germ $f(x, y) = x^4 + 2x^2 y + 2y^3 \in \mathcal{E}_2$. Find the corank of f. Show that $xy \in Jf$ and hence find the codimension of f. Use this information to determine the type of singularity represented by f. (†)

6.4 Provide the details of the argument that the parabolic umbilic with $e \neq 0$ in (6.2) is equivalent to $f_0 = x^2 y \pm y^4$ as follows. Fix (a, b, c, d, e) with $e \neq 0$, and let

$$f = x^2 y + ax^4 + bx^3 y + cx^2 y^2 + dxy^3 + ey^4.$$

(i). Use Nakayama's lemma to show that $\mathfrak{m}_2^4 \subset \mathfrak{m}_2 Jf$. Deduce that f is 4-determined.

(ii). Change coordinates so that $x = X + \alpha(X, Y)$ and $Y = Y + \beta(X, y)$, where $\alpha, \beta \in \mathfrak{m}_2^2$. Show that α and β can be chosen in order to eliminate the degree 4 terms except for ey^4, and use the 4 determinacy of f to reduce to $f_1 = x^2 y + ey^4$ with $e \neq 0$.

(iii). Rescale y to make $e = \pm 1$ and then x so that coefficient of $x^2 y$ returns to 1. (†)

6.5 Consider the $\mathsf{T}_{3,3,3}$ singularity, $f = x^3 + y^3 + z^3 + \lambda xyz$ (with $\lambda \in \mathbb{R}$). Show that $\mathfrak{m}_3 Jf \neq \mathfrak{m}_2^3$, but $\mathfrak{m}_3^2 Jf = \mathfrak{m}_3^4$ provided $\lambda \neq -3$. Find the set of critical points of f when $\lambda = -3$ and show that in this case the origin is not an isolated critical point, and compare with the geometric criterion of Theorem 5.20.

6.6 Show that of the singularities $\mathsf{T}_{p,q,r}$ given in Equation (6.3), the three parabolic ones are weighted homogeneous, while the hyperbolic ones are not. (†)

6.7 Consider the binary cubic form $C = ax^3 + bx^2y + cxy^2 + dy^3$, and form its ***Hessian quadratic form***,

$$h(x, y) = \det \begin{pmatrix} \frac{\partial^2 C}{\partial x^2} & \frac{\partial^2 C}{\partial x \partial y} \\ \frac{\partial^2 C}{\partial x \partial y} & \frac{\partial^2 C}{\partial y^2} \end{pmatrix}.$$

Show that the classification of C into elliptic, hyperbolic and parabolic corresponds to the same classification of the quadratic form, and that C is a perfect cube if and only if h is zero. (Further details can be found in [97].)

7

Unfoldings and catastrophes

WHEN A FUNCTION WITH a degenerate critical point is deformed (perturbed), the new function may have several critical points nearby the old degenerate one. We saw examples of this in Chapter 2, for example in the family $x^3 + ux$, which deforms the degenerate critical point $f_0(x) = x^3$, producing either two or no real critical points, according to the sign of u. In this chapter we study this in greater detail, using the classification of elementary singularities from the previous chapter, which leads to René Thom's *elementary catastrophes*.

Recall from Definition 5.1 that a smooth family of functions is a smooth map

$$F\colon \mathbb{R}^n \times \mathbb{R}^a \rightarrowtail \mathbb{R}$$

and we write $F(x; u) = f_u(x)$, with $x \in \mathbb{R}^n$ and $u \in \mathbb{R}^a$, the latter being the *parameter*. In Chapter 2 we looked in some detail at two examples: the fold family

$$F(x; u) = x^3 + ux$$

and the cusp family

$$F(x; u, v) = x^4 + ux^2 + vx.$$

(These differ from the expressions used in Chapter 2 by some rational coefficients introduced to simplify the calculations, but they make little difference). In both families, when the parameters are zero the critical point is at its most degenerate (an A_2 in the first and an A_3 in the second). Such families, in a neighbourhood of a degenerate critical point, are said to be deformations or *unfoldings* of that degenerate critical point. Thus the fold family is an unfolding of x^3 and the cusp family an unfolding of x^4.

In the previous chapter we found the list of all critical points of codimension up to 4. In this one we will study their unfoldings. First we recall briefly the main characteristics of families/unfoldings; this is a repeat of definitions from Chapter 2.

7.1 Geometry of families of functions

A smooth a–parameter ***family of functions*** on $\mathbb{R}^n$ is a smooth map (whose domain we assume includes the origin)

$$\begin{aligned} F\colon \mathbb{R}^n \times \mathbb{R}^a &\rightarrowtail \mathbb{R} \\ (x,\, u) &\mapsto f_u(x). \end{aligned}$$

The notations $f_u(x)$ and $F(x;u)$ reflect the interpretation that u is a parameter and in applications x is often called the ***state variable***. The parameter space $\mathbb{R}^a$ is often called the ***base space*** of the family.

Such a family is called an ***unfolding*** or ***deformation*** of a given germ $f_0 \in \mathcal{E}_n$, where $f_0(x) = F(x,0)$. In this case one usually only considers the family for u in a neighbourhood of 0 (or even the germ at $u = 0$).

Recall the definitions made in Chapter 2. Given a family F as above, then the ***catastrophe set*** of F is

$$C_F = \{(x,u) \in \mathbb{R}^n \times \mathbb{R}^a \mid \mathsf{d}(f_u)_x = 0\},$$

where $\mathsf{d}(f_u)$ means the differential with respect to x only. In many important cases, this will be a submanifold of dimension a (the number of parameters – see Appendix B for the definition of submanifold). Next we defined the ***singular set*** Σ_F of the family as the set of points in C_F where the critical point is degenerate, and the ***discriminant*** Δ_F to be the image of Σ_F in the base space, under the projection

$$\begin{aligned} \pi_F\colon C_F &\longrightarrow \mathbb{R}^a \\ (x,u) &\longmapsto u. \end{aligned}$$

so it is the set of points u in the base (or parameter) space $\mathbb{R}^a$ for which f_u has a degenerate critical point. The discriminant is also known as the ***bifurcation set***.

There is a close relationship between the geometry of the projection π_F and the type of critical point. We begin by stating what happens near a nondegenerate critical point.

Proposition 7.1. *If* $(x_0,u_0) \in C_F \setminus \Sigma_F$ *then there is a neighbourhood* U *of* (x_0,u_0) *in* $\mathbb{R}^n \times \mathbb{R}^a$ *such that* $U \cap C_F$ *is a submanifold of* $\mathbb{R}^n \times \mathbb{R}^a$ *of dimension* a*, and such that the restriction of* π_F *to* U *is a diffeomorphism.*

This statement implies in particular that if C_F fails to be a submanifold at (x_0,u_0), then x_0 is necessarily a degenerate critical point of f_{u_0}.

PROOF: Suppose that $(x_0, u_0) \in C_F$ and f_{u_0} has a nondegenerate critical point at x_0. Then by the splitting lemma (see p. 55), there is a neighbourhood U of (x_0, u_0) and a change of coordinates of the form $(x, u) \mapsto (X(x, u), u)$ such that in these coordinates

$$F(X, u) = Q(X) + h(u).$$

Here $Q(X) = \frac{1}{2}X^T HX$ where H is the Hessian matrix of f_{u_0} at x_0. In these coordinates, $C_F \cap U = \{(X, u) \mid X = 0,\ (0, u) \in U\}$. It follows that $C_F \cap U$ is a submanifold of dimension a of $\mathbb{R}^n \times \mathbb{R}^a$ (and the change of coordinates is the straightening map).

Furthermore, continuing with the same coordinates, the projection π_F on U is given by

$$\pi_F(0, u) = u$$

which is clearly invertible and a diffeomorphism. ✔

The reason Δ_F is called the *bifurcation set* is that it is where transitions, or bifurcations, take place. If $u_0 \notin \Delta_F$ then there is a neighbourhood V of u_0 such that $\forall v \in V$, the number of critical points of f_v is the same as for f_{u_0}, as follows from the proposition above (the V here is the projection to the base space of the U in the proposition). Thus Δ_F divides $\mathbb{R}^a$ into a number of connected components, and in each of these the number of critical points is constant. We already saw this in the examples of Chapter 2.

Example 7.2. To illustrate the possibility that the catastrophe set may not be submanifold, consider the family $F\colon \mathbb{R} \times \mathbb{R}^2 \to \mathbb{R}$ given by $f(x, u, v) = \frac{1}{4}x^4 + \frac{1}{3}ux^3 + \frac{1}{2}vx^2$. Then $\frac{\partial f}{\partial x} = x^3 + ux^2 + vx$ so that

$$C_F = \left\{(x, u, v) \in \mathbb{R}^3 \mid x(x^2 + ux + v) = 0\right\}.$$

This is not a submanifold, although it is made up of the union of two submanifolds (smooth surfaces): the plane $\{x = 0\}$ and the curved surface $\{v = -x^2 - ux\}$. These two surfaces intersect along the u-axis. Thus

$$C_F = \left\{(0, u, v) \mid u, v \in \mathbb{R}\right\} \cup \left\{(x, u, v) \mid v = -x^2 - ux\right\}.$$

Now, $\Sigma_F = \{(x, u, v) \in C_F \mid 3x^2 + 2ux + v = 0\}$. Treating each component in turn gives

$$\begin{aligned}\Sigma_F &= \left\{(0, u, 0)\right\} \cup \left\{(x, u, v) \mid v = -x^2 - ux = -3x^2 - 2ux\right\} \\ &= \left\{(0, u, 0) \mid u \in \mathbb{R}\right\} \cup \left\{(x, -2x, x^2) \mid x \in \mathbb{R}\right\}.\end{aligned}$$

This is the union of two curves (one the u-axis, the other a parabola), and as we have seen the u-axis is where the catastrophe set fails to be a submanifold. On the other hand, it is a submanifold in a neighbourhood of the other points in Σ_F.

The discriminant is the projection of these curves to the parameter plane:

$$\Delta_F = \left\{(u,0)\right\} \cup \left\{(u,v) \mid u^2 = 4v\right\}.$$

The curves are tangent at the origin, as shown in the figure below.

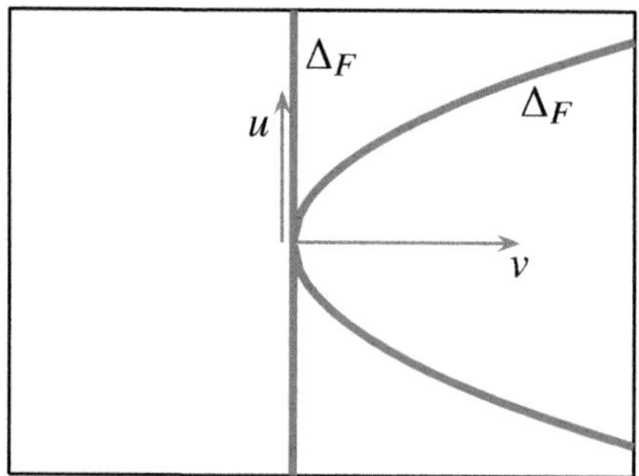

Now consider the geometry of the map π_F. This map is given as usual by $\pi_F(x,u,v) = (u,v)$. The component $\{x = 0\}$ can be parametrized by (u,v), and this map becomes $\pi_F(u,v) = (u,v)$ which is of course a diffeomorphism. The other component of Σ_F satisfies $v = -x^2 - ux$ so can be parametrized by (u,x), and the projection is

$$\pi_F(u,x) = (u, -x^2 - ux).$$

This has differential

$$\mathrm{d}\pi_F = \begin{pmatrix} 1 & 0 \\ -x & -2x - u \end{pmatrix},$$

which fails to be invertible when $u = -2x$. Of the two components of the singular set Σ_F, one is therefore the intersection of the two components of C_F (so points where C_F is not a submanifold) and the other is the set of singular points of π_F. See Problem 7.9.

7.2 Change of parameter and induced unfoldings

We wish to define a notion of *equivalent unfoldings*, which will involve changes of parameter. We begin with an important definition which will be central to further developments.

Definition 7.3. Let $f\colon \mathbb{R}^n \rightarrowtail \mathbb{R}$ be a given function, and let

$$F\colon \mathbb{R}^n \times \mathbb{R}^a \rightarrowtail \mathbb{R}$$

be a given unfolding of f. Now let $\phi\colon \mathbb{R}^b \to \mathbb{R}^a$ be a given map with target the base of the given unfolding, and with $\phi(0) = 0$. One defines the unfolding of f ***induced from F by ϕ***, written ϕ^*F to be the family,

$$(\phi^*F)(x; w) = F(x; \phi(w)).$$

Thus $\phi^*F\colon \mathbb{R}^n \times \mathbb{R}^b \to \mathbb{R}$. ✯

Note that $\phi^*F(x;0) = F(x;0)$, so F and ϕ^*F are both unfoldings of the same function f. It is useful to see it as a diagram:

$$\begin{array}{ccc} \mathbb{R}^n \times \mathbb{R}^b & & \\ {\scriptstyle \mathsf{Id}\times\phi}\downarrow & \searrow {\scriptstyle \phi^*F} & \\ \mathbb{R}^n \times \mathbb{R}^a & \xrightarrow{F} & \mathbb{R} \end{array} \tag{7.1}$$

(The dashed line here means ϕ^*F is constructed using the other maps: $\phi^*F = F \circ (\mathsf{Id} \times \phi)$, where of course $(\mathsf{Id} \times \phi)(x, w) = (x, \phi(w))$.)

This can all be stated in terms of germs at 0: if F, ϕ are both germs, then so is ϕ^*F. The reader may check that if indeed F and ϕ are replaced by germs, then the induced germ ϕ^*F is well-defined.

Example 7.4. Consider the family $G\colon \mathbb{R}\times\mathbb{R} \to \mathbb{R}$ given by $G(x; w) = \frac{1}{4}x^4 + \frac{1}{2}wx^2$. This can be induced from the unfolding given in Section 2.4, namely $F(x; u, v) = x^4 + ux^2 + vx$ by the map $\phi(w) = (w, 0)$. Notice that

$$C_G = \left\{(x, w) \mid x^3 + wx = 0\right\}.$$

and therefore $C_G = \{(x, w) \mid (x, \phi(w)) \in C_F\}$ – see Problem 7.3 for more in this direction. ✐

7.3 Equivalence of unfoldings

Similar to the right equivalence of germs, there is a useful notion of equivalence of unfoldings, illustrated and motivated by the following example.

Example 7.5. Consider the 2–parameter unfolding of $f_0(x) = x^3$ given by

$$G(x, r, s) = x^3 + 3rx^2 + sx.$$

We wish to compare this to the 1–parameter unfolding of the fold catastrophe, given in Example 2.7, namely $F(x; u) = x^3 + ux$. We begin by 'completing the cube', $(x + r)^3 = x^3 + 3rx^2 + 3r^2x + r^3$, so that

$$\begin{aligned} G(x; r, s) &= (x + r)^3 + (s - 3r^2)x - r^3 \\ &= (x + r)^3 + (s - 3r^2)(x + r) - r(s - 3r^2) - r^3 \\ &= y^3 + (s - 3r^2)y + (2r^3 - rs). \end{aligned}$$

where $y = x + r$. The relation between this family and F involves several steps:
(i) replace x with $y = x + r$ (this is a change of coordinates depending on r)
(ii) replace u with $s - 3r^2$
(iii) add the constant term $(2r^3 - rs)$ (constant for each value of the parameters)
To put these together, number (ii) is inducing an unfolding using the map $\phi\colon \mathbb{R}^2 \to \mathbb{R}$, $\phi(r, s) = u = s - 3r^2$, while (i) is an $\mathcal{R}$-equivalence (depending on the parameter r), and (iii) makes it $\mathcal{R}^+$-equivalence (depending on both parameters).

In conclusion, $\phi^* F(x; r, s) = x^3 + (s - 3r^2)x$ is 'unfolding-equivalent' to G, as made precise in the following definition. ✐

Note that the change of coordinates (diffeomorphism) $x \mapsto x+r$ in this example does not preserve the origin (if $r \neq 0$), whereas previously the $\mathcal{R}$-equivalence we have considered has always preserved the origin. When we consider unfoldings, preserving the origin is not required.

This example suggests the following definition of equivalent unfoldings:

Definition 7.6. Two families $F, G\colon \mathbb{R}^n \times \mathbb{R}^a \to \mathbb{R}$ are ***equivalent*** if there is a diffeomorphism $\Phi\colon \mathbb{R}^n \times \mathbb{R}^a \to \mathbb{R}^n \times \mathbb{R}^a$ of the form

$$\Phi(x, u) = (\phi(x, u), \psi(u))$$

and a function-germ $C\colon \mathbb{R}^a \to \mathbb{R}$ such that

$$F(x, u) = G(\phi(x, u), \psi(u)) + C(u).$$

This equivalence is called $\mathcal{R}^+_{\mathrm{un}}$***-equivalence***. ☆

The subscript 'un' in $\mathcal{R}^+_{\mathrm{un}}$ is of course for 'unfolding'. In essence therefore, for each value of u, the function $G_{\psi(u)}$ is $\mathcal{R}^+$-equivalent to F_u, via the change of coordinates ϕ_u (here $\phi_u(x) = \phi(x, u)$), and the constant $C(u)$.

It is important to emphasize that the change in parameters does not involve the state variable x.

7.4 Versal unfoldings

One of the important ideas introduced by Thom is the notion of versal unfolding [111]. The idea is that a versal unfolding of a germ f_0 contains all the information of all possible unfoldings of f_0. More precisely, the definition is the following.

Definition 7.7. Let $f_0 \in \mathcal{E}_n$ be a function-germ, and $F\colon (\mathbb{R}^n \times \mathbb{R}^a, (0,0)) \to \mathbb{R}$ be an unfolding of f_0. The unfolding F is ***versal*** if given any other unfolding $G\colon \left(\mathbb{R}^n \times \mathbb{R}^b, (0,0)\right) \to \mathbb{R}$ of f_0 there is a map germ $\phi\colon (\mathbb{R}^b, 0) \to (\mathbb{R}^a, 0)$ such that G is equivalent to $\phi^* F$. ✯

The equivalence here is of course $\mathcal{R}^+_{\mathrm{un}}$-equivalence.

Not only did Thom introduce the notion of versal unfolding, he provided an easily computable way to recognize whether a given unfolding is versal, and indeed to construct a versal unfolding of any germ of finite codimension.

Given an unfolding $F(x,u)$ (with $u \in \mathbb{R}^b$), the ***initial speeds*** of the unfolding are defined to be,

$$\dot{F}_j(x) = \frac{\partial F}{\partial u_j}(x,0), \quad (j = 1, \dots, b).$$

These are elements of $\mathcal{E}_n$. Let $\dot{F} \subset \mathcal{E}_n$ be the vector subspace spanned by the initial speeds:

$$\dot{F} = \mathbb{R}\left\{\dot{F}_1, \dots, \dot{F}_b\right\}.$$

For example, if $F(x,y;u_1,u_2) = x^3 - y^2 + u_1 x - u_2^2 y - u_2 xy$ then $\dot{F}_1 = x$ and $\dot{F}_2 = -xy$, and so $\dot{F} = \mathbb{R}\{x,\ xy\} \subset \mathcal{E}_2$, a 2–dimensional subspace.

Theorem 7.8 (Versal unfoldings). *Let $F(x;u)$ be an a–parameter unfolding of $f \in \mathcal{E}_n$ (so $u \in \mathbb{R}^a$ and $f(x) = F(x,0)$). Then F is versal if and only if*

$$\mathcal{E}_n = Jf + \mathbb{R} + \dot{F}.$$

It is usual to take the initial speeds such that all $\dot{F}_j \in \mathfrak{m}_n$ and in that case the condition for versality can be replaced by

$$\mathfrak{m}_n = Jf + \dot{F}.$$

In other words, for F to be versal one requires that (some of) the initial speeds form a cobasis for Jf in $\mathfrak{m}_n$.

This theorem is a consequence of the Malgrange preparation theorem, and is proved in Chapter 16.

Corollary 7.9. *If* $F\colon \mathbb{R}^n \times \mathbb{R}^a \to \mathbb{R}$ *is a versal unfolding of* f*, then* $a \geq \operatorname{codim}(f)$.

Versal unfoldings with exactly $\operatorname{codim}(f)$ parameters (the least possible number), are known as ***miniversal unfoldings***, and for a miniversal unfolding the set of initial speeds indeed forms a cobasis for Jf in $\mathfrak{m}_n$. The reader should check that the fold and cusp catastrophes described in Chapter 2 are miniversal.

The theorem also provides a simple way of constructing versal unfoldings of a given germ:

Corollary 7.10. *Let* $f \in \mathcal{E}_n$ *be a germ of finite codimension* r*, and let* $h_1, \dots, h_r$ *be a cobasis for* Jf *in* $\mathfrak{m}_n$*. Then*

$$F(x; u) = f(x) + u_1 h_1(x) + \cdots + u_r h_r(x)$$

is a versal unfolding of f.

Example 7.11. Let $f = x^2 + xy^3$. Then $Jf = \langle 2x + y^3,\ xy^2 \rangle$. This ideal contains the monomials x^2 and y^5. Its Newton diagram is

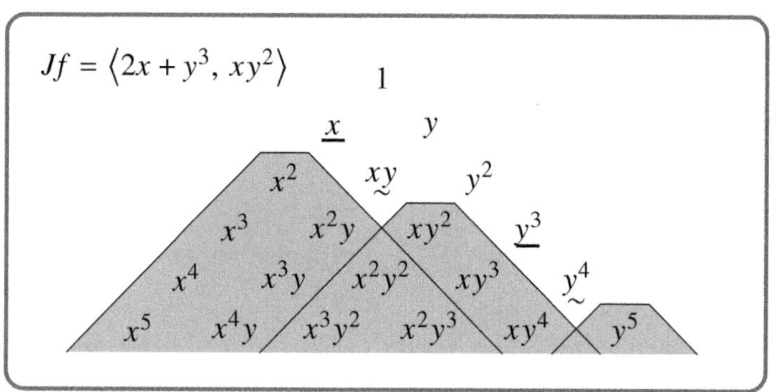

And both $2x + y^3$ and $2xy + y^4$ are in Jf. So as a cobasis for Jf we can take,

$$\{y,\ y^2,\ y^3,\ y^4\}.$$

One versal unfolding of f is therefore

$$F(x, y; a, b, c, d) = x^2 + xy^3 + ay + by^2 + cy^3 + dy^4.$$

If we choose $\{x,\ y,\ xy,\ y^2\}$ as an alternative cobasis for Jf we would have the versal unfolding

$$F_1(x, y; a, b, c, d) = x^2 + xy^3 + ax + by + cxy + dy^2.$$

The versality theorem tells us that F and F_1 are equivalent unfoldings of f, as each can be induced from the other.

Note that under the change of coordinates, $x' = x + \frac{1}{2}y^3$, $y' = y$, one gets $f = (x')^2 - \frac{1}{4}(y')^6$ which is a critical point of type A_5^-.

Example 7.12. Show that $G(x, y; r, s) = x^2 + y^4 + ry^2 + r^2 y + s(x + y)$ is a versal unfolding of $g_0(x, y) = x^2 + y^4$.
Solution: We have $Jg_0 = \langle x, y^3 \rangle$. The initial speeds of G are,

$$\dot{G}_1 = \frac{\partial G}{\partial r}(x, y, 0, 0) = y^2, \quad \dot{G}_2 = \frac{\partial G}{\partial s}(x, y, 0, 0) = x + y.$$

Thus,

$$\dot{G} = \mathbb{R}\left\{y^2, x + y\right\}.$$

It is clear then that $\mathfrak{m}_2 = Jg_0 + \dot{G}$ so that G is versal.

Example 7.13. Let $f(x, y) = x^5 + y^5$ (see Example 5.11 on p. 68).
The Newton diagram for $Jf = \langle x^4, y^4 \rangle$ is shown on the right. A cobasis for Jf in $\mathfrak{m}_2$ consists of the 15 monomials not shaded (excluding 1, which is not in $\mathfrak{m}_2$), so $\operatorname{codim}(f) = 15$. If we call these monomials h_1 to h_{15} (with $h_1 = x$, $h_2 = y$, $\ldots$, $h_{15} = x^3y^3$), then a versal unfolding of f is given by

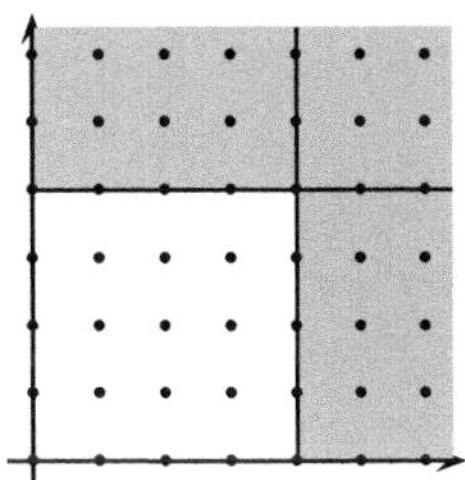

$$F(x, y; u_1, \ldots, u_{15}) = x^5 + y^5 + \sum_{j=1}^{15} u_j h_j.$$

In each of the examples above and in those of Chapter 2, the catastrophe set is a submanifold. In fact this is true in general for versal unfoldings:

Proposition 7.14. *Let* $F\colon (\mathbb{R}^n \times \mathbb{R}^a, (0, 0)) \to \mathbb{R}$ *be a versal unfolding of a germ* $f \in \mathcal{E}_n$. *Then,*
(i) C_F *is a submanifold of* $\mathbb{R}^n \times \mathbb{R}^a$ *of dimension* a, *and*
(ii) $(x, u) \in \Sigma_F$ *if and only if* π_u *fails to be a local diffeomorphism at* (x, u).

PROOF: (i) Suppose $f \in \mathfrak{m}_n^3$ (if not begin by applying the splitting lemma). Then the monomials $x_1, \ldots, x_n$ are part of the cobasis for Jf, so one versal unfolding is of the form

$$G(x; u, v) = G_1(x; u) + \sum_{j=1}^{n} v_j x_j,$$

where $v \in \mathbb{R}^n$ and $u \in \mathbb{R}^{a-n}$ and G_1 involves the remaining unfolding terms. Then

$$C_G = \left\{(x, u, v) \in \mathbb{R}^n \times \mathbb{R}^{a-n} \times \mathbb{R}^n \mid v_j = -\frac{\partial G_1}{\partial x_j}(x, u),\ (j = 1.\ldots, n)\right\}.$$

This is a submanifold parametrized by (x, u); indeed, it is the graph of a map $\mathbb{R}^n \times \mathbb{R}^{a-n} \to \mathbb{R}^n$, so is of dimension a. For the given versal unfolding F, there is a diffeomorphism of $\mathbb{R}^n \times \mathbb{R}^a$ (as per the definition of equivalent unfoldings) which maps C_G to C_F, hence since C_G is a submanifold so is C_F.

(ii) One implication is the subject of Proposition 7.1, though we give a different proof here, based on (i). Using the expression for C_G above (as a graph $v_j = -\frac{\partial G_1}{\partial x_j}(x, u)$) we can parametrize C_G by (x, u) and we have

$$\pi_G(x, u) = (v, u) = \left(-\frac{\partial G_1}{\partial x}, u\right),$$

(we have swapped the order of u, v to make the next bit clearer). The Jacobian matrix is then

$$\mathrm{d}\pi_G(x, u) = \begin{pmatrix} H & A \\ 0 & I \end{pmatrix},$$

where $H = -\left(\frac{\partial^2 G_1}{\partial x_i \partial x_j}\right)$ and $A = \frac{\partial^2 G_1}{\partial x_i \partial u_j}$. Now π_G is a diffeomorphism in a neighbourhood of the point (x, u, v) if and only if this Jacobian matrix is invertible, and this is the case if and only if H is invertible. But the matrix H is the Hessian of $f_{(u,v)}$ at x, so this is equivalent to $f_{(u,v)}$ having a nondegenerate critical point at x. The equivalence of unfoldings again shows the same would be true for the map π_F. ✔

7.5 The elementary catastrophes

We record here in Table 7.1 a list of versal unfoldings of the elementary singularities given in Chapter 6. Thom called these the *elementary catastrophes*. The reader will

TABLE 7.1 Versal unfoldings of the elementary catastrophes.

Label	Normal form	with miniversal unfolding	Name
A_1	$\pm x^2 \pm y^2$		Morse
A_2	$x^3 + y^2$	$+\ ax$	fold
A_3	$\pm x^4 \pm y^2$	$+\ ax^2 + bx$	cusp
A_4	$x^5 \pm y^2$	$+\ ax^3 + bx^2 + cx$	swallowtail
A_5	$\pm x^6 \pm y^6$	$+\ ax^4 + bx^3 + cx^2 + dx$	butterfly
D_4^-	$x^2y - y^3$	$+\ a(x^2 + y^2) + bx + cy$	elliptic umbilic
D_4^+	$x^2y + y^3$	$+\ a(x^2 - y^2) + bx + cy$	hyperbolic umbilic
D_5	$x^2y + y^4$	$+\ ax^2 + by^2 + cx + dy$	parabolic umbilic

recognize the fold and cusp families from Chapter 2. The swallowtail is the subject of Problem 7.14. The discriminant of the swallowtail is illustrated in Figures 7.1 and 7.3, while that of the elliptic and hyperbolic umbilics are shown in Figures 7.3 and 7.4 respectively.

The families in the table are all presented as functions of 2 variables x and y, even though in Chapter 2 and in Problem 7.14 there is only 1 variable. Notice however that in the table, the A_k families all contain y^2 as the only term involving y, so the critical points all occur at $y = 0$. In this way they are very similar to their 1-variable counterparts (see Problem 7.7). The D_k critical points were called ***umbilics*** by Thom, because of their association with umbilics on surfaces in differential geometry (see the end of Chapter 15).

7.6 Simple singularities

A ***simple singularity*** is one for which, in its versal unfolding, there occur only finitely many distinct types of critical point. For example, in the versal unfolding of a D_4 singularity, only critical points of types A_1, A_2 and A_3 occur. One sees from the list that the elementary catastrophes are all simple singularities; the full list is given in Table 6.2. The adjacencies between the different simple singularities is shown in Figure 7.2 (i.e. which one occurs in the unfolding of which).

On the other hand, consider the germ $f_0 = x^4 + y^4$. The versal unfolding includes

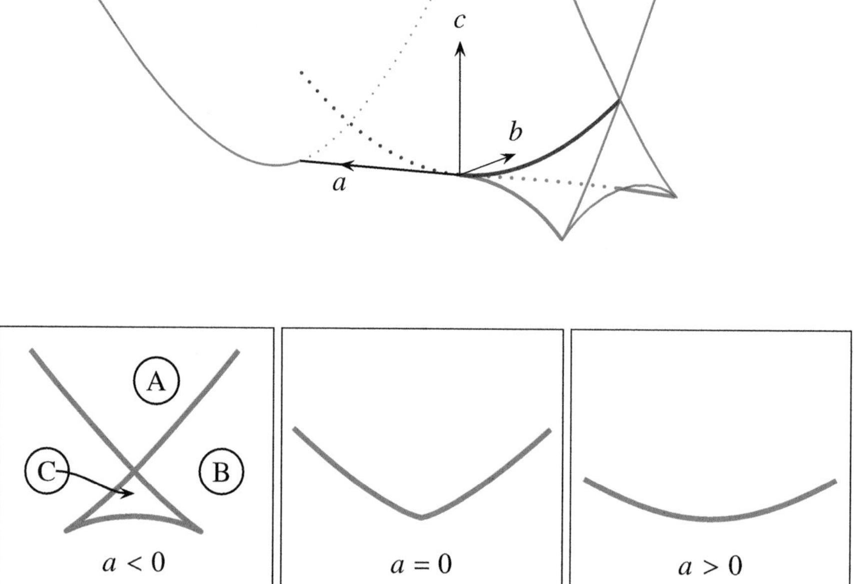

FIGURE 7.1 (ABOVE) Line drawing of the swallowtail surface. (BELOW) Three sections of the surface with a constant (the left-hand section gives the swallowtail its name). For future reference, A is the region above the surface, B the region below it and C the region in the 'pocket'. See text for discussion, as well as Figure 7.3

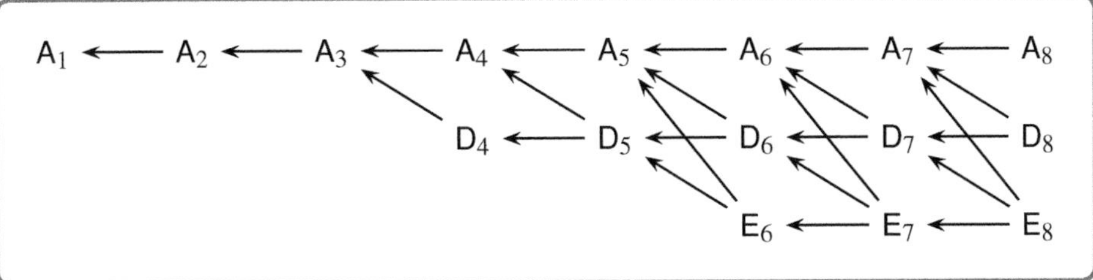

FIGURE 7.2 Adjacency between the simple singularities: an arrow from A to B means that B occurs in the versal unfolding of A. The pattern with the A_k and D_k continues to the right: they are all simple.

the term x^2y^2, and for different values of λ near 0 the germs $f_\lambda = x^4 + \lambda x^2y^2 + y^4$ are all inequivalent. It follows that f_0 is not a simple singularity (nor is f_λ, for any value of λ). In such a setting, the parameter λ is called a ***modulus*** or ***modal parameter***. Since there is only one modal parameter, these are examples of ***unimodal*** singularities.

A similar analysis for the more general singularities $T_{p,q,r}$ (see (6.3)) show these are also not simple, and are also unimodal. An example of a bimodal critical point is $f = x^4 + ax^2y^3 + bx^2y^4 + y^6$ (for $a \neq \pm 2$). See the classification by Arnold reported in [6, Chapter 15] for many more.

7.7 A variational bifurcation problem

In Chapter 1 we saw several bifurcation problems which are of a variational nature. For example, the Zeeman catastrophe machine in Section 1.3 and the evolute of an ellipse in Section 1.4. Let us consider again the second of these, and in particular Example 1.1. The family of distance squared functions is $f_{\mathbf{c}}$, with parameter $\mathbf{c} \in \mathbb{R}^2$. The discriminant is given in Equation (1.6). Consider one of the cusp points, say $\mathbf{c}_0 = (u_0, v_0) = (5/3, 0)$. As stated in that example, the degenerate critical point (at $t = 0$) is of type A_3. I claim that the unfolding of this critical point is versal. To see this, we note that (ignoring the constant)

$$F(t, \mathbf{c}_0) = f_{\mathbf{c}_0}(t) = 5\cos t - \tfrac{5}{2}\cos^2 t = \tfrac{5}{2} - \tfrac{5}{8}t^4 + O(t^6),$$

which is clearly of type A_3 at $t = 0$. This critical point is of codimension 2, and the initial speeds of the unfolding at $\mathbf{c}_0$ are

$$\dot{F}_1 = 3\cos t, \quad \text{and} \quad \dot{F}_2 = \sin t,$$

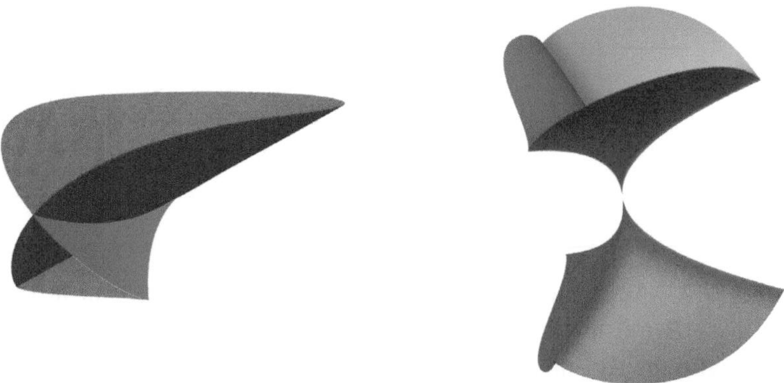

FIGURE 7.3 Swallowtail and elliptic umbilic discriminants.

where $\dot{F}_1 = \frac{\partial}{\partial u}F$ and $\dot{F}_2 = \frac{\partial}{\partial v}F$ evaluated at $\mathbf{c}_0$. In $\mathfrak{m}_1/Jf$ these two functions give a cobasis for Jf as required for a (mini-)versal unfolding. The same analysis can be carried out at any of the other 3 cusps, and explains the cuspidal shape of the discriminant near the 4 points where the family has an A_3 critical point.

A similar analysis can be carried out for the Zeeman catastrophe machine, given an explicit form of the potential.

Problems

7.1 Write down the unfolding $G(x;\lambda)$ induced from the cusp family $F(x;u,v) = \frac{1}{4}x^4 + \frac{1}{2}ux^2 + vx$ by the map $\phi(\lambda) = (\lambda, 2\lambda)$. Draw a sketch of the discriminant of the cusp family together with the image of ϕ (you may like to refer to Chapter 2).

7.2 Show by constructing explicit changes of coordinates that the families

$$F(x;u,v) = x^4 + ux^2 + vx, \quad \text{and} \quad G(X;s,t) = 16X^4 + sX^2 + tX$$

are equivalent.

7.3 Suppose an unfolding $G(x,v)$ is induced from an unfolding $F(x,u)$ via a map ϕ (so $u = \phi(v)$). Show that $\Delta_G = \phi^{-1}(\Delta_F)$; that is, show

$$v \in \Delta_G \iff \phi(v) \in \Delta_F. \qquad (\dagger)$$

7.4 Let $F,\ G\colon \mathbb{R}^n \times \mathbb{R}^a \to \mathbb{R}$ be two equivalent unfoldings. Using the notation of Definition 7.6 (see p. 96), show that the diffeomorphism Φ maps C_F to C_G, and Σ_F to Σ_G and that ϕ maps Δ_F to Δ_G

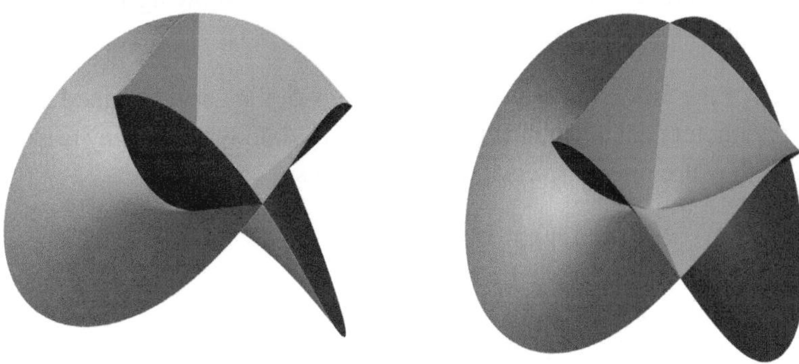

FIGURE 7.4 Two views of the hyperbolic umbilic discriminant, showing the curve of self-intersection and the cuspidal edge, which cross each other at the origin.

7.5 Show that the unfoldings given in Section 7.5 are indeed versal unfoldings of each of the germs in the tables on p. 86. Find versal unfoldings for the E_6, E_7 and E_8 germs given on p. 87.

7.6 Consider the (trivial) family of functions given by $F\colon \mathbb{R}\times\mathbb{R} \to \mathbb{R}$, $F(x;u) = (x-u)^3$. Find C_F, and show that $\Sigma_F = C_F$. Comment on the relevance of this example to a possible converse of Proposition 7.1. (†)

7.7 Let $F\colon \mathbb{R}^n \times \mathbb{R}^a \to \mathbb{R}$ be a family of functions, and let $G\colon \mathbb{R}^{n+k} \times \mathbb{R}^a \to \mathbb{R}$ be given by $G(x,y;u) = F(x;u) + \sum_{i=1}^k \pm y_i^2$ (here $x \in \mathbb{R}^n$ and $y \in \mathbb{R}^k$). Show that $\Delta_F = \Delta_G$ (first you will need to relate C_F and C_G).

7.8 Let $f \in \mathcal{E}_n$ be a germ of corank k and finite codimension, and suppose there are coordinates so that $f(x,y) = Q(x) + h(y)$ with Q a nondegenerate quadratic form and $h \in \mathfrak{m}_k^3$ (eg, after applying the splitting lemma). Let $H\colon \mathbb{R}^k \times \mathbb{R}^a \to \mathbb{R}$ be a versal unfolding of h. Show that $F\colon \mathbb{R}^n \times \mathbb{R}^a \to \mathbb{R}$ given by

$$F(x,y;\,u) = Q(x) + H(y;u)$$

is a versal unfolding of f.

7.9 For the family in Example 7.2, find the number and type of critical points of $f_{(u,v)}$ for (u,v) in each region of the parameter plane.

7.10 Consider the family $F\colon \mathbb{R}\times\mathbb{R} \to \mathbb{R}$ given by $F(x,u) = ux^3 + x^2$. Show that for $u = 0$ this has a single nondegenerate critical point, while for $u \neq 0$ it has

2 nondegenerate critical points. Draw a diagram showing C_F, and explain what happens to the 'extra' critical point as $u \to 0$.

7.11 Which of the following are versal unfoldings of the germ obtained by putting the parameters equal to 0? If they are versal, state whether they are miniversal. If they are not miniversal, modify them so that they are.

(i) $F(x; u, v) = x^3 + ux^2 + vx$
(ii) $F(x; u, v) = x^4 + ux^3 + vx$
(iii) $F(x, y; u, v) = x^2 + y^3 + ux + vy$
(iv) $F(x, y; u, v) = x^2 + y^4 + ux + vy$
(v) $F(x, y, u, v, w) = x^3 + y^3 + u(x^2 + y^2) + vx + wy$
(vi) $F(x, y, u, v, w) = x^3 - xy^2 + u(x^2 + y^2) + vx + wy.$

7.12 For each of the following pairs of families, F is versal and G is therefore equivalent to a family induced from F. In each case, find the equivalence and inducing map.
(i) $F(x;\, u) = x^3 + ux, \quad G(y;\, \lambda) = y^3 - \lambda y + \lambda^2$
(ii) $F(x, y;\, u) = x^2 + y^3 - uy, \quad G(x, y; \lambda, \mu) = x^3 + y^2 + \lambda x + \mu y$
(iii) $F(x; u) = x^3 + ux, \quad G(y;\, \lambda) = y^3 - \lambda y^2$
(iv) $F(x, y;\, u, v, w) = x^3 + y^3 + uxy + vx + wy,$
$G(x, y;\, \lambda) = x^3 + y^3 + \lambda(x^2 - y^2)$ (†)

7.13 Let $f \in \mathfrak{m}_n^2$ have a critical point of type A_2, and let $F(x; u)$ be a versal unfolding of f. Show that the two (real) critical points appearing for $F(x; u)$ with $u \neq 0$ have (Morse) indices that differ by 1.

7.14 **The swallowtail catastrophe** Consider the family (based on the A_4 critical point $x \mapsto \frac{1}{5}x^5$)

$$F(x; a, b, c) = \tfrac{1}{5}x^5 + \tfrac{1}{3}ax^3 + \tfrac{1}{2}bx^2 + cx.$$

Refer to Figure 7.1.

(i). Show this is a versal unfolding of $f(x) = \frac{1}{5}x^5$, and show directly that C_F is a submanifold of $\mathbb{R}^4$.

(ii). Consider the 1–parameter unfolding $G(x, u)$ induced from F by $\phi(u) = (2u, 0, u^2)$. Show that for each $u \neq 0$, g_u has 2 degenerate critical points (real for $u < 0$ and imaginary for $u > 0$). [For $u < 0$ this is the curve of self-intersection on the swallowtail surface. This curve continues for $u > 0$ and this part is sometimes called the whisker, or thread, of the swallowtail surface; see the blue curve in Figure 7.1.]

(iii). Consider the restriction $H(x,a)$ of the unfolding F to the a-axis (that is, the unfolding induced by $\phi(a) = (a,0,0)$). What can you say about the critical points of the functions in this family?

(iv). Let h_c be given by putting $a = -1$ and $b = 0$ in F. Sketch the graph of h_c for relevant values, or ranges of values, of c, making clear the numbers of critical points, and what happens to them as c is varied.

(v). Referring to Figure 7.1, deduce the number of (real) critical points when (a,b,c) lies in each of the regions marked A, B and C.

(vi). Show that $f_{(a,b,c)}$ has an A_3-critical point if and only if there is a t such that $a = -6t^2$, $b = 8t^3$, $c = -3t^4$ (the A_3-locus on the discriminant is therefore a curve: it is the cuspidal edge in the figure). (†)

For the record, though we do not use it, the equation for the discriminant of the swallowtail family in this form is

$$b^2(4a^3 + 27b^2 - 144ac) - 16c(a^2 - 4c)^2 = 0. \tag{7.2}$$

(vii). Check that indeed the parabola found in part (ii) and the cuspidal edge in part (vi) satisfy this equation.

7.15 **Elliptic umbilic** Consider the family (from the D_4^- singularity)

$$F(x,y;a,b,c) = \tfrac{1}{3}x^3 - xy^2 + a(x^2+y^2) - bx - cy.$$

(i). Show this is a versal unfolding of f_0, and that C_F is a submanifold of $\mathbb{R}^5$ of dimension 3.

(ii). Using the natural parametrization of C_F by (x,y,a) show that for each $a \neq 0$ the singular set Σ_F is a circle – what is its radius? What is the overall shape of $\Sigma_F \subset \mathbb{R}^3$?

(iii). Without computing an equation for the discriminant, show that for $a \neq 0$, the point $(a,b,c) = (a,0,0) \notin \Delta_F$. Find the critical points of $f_{(a,0,0)}$, and show they are nondegenerate. What are their indices?

(iv). Find the locus of points $(a,b,c) \in \Delta_F$ such that the tangent vector to the Σ_F circle from part (ii) is contained in the kernel of the Hessian matrix. You should find three curves (or lines), each parametrized by a. (It turns out these are precisely the points for which $f_{(a,b,c)}$ has an A_3-critical point.) See Figure 7.3 for the discriminant of the elliptic umbilic, showing the three cuspidal edges. (†)

Again, for the record, the equation of the discriminant of this elliptic umbilic is

$$27a^8 - 18a^4(b^2 + c^2) - (b^2 + c^2)^2 + 8a^2(b^3 - 3bc^2) = 0. \tag{7.3}$$

7.16 **Hyperbolic umbilic** Repeat all 4 parts of the previous question, but for the family (the D_4^+ singularity)

$$F(x, y; a, b, c) = \tfrac{1}{3}x^3 + \tfrac{1}{3}y^3 + axy - bx - cy.$$

Note that in (ii) the relevant curve is no longer a circle. What shape is it? And in part (iv) there is only a single curve of points with the property described. There is also a curve of double points in the discriminant (see Figure 7.4).

The equation of the discriminant for this version of the hyperbolic umbilic is

$$27a^8 - 288a^4bc - 256b^2c^2 + 256a^2(b^3 + c^3) = 0. \tag{7.4}$$

7.17 Let $f_u(x)$ be a 1–parameter family of smooth germs in 1 variable, $x \in \mathbb{R}$. Suppose $f_u'(0) = 0$ for all u in a neighbourhood of $u = 0$. Suppose that $f_0''(0) = 0$ (degenerate critical point) and that for $u \neq 0$ the critical point is nondegenerate, with $f_u''(0) > 0$ if $u > 0$ and $f_u''(0) < 0$ if $u < 0$. Show that there is a neighbourhood U of 0 in $\mathbb{R}$ such that for each $x \in U$ there is a u (not necessarily unique) such that x is a critical point of f_u. [*Note*: *there are no assumptions of finite determinacy.*] (†)

8

Singularities of plane curves

THERE ARE TWO WAYS TO DEFINE a curve in the plane: either as the image of a map $\gamma\colon \mathbb{R} \to \mathbb{R}^2$ (the parametric representation), or as the set of solutions of an equation $f(x, y) = 0$, where $f\colon \mathbb{R}^2 \to \mathbb{R}$ (the Cartesian representation). The singularity theory of the two is somewhat different.

For example, consider a figure-8 curve in the plane. If we see it as the image of a map γ (ie, as a parametric curve) then under small perturbation it will remain very similar to a figure-8 curve, a little distorted perhaps, but it will retain its characteristic self-crossing point. On the other hand, if this same figure-8 curve is considered as the set of solutions of an equation, then a small perturbation of the equation can smooth out the self-crossing, as we will see below.

Of these two types of curve (parametric and Cartesian), the former requires some new theory – namely, left equivalence – but the latter falls into what we have been developing, but with one minor modification discussed below.

For the remainder of this chapter, we consider curves in Cartesian form. A ***plane curve singularity*** at the origin is a smooth germ $f\colon (\mathbb{R}^2, 0) \to (\mathbb{R}, 0)$. The curve in question is the set $f^{-1}(0)$, and a plane curve singularity where f is not singular is said to be a *smooth curve*. We say two plane curve singularities f_1, f_2 are ***equivalent*** if there is a change of coordinates, or diffeomorphism germ, $\phi\colon (\mathbb{R}^2, 0) \to (\mathbb{R}^2, 0)$ such that $f_1 = f_2 \circ \phi$. In other words the singularities are right equivalent. Note that they should not be $\mathcal{R}^+$-equivalent, as that would prevent $f(0, 0) = 0$; this is the minor modification required in our equivalence.

Remark 8.1. Since it is just the set $f^{-1}(0)$ that is of interest, one could extend the equivalence to include $f_1(x, y) = h(x, y) f_2(x, y)$ with $h(0, 0) \neq 0$, because one can divide by h and $f_1^{-1}(0) = f_2^{-1}(0)$. This is an example of *contact equivalence* which we meet in Part II. It turns out that for singularities of low codimension, extending the equivalence by incorporating such an h does not make any difference to the classification or unfoldings of plane curve singularities. ❞

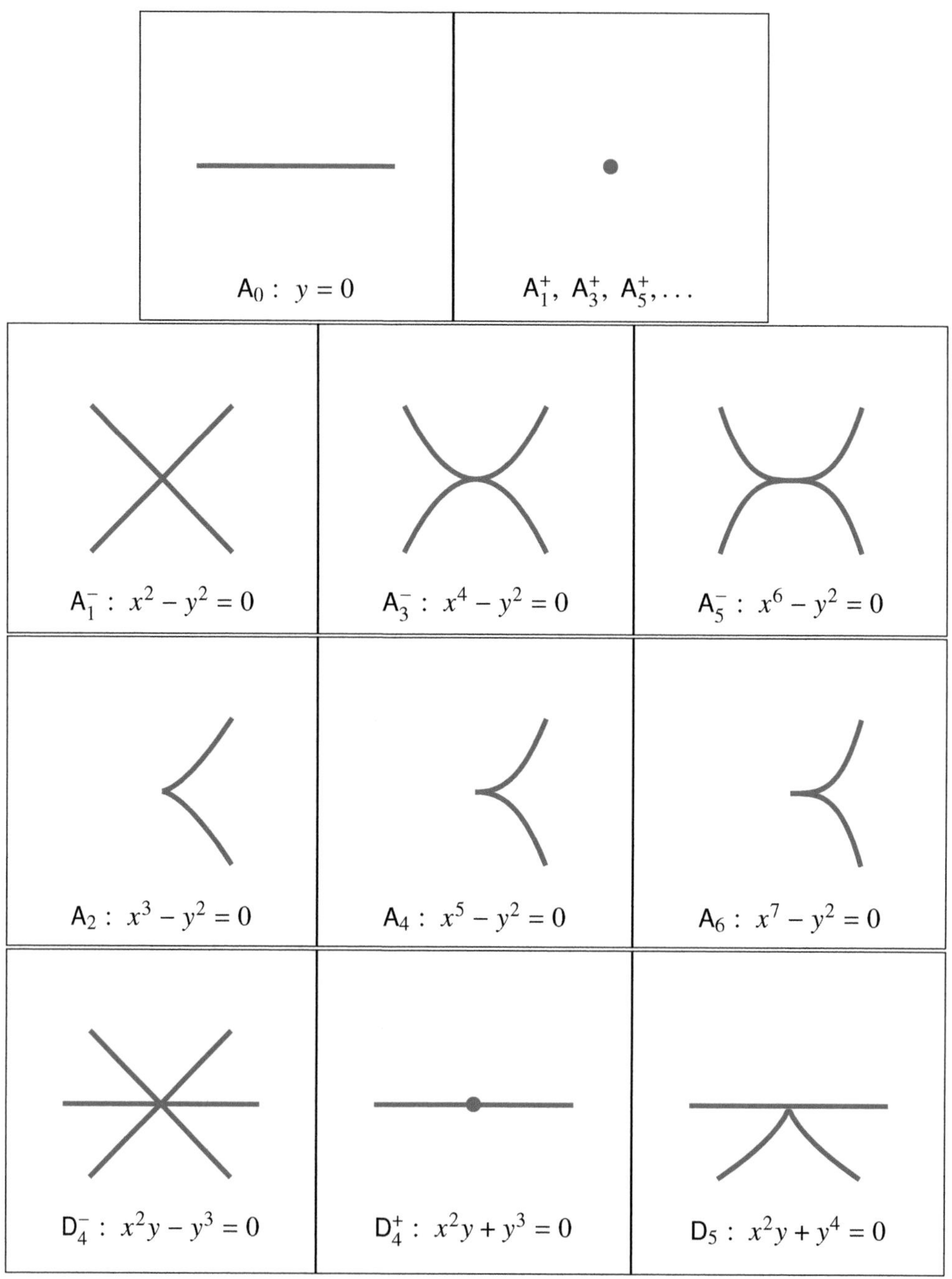

FIGURE 8.1 Plane curve singularities.

8.1 Classification

Let $f\colon (\mathbb{R}^2,0) \to (\mathbb{R},0)$, and denote the curve germ $f^{-1}(0)$ by $\mathcal{C}$. Because the equivalence is $\mathcal{R}$ rather than $\mathcal{R}^+$ we define the ***codimension*** of the curve singularity to be

$$\operatorname{codim}\mathcal{C} = \dim \mathcal{E}_2/_{Jf}.$$

Thus $\operatorname{codim}\mathcal{C} = 1 + \operatorname{codim} f$.

The first case is where f is non-singular. Then by the submersion theorem (Theorem B.13), there is a change of coordinates so that $f(x,y) = y$, and $\mathcal{C}$ is the x-axis. Since in this case $Jf = \mathcal{E}_2$, the non-singular curve has codimension 0.

Now suppose f is singular, and refer to Table 6.1 on p. 87. Note that some entries in that table are not what one would really call (real) curves, such as $x^2 + y^2 = 0$. However, the set of complex solutions to the equations are curves, and moreover, even as real sets, their deformations do give rise to (real) curves: for example $x^2 + y^2 = u$ is a circle when $u > 0$. For these reasons we keep such 'non-curves' in the classification.

All curve singularities up to codimension 5 are shown in Figure 8.1, as well as A_6 which has codimension 6 (the A_k and D_k curves have codimension k).

8.2 Unfoldings

Since the equivalence is $\mathcal{R}$-equivalence, we should now consider $\mathcal{R}$-versal unfoldings of the function defining the curve. The effect is to require an extra parameter in the unfolding compared to the unfoldings given in Table 7.1. Thus for example, the $\mathcal{R}$-versal unfolding of $f = x^3 - y^2$ (the A_2 curve singularity) would be

$$F(x,y;u,v) = x^3 - y^2 + ux + v \ (= 0).$$

The curve singularity is now of codimension 2, rather than the codimension 1 for the critical point itself. The number of unfolding parameters needed is equal to the codimension of the curve singularity as defined above. And for the (mini-versal) unfolding terms one chooses a cobasis for Jf in $\mathcal{E}_2$.

The ***discriminant*** $\Delta_{\mathcal{C}}$ of an unfolding $F(x;u)$ with base $\mathbb{R}^a$, is the set of parameter values $u \in \mathbb{R}^a$ for which the curve $f_u(x,y) = 0$ has a singular point: that is, there is an $(x,y) \in \mathbb{R}^2$ such that $f_u(x,y) = 0$ and $\mathsf{d}f_u(x,y) = 0$

$$\Delta_{\mathcal{C}} = \left\{ u \in \mathbb{R}^a \mid f_u(x,y) = \frac{\partial}{\partial x} f_u(x,y) = \frac{\partial}{\partial y} f_u(x,y) = 0 \right\}. \tag{8.1}$$

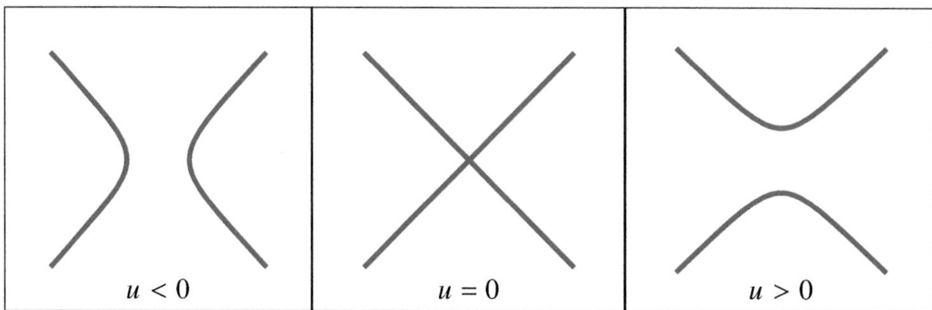

FIGURE 8.2 Versal unfolding of the crunode: $F(x, y; u) = x^2 - y^2 - u$.

This is different from the definition of discriminant of a family of functions that we have been using hitherto, and is more in keeping with the definition used in Chapter 10 below. This is because we are not interested primarily in critical points of the functions but in the structure of their level sets.

Example 8.2. For the A_2 cusp singularity with $f_{(u,v)}(x, y) = x^3 - y^2 - ux - v$, the discriminant is derived from the equations

$$x^3 - y^2 - ux - v = 0, \quad 3x^2 - u = 0, \quad 2y = 0.$$

Eliminating x and y from these gives the discriminant

$$\Delta = \{(u, v) \in \mathbb{R}^2 \mid (\exists x)\ u = 3x^2,\ v = -2x^3\}.$$

This is the semicubical parabola with equation $27v^2 = 4u^3$. See Figure 8.3 which shows some typical deformations of the curve singularity. ✐

Brief descriptions

- For A_0 (non-singular curve) the codimension is zero, and there is no unfolding to consider.

- For A_1^- (called a *crunode*) the codimension is now 1, and the versal unfolding is given by $F(x, y; u) = x^2 - y^2 - u$. For each $u \neq 0$ this is the equation of a hyperbola; see Figure 8.2.

- For A_1^+ (called an *acnode*) the unfolding is $F(x, y; a) = x^2 + y^2 + u$. For $u < 0$ the zero-set (i.e. the points (x, y) where $F(x, y; u) = 0$) is a circle, while for $u > 0$ it is empty.

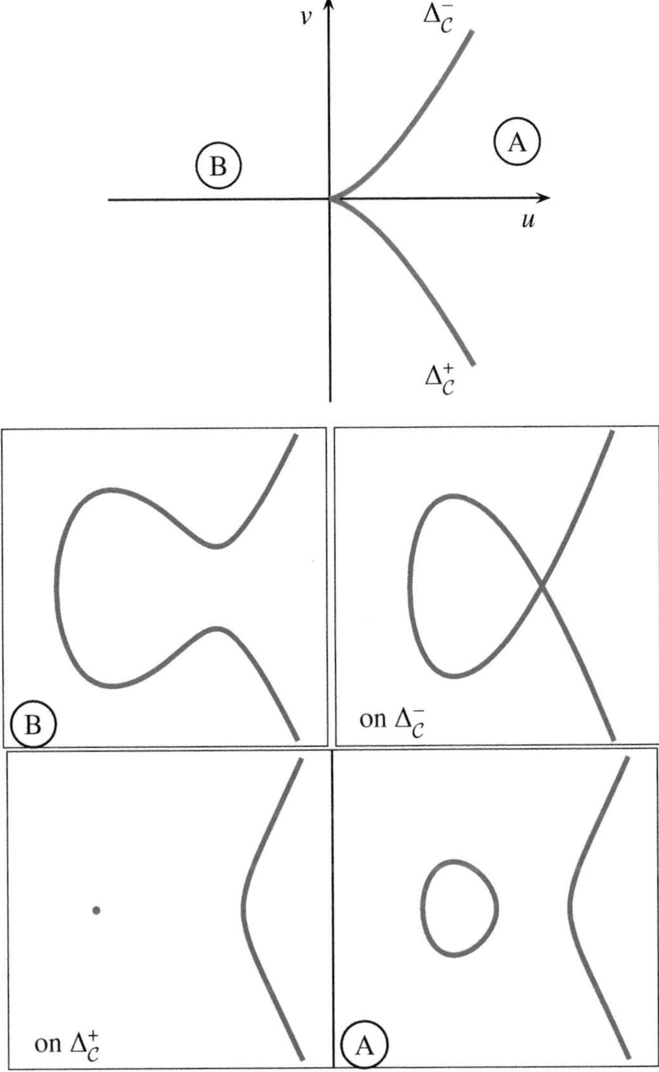

FIGURE 8.3 Versal unfolding $F(x, y; u, v) = x^3 - y^2 - ux - v$ of the A_2 singular curve $x^3 - y^2 = 0$ depicted in Figure 8.1. The curve with the self-crossing (double point) corresponds to a choice of (a, b) on the discriminant. (It is a coincidence that the central curve $x^3 - y^2 = 0$ (shown in Figure 8.1) is diffeomorphic to the discriminant.)

- For A_2 (the simple cusp, or *spinode*) the unfolding is as above $F(x, y; a, b) = x^3 - y^2 + ux + v$. The discriminant is the set $27v^2 = 4u^3$. See Figure 8.3 for the unfoldings.

- For $\mathsf{A}_3^\pm$ (A_3^- is called the *tacnode*) the unfoldings are $F(x, y; u, v, w) = x^4 \pm y^2 + ux^2 + vx + w$. The discriminant is the swallowtail surface. See Figure 8.4 for the unfoldings.

The names *acnode, crunode, spinode* and *tacnode* are terms from classical nineteenth century algebraic geometry, and barely used any more.

8.3 Higher dimensions

Surfaces in $\mathbb{R}^3$ can also be defined as level sets of real-valued functions, and more generally hypersurfaces in $\mathbb{R}^n$. Indeed, one defines a ***hypersurface singularity*** to be a smooth germ $f\colon (\mathbb{R}^n, 0) \to (\mathbb{R}, 0)$, where the hypersurface itself is $f^{-1}(0)$.

The analysis of these singularities is analogous to the plane curve singularities above, only with n variables instead of 2. We show some figures for the case of surfaces in $\mathbb{R}^3$, where the singularity is type A_2. See Figs 8.5 and 8.6: these illustrate 1–parameter unfoldings, while the versal unfolding has 2 parameters (as in Example 8.2).

Note that the family of Figure 8.6, namely $F(x, y, z) = x^3+y^2-z^2+ax = 0$, has an interesting symmetry. If one performs the rotation by π about the line $x = 0,\ y = z$, which is given by $(x, y, z) \mapsto (-x, z, y)$, then $f(-x, z - y) = -f(x, y, z)$ showing that the side of the surface where $f > 0$ is rotated to the side where $f < 0$ and vice versa.

Problem

8.1 Consider the A_3^+ family of plane curves,

$$x^2 + y^4 + ay^2 + by + c = 0.$$

The discriminant is the same swallowtail surface as for the A_3^- family. For parameter values in region A of Figure 8.4, the curve is empty. Sketch the curve for other regions of parameter space, paying particular attention to the transitions as the parameter crosses different parts of the discriminant. (†)

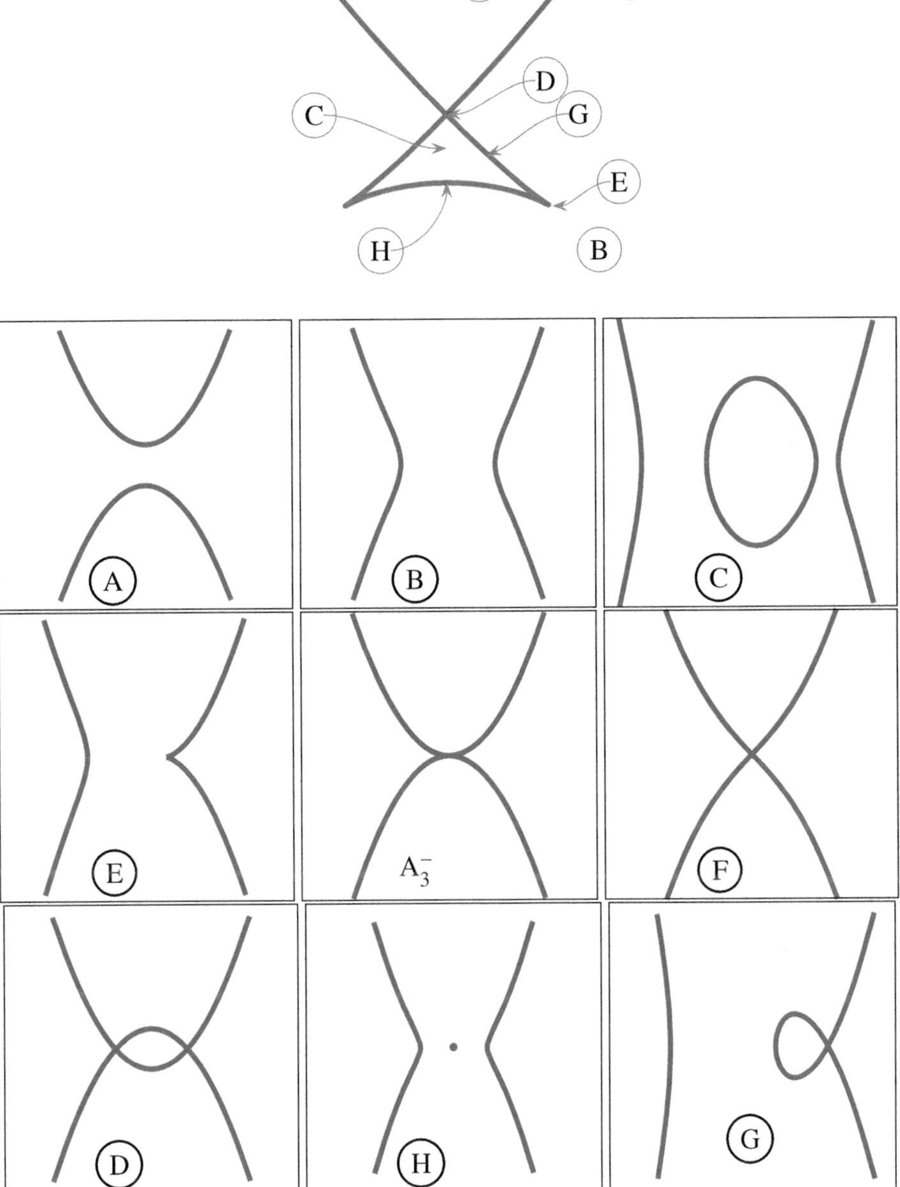

FIGURE 8.4 Representative deformations of the A_3^- singular curve occurring in the versal unfolding: $F(x, y; u, v, w) = x^4 - y^2 + ux^2 + vx + w = 0$. The discriminant is the swallowtail, and the curve at the top is a slice through it (with $u < 0$ fixed) (see also Figure 7.1).

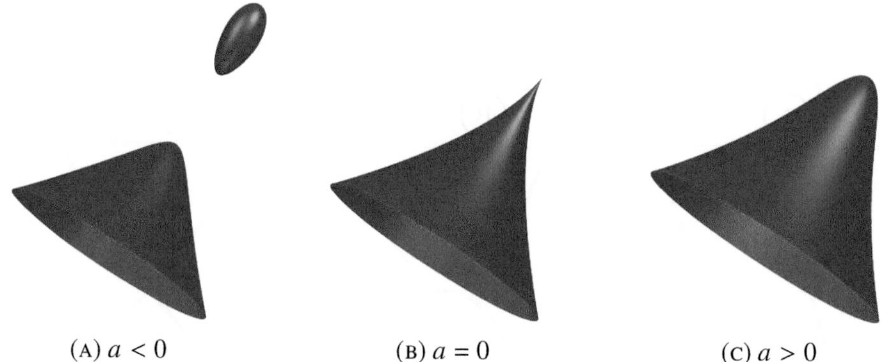

(A) $a < 0$ (B) $a = 0$ (C) $a > 0$

FIGURE 8.5 The family $x^3 + y^2 + z^2 + ax = 0$ of deformations of one of the A_2 surface singularities. It is a surface of revolution based on the curve family in Figure 8.3

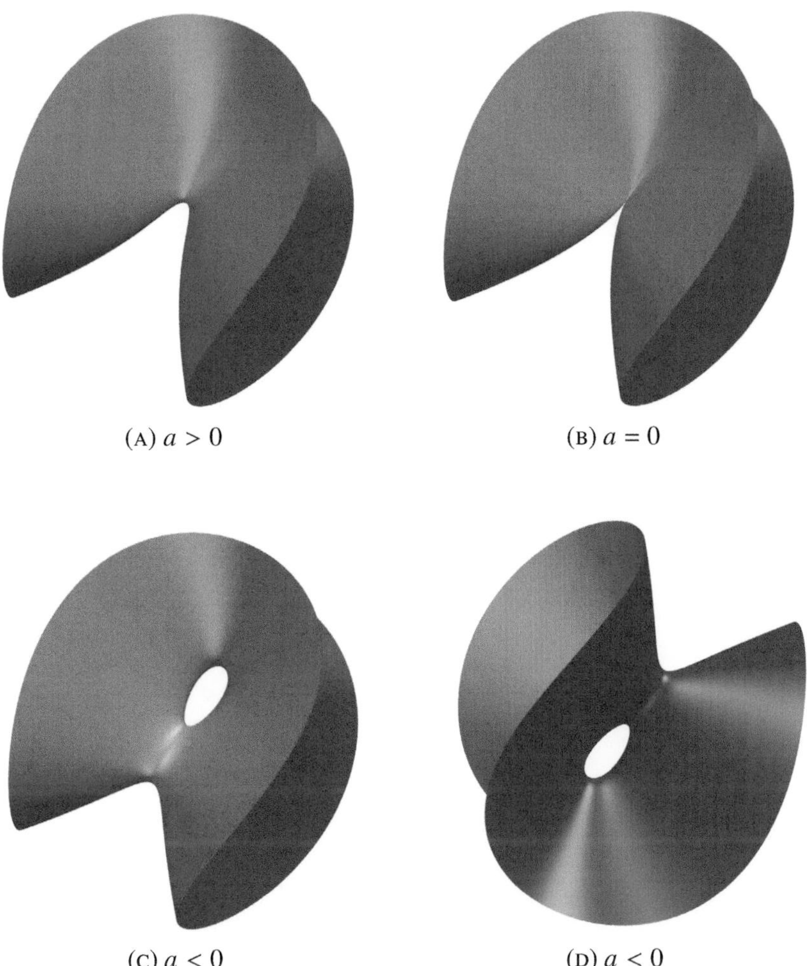

(A) $a > 0$ (B) $a = 0$

(C) $a < 0$ (D) $a < 0$

FIGURE 8.6 The family $x^3 + y^2 - z^2 + ax = 0$ of deformations of the other possible A_2 surface singularity. The two images with $a < 0$ are different views of the same surface, illustrating a pleasing symmetry between the two sides of the surface: the rotation taking (x, y, z) to $(-x, z, y)$ changes the sign of the function.

9

Even functions

HIS CHAPTER DEMONSTRATES HOW the general theory of right equivalence might be adapted to specific requirements. Here we consider the class of ***even functions***, that is, functions $f\colon \mathbb{R}^n \rightarrowtail \mathbb{R}$ such that

$$f(-x) = f(x),$$

and address similar questions to before, in particular to find criteria for finite determinacy and versal unfoldings. Such functions arise naturally in certain problems with symmetry.

Write $\mathcal{E}_n^+$ for the set (ring) of all germs at 0 of even functions. This ring has a unique maximal ideal $\mathfrak{m}_n^+$ consisting of even functions satisfying $f(0) = 0$, and one can show that

$$\mathfrak{m}_n^+ = \langle x_1^2, x_1x_2, \dots, x_n^2 \rangle = \mathfrak{m}_n \cap \mathcal{E}_n^+$$

(the same generators as $\mathfrak{m}_n^2$, but the two are ideals in different rings, so are not equal – for example, $x_1^3 \notin \mathfrak{m}_n^+$ as it's not even).

Similarly, we write $\mathcal{E}_n^-$ for the set of all ***odd*** functions; that is, those satisfying $h(-x) = -h(x)$. This set $\mathcal{E}_n^-$ is not a ring, but it is a *module* over $\mathcal{E}_n^+$ – see Problem 9.3

In one variable, a polynomial is even if it is of the form[1]

$$f = a_0 + a_1x^2 + a_2x^4 + a_3x^6 + \cdots + a_nx^{2n}.$$

Notice that this polynomial of degree $2n$ can be written as a polynomial (of degree n) in x^2. A fundamental theorem of Whitney extends this to all C^∞ even functions of one variable: a smooth function $f\colon \mathbb{R} \rightarrowtail \mathbb{R}$ is even if and only if there is another smooth function $g\colon \mathbb{R} \rightarrowtail \mathbb{R}$ such that,

$$f(x) = g(x^2).$$

While this is reasonably clear for polynomials and even analytic functions, it is not at all obvious for smooth functions. (See Chapter 16 for a proof using the Preparation Theorem.)

[1]The fact that all the powers appearing are even explains why they are known as *even* functions

9.1 Right equivalence

In order to preserve the evenness of functions, it is natural to assume that a change of coordinates ϕ respects the symmetry $x \leftrightarrow -x$, which requires it to satisfy

$$\phi(-x) = -\phi(x), \tag{9.1}$$

for then if f is even, so is $f \circ \phi$:

$$f \circ \phi(-x) = f(\phi(-x)) = f(-\phi(x)) = f(\phi(x)) = f \circ \phi(x).$$

For $f, g \in \mathcal{E}_n^+$ we write the right equivalence via ϕ with property (9.1) as $\mathcal{R}^{\mathrm{ev}}$-equivalence ($\mathcal{R}^+$ already having another meaning). The infinitesimal version of such diffeomorphisms (obtained by differentiating (9.1)) consists of vector fields $\mathbf{v}(x)$ satisfying $\mathbf{v}(-x) = -\mathbf{v}(x)$. We write the set of such vector fields as θ_n^- (the *odd* vector fields). Notice that $\mathbf{v}(-x) = -\mathbf{v}(x)$ implies $\mathbf{v}(0) = 0$, so $\mathbf{v} \in \theta_n^- \Rightarrow \mathbf{v} \in \mathfrak{m}_n\theta_n$.

For $f \in \mathcal{E}_n^+$, define the ***even Jacobian ideal*** to be

$$J^+f = \mathrm{t}f(\theta_n^-) = \left\{ \sum_j \frac{\partial f}{\partial x_j} v_j \mid v_j \in \mathcal{E}_n^- \right\}.$$

Note that as f is even, $\frac{\partial f}{\partial x_j}$ is odd, and since v_j is also odd, their product is indeed even; see also Problem 9.1. The generators of J^+f as an ideal in $\mathcal{E}_n^+$ are the n^2 terms, $x_i \frac{\partial f}{\partial x_j}$.

Theorem 9.1. *$f \in \mathcal{E}_n^+$ is $2k$-determined with respect to $\mathcal{R}^{\mathrm{ev}}$-equivalence if*

$$(\mathfrak{m}_n^+)^{k+1} \subset \mathfrak{m}_n^+ J^+f$$

The left-hand side is $(\mathfrak{m}_n^+)^{k+1} = \mathfrak{m}_n^{2k+1} \cap \mathcal{E}_n^+$, and the right-hand side is equal to $\mathfrak{m}_n^2 Jf \cap \mathcal{E}_n^+$. The proof of this theorem is the same as the standard finite determinacy theorem for right equivalence in Chapter 5, see Problem 9.4 for details.

9.2 Unfoldings and versality

Let $f_0 \in \mathfrak{m}_n^+$ be an even function germ. We define the ***even-codimension*** of f to be

$$\operatorname{codim}^+(f) = \dim\left(\mathfrak{m}_n^+/J^+f\right).$$

For example, if $f = \sum_j \pm x_j^2$ (which is even) then $J^+f = \mathfrak{m}_n^+$ so $\operatorname{codim}^+(f) = 0$. And for a degenerate example, if $f = x^4 + y^4$ one has

$$J^+f = \langle x^4,\, x^3y,\, xy^3,\, y^4 \rangle\,.$$

If follows that $\operatorname{codim}^+(f) = 4$, as the monomials of $\mathfrak{m}_2^+$ missing from J^+f are x^2, xy, y^2 and x^2y^2.

Now suppose $F\colon \mathbb{R}^n \times \mathbb{R}^a \to \mathbb{R}$ is a smooth family of even functions with base $\mathbb{R}^a$; that is for each $u \in \mathbb{R}^a$, f_u is an even function of $x \in \mathbb{R}^n$. The initial speeds of the family $\dot{F}_i$ are also even functions. The vector space $\dot{F}$ spanned by the initial speeds is a subspace of $\mathcal{E}_n^+$.

Theorem 9.2. *The family F is versal amongst the class of all smooth families of even functions if and only if*

$$J^+f + \mathbb{R} + \dot{F} \;=\; \mathcal{E}_n^+.$$

Example 9.3. Consider the germ $f_0(x,y) = x^6 + y^4 \in \mathcal{E}_2^+$. Then

$$J^+f_0 = \langle x^6,\, x^5y,\, xy^3,\, y^4 \rangle \lhd \mathcal{E}_2^+.$$

The Newton diagram with just monomials of even degree is shown below. Now $(\mathfrak{m}_2^+)^4 = \langle x^8, x^7y, \ldots, xy^7, y^8 \rangle$ and this is contained in J^+f_0. It follows from Theorem 9.1 that f_0 is 8-determined with respect to $\mathcal{R}^{\text{ev}}$-equivalence.

Moreover, $\operatorname{codim}^+(f) = 7$ and

$$F(x,y;u_1,\ldots,u_7) = x^6 + y^4 + u_1x^4y^2 + u_2x^4 + u_3x^3y + u_4x^2y^2 + u_5x^2 + u_6xy + u_7y^2$$

is an $\mathcal{R}^{\text{ev}}$-versal unfolding of f.

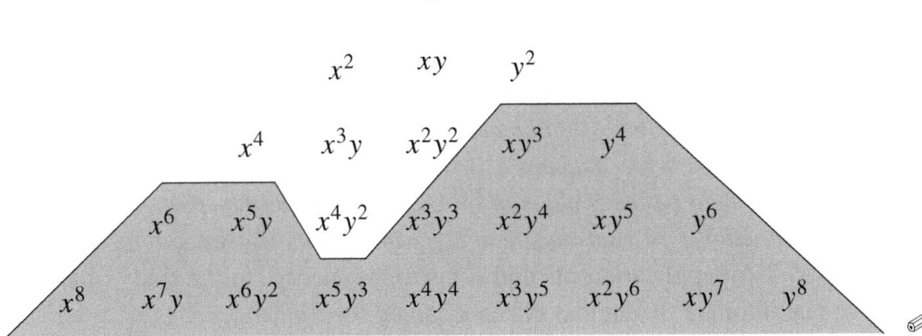

Problems

9.1 Let $f \in \mathcal{E}_n^+$. Show that $J^+f = Jf \cap \mathcal{E}_n^+$, and deduce that J^+f is an ideal in $\mathcal{E}_n^+$. [*Hint: the second part follows from the first by a general result, since $\mathcal{E}_n^+$ is a subring of $\mathcal{E}_n$, namely, show that if $I \lhd R$ and S is a subring of R then $(I \cap S) \lhd S$.*]

9.2 Find $\operatorname{codim}^+(f)$ for the even function $f(x, y, z) = x^4 + y^4 + z^2$ and find a versal unfolding. (†)

9.3 Write $\mathcal{E}_n^-$ for the set (not a ring) of *odd functions*: those $f \in \mathcal{E}_n$ satisfying $f(-x) = -f(x)$.

(i). Show by an example that $\mathcal{E}_n^-$ is not a ring.

(ii). Show that $\mathcal{E}_n^-$ is a module over the ring $\mathcal{E}_n^+$ (see Section D.5 on p. 374 for the definition of a module over a ring).

(iii). Let $f \in \mathcal{E}_n$, and define $f_+(x) := \frac{1}{2}(f(x) + f(-x))$ and $f_-(x) := \frac{1}{2}(f(x) - f(-x))$. Show that f_+ is even, f_- is odd and deduce that $\mathcal{E}_n = \mathcal{E}_n^+ \oplus \mathcal{E}_n^-$.
(This is similar to showing that any matrix can be written uniquely as the sum of a symmetric matrix and a skew-symmetric matrix.)

9.4 Denote by θ_n^+ the set of germs of *even* vector vector fields, those satisfying $\mathbf{v}(-x) = \mathbf{v}(x)$.

(i). Show that both θ_n^- and θ_n^+ are modules over $\mathcal{E}_n^+$, and that $\theta_n = \theta_n^+ \oplus \theta_n^-$ (cf. previous problem). Write correspondingly, $\mathbf{v} = \mathbf{v}_+ + \mathbf{v}_-$.

(ii). Show that a diffeomorphism obtained by integrating an odd or even vector field is itself odd or even, respectively.

(iii). Suppose $f, h \in \mathcal{E}_n^+$. Show that if there is a vector field $\mathbf{v}$ satisfying the infinitesimal homotopy equation (5.9) then $\mathbf{v}_-$ also satisfies the same equation.

(iv). Use this to prove Theorem 9.1.

9.5 How would this chapter be modified if one were interested in critical points of *odd* functions? [*Hint: we still require changes of coordinates with $\phi(-x) = -\phi(x)$ (in order to preserve 'oddness'), so in the homotopy equation, the expression $\frac{\partial f}{\partial x_i} v_i(x)$ would be the product of an even and an odd function – show that the resulting set J^-f is a submodule of $\mathcal{E}_n^-$, with the module structure from question 9.3(ii).*] An application of 'odd catastrophe theory' to the study of caustics in Hamiltonian dynamical systems can be seen in [84].

Part II

Singularity theory

10

Families of maps and bifurcations

N CONTRAST TO THE FIRST part of the book, we are now interested in systems of equations whose solutions are not given as critical points of functions; in other words, the equations are not necessarily variational.

Systems of equations can be interpreted as maps, and their solutions as the zero-set of the map. We saw some examples in Chapter 1, but they were all with a single state variable (in one variable, every equation is variational). In this short chapter, we discuss a few examples with two state variables, paving the way for later developments and in particular *bifurcation theory*, which provides methods for studying solutions to systems of equations and how they might vary, or bifurcate, as external parameters are varied. As usual, our study will be entirely local.

10.1 Bifurcation problems

A bifurcation problem usually refers to a set of equations depending on one or more parameters. As we said above, these can be interpreted as families of maps, as illustrated in the following simple example. Consider the 1–parameter system of two equations in the plane (that is, with two state variables),

$$\begin{cases} x^2 + y^2 &= u \\ xy &= 0, \end{cases}$$

where u is the parameter. If we define the 1–parameter family of maps,

$$G(x, y; u) = (x^2 + y^2 - u, xy).$$

where $G\colon \mathbb{R}^2 \times \mathbb{R} \to \mathbb{R}^2$, we see that solutions of the equations coincide with zeros of G. In this example, there is a bifurcation when $u = 0$, for when $u < 0$ there are no solutions, when $u = 0$ there is one (at the origin) and when $u > 0$ there are four (on the axes).

Definition 10.1. A ***bifurcation problem*** is a smooth map $G\colon \mathbb{R}^n \times \mathbb{R}^k \rightarrowtail \mathbb{R}^p$ (defined on an open subset of $\mathbb{R}^n \times \mathbb{R}^k$), or it is the germ at a point of such a map. For each $u \in \mathbb{R}^k$ (the *base* space, or parameter space) we denote by $g_u\colon \mathbb{R}^n \rightarrowtail \mathbb{R}^p$ the map

$$g_u(x) := G(x; u).$$

As usual, we refer to the g_u as a smooth family of maps with parameter u; the variables denoted by x are called ***state variables***. Given such a bifurcation problem, we are interested in the solutions of the corresponding equations, so we denote the set of zeros of G by

$$Z_G := \left\{(x, u) \in \mathbb{R}^n \times \mathbb{R}^k \mid G(x, u) = 0\right\}.$$

This is called the ***zero-set*** of G. The bifurcation problem (or family of maps) G is said to be ***regular*** if as a map (germ), G is a submersion. ✩

Note that for regular bifurcation problems, Z_G is a submanifold.

By far the most important setting in applications is where $n = p$ (the same number of equations as state variables), though we will not make this an assumption for now.

Notice that if $F\colon \mathbb{R}^n \times \mathbb{R}^k \rightarrow \mathbb{R}$ is a family of *functions* (as defined in Chapter 2 and Chapter 7), and we put $G = \nabla_x F = (\frac{\partial F}{\partial x_1}, \dots, \frac{\partial F}{\partial x_n})$ then Z_G is the catastrophe set C_F of F. The following definition is also analogous to the corresponding definition for a family of functions from Chapter 2.

Definition 10.2. The ***singular set*** of a bifurcation problem G is

$$\Sigma_G = \{(x, u) \in Z_G \mid \mathsf{d}(g_u)(x) \text{ does not have rank } p\}.$$

In most applications $p = n$ and in that case,

$$\Sigma_G = \{(x, u) \in Z_G \mid \det(\mathsf{d}g_u(x)) = 0\}.$$

The ***discriminant*** or ***bifurcation set*** is the subset of the base space,

$$\Delta_G = \pi_G(\Sigma_G)$$

where $\pi_G\colon Z_G \rightarrow \mathbb{R}^k$ is the projection $\pi_G(x, u) = u$. That is,

$$\Delta_G = \{u \in \mathbb{R}^k \mid \exists x, (x, u) \in \Sigma_F\}.$$

✩

Example 10.3. Continuing the example above, with $G(x, y; u) = (x^2 + y^2 - u, xy)$, one has

$$\begin{aligned} Z_G &= \{(x, y, u) \in \mathbb{R}^3 \mid xy = 0,\ u = x^2 + y^2\} \\ &= \{(x, 0, x^2) \mid x \in \mathbb{R}\} \bigcup \{(0, y, y^2) \mid y \in \mathbb{R}\}, \end{aligned}$$

which is the union of two parabolae in $\mathbb{R}^3$ and is singular at the origin. Next, $\mathsf{d}g_u = \begin{pmatrix} 2x & 2y \\ y & x \end{pmatrix}$, so that

$$\Sigma_G = \{(x, y, u) \in Z_G \mid 2x^2 - 2y^2 = 0\} = \{(0, 0, 0)\}.$$

Consequently, Δ_G is just the origin $\{u = 0\}$.

We see in this example that the origin is a singular point of Z_G. This is a particular case of a general fact: if G is not regular then Z_G will be singular, and these singular points belong to Σ_G. Indeed, a point $(x, u) \in Z_G$ is singular if $[\mathsf{d}_x G,\ \mathsf{d}_u G]$ has rank less than p there, and therefore a fortiori the rank of $\mathsf{d}_x G$ is less than p at that point.

A typical source of such problems is in finding equilibrium points of systems of first order differential equations, such as described in Section 1.1 but with perhaps more variables: see Problem 10.3 for an example.

In the next chapter we address the question of what the appropriate equivalence is for studying zero-sets of systems of equations, which is called *contact equivalence*, analogous to right equivalence for critical points of functions. We will later study versal unfoldings for such systems of equations, and then (in Part III) discuss bifurcation problems from this point of view.

This part of the book is entitled 'Singularity theory', though it should be pointed out that we are only considering that part of singularity theory dealing with systems of equations. There are many other interesting areas, such as how the geometry of singular maps can deform in unfoldings, relying instead on left–right equivalence of maps. We discuss a little of this in Chapter 17, but not in any great detail.

We end this chapter with a definition for maps (or map germs), as distinct from bifurcation problems, which will be used in later chapters.

Definition 10.4. Given a map $f\colon \mathbb{R}^n \rightarrowtail \mathbb{R}^p$, one defines the ***singular set*** (or critical set) of f to be

$$\Sigma_f := \{x \in \mathbb{R}^n \mid \mathrm{rk}(\mathsf{d}f_x) < p\},$$

and its ***discriminant*** Δ_f is the image $f(\Sigma_f) \subset \mathbb{R}^p$. ☆

Note that if $p > n$ then $\Sigma_f = \operatorname{dom} f$ and $\Delta_f = \operatorname{image}(f)$.

A word of warning is perhaps worthwhile: the definitions of the singular set of a bifurcation problem and the singular set of the underlying map are not the same. Let G be a bifurcation problem. If we denote by Σ'_G the singular set of G as a map, then $\Sigma'_G \subset \Sigma_G$; this follows from the discussion in the paragraph following Example 10.3.

Problems

10.1 Consider the 1–parameter family $G\colon \mathbb{R}\times\mathbb{R} \to \mathbb{R}$ given by $G(x;u) = x^3 - ux$. Sketch $Z_G \subset \mathbb{R}^2$ and discuss the number of solutions to $g_u = 0$ for different values of u. This is the ***pitchfork bifurcation*** discussed in Chapter 1.

10.2 Find Z_G, Σ_G and Δ_G for the (regular) bifurcation problem

$$G(x, y; u, v) = (x^2 + y^2 - u,\ xy - v).$$

Find the number of zeros of $g_{(u,v)}$ for each $(u, v) \in \mathbb{R}^2$.

10.3 Consider the 1–parameter family of ODEs in the plane $\mathbb{R}^2$,

$$\begin{cases} \dot{x} &= x^3 - \lambda x \\ \dot{y} &= y - x^2. \end{cases}$$

Find the number of equilibria of this system for all values of λ. Sketch 'phase diagrams' for x (ignoring y) for different values of λ, analogous to Figure 1.1B. (†)

10.4 Find Z_G, Σ_G, Δ_G for the family $G\colon \mathbb{R}\times\mathbb{R}^2 \to \mathbb{R}$ given by

$$G(x; u, v) = x^3 + ux + v,$$

and sketch the discriminant curve Δ_G. How many solutions does $g_{(u,v)}(x) = 0$ have, for each possible value (u, v) of the parameters?

10.5 Let $g(x;u) = 0$ be a bifurcation problem, and let G be the k–dimensional suspension of g given by $G(x;u,v) = (g(x;u), v)$ (see Definition B.17). Show that $Z_G = Z_g \times \{0\}$, and using this to identify the two, then $\pi_g = \pi_G$, $\Sigma_g = \Sigma_G$ and $\Delta_g = \Delta_G$.

10.6 Let $f\colon \mathbb{R}^n \rightarrowtail \mathbb{R}^p$ be a smooth map. What is the relation between the singular set of f and the singular set of the bifurcation problem

$$F(x;u) := f(x) - u = 0,$$

where $u \in \mathbb{R}^p$? (†)

11

Contact equivalence

CONTACT EQUIVALENCE IS the equivalence relation on map germs suitable for studying bifurcation problems. The basic equivalence is defined for a single map or equation, but it is the unfolding version of contact equivalence that is suitable for bifurcation problems (families of equations, as discussed in the previous chapter). Contact equivalence is also called $\mathcal{K}$-equivalence, or V-equivalence.

This equivalence was first introduced by John Mather as a technical tool to aid with the classification of singularities of maps $\mathbb{R}^n \rightarrowtail \mathbb{R}^p$ up to what is called *left–right equivalence* (see Chapter 12), but it soon became clear that it was an important concept in its own right and now lies behind some of the most important applications of singularity theory. In this chapter we define the equivalence and later we address questions of finite determinacy and of $\mathcal{K}$-codimension of map germs; after that we consider their classification and unfoldings. The results of this chapter are entirely due to Mather [76, III]. Contact equivalence is also called V-equivalence by some authors [75], [6].

11.1 $\mathcal{K}$-equivalence

The idea of contact equivalence of two map germs $f, g\colon (\mathbb{R}^n, 0) \to (\mathbb{R}^p, 0)$ is that the 'solution sets' $f^{-1}(0)$ and $g^{-1}(0)$ should be diffeomorphic, but that is not the definition; after all $x = 0$ and $x^2 = 0$ have the same solution set but their deformations or unfoldings (and resulting bifurcations) are quite different.

Definition 11.1. Two map germs $f, g\colon (\mathbb{R}^n, 0) \to (\mathbb{R}^p, 0)$ are ***contact equivalent*** or ***$\mathcal{K}$-equivalent***, if there exist,
(i) a diffeomorphism ϕ of the source $(\mathbb{R}^n, 0)$, and
(ii) a matrix $M \in \mathsf{GL}_p(\mathcal{E}_n)$ such that

$$f \circ \phi(x) = M(x)g(x),$$

where $f(x)$ and $g(x)$ are written as column vectors, and $M(x)g(x)$ is the usual product of matrix times vector.

If ϕ is the identity, one says f and g are ***$\mathcal{C}$-equivalent***: in this case $f(x) = M(x)g(x)$. ✯

The definition states that the matrix $M \in \mathsf{GL}_p(\mathcal{E}_n)$. This means that M is a $p \times p$ matrix whose entries are in $\mathcal{E}_n$, and which is moreover *invertible*, with the entries of $M^{-1}(x)$ also belonging to $\mathcal{E}_n$. In practice, one just needs to check that $\det(M(0)) \neq 0$ in order to deduce that M is invertible in a neighbourhood of 0, and that the inverse is smooth. See Lemma D.18.

Example 11.2. Let $g(x, y) = (x^2, y^2)$ and $f(x, y) = (x^2 + y^2, x^2 - y^2 + y^3)$, considered as germs at the origin. Then f and g are $\mathcal{C}$-equivalent, with $M(x, y) = \begin{pmatrix} 1 & 1 \\ 1 & y-1 \end{pmatrix}$ as the reader can readily check. Note that $M(x, y)$ is invertible for (x, y) in a neighbourhood of $(0, 0)$. ✎

If f and g are $\mathcal{C}$-equivalent, and because $M(x)$ is invertible, $f(x) = 0 \Leftrightarrow g(x) = 0$; that is, f and g have the same zero-set. In the example above the zero-set is just the origin $x = y = 0$.

More generally, and still because $M(x)$ is invertible, if f and g are $\mathcal{K}$-equivalent then

$$g(x) = 0 \Longleftrightarrow f(\phi(x)) = 0$$

so that $\phi(g^{-1}(0)) = f^{-1}(0)$. This proves the following.

Proposition 11.3. *If f and g are contact equivalent then their zero-sets are diffeomorphic subsets of $\mathbb{R}^n$.*

Note however that the converse does not hold, as demonstrated by the example above with x^2 and x. This proposition explains why some authors call this V-equivalence of map germs, rather than $\mathcal{K}$-equivalence, with 'V' for variety, and varieties being defined usually as zero-sets of maps (or of ideals).

On the other hand, the *images* of contact equivalent map germs can be quite different. For example the maps $\mathbb{R} \to \mathbb{R}^2$ given by $f(t) = (t^2, 0)$ and $g(t) = (t^2, t^3)$ are $\mathcal{K}$-equivalent (see Problem 11.1), but their images are not diffeomorphic, as shown in Figure 11.1. If one is interested in images of maps then $\mathcal{K}$-equivalence is not relevant and one needs to use left–right equivalence.

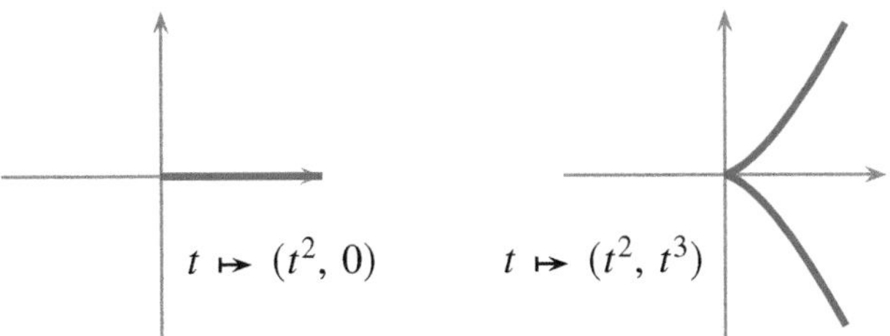

FIGURE 11.1 Images of two maps which are $\mathcal{K}$-equivalent.

11.2 Local algebra of a map germ

Let $f\colon (\mathbb{R}^n, 0) \to (\mathbb{R}^p, 0)$ be a smooth map germ, and denote[1] by I_f the ideal in $\mathcal{E}_n$ generated by the components of f. Define the ***local algebra of*** f to be

$$Q(f) := \mathcal{E}_n/I_f\,.$$

For example, if $f(x, y) = (x^2, y^2)$ then $I_f = \langle x^2, y^2\rangle$ and

$$Q(f) = \mathcal{E}_2/\langle x^2, y^2\rangle \simeq \mathbb{R}\{1, x, y, xy\}.$$

Here the isomorphism is as vector spaces, but the local algebra $Q(f)$ has the structure of a ring (or algebra[2]) inherited from that of $\mathcal{E}_n$. This maps to a ring structure on $\mathbb{R}\{1, x, y, xy\}$, with $x.y = xy$ while $x.xy = 0$ because $x^2y \in I_f$.

Example 11.4. If $h \in \mathcal{E}_n$ is the germ of a smooth function then with $f = \operatorname{grad} h$ we have

$$Q(f) = \mathcal{E}_n/Jh,$$

which is a familiar object from Chapter 4. ✐

The local algebra, and indeed the ideal I_f, is an important invariant for $\mathcal{C}$-equivalence, as observed by Mather.

Theorem 11.5. *Let $f, g\colon (\mathbb{R}^n, 0) \to (\mathbb{R}^p, 0)$ be two map germs. The following are equivalent:*

(i). f and g are $\mathcal{C}$-equivalent,

[1] Other authors often denote this ideal I_f by $f^*\mathfrak{m}_p$

[2] An algebra is a ring which also has 'scalar multiplication' by elements of a field – here the field is $\mathbb{R}$, and indeed $\mathcal{E}_n$ itself is an algebra.

(ii). *their ideals* I_f *and* I_g *are equal,*

(iii). *their local algebras are equal.*

We prove this below.

Example 11.6. Consider $f(x,y) = (x^2, y^2)$, and $g(x,y) = (x^2+y^3, y^2+x^3)$. By Nakayama's lemma, $\langle x^2, y^2\rangle = \langle x^2+y^3, y^2+x^3\rangle$ so consequently $I_f = I_g$ and the two maps are $\mathcal{C}$-equivalent. The reader can also show this directly by constructing an appropriate matrix M.

This theorem extends to $\mathcal{K}$-equivalence as follows. The difference between $\mathcal{C}$-equivalence and $\mathcal{K}$-equivalence is a change of coordinates in the source. It follows that if f, g are two $\mathcal{K}$-equivalent map germs there is a diffeomorphism ϕ of $(\mathbb{R}^n, 0)$ such that f and $g\circ\phi$ are $\mathcal{C}$-equivalent. By the theorem this is equivalent to $I_f = I_{g\circ\phi}$. Now I_g and $I_{g\circ\phi}$ are themselves related by the change of coordinates ϕ, that is $I_{g\circ\phi} = \phi^* I_g$. It follows that $I_f = \phi^* I_g$.

This motivates the definition: we say two ideals I and J in $\mathcal{E}_n$ are ***diffeomorphic ideals*** if there is a diffeomorphism germ ϕ such that $I = \phi^* J$. The argument of the previous paragraph shows the following.

Corollary 11.7. *Let* $f, g\colon (\mathbb{R}^n, 0) \to (\mathbb{R}^p, 0)$ *be two map germs. The following are equivalent:*

(i). f *and* g *are* $\mathcal{K}$*-equivalent,*

(ii). *their ideals* I_f *and* I_g *are diffeomorphic ideals in* $\mathcal{E}_n$*,*

(iii). *their local algebras are diffeomorphic.*

For example, the local algebras

$$Q_1 = \mathcal{E}_2/\langle x^2, y^3\rangle, \qquad \text{and} \qquad Q_2 = \mathcal{E}_2/\langle x^3, y^2\rangle$$

are clearly isomorphic via the change of coordinates $(x,y) \mapsto (y,x)$, so are diffeomorphic.[3]

If f, g are finitely determined with respect to $\mathcal{K}$-equivalence (as described in Chapter 13), then the word 'diffeomorphic' can be replaced simply by 'isomorphic', so for example $f \sim_{\mathcal{K}} g$ iff $Q(f) \simeq Q(g)$, where $\simeq$ means isomorphic. We do not give the proof, but it can be found, along with almost everything else in this chapter, in Mather's paper [76, III].

[3] This notion of diffeomorphic ideals or algebras is sometimes called 'induced isomorphic'.

Proof of Theorem 11.5: (ii) $\iff$ (iii): R/I is the set of equivalence classes in the ring R modulo the ideal I, and one of these classes (the zero of multiplication in Q_I) is just I. So if $R/I = R/J$ then $I = J$. And conversely.

(i) $\Longrightarrow$ (ii): Since $f(x) = M(x)g(x)$ we have $f_i = \sum m_{ij}g_j$ so that each $f_i \in I_g$, and conversely using M^{-1} we have $g_j \in I_f$.

(ii) $\Longrightarrow$ (i): Since $f_i \in I_g$ there exist functions $a_{ij}(x)$ such that $f_i = \sum_j a_{ij}g_j$, and similarly there are functions b_{ji} such that $g_j = \sum_i b_{ji}f_i$. It follows that $f_i = \sum_j\sum_k a_{ij}b_{jk}f_k$. Let $A = (a_{ij})$ and $B = (b_{ij})$ which are both $p \times p$ matrices. However, A and B may not be invertible – if either was we would be done for it would define a $\mathcal{C}$-equivalence. Let A_0, B_0 be the matrices A, B at $x = 0$. The lemma below tells us that there is a matrix C such that $C(\mathsf{Id}_p - A_0B_0) + B_0$ is invertible. Let

$$M(x) = C(\mathsf{Id}_p - A(x)B(x)) + B(x).$$

Since $M(0)$ is invertible (by the lemma), so $M(x)$ must be invertible for x in a neighbourhood of $x = 0$ (by continuity of the determinant). Now

$$\begin{aligned} M(x)f &= C(I - AB)f + Bf \\ &= C(f - Ag) + g \\ &= C(f - f) + g \;=\; g. \end{aligned}$$

Thus f and g are $\mathcal{C}$-equivalent. ✔

Lemma 11.8. *Given any pair of $p \times p$ matrices A, B with real entries, there is a matrix C (depending only on B) such that $M := C(\mathsf{Id}_p - AB) + B$ is invertible.*

Proof: If B has rank p then it is invertible and we can take $C = 0$. Suppose therefore that $\operatorname{rk} B = r < p$. Then we can choose linearly adapted bases in the source and target so that B takes the form

$$B = \begin{pmatrix} I_r & 0 \\ 0 & 0 \end{pmatrix},$$

as discussed in Appendix A (Proposition A.2). Now with respect to these bases, let

$$C = \begin{pmatrix} 0 & 0 \\ 0 & I_{p-r} \end{pmatrix}.$$

Then with any matrix A,

$$A = \begin{pmatrix} A_{11} & A_{12} \\ A_{21} & A_{22} \end{pmatrix}$$

one finds

$$M = C(\mathsf{Id}_p - AB) + B = \begin{pmatrix} I_r & 0 \\ A_{21} & I_{p-r} \end{pmatrix}$$

which is clearly invertible (being lower triangular with 1s on the diagonal). ✔

Multiplicities There are two notions of multiplicity of equidimensional map germs: one algebraic and one geometric.

Definition 11.9. Let $f\colon (\mathbb{R}^n, 0) \to (\mathbb{R}^n, 0)$ be a map germ. Define the ***algebraic multiplicity*** of f to be $m_A(f) = \dim Q(f)$. ✯

For the map $f(x, y) = (x^2, y^2)$, we have $Q(f) \simeq \mathbb{R}\{1, x, y, xy\}$ so $m_A(f) = 4$. Notice that in this example $f^{-1}(a, b)$ consists of four points if $a, b > 0$. The geometric multiplicity $m_G(f)$ is defined to be the maximal number of pre-image points for a representative function on an arbitrarily small neighbourhood, so in this example the geometric multiplicity is also equal to four.

Indeed, one can show in general that if f is finitely determined then $m_G(f) \leq m_A(f)$. On the other hand, for complex analytic maps, one can show $m_G(f) = m_A(f)$ – see Theorem 13.12.

It follows from the corollary above that $\mathcal{K}$-equivalent germs have equal algebraic multiplicity. However the same is not true for geometric multiplicity (a simple example is given in Problem 11.6).

11.3 The alternative definition

There is another definition of contact equivalence which appears to be coarser (i.e. more maps become contact equivalent), but is in fact equivalent. Because it allows more than just matrix multiplication on the left by using general diffeomorphisms of $\mathbb{R}^p$ depending on $x \in \mathbb{R}^n$, this version lends itself to interesting and useful extensions (obtained by modifying condition (i) below). For the moment, we call this $\mathcal{K}'$-equivalence.

By definition, $f \sim_{\mathcal{K}'} g$ if there exists a diffeomorphism germ Ψ of $\mathbb{R}^n \times \mathbb{R}^p$, $(0, 0)$ of the form

$$\begin{array}{rccc} \Psi\colon & \mathbb{R}^n \times \mathbb{R}^p & \longrightarrow & \mathbb{R}^n \times \mathbb{R}^p \\ & (x, y) & \longmapsto & (\phi(x), \psi(x, y)), \end{array} \tag{11.1}$$

such that,

(i). $\Psi(\Gamma_0) = \Gamma_0$, and

(ii). $\Psi(\Gamma_f) = \Gamma_g$.

Recall that Γ_f denotes the graph of f, so Γ_0 is the graph of the zero map: $\Gamma_0 = \mathbb{R}^n \times \{0\}$. More explicitly, these two conditions are equivalent to,

(i)′. $\psi(x,0) = 0$,

(ii)′. $g \circ \phi(x) = \psi(x, f(x))$.

To see that (ii)′ is equivalent to (ii), we have $(x,y) \in \Gamma_f \Leftrightarrow y = f(x)$, and $\Psi(x,y) \in \Gamma_g \Leftrightarrow \psi(x,y) = g(\phi(x))$.

The fact that Ψ is a diffeomorphism of the form (11.1) implies firstly that ϕ is a diffeomorphism of $(\mathbb{R}^n, 0)$, and secondly that for each $x \in \mathbb{R}^n$ the resulting map $\mathbb{R}^p \to \mathbb{R}^p$, $y \mapsto \psi(x,y)$ is a diffeomorphism. This is because (11.1) forces the Jacobian matrix of Ψ to take the following block form:

$$\mathsf{d}\Psi_{(x,y)} = \begin{pmatrix} \mathsf{d}\phi_x & 0 \\ \mathsf{d}_x\psi_{(x,y)} & \mathsf{d}_y\psi_{(x,y)} \end{pmatrix},$$

where $\mathsf{d}_x\psi$ means the differential with respect to the x-variables, and $\mathsf{d}_y\psi$ similarly. Indeed, as Ψ is a diffeomorphism, the Jacobian matrix is invertible, so that because of the zero in the top-right corner, the top-left and bottom-right blocks must each be invertible, and so each is a diffeomorphism by the inverse function theorem.

Proposition 11.10. *$\mathcal{K}$ and $\mathcal{K}'$ equivalences are identical.*

In other words, two germs are $\mathcal{K}$-equivalent if and only if they are $\mathcal{K}'$-equivalent.

Proof: One way is straightforward: if $f \sim_{\mathcal{K}} g$ then $f \sim_{\mathcal{K}'} g$. This follows from the fact that a given $\mathcal{K}$-equivalence $g \circ \phi(x) = M(x) f(x)$, defines a diffeomorphism Ψ by

$$\Psi(x,y) = (\phi(x),\ M(x)y).$$

To see this is a diffeomorphism it suffices to note that its inverse is given by,

$$\Psi^{-1}(u,v) = (\phi^{-1}(u),\ M(\phi^{-1}(u))^{-1}v),$$

(as ϕ is a diffeomorphism and M is invertible), and it is easily checked it has the required properties for $\mathcal{K}'$-equivalence.

For the converse we rely on Hadamard's lemma. Suppose then that $f \sim_{\mathcal{K}'} g$. Then there is a diffeomorphism Ψ of the form (11.1) such that

$$g \circ \phi(x) = \psi(x, f(x)). \tag{11.2}$$

Since each component ψ_j of ψ satisfies $\psi_j(x,0) = 0$ for $j = 1, \dots, p$ (see (i)′ above), we can apply Hadamard's lemma to write

$$\psi_i(x,y) = \sum_j \chi_{ij}(x,y)\, y_j,$$

for functions $\chi_{ij} \in \mathcal{E}_n$ $(i, j = 1, \dots, p)$. Then for each i,

$$\psi_i(x, f(x)) = \sum_j \chi_{ij}(x, f(x))\, f_j(x).$$

Let $M(x)$ be the matrix whose i, j entry is $\chi_{ij}(x, f(x))$. Then the previous equation can be rewritten, using also (11.2), as

$$g \circ \phi(x) = M(x) f(x),$$

so that $g \sim_{\mathcal{K}} f$. ✔

Remark 11.11. In the definition of $\mathcal{K}'$-equivalence, Ψ is a diffeomorphism which takes the pair of submanifolds (Γ_0, Γ_f) to the pair (Γ_0, Γ_g), which means it is respecting the "contact" between the graphs; this is the origin of the name of contact equivalence. Indeed, in Mather's paper [76, III] where he introduces contact equivalence, $\mathcal{K}'$-equivalence is called contact equivalence, and the equivalence involving the matrix $M(x)$ is called $\mathcal{K}$-equivalence.

This idea of contact between two graphs was extended to a general notion of contact between pairs of submanifolds of $\mathbb{R}^n$ of the same dimension in [45], and then to general pairs of submanifolds [83] as follows.

Let X, Y and X', Y' be two pairs of submanifold-germs of $(\mathbb{R}^n, 0)$. Suppose X, X' are given as the images of immersion-germs g, g': $(\mathbb{R}^k, 0) \to (\mathbb{R}^n, 0)$, respectively, and Y, Y' are the zero-sets of two submersion-germs $f, f' : (\mathbb{R}^n, 0) \to (\mathbb{R}^p, 0)$, respectively. Then *there is a diffeomorphism-germ of* $(\mathbb{R}^n, 0)$ *taking* X *to* X' *and* Y *to* Y' *if and only if* $f \circ g$ *and* $f' \circ g'$ *are* $\mathcal{K}$*-equivalent.* ❞

Problems

11.1 Show that the following pairs of maps are contact equivalent:

(i). $(\mathbb{R} \to \mathbb{R}^2)$ $f(t) = (t^2, 0)$ and $g(t) = (t^2, t^a)$, with $a \geq 2$.

(ii). $(\mathbb{R}^2 \to \mathbb{R}^2)$ $f(x,y) = (x^2, y^2)$, $g(x,y) = (x^2 + y^2,\, x^2 - y^2)$.

(iii). ($\mathbb{R}^2 \to \mathbb{R}^2$) $f(x, y) = (x^2, y^2)$, $g(x, y) = (x^2 + y^2, xy)$
[*Hint*: *use* $\phi(x, y) = (x + y, x - y)$.]

(iv). ($\mathbb{R}^2 \to \mathbb{R}^2$) $f(x, y) = (x^2, y^2)$, $g(x, y) = (x^2 + y^3, y^2 + x^3)$.

(v). ($\mathbb{R}^2 \to \mathbb{R}^2$) $f(x, y) = (x, y^3 + xy)$, $g(x, y) = (x, y^3)$
[*Note*: *in* (v), $f(x, y)$ *is the* cusp *map discussed in Chapter* 7.] (†)

11.2 Extending (i) of the previous problem, let $\gamma\colon (\mathbb{R}, 0) \to (\mathbb{R}^n, 0)$ be a curve germ, and suppose that the $(k-1)$-jet of γ is zero, but not the k-jet. Show that γ is $\mathcal{K}$-equivalent to $\sigma : (\mathbb{R}, 0) \to (\mathbb{R}^n, 0)$ defined by $\sigma(t) = (t^k, 0, 0, \ldots, 0)$.

11.3 Show that if two function-germs $f, g\colon (\mathbb{R}^n, 0) \to \mathbb{R}$ are right equivalent, then their gradient maps ∇f, $\nabla g\colon (\mathbb{R}^n, 0) \to \mathbb{R}^n$ are contact equivalent. [*Note*: *the converse is false*; *for example* $f = x^2 + y^2$ *and* $g = x^2 - y^2$.]

11.4 Show contact equivalence is an equivalence relation on $\mathcal{E}_n^p$. Deduce that the set of contact equivalences (i.e., of pairs (ϕ, M) as in the definition) forms a group under composition.

11.5 Let $f\colon (\mathbb{R}^n, 0) \to (\mathbb{R}^p, 0)$. Show that $I_f = \mathfrak{m}_n$ if and only if f is an immersion germ.

11.6 Consider the germ $f(x, y) = (x, y^3 + xy)$. Calculate the algebraic multiplicity $m_A(f)$ and the geometric multiplicity $m_G(f)$ of f. Compare with the same multiplicities for the $\mathcal{K}$-equivalent germ $g(x, y) = (x, y^3)$. (†)

11.7 Let $f, g \in \mathcal{E}_n^p$ be $\mathcal{K}$-equivalent map germs.
(i) Show that $j^k f = 0 \Leftrightarrow j^k g = 0$,
(ii) Suppose in particular $p = 1$ and $f(0) = g(0) = 0$. Show that f has a nondegenerate critical point of index i then g also has a nondegenerate critical point, but of index either i of $n - i$. (Recall that the index is the number of negative signs when f is written $f = \sum_i \pm x_i^2$.)

11.8 The following are a variation and an extension of the Lagrange multiplier theorem (for which, see Problem B.12).
(i) Let $g\colon (\mathbb{R}^n, 0) \to (\mathbb{R}^{n-k}, 0)$ be a submersion (germ), so that $M := g^{-1}(0)$ is a submanifold germ of $\mathbb{R}^n$ of dimension k. Now let $f \in \mathcal{E}_n$. Show that the restriction $f|_M$ has a critical point at 0 if and only if the map germ $(f, g)\colon (\mathbb{R}^n, 0) \to (\mathbb{R}^{n-k+1}, 0)$ is singular at 0. [*Hint*: *use linearly adapted coordinates for* g.]

(ii) With the same notation, let $j\colon (\mathbb{R}^k, 0) \to (\mathbb{R}^n, 0)$ be a regular parametrization of M (regular here means that j has rank k), and let $j^*\colon \mathcal{E}_n \to \mathcal{E}_k$ be

the 'pull-back' homomorphism, $j^*(h) = h \circ j$.
Show that j^* induces an isomorphism of local algebras $j^*: Q_{(f,g)} \to Q_{f \circ j}$.

[*Remark: this result can be used to find the local algebra of* $f|_M$ *(which is the same as* $Q_{f \circ j}$*) without explicitly knowing* j. (†)

11.9 Part (ii) of the problem above was for $f \in \mathcal{E}_n$. Show the corresponding result for arbitrary $f \in \mathcal{E}_n^p$ with $p > 1$.

11.10 Use Problem 11.8 to show that if M is the sphere $g(x, y, z) = 0$, where $g(x, y, z) = x^2 + y^2 + (z-1)^2 - 1$, and $f(x, y, z) = x^2 + y^2 + z^2$, then $f|_M$ has a nondegenerate critical point at the origin, and that it is a local minimum. [*Hint*: *you will probably need to use Problem* 11.7.]

12

Tangent spaces

LET $\gamma(s)$ BE A SMOOTH CURVE in the vector space $\mathbb{R}^N$ then $\dot\gamma(s) = \frac{\mathrm{d}}{\mathrm{d}s}\gamma(s)$ is a vector in $\mathbb{R}^N$ tangent to the curve at the point $\gamma(s)$. If the curve lies in a submanifold X of $\mathbb{R}^N$, then one says that $\dot\gamma(s)$ is tangent to X at $\gamma(s)$, and indeed one can define the tangent space T_xX to be the set in $\mathbb{R}^N$ of all vectors tangent at x to smooth curves lying in X (see p. 348).

Now consider the set of all map germs $(\mathbb{R}^n, 0) \longrightarrow (\mathbb{R}^p, 0)$, that is, the vector space $\mathcal{E}_n^p$ (though now infinite–dimensional), and suppose $\gamma(s)$ is a curve in $\mathcal{E}_n^p$. We can still think of $\frac{\mathrm{d}}{\mathrm{d}s}\gamma(s)$ as a tangent vector, but now in $\mathcal{E}_n^p$. And if $\gamma(0) = f$ then the set of all tangent vectors at f to such curves in $\mathcal{E}_n^p$ defines the tangent space to $\mathcal{E}_n^p$ at f. In Section A.4 in Appendix A, it is explained that such tangent vectors are vector fields along the map. We denote the set of such germs of vector fields along f by $\theta(f)$.

In this way, we think of $\theta(f)$ as the tangent space at f to $\mathcal{E}_n^p$ (it consists of p-tuples of germs in $\mathcal{E}_n$, so is really the same as $\mathcal{E}_n^p$, just as a tangent space of $\mathbb{R}^N$ can be identified with $\mathbb{R}^N$).

For each of the singularity theory equivalences (we have already seen right and contact equivalences, and others will follow shortly), consider the subset of $\mathcal{E}_n^p$ consisting of map germs equivalent to f. One can let $\gamma(s)$ be a curve in that set, and then its tangent vector $\dot\gamma$ can justifiably be called a tangent vector to the equivalence class in question.

This is the same approach we used in Chapter 5 for right equivalence, where we let $\gamma(s) = f \circ \phi_s$, and then, using the chain rule,

$$\dot\gamma(s)(x) = \frac{\mathrm{d}}{\mathrm{d}s} f(\phi_s(x)) = \sum_{j=1}^{n} \frac{\partial f}{\partial x_j} v_j(y) = \mathrm{t}f(\mathbf{v}_s)(y), \tag{12.1}$$

where $y = \phi_s(x)$ and $\mathbf{v}_s$ is the vector field $\mathbf{v}_s(y) = \frac{\mathrm{d}}{\mathrm{d}s}\phi_s(x)$. Then $\dot\gamma(0)$ is a tangent vector to the equivalence class of functions $\mathcal{R}$-equivalent to f, written $T\mathcal{R}{\cdot}f$.

For different families ϕ_s of diffeomorphisms in the expression above, one obtains different tangent vectors. For example, if we consider diffeomorphisms whose linear part at the origin is equal to the identity ($\mathcal{R}_1$-equivalence), then the tangent space to the set of equivalent functions would be smaller, and denoted $T\mathcal{R}_1{\cdot}f$. Moreover, when dealing with families it is important to allow diffeomorphisms that do not fix the origin ($\phi_s(0) \neq 0$ for $s \neq 0$) and then the tangent space to the set of equivalent functions would be larger, and in this case it is denoted $T_e\mathcal{R}{\cdot}f$, called the *extended* right tangent space, and which is equal to Jf.

A similar approach, with different equivalence relations on map germs, gives rise to different tangent spaces, and we will see that some tangent spaces are easier to compute than others.

Technical aside One should be aware that the discussion above is heursitic. What does it mean to differentiate a curve in $\mathcal{E}_n$? Differentiation involves limits, and we don't have a definition for convergence in $\mathcal{E}_n$, so we don't even have a well-defined notion of continuous curve in $\mathcal{E}_n$, let alone differentiable ones. However, the curves we consider are all of the form where $\gamma(s) = \gamma_s$, which is a function of x, are such that $\gamma_s(x) = F(x, s)$ is a *smooth* function of x and s, and then we can *define the derivative* $\dot\gamma(s)$ to be the function of x obtained by differentiating F with respect to s:

$$\left(\frac{\mathrm{d}}{\mathrm{d}s}\gamma_s\right)(x) := \frac{\partial F}{\partial s}(x, s). \tag{12.2}$$

For example, if $\gamma_s(x) = \mathrm{e}^{xs}$ then $\dot\gamma_s(x) = x\mathrm{e}^{xs}$. One observation which justifies this heuristic, is that if we work with jets of a fixed order k say (which form a finite–dimensional vector space) then we do know what differentiation means and it follows from the fact that partial derivatives commute that

$$\frac{\mathrm{d}}{\mathrm{d}s}(\mathrm{j}^k_x\gamma_s)(x) := \mathrm{j}^k_x\left(\frac{\partial F}{\partial s}(x, s)\right),$$

where j^k_x means the k-jet with respect to the x-variables, and this is compatible with (12.2).

Much of this chapter is based on the survey paper of Wall [115] and the monograph of Damon [27]. The first section below is an introduction to modules: while $\mathcal{E}_n$ is a ring and the subsets of interest are ideals, $\mathcal{E}_n^p$ is no longer a ring (for $p > 1$), but a module, and ideals are replaced by submodules. Readers already familiar with modules and submodules may skip this section, referring back to it perhaps for notation. Subsequently, we apply the above heuristic argument to define the various notions of tangent space arising in singularity theory.

12.1 Modules

Before proceeding with the various tangent spaces, we need to introduce the algebraic notion of modules over the ring $\mathcal{E}_n$. Readers familiar with this notion can skip this (or better, skim it to familiarize themselves with the notation) and proceed directly with Section 12.2. See Section D.5 for definitions and examples of modules.

A germ $f\colon (\mathbb{R}^n, 0) \to \mathbb{R}$ is an element of $\mathcal{E}_n$. If instead we consider $f\colon (\mathbb{R}^n, 0) \to \mathbb{R}^p$ then each component of f is an element of $\mathcal{E}_n$ so that

$$f \in \mathcal{E}_n^p = \mathcal{E}_n \times \cdots \times \mathcal{E}_n \quad (p \text{ copies}).$$

That is $\mathcal{E}_n^p$ consists of column vectors with a smooth germ at 0 in each component.[1] Two such vectors can be added, and $f \in \mathcal{E}_n^p$ can be multiplied by a 'scalar' $h \in \mathcal{E}_n$:

$$h\begin{pmatrix} f_1 \\ f_2 \\ \vdots \\ f_p \end{pmatrix} = \begin{pmatrix} hf_1 \\ hf_2 \\ \vdots \\ hf_p \end{pmatrix}.$$

This gives $\mathcal{E}_n^p$ the structure of a module over the ring $\mathcal{E}_n$.

Instead of ideals in a ring, in a module one has submodules and if A is a submodule of B we write $A \lhd B$, as for ideals. Thus for example

$$\{f \in \mathcal{E}_n^p \mid f_1(0) = \cdots = f_p(0)\}$$

is a submodule of $\mathcal{E}_n^p$, usually written $\mathfrak{m}_n\mathcal{E}_n^p \lhd \mathcal{E}_n^p$. Note that if different powers of the maximal ideal (or different ideals) are used in different components of a module, one needs to use Cartesian product notation, as in the following example.

Example 12.1. The Cartesian product

$$\mathfrak{m}_n^2 \times \mathfrak{m}_n^3 = \{(f, g) \in \mathcal{E}_n^2 \mid f \in \mathfrak{m}_n^2,\ g \in \mathfrak{m}_n^3\}$$

is a submodule of $\mathcal{E}_n^2$. And more generally, if $I_1, I_2, \ldots, I_p$ are ideals in $\mathcal{E}_n$ then $I_1 \times I_2 \times \cdots \times I_p \lhd \mathcal{E}_n^p$ is the submodule consisting of elements $(v_1, \ldots, v_p) \in \mathcal{E}_n^p$ with $v_j \in I_j$. Checking that these indeed form submodules is left to the reader. ✐

[1] One pronounces $\mathcal{E}_n^p$ as ee-n-p, like $\mathbb{R}^p$ is r-p.

Many of the definitions and results from Part I can be extended to this multi-component setting. For example, a submodule $M \lhd \mathcal{E}_n^p$ is of ***finite codimension*** if there is a finite–dimensional vector subspace V of $\mathcal{E}_n^p$ such that $M + V = \mathcal{E}_n^p$. If moreover $M \cap V = 0$ then one writes $\mathcal{E}_n^p = M \oplus V$, and in this case any basis for V is called a ***cobasis*** for M in $\mathcal{E}_n^p$. Furthermore, Proposition 3.13 extends to saying that a submodule $M \lhd \mathcal{E}_n^p$ has finite codimension if and only if there is an $r > 0$ such that $\mathfrak{m}_n^r \mathcal{E}_n^p \subset M$. Note that if $\mathcal{E}_n^p = M \oplus V$, then as vector spaces $\mathcal{E}_n/M \simeq V$. See Problem 3.15. In particular, if $\{h_1, \ldots, h_r\}$ is a cobasis for M in $\mathcal{E}_n^p$ then given $f \in \mathcal{E}_n^p$ there are unique real numbers $a_1, \ldots, a_r$ and $m \in M$ such that

$$f = a_1 h_1 + \cdots + a_r h_r + m.$$

Finally, it should be noted that Nakayama's lemma is precisely the same statement as before, but where I, J are submodules of a given module with, as before, I being finitely generated. See Section D.7 in Appendix D for details.

Example 12.2. Let M be the submodule of $\mathcal{E}_2^2$ generated by

$$\alpha = \begin{pmatrix} x^2 \\ 0 \end{pmatrix}, \quad \beta = \begin{pmatrix} y^2 \\ 0 \end{pmatrix}, \quad \gamma = \begin{pmatrix} xy \\ xy \end{pmatrix}, \quad \delta = \begin{pmatrix} 0 \\ x^2 - y^2 \end{pmatrix}.$$

Show that M is of finite codimension, and find a cobasis in $\mathcal{E}_2^2$.
Solution: The first step is to find as many monomial elements as possible. From α and β one obtains $\mathfrak{m}_2^3 \times \{0\} \subset M$, for example $\begin{pmatrix} x^2 y \\ 0 \end{pmatrix} = y\alpha$. For the bottom row (the second component), the same is true, but requires a little more work. For example,

$$\begin{pmatrix} 0 \\ x^3 \end{pmatrix} = x\delta + \begin{pmatrix} 0 \\ xy^2 \end{pmatrix} = x\delta - y\gamma + \begin{pmatrix} xy^2 \\ 0 \end{pmatrix}.$$

Since we have already shown the final term is in M (it is $x\beta$), we see $\begin{pmatrix} 0 \\ x^3 \end{pmatrix} \in M$. The other generators of $\{0\} \times \mathfrak{m}_2^3$ follow similarly. Thus

$$\mathfrak{m}_2^3 \mathcal{E}_2^2 \subset M,$$

whence M is of finite codimension. For a cobasis of M in $\mathcal{E}_2^2$, we can take

$$\left\{ \begin{pmatrix} 1 \\ 0 \end{pmatrix}, \begin{pmatrix} 0 \\ 1 \end{pmatrix}, \begin{pmatrix} x \\ 0 \end{pmatrix}, \begin{pmatrix} 0 \\ x \end{pmatrix}, \begin{pmatrix} y \\ 0 \end{pmatrix}, \begin{pmatrix} 0 \\ y \end{pmatrix}, \begin{pmatrix} xy \\ 0 \end{pmatrix}, \begin{pmatrix} 0 \\ x^2 \end{pmatrix} \right\},$$

and so M is of codimension 8 in $\mathcal{E}_2^2$. One can also use Newton diagrams for submodules of $\mathcal{E}_n^p$, with one diagram for each component. For this example the diagram would be as follows:

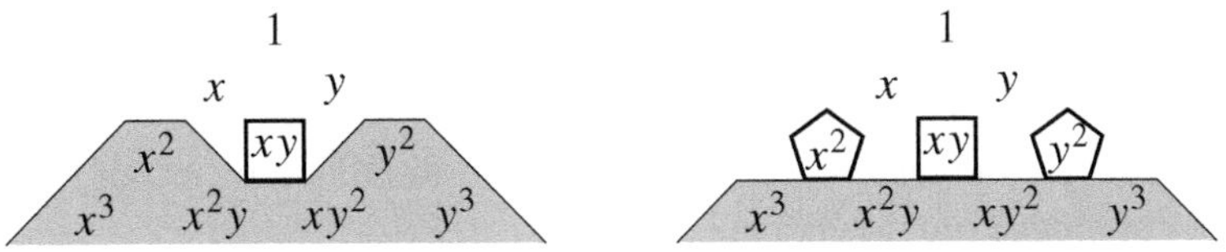

Top component	Bottom component
1	1
$x \quad y$	$x \quad y$
$x^2 \quad \boxed{xy} \quad y^2$	$x^2 \quad \boxed{xy} \quad y^2$
$x^3 \quad x^2y \quad xy^2 \quad y^3$	$x^3 \quad x^2y \quad xy^2 \quad y^3$

As before, the greyed out region represents the monomial terms in the module, while the 'decorated' terms correspond to relations between the remaining monomials.

If insufficiently many of the generators are monomials one usually needs a slightly more sophisticated approach, as in the following example.

Example 12.3. Let M be the submodule of $\mathcal{E}_2^2$ generated by

$$\alpha = \begin{pmatrix} x \\ 2y \end{pmatrix}, \quad \beta = \begin{pmatrix} y \\ -x \end{pmatrix}, \quad \gamma = \begin{pmatrix} x^2 + y^2 \\ 3xy \end{pmatrix}, \quad \delta = \begin{pmatrix} xy \\ x^2 - y^2 \end{pmatrix}.$$

Show that M is of finite codimension, and find its codimension in $\mathcal{E}_2^2$.

Solution: One begins as before by finding as many monomial elements as possible: the first hope is that all 6 of the quadratic monomials are in M, and indeed this is the case. To see this, consider the following table (matrix):

	$\begin{pmatrix} x^2 \\ 0 \end{pmatrix}$	$\begin{pmatrix} xy \\ 0 \end{pmatrix}$	$\begin{pmatrix} y^2 \\ 0 \end{pmatrix}$	$\begin{pmatrix} 0 \\ x^2 \end{pmatrix}$	$\begin{pmatrix} 0 \\ xy \end{pmatrix}$	$\begin{pmatrix} 0 \\ y^2 \end{pmatrix}$
$x\alpha$	1	0	0	0	2	0
$y\alpha$	0	1	0	0	0	2
$x\beta$	0	1	0	-1	0	0
$y\beta$	0	0	1	0	-1	0
γ	1	0	1	0	3	0
δ	0	1	0	1	0	-1

The first row of this 6×6 matrix says that $x\alpha = 1\begin{pmatrix} x^2 \\ 0 \end{pmatrix} + 2\begin{pmatrix} 0 \\ xy \end{pmatrix}$. The other rows have similar interpretations. This 6×6 matrix is invertible (as can be checked by doing some row/column operations). It follows that each monomial vector in $\mathfrak{m}_2^2\mathcal{E}_2^2$ can be obtained from linear combinations of the 6 terms $x\alpha$, $y\alpha$, $x\beta$, $y\beta$, γ and δ.

For example,

$$\begin{pmatrix} xy \\ 0 \end{pmatrix} = \tfrac{1}{5}(y\alpha + 2x\beta + 2\delta).$$

Thus $\mathfrak{m}_2^2 \mathcal{E}_2^2 \subset M$, so M is of finite codimension. Now $\mathcal{E}_2^2/\mathfrak{m}_2^2\mathcal{E}_2^2$ is isomorphic (as a vector space) to $\mathcal{E}_2/\mathfrak{m}_2^2 \times \mathcal{E}_2/\mathfrak{m}_2^2$ which has dimension $3+3=6$. Since α, β are non-zero and linearly independent in this space, it follows that $\dim(\mathcal{E}_2^2/M) = 6-2 = 4$. The reader should write down a cobasis for M in $\mathcal{E}_2^2$. ✐

An important module we make much use of is the ***module of germs of vector fields*** on $(\mathbb{R}^n, 0)$, denoted θ_n (introduced in Chapter 5). A representative of an element $\xi \in \theta_n$ is a vector field defined in some neighbourhood U of 0 in $\mathbb{R}^n$. Since such a vector field germ has n components, each of which is a smooth function (germ), it is simply a map (germ) $(\mathbb{R}^n, 0) \to \mathbb{R}^n$, so θ_n can be identified with $\mathcal{E}_n^n = \mathcal{E}_n \times \cdots \times \mathcal{E}_n$. Similarly, for $f \in \mathcal{E}_n^p$ the set $\theta(f) = \mathcal{E}_n^p$ of vector fields along f is an $\mathcal{E}_n$-module. The submodule $\mathfrak{m}_n\theta_n$ consists of all those vector fields that vanish at the origin: $\mathbf{v}(0) = 0$, so $\mathfrak{m}_n\theta_n$ can be identified with $\mathfrak{m}_n \times \cdots \times \mathfrak{m}_n$.

Recall that for a function germ $f \in \mathcal{E}_n$ the Jacobian ideal is the ideal in $\mathcal{E}_n$ generated by its partial derivatives. If instead $f \in \mathcal{E}_n^p$ is a map germ (with $p > 1$), one defines the Jacobian module in the same way, but now the partial derivatives of f are (column) vectors.

Definition 12.4. Let $f\colon (\mathbb{R}^n, 0) \to \mathbb{R}^p$ be a smooth map germ. The ***Jacobian module*** Jf of f is the $\mathcal{E}_n$-submodule of $\theta(f)$ generated by the partial derivatives of f (the columns of the Jacobian matrix). ☆

Example 12.5. (i) Let $f(x, y) = (x^2, xy, y^2) \in \mathcal{E}_2^3$, then

$$\frac{\partial f}{\partial x} = \begin{pmatrix} 2x \\ y \\ 0 \end{pmatrix}, \quad \frac{\partial f}{\partial y} = \begin{pmatrix} 0 \\ x \\ 2y \end{pmatrix},$$

so

$$Jf = \mathcal{E}_2 \left\{ \begin{pmatrix} 2x \\ y \\ 0 \end{pmatrix}, \begin{pmatrix} 0 \\ x \\ 2y \end{pmatrix} \right\}.$$

(ii) Let f be the submersion $f(x_1, \ldots, x_n) = (x_1, \ldots, x_p)$ (as in Theorem B.13). Then

$$Jf = \mathcal{E}_n \left\{ \begin{pmatrix} 1 \\ \vdots \\ 0 \end{pmatrix}, \ldots, \begin{pmatrix} 0 \\ \vdots \\ 1 \end{pmatrix} \right\} = \theta(f).$$

✐

12.2 Right tangent space

Returning to equation (12.1), for $f \in \mathcal{E}_n^p$ we see that the ***right tangent space*** is the set of all possible $\mathrm{t}f(\mathbf{v})$, where $\mathbf{v}$ is any vector field with $\mathbf{v}(0) = 0$, so that each component $v_j \in \mathfrak{m}_n$. That is,

$$T\mathcal{R}{\cdot}f = \mathfrak{m}_n Jf = \mathrm{t}f(\mathfrak{m}_n\theta_n) \lhd \mathfrak{m}_n\theta(f) = \mathfrak{m}_n\mathcal{E}_n^p.$$

This right tangent space was used in the finite determinacy theorem (Section 5.2). On the other hand, in the versal unfolding theorem we do not require that $\phi(0) = 0$ so that each component $v_j \in \mathcal{E}_n$ and we implicitly use the ***extended right tangent space***,

$$T_e\mathcal{R}{\cdot}f = \mathrm{t}f(\theta_n) = Jf.$$

We discussed $\mathcal{R}^+$-equivalence with $p = 1$ in Chapter 4, which allowed equivalences where we add a constant to f. Thus a 1–parameter family of $\mathcal{R}^+$-equivalent germs would be

$$\gamma(s) = f \circ \phi_s + C_s$$

where $C_s \in \mathbb{R}$ is a 'constant' depending smoothly on s, which means $C\colon \mathbb{R} \to \mathbb{R}$, $s \mapsto C_s$ is a smooth function. Differentiating this gives a sum of the part coming from ϕ_s and the part from C_s. Thus

$$T\mathcal{R}^+{\cdot}f = \mathfrak{m}_n Jf + \mathbb{R}, \quad \text{and} \quad T_e\mathcal{R}^+{\cdot}f = Jf + \mathbb{R}.$$

Notice then that, as vector spaces,

$$\mathcal{E}_n/T_e\mathcal{R}^+{\cdot}f \simeq \mathfrak{m}_n/Jf,$$

and it's really the first that defines the codimension of f, although it is customary to use the second:

$$\operatorname{codim}(f, \mathcal{R}^+) := \dim\left(\mathcal{E}_n/T_e\mathcal{R}^+{\cdot}f\right).$$

We now proceed to adapt this approach to determine the tangent spaces coming from several other equivalence relations defined on map germs.

12.3 Left tangent space

Two map germs $f, g\colon (\mathbb{R}^n, 0) \to (\mathbb{R}^p, 0)$ are said to be ***left equivalent*** if there is a diffeomorphism germ $\psi\colon (\mathbb{R}^p, 0) \to (\mathbb{R}^p, 0)$ such that

$$g = \psi \circ f.$$

If ψ_s is a 1–parameter family of such diffeomorphisms, then $\frac{\mathsf{d}}{\mathsf{d}s}\psi_s(y) = \mathbf{w}_s(\psi_s(y))$, where $\mathbf{w}_s$ is a vector field on the target $\mathbb{R}^p$. Moreover,

$$\frac{\mathsf{d}}{\mathsf{d}s}(\psi_s \circ f)(x) = \mathbf{w}_s(\psi_s(f(x))).$$

At $s = 0$, this becomes, assuming $\psi_0 = \mathsf{Id}_p$,

$$\frac{\mathsf{d}}{\mathsf{d}s}(\psi_s \circ f)(x)\big|_{s=0} = \mathbf{w}_0(f(x)).$$

Note that for each $x \in \mathbb{R}^n$, the vector $\mathbf{w}_0 \circ f(x)$ is a vector at $f(x)$ on $\mathbb{R}^p$. The map $x \mapsto \mathbf{w}(f(x))$ therefore defines a ***vector field along*** f. We write the vector field $\mathbf{w} \circ f$ as $(\omega f)(\mathbf{w})$. Thus $\omega f : \theta_p \to \theta(f)$. This is a homomorphism of $\mathcal{E}_p$-modules, where $\theta(f)$ is given the structure of an $\mathcal{E}_p$ module by $(h \cdot \mathbf{v})(x) = h(f(x))\,\mathbf{v}(x)$, for $h \in \mathcal{E}_p$ and $\mathbf{v} \in \theta(f)$ (see also Section D.5A).

As before with right equivalence, we begin by requiring the diffeomorphisms in question to preserve the origin, so that $\mathbf{w}(0) = 0$, and each component $w_i \in \mathfrak{m}_p$. One therefore defines the ***left tangent space*** to be

$$T\mathcal{L}{\cdot}f = \omega f(\mathfrak{m}_p\theta_p),$$

where

$$\omega f(\mathfrak{m}_p\theta_p) = \left\{\mathbf{w} \circ f \mid \mathbf{w} \in \mathfrak{m}_p\theta_p\right\} \subset \mathfrak{m}_n\theta(f).$$

And if we don't require $\psi(0) = 0$, then we obtain the ***extended left tangent space***,

$$T_e\mathcal{L}{\cdot}f = \omega f(\theta_p).$$

It is important to remember that these tangent spaces are $\mathcal{E}_p$-modules, rather than $\mathcal{E}_n$-modules.

If we let L_f be the $\mathcal{E}_p$-module

$$L_f := f^{-1}(\mathcal{E}_p) = \{f \circ h \mid h \in \mathcal{E}_p\},$$

then one observes that $\omega f(\theta_p)$ consists of a copy of L_f in each of the p components.[2] That is,

$$T_e\mathcal{L}{\cdot}f = L_f \times L_f \times \cdots \times L_f \quad (p \text{ copies}).$$

Similarly for $T\mathcal{L}{\cdot}f$: let $L^1_f = f^{-1}(\mathfrak{m}_p)$, then

$$T\mathcal{L}{\cdot}f = L^1_f \times L^1_f \times \cdots \times L^1_f \quad (p \text{ copies}).$$

[2]The maps ωf and f^{-1} are closely related, but ωf applies to vector fields only, and consists of f^{-1} in each component

Analogous to the right codimension discussed in the previous section, one defines the ***left codimension*** of a map germ to be

$$\operatorname{codim}(f, \mathcal{L}) = \dim\left(\theta_n / T_e\mathcal{L}\cdot f\right) = p \dim\left(\mathcal{E}_n / L_f\right).$$

Example 12.6. Let $f\colon (\mathbb{R},0) \to (\mathbb{R}^2,0)$, $f(t) = (t^2, t^3)$, whose image is the semicubical parabola. Write y_1, y_2 for the coordinates on $\mathbb{R}^2$. Clearly, any function of y_1, y_2 lying in $\mathfrak{m}_2$ becomes an element of $\mathfrak{m}_1^2$ under f^{-1}. Conversely any function in $h \in \mathfrak{m}_1^2$ can be written as a function of t^2 and t^3, so as a function of y_1 and y_2 (this is fairly clear for polynomials and analytic functions, but not obvious for general smooth functions; it can be proved using the preparation theorem, see Example 16.3). It follows that

$$T\mathcal{L}\cdot f = \begin{pmatrix} \mathfrak{m}_1^2 \\ 0 \end{pmatrix} \oplus \begin{pmatrix} 0 \\ \mathfrak{m}_1^2 \end{pmatrix} = \mathfrak{m}_1^2\mathcal{E}_1^2.$$

In other words, $L_f^1 = \mathfrak{m}_1^2$. It follows from the theorem below that f is 3-$\mathcal{L}$-determined. ✐

Using the homotopy method analogous to the approach for right-determinacy, one can prove that if $f \in \mathcal{E}_n^p$ satisfies

$$\mathfrak{m}_n^{k+1} \subset f^{-1}\mathcal{E}_p$$

then it is $(2k+1)$-determined w.r.t. left equivalence. See the survey paper of C.T.C. Wall [115] for details.

However, a stronger statement can be derived using different (and simpler) methods as discovered by Bruce, Gaffney and du Plessis [17]. They show that two germs $f, g \in \mathcal{E}_n^p$ are left equivalent if and only if $L_f = L_g$.

Indeed, if f and g are left equivalent then $f = \phi \circ g$. Now if $h \in L_f$ then $h = v \circ f$ for some $v \in \mathcal{E}_p$, and then $h = (v \circ \phi) \circ g \in L_g$, so that $L_f \subset L_g$. A similar argument shows $L_g \subset L_f$.

The converse argument, starting from $L_f = L_g$, is only slightly more involved. This then implies the following greatly improved finite determinacy theorem for left equivalence; see [17] for details.

Theorem 12.7 (Finite determinacy for left equivalence)**.** *The germ $f \in \mathcal{E}_n^p$ is k-determined w.r.t. left equivalence if and only if*

$$\mathfrak{m}_n^{k+1} \subset f^{-1}\mathfrak{m}_p^2.$$

Here $f^{-1}\mathfrak{m}_p^2 = \{h \circ f \mid h \in \mathfrak{m}_p^2\}$.

For example, the plane curve germ $f(t) = (t^2, t^{2r+1})$ satisfies $\mathfrak{m}_1^{2r+2} \subset f^{-1}\mathfrak{m}_2^2$, whence f is $(2r+1)$ left determined (the proof of this requires the Preparation Theorem; see for example Problem 16.3).

Remark 12.8 (Geometric criterion)**.** Analogous to the geometric criterion for finite codimension for $\mathcal{R}$-equivalence (see Chapter 5) there is a criterion for left equivalence, as follows. Let $f\colon (\mathbb{C}^n, 0) \to (\mathbb{C}^p, 0)$ be a holomorphic map germ. Then *f is of finite $\mathcal{L}$-codimension iff there is a neighbourhood U of the origin in $\mathbb{C}^n$ on which a representative is defined and is a 1–1 immersion outside the origin.*

That is, if $0, x, y$ are any 3 distinct points in U then $f(x) \neq f(y)$ and $\mathrm{d}f_x$ is of rank n. It follows from this that for a singular map germ to be left-finite one must have $p \geq 2n$. Details can be found in [115]. ❞

12.4 Left-right tangent space

Combining the definitions of right and left equivalence leads to the equivalence defined below. It is the most natural equivalence respecting properties of maps which are not dependent on particular coordinate systems; however, it has the disadvantage of being technically difficult to deal with. This equivalence relation is not central to this text, but we do discuss its relation with contact equivalence in Chapter 17.

Definition 12.9. Two map germs $f, g\colon (\mathbb{R}^n, 0) \to (\mathbb{R}^p, 0)$ are ***left–right equivalent*** if there are diffeomorphism germs ϕ of $(\mathbb{R}^n, 0)$ and ψ of $(\mathbb{R}^p, 0)$ such that

$$g = \psi \circ f \circ \phi.$$

This is often called ***$\mathcal{A}$-equivalence.*** ✯

An example of a left–right equivalence class is the set of map germs of constant rank, whose normal form is given in Theorem B.26 on p. 350, as this theorem involves changing coordinates in both source and target.

Combining the infinitesimal calculations for right and left equivalence, one defines

$$T\mathcal{A}{\cdot}f = \mathrm{t}f(\mathfrak{m}_n\theta_n) + \omega f(\mathfrak{m}_p\theta_p)$$

and the extended version

$$T_e\mathcal{A}{\cdot}f = \mathrm{t}f(\theta_n) + \omega f(\theta_p).$$

The ***codimension*** is defined to be

$$\operatorname{codim}(f,\mathcal{A}) = \dim\left(\theta(f)/T_e\mathcal{A}{\cdot}f\right).$$

$\mathcal{A}$-equivalence is harder to deal with than either $\mathcal{R}$- or $\mathcal{L}$-equivalence because of the module structure of $T\mathcal{A}{\cdot}f$. The first component $\mathrm{t}f(\theta_n)$ is a finitely generated (f.g.) $\mathcal{E}_n$-module, while θ_p is a f.g. $\mathcal{E}_p$-module, and hence so is its image $\omega f(\theta_p)$ in $\theta(f)$. Thus $T\mathcal{A}{\cdot}f$ is a sum of a f.g. $\mathcal{E}_n$-module and a f.g. $\mathcal{E}_p$-module, and this mixed module structure causes the difficulty and means one cannot use Nakayama's lemma (one needs instead the much deeper 'preparation theorem' of Malgrange, which is discussed Chapter 16). Below we see that the tangent space for contact equivalence effectively extends the $\mathcal{A}$-tangent space into an $\mathcal{E}_n$-module, making it far easier to compute. For the record, we state the optimal finite determinacy theorem for left–right equivalence, which is due to T. Gaffney [40].

Theorem 12.10 (Gaffney). *If $f \in \mathcal{E}_n^p$ is such that*

- $\mathfrak{m}_n^k\theta(f) \subset T_e\mathcal{A}{\cdot}f$, *and*
- $\mathfrak{m}_n^\ell\theta(f) \subset T_e\mathcal{K}{\cdot}f$ *(defined in Def. 12.11 below)*

then f is $(k+\ell)$-determined w.r.t. $\mathcal{A}$-equivalence.

12.5 Contact tangent space

The main aim of this chapter is to extend these ideas to contact equivalence. To this end, suppose there is a 1–parameter family of map germs $f_s\colon (\mathbb{R}^n,0) \to (\mathbb{R}^p,0)$ which are all $\mathcal{K}$-equivalent to $f = f_0$. That is, there are invertible matrices M_s and diffeomorphism germs ϕ_s (all depending smoothly on s) such that

$$f_s(x) = M_s(x)\, f \circ \phi_s(x),$$

where $M_0(x) = \mathsf{Id}_p$ and $\phi_0 = \mathsf{Id}_n$. Differentiating this with respect to s yields (using the product and chain rules), one obtains

$$\frac{\mathsf{d}}{\mathsf{d}s} f_s(x) = M_s(x)\,\mathrm{t}f(\mathbf{v}_s)(\phi_s(x)) + N_s(x) f \circ \phi_s(x),$$

where $\mathbf{v}_s(y) = \frac{\mathsf{d}}{\mathsf{d}s}\phi_s(x)$ as usual, and $N_s(x)$ is the square matrix equal to $\frac{\mathsf{d}}{\mathsf{d}s}M_s(x)$ (which is not necessarily invertible). Putting $s = 0$ gives

$$\frac{\mathsf{d}}{\mathsf{d}s} f_s\big|_{s=0}(x) = \mathrm{t}f(\mathbf{v}_0)(x) + N_0(x) f(x).$$

The first summand $\mathrm{t}f(\mathbf{v}_0)$ we understand from previous chapters, but the second is new and requires some explanation. Let $N(x)$ be the matrix with entries $a_{ij}(x)$ $(1 \leq i, j \leq p)$. Then the ith component of $N(x)f(x)$ is of course

$$[N(x)f(x)]_i = \sum_{j=1}^{p} a_{ij}(x)f_j(x).$$

If we take N to be the 'elementary' matrix with just one 1 in position (i, j) and 0 elsewhere, then

$$Nf(x) = f_j(x)\,\mathbf{e}_i,$$

where $\mathbf{e}_1, \ldots, \mathbf{e}_p$ are the usual basis vectors of $\mathbb{R}^p$. In other words, by using the different elementary matrices, we have p^2 generators, consisting of each of the p components of f in each of the p components of the vector field. For example, if $f\colon \mathbb{R}^3 \to \mathbb{R}^2$ is given by $f(x, y, z) = (xz, xy^2)$ then the four elementary 2×2 matrices give rise to the four elements of $\theta(f)$,

$$\begin{pmatrix} xz \\ 0 \end{pmatrix}, \begin{pmatrix} 0 \\ xz \end{pmatrix}, \begin{pmatrix} xy^2 \\ 0 \end{pmatrix}, \begin{pmatrix} 0 \\ xy^2 \end{pmatrix}.$$

It follows (in general) that

$$\left\{Nf \mid N \in \mathsf{Mat}_p(\mathcal{E}_n)\right\} = I_f\,\theta(f),$$

where $I_f = \langle f_1, \ldots, f_p \rangle \lhd \mathcal{E}_n$ and we write $\mathsf{Mat}_p(\mathcal{E}_n)$ for the set of all $p \times p$ matrices with entries in $\mathcal{E}_n$. This motivates the following definition.

Definition 12.11. The ***tangent space*** for contact equivalence of a smooth map germ $f\colon (\mathbb{R}^n, 0) \to (\mathbb{R}^p, 0)$ is defined to be

$$T\mathcal{K}{\cdot}f = \mathrm{t}f(\mathfrak{m}_n\theta_n) + I_f\,\theta(f).$$

This is an $\mathcal{E}_n$-module, a submodule of $\theta(f) \simeq \mathcal{E}_n^p$.

The ***extended tangent space*** for contact equivalence, which will be important for unfolding theory, is defined to be

$$T_e\mathcal{K}{\cdot}f = \mathrm{t}f(\theta_n) + I_f\theta(f).$$

☆

Notice that only the 'source-terms' $\mathrm{t}f(\theta_n)$ and $\mathrm{t}f(\mathfrak{m}_n\theta_n)$ are different between the two tangent spaces. This is because the target value of 0 is essential for contact equivalence, and one does not allow diffeomorphisms of the target that move the origin.

Example 12.12. Let $f(x,y) = (x^2, y^2)$ (the so-called *folded handkerchief* singularity). Then

$$\mathrm{t}f\begin{pmatrix}u\\v\end{pmatrix} = \begin{pmatrix}2xu\\2yv\end{pmatrix}$$

so that, with $u, v \in \mathfrak{m}_2$

$$\mathrm{t}f(\mathfrak{m}_2\theta(2)) = \mathfrak{m}_2\left\{\begin{pmatrix}2x\\0\end{pmatrix}, \begin{pmatrix}0\\2y\end{pmatrix}\right\} = \left\langle\begin{pmatrix}x^2\\0\end{pmatrix}, \begin{pmatrix}xy\\0\end{pmatrix}, \begin{pmatrix}0\\yx\end{pmatrix}, \begin{pmatrix}0\\y^2\end{pmatrix}\right\rangle.$$

The first two are obtained from $v = 0$ with $u = x$ and y in turn, while the last two are from $u = 0$ with $v = x$ and y in turn. Moreover, $I_f = \langle x^2, y^2\rangle$ so

$$I_f\,\theta(f) = \langle x^2, y^2\rangle\,\theta(f) = \left\langle\begin{pmatrix}x^2\\0\end{pmatrix}, \begin{pmatrix}y^2\\0\end{pmatrix}, \begin{pmatrix}0\\x^2\end{pmatrix}, \begin{pmatrix}0\\y^2\end{pmatrix}\right\rangle \subset \theta(f).$$

Putting these together shows that

$$\begin{aligned}
T\mathcal{K}{\cdot}f &= \mathrm{t}f(\mathfrak{m}_2\,\theta(2)) + I_f\,\theta(f)\\
&= \left\langle\begin{pmatrix}x^2\\0\end{pmatrix}, \begin{pmatrix}xy\\0\end{pmatrix}, \begin{pmatrix}y^2\\0\end{pmatrix}, \begin{pmatrix}0\\x^2\end{pmatrix}, \begin{pmatrix}0\\xy\end{pmatrix}, \begin{pmatrix}0\\y^2\end{pmatrix}\right\rangle\\
&= \mathfrak{m}_2^2\,\theta(f) \subset \theta(f).
\end{aligned}$$

For the extended tangent space,

$$T_e\mathcal{K}{\cdot}f = \langle x, y^2\rangle \times \langle x^2, y\rangle = \langle x, y^2\rangle\begin{pmatrix}1\\0\end{pmatrix} \bigoplus \langle x^2, y\rangle\begin{pmatrix}0\\1\end{pmatrix}.$$

A cobasis for this submodule of $\theta(f)$ is given by

$$\left\{\begin{pmatrix}1\\0\end{pmatrix}, \begin{pmatrix}y\\0\end{pmatrix}, \begin{pmatrix}0\\1\end{pmatrix}, \begin{pmatrix}0\\x\end{pmatrix}\right\}.$$

In a similar manner to the equivalence relations discussed above, one defines the contact-codimension to be

$$\operatorname{codim}(f,\mathcal{K}) := \dim\left({}^{\theta(f)}\big/_{T_e\mathcal{K}\cdot f}\right). \tag{12.3}$$

A map germ f is therefore said to be of ***finite*** $\mathcal{K}$***-codimension*** if $\operatorname{codim}(f, \mathcal{K}) < \infty$. And following the proof of Proposition 3.13 one shows that this is equivalent to the existence of $r > 0$ such that

$$\mathfrak{m}_n^r\theta(f) \subset T\mathcal{K}{\cdot}f.$$

In the folded handkerchief example above, $\operatorname{codim}(f, \mathcal{K}) = 4$. This condition of finite codimension also implies finite determinacy, but we leave precise statements to the next chapter.

The analogue of the result of Problem 4.9 holds for contact equivalence.

Proposition 12.13. *If f, g are contact equivalent then they have the same codimension:* $\operatorname{codim}(f, \mathcal{K}) = \operatorname{codim}(g, \mathcal{K})$.

If one defines a variety X not just as the set $f^{-1}(0)$, but the set together with the ideal I_f, then this proposition shows the $\mathcal{K}$-codimension is a property of the variety (rather than the map); it is called the *Tjurina* number of the variety, denoted $\tau(X)$. Note that if $f\colon (\mathbb{R}^n, 0) \to (\mathbb{R}, 0)$, then (from the definitions), $\operatorname{codim}(f, \mathcal{K}) \leq \operatorname{codim}(f, \mathcal{R})$, or $\tau(X) \leq \mu(X)$ (the Milnor number), and using the Euler vector field, one can show that for weighted homogeneous functions, the two are in fact equal.

A proof of the proposition can be found in [41, p. 154].

12.6 All together

Let $\mathcal{G}$ be any of the singularity equivalences considered thus far. The tangent space for each one is listed in Table 12.1. In each case, one has $T\mathcal{G}\cdot f \subset \mathfrak{m}_n\theta(f)$, while $T_e\mathcal{G}\cdot f \subset \theta(f)$. For each equivalence, there is the corresponding notion of codimension:

$$\operatorname{codim}(f, \mathcal{G}) = \dim \left(\theta(f) \big/ T_e\mathcal{G}\cdot f \right).$$

To show the uniform results that are possible, we quote a result, which can be found together with more details in the survey by Wall [115].

Theorem 12.14. *For a germ $f \in \mathcal{E}_n^p$, the following are equivalent:*

(i). f is finitely determined with respect to $\mathcal{G}$,

(ii). $\operatorname{codim}(f, \mathcal{G}) < \infty$.

(iii). f possesses a $\mathcal{G}$-versal unfolding.

Note that being of finite codimension is equivalent to the existence of an $r \geq 1$ for which $\mathfrak{m}_n^r\theta(f) \subset T\mathcal{G}\cdot f$.

TABLE 12.1 Tangent spaces for the basic singularity equivalences

$\mathcal{G}$	$T\mathcal{G}{\cdot}f$	$T_e\mathcal{G}{\cdot}f$
$\mathcal{R}$	$\mathfrak{m}_n Jf$	Jf
$\mathcal{R}^+$	$\mathfrak{m}_n Jf + \mathbb{R}$	$Jf + \mathbb{R}$
$\mathcal{L}$	$\omega f(\mathfrak{m}_p\theta_p)$	$\omega f(\theta_p)$
$\mathcal{C}$	$I_f\theta(f)$	$I_f\theta(f)$
$\mathcal{A}$	$\mathfrak{m}_n Jf + \omega f(\mathfrak{m}_p\theta_p)$	$Jf + \omega f(\theta_p)$
$\mathcal{K}$	$\mathfrak{m}_n Jf + I_f\theta(f)$	$Jf + I_f\theta(f)$

The definition of finite determinacy with respect to any $\mathcal{G}$ is as one would expect: $f \in \mathcal{E}_n^p$ is k-$\mathcal{G}$-determined if, for any $h \in \mathfrak{m}_n^{k+1}\mathcal{E}_n^p$, the germ $f + h$ is $\mathcal{G}$-equivalent to f. In other words, the $\mathcal{G}$-equivalence class is determined by the k-jet of f. We will be looking at the case of contact equivalence in detail in the next chapter.

We have discussed the most usual equivalence relations used and studied in singularity theory; however, there are many other possibilities. In [27] J. Damon introduces a general notion of *geometric subgroup* of $\mathcal{A}$ and $\mathcal{K}$, and Theorem 12.14 holds for all these. Among these are the unfolding equivalence relations, such as $\mathcal{K}_{\rm un}$, $\mathcal{R}_{\rm un}$ and $\mathcal{A}_{\rm un}$ discussed below. In later chapters, we will be interested in another particular geometric subgroup of $\mathcal{K}$, namely $\mathcal{K}_V$-equivalence, for which Theorem 12.14 also holds. See Chapter 20.

Remark 12.15. In the literature (e.g. [115]), one often finds two notions of codimension: $d_e(f,\mathcal{G}) = \operatorname{codim}(f,\mathcal{G})$ defined above and

$$d(f,\mathcal{G}) = \dim\left(\mathfrak{m}_n\theta(f)\big/T\mathcal{G}{\cdot}f\right).$$

For example, $d(f,\mathcal{R}) = \dim\left(\mathfrak{m}_n/\mathfrak{m}_n Jf\right)$. We make no use of $d(f,\mathcal{G})$ in this book. On the other hand, $\operatorname{codim}(f,\mathcal{G})$ arises naturally in the study of unfoldings, and we make considerable use of that.

The heuristic interpretation of the two notions of codimension is as follows. If we consider for simplicity the vector space $C^\infty(\mathbb{R}^n,\mathbb{R}^p)$, and ask for the set of pairs (f,x) with $x \in \operatorname{dom}(f)$ for which the germ $[f]_x$ of f at x has a particular singularity type T, then $\operatorname{codim}([f]_x,\mathcal{G})$ is the codimension of this set in $\mathbb{R}^n \times C^\infty(\mathbb{R}^n,\mathbb{R}^p)$.

On the other hand, if we fix $x_0 = 0 \in \mathbb{R}^n$ and $y_0 = 0 \in \mathbb{R}^p$ then $d([f]_{x_0}, \mathcal{G})$ is the codimension of the set of all germs in $\mathfrak{m}_n\mathcal{E}_n^p$ of type T. Since germs of finite codimension are finitely determined, this heuristic discussion can be made more precise by using jets and jet spaces. ❞

12.7 Tangent spaces of unfoldings

In Chapter 7 we defined equivalence of unfoldings of function germs, which was called $\mathcal{R}^+_{\mathrm{un}}$-equivalence (Definition 7.6). This definition can be adapted for each of the other equivalence relations $\mathcal{L}$, $\mathcal{A}$, $\mathcal{C}$ and $\mathcal{K}$ described above to obtain $\mathcal{L}_{\mathrm{un}}$, $\mathcal{A}_{\mathrm{un}}$, $\mathcal{C}_{\mathrm{un}}$ and $\mathcal{K}_{\mathrm{un}}$ equivalences. These equivalence relations on families also have tangent spaces, and for $\mathcal{R}^+_{\mathrm{un}}$ this is defined as follows.

It will be useful in what follows, when dealing with families of maps $F(x,u)$ with $x \in \mathbb{R}^n$, $u \in \mathbb{R}^k$, to change notation for the rings of germs, and use the name of the variable in the index; thus, $\mathcal{E}_n$ becomes $\mathcal{E}_x$, and $\mathcal{E}_{n+k}$ becomes $\mathcal{E}_{x,u}$ and θ_k becomes θ_u, and so on.

Recall Definition 7.6: two germs of k–parameter families $F, G \in \mathcal{E}_{x,u}$ are $\mathcal{R}^+_{\mathrm{un}}$-equivalent if there exist, the germ of a diffeomorphism $\Phi\colon (\mathbb{R}^n \times \mathbb{R}^k, (0,0)) \to (\mathbb{R}^n \times \mathbb{R}^k, (0,0))$ of the form $\Phi(x,u) = (\phi(x,u), \psi(u))$ and the germ of a function $C \in \mathcal{E}_u$ such that

$$G(x,u) = F(\phi(x,u),\ \psi(u)) + C(u).$$

To derive the tangent space for this equivalence relation on $\mathcal{E}_{x,u}$, suppose Φ_s and C_s are smooth 1–parameter families of such maps, with $\Phi_s(x,u) = (\phi_s(x,u),\ \psi_s(u))$.

- First, ψ_s is an arbitrary 1–parameter family of diffeomorphisms, and (as earlier in the chapter) $\frac{\mathrm{d}}{\mathrm{d}s}\psi_s(u) = \mathbf{w}_s(\psi_s(u))$ for some family of vector fields $\mathbf{w}_s$ on $\mathbb{R}^k$, whence for each s, $\mathbf{w}_s \in \theta_u$.

- On the other hand, $\phi_s\colon \mathbb{R}^n \times \mathbb{R}^k \to \mathbb{R}^n$, so that $\frac{\mathrm{d}\phi_s}{\mathrm{d}s}$ is a vector field $\mathbf{v}_s$ on $\mathbb{R}^n$ which in general depends on $u \in \mathbb{R}^k$, not just on x. We denote the $\mathcal{E}_{x,u}$ module of such vector fields as $\theta_{x/u}$; thus,

  $$\theta_{x/u} = \mathcal{E}_{x,u}\theta_x.$$

  See Definition D.14 for extending a finitely generated module over one ring to a larger ring.

- Finally differentiating $C_s(u)$ gives rise to an arbitrary element of $\mathcal{E}_u$.

Putting these together, we define

$$T_e\mathcal{R}^+_{\text{un}}\cdot F \;=\; \mathrm{t}_x F(\theta_{x/u}) \;+\; \mathrm{t}_u F(\theta_u) \;+\; \mathcal{E}_u. \tag{12.4}$$

More explicitly,

$$T_e\mathcal{R}^+_{\text{un}}\cdot F \;=\; \mathcal{E}_{x,u}\left\{\frac{\partial F}{\partial x_1},\ldots,\frac{\partial F}{\partial x_n}\right\} \;+\; \mathcal{E}_u\left\{\frac{\partial F}{\partial u_1},\ldots,\frac{\partial F}{\partial u_k}\right\} \;+\; \mathcal{E}_u.$$

There is also a restricted version, where the diffeomorphisms are required to fix the origin in $\mathbb{R}^n\times\mathbb{R}^k$, giving

$$T\mathcal{R}^+_{\text{un}}\cdot F \;=\; \mathrm{t}_x F(\mathfrak{m}_{x,u}\theta_{x/u}) \;+\; \mathrm{t}_u F(\mathfrak{m}_u\theta_u) \;+\; \mathcal{E}_u.$$

In these expressions, t_x refers to the partial derivatives of F with respect to x, while t_u those with respect to u. Thus, for $\mathbf{v}(x,u)\in\theta_{x/u}$ we have

$$\mathrm{t}_x(\mathbf{v}) = \sum_{j=1}^{n} v_j(x,u)\frac{\partial F}{\partial x_j}.$$

In the case of an unfolding of the form $F(x,u) = f(x) + \sum_j u_j h_j(x)$, we have

$$\mathrm{t}_u F(\theta_u) = \mathcal{E}_u\dot{F} = \mathcal{E}_u\{h_1,\ldots,h_k\} \subset \mathcal{E}_{x,u}.$$

Note the complication in $T_e\mathcal{R}^+\cdot F$ (similar to that for the $\mathcal{A}$-tangent space): the first term is an $\mathcal{E}_{x,u}$-submodule of $\theta(F)$, while the other two terms are only $\mathcal{E}_u$-submodules.

A similar analysis for unfoldings up to contact equivalence (discussed in detail in Chapter 14), namely $\mathcal{K}_{\text{un}}$-equivalence, leads to

$$T_e\mathcal{K}_{\text{un}}\cdot F \;=\; \mathrm{t}_x F(\theta_{x/u}) \;+\; I_F\,\theta(F) \;+\; \mathrm{t}_u F(\theta_u), \tag{12.5}$$

where $F(x;u)$ is an unfolding of $f\in\mathcal{E}_x^p$, I_F is the ideal in $\mathcal{E}_{x,u}$ generated by the components of F, and as usual $\theta(F)$ are germs of vector fields along F.

It is left to the reader to find expressions for $T_e\mathcal{L}_{\text{un}}\cdot F$ and $T_e\mathcal{A}_{\text{un}}\cdot F$.

Example 12.16. Consider the unfolding $F(x,u) = x^3 + ux$ of $f(x) = x^3$. Then

$$T_e\mathcal{R}^+_{\text{un}}\cdot F \;=\; \mathcal{E}_{x,u}\left\{3x^2+u\right\} + \mathcal{E}_u\left\{x,1\right\}.$$

On the other hand,

$$\begin{aligned} T_e\mathcal{K}_{\text{un}}\cdot F &= \mathcal{E}_{x,u}\left\{3x^2+u, x^3+ux\right\} + \mathcal{E}_u\left\{x\right\} \\ &= \mathcal{E}_{x,u}\left\{3x^2+u, x^3\right\} + \mathcal{E}_u\left\{x\right\}. \end{aligned}$$

Note that these tangent spaces are modules over $\mathcal{E}_u$, but not over $\mathcal{E}_{x,u}$; however, as $\mathcal{E}_u$-modules they are not finitely generated. It turns out that, in this example, $T\mathcal{R}^+_{\text{un}} \cdot F$ is equal to $\mathcal{E}_{x,u}$ (the proof involves the preparation theorem – see Problem 16.6). On the other hand $T\mathcal{K}_{\text{un}} \cdot F$ is of infinite codimension in $\mathcal{E}_{x,u}$: it contains no function of the form $h(u)$. ✐

12.8 Thom–Levine principle

One advantage of this general point of view, with $\mathcal{G}$ representing any of the singularity theoretic equivalence relations we have met so far, is that one can state a uniform Thom–Levine principle for $\mathcal{G}$.

Recall (Definition 5.3) that a smooth family f_s of germs is $\mathcal{R}$-trivial if there is a smooth family of diffeomorphisms ϕ_s such that $f_s \circ \phi_s = f_0$. Likewise, for any of the equivalence relations $\mathcal{G}$ on the space of germs, we say a family f_s of germs is $\mathcal{G}$-trivial if there are smooth family of appropriate diffeomoprhisms, or matrices in the case that $\mathcal{G} = \mathcal{K}$ or $\mathcal{C}$, which transform f_s into f_0.

Theorem 12.17 (A general Thom–Levine theorem). *Let $\mathcal{G}$ be any of the equivalence relations $\mathcal{R}$, $\mathcal{L}$, $\mathcal{A}$, $\mathcal{C}$ or $\mathcal{K}$, and let f_s be a smooth family of germs in $\mathcal{E}_n^p$. Then f_s is a $\mathcal{G}$-trivial family if and only if $\dot{f}_s \in T\mathcal{G} \cdot f_s$, smoothly in s.*

This theorem is a key part of the homotopy method described in Chapter 5, but applied to all the equivalence relations.

Proof: Each equivalence requires its own argument. In each case, the tangent space $T\mathcal{G} \cdot f$ was defined by deriving the infinitesimal condition for the family being trivial, which shows that one direction ('only if') of the theorem follows from the definitions of $T\mathcal{G} \cdot f$. However, the converse does need proving. Note that the statements of the theorem are for germs of maps, but existence theorems for differential equations require representative neighbourhoods; we take the passages from germs to neighbourhoods and back to germs for granted with no further mention.

We have already proved this theorem for $\mathcal{G} = \mathcal{R}$ (and $\mathcal{R}_1$) in Chapter 5. The argument extends easily to $\mathcal{R}^+$. We now consider the other equivalence relations in turn.

Consider first $\mathcal{G} = \mathcal{L}$, and let f_s be a family for which for which $\dot{f}_s(x) \in T\mathcal{L} \cdot f_s$, smoothly in s; that is, there is a smooth vector field $\mathbf{w}_s(y)$ on $\mathbb{R}^p$ satisfying

$\mathbf{w}_s(0) = 0$ and

$$\dot{f}_s(x) = -\mathbf{w}_s(f_s(x)). \tag{12.6}$$

Since $\mathbf{w}_s(0) = 0$ there is an integral flow ψ_s defined on a neighbourhood of the origin in the target, for which $\frac{\mathrm{d}}{\mathrm{d}s}(\psi_s)(y) = \mathbf{w}_s(\psi_s(y))$. Then using (12.6) one finds $\frac{\mathrm{d}}{\mathrm{d}s}(\psi_s \circ f_s)(x) = 0$, showing that indeed $\psi_s \circ f_s$ is constant and the family is left trivial.

Consider now the case $\mathcal{G} = \mathcal{C}$. Let f_s be a smooth family of germs in $\mathcal{E}_n^p$, such that $\dot{f}_s \in T\mathcal{C} \cdot f_s = I_{f_s}\theta(f_s)$, smoothly in s. That is, there are smooth functions $g_{ij}(x, s)$ such that

$$(\dot{f}_s)_i(x) = -\sum_j g_{ij}(x, s)(f_s)_j(x),$$

or as vectors and matrices, $\dot{f}_s(x) = G_s(x)f_s(x)$. We wish to show there is a family of matrices $M_s(x) \in \mathsf{GL}_p(\mathcal{E}_{n,I})$ (germs at $x = 0$) satisfying $f_s(x) = M_s(x)f_0(x)$ with $M_0(x) = \mathsf{Id}_p$. For this, let $M_s(x)$ be the unique solution of the initial value problem

$$\frac{\mathrm{d}}{\mathrm{d}s}M_s = M_s G_s, \quad M_0 = \mathsf{Id}_p,$$

whose existence is guaranteed by Theorem C.9 in Appendix C (where t plays the role of s and L that of G). Then

$$\frac{\mathrm{d}}{\mathrm{d}s}(M_s f_s) = (\dot{M}_s)f_s + M_s\dot{f}_s = M_s\left(G_s f_s + \dot{f}_s\right) = 0.$$

Since $M_0 = \mathsf{Id}_p$ if follows that $M_s f_s = f_0$ as required.

The cases $\mathcal{G} = \mathcal{A}$ and $\mathcal{G} = \mathcal{K}$ are combinations of $\mathcal{R}$ and either $\mathcal{L}$ or $\mathcal{C}$ respectively, and the details are left to the reader. ✔

Remark 12.18. Above we defined $\mathcal{R}_{\mathrm{un}}$ and $\mathcal{K}_{\mathrm{un}}$ tangent spaces, and there are similar $\mathcal{L}_{\mathrm{un}}$ and $\mathcal{A}_{\mathrm{un}}$ equivalence relations and tangent spaces, and the Thom–Levine principle holds for these too. In other chapters we discuss refinements of the equivalence relations $\mathcal{G}$ discussed above, in particular where the linear part at the origin is the identity. We have already met $\mathcal{R}_1$-equivalence in Chapter 5, and in Chapter 13 we define $\mathcal{K}_1$-equivalence, and one can define $\mathcal{A}_1$-equivalence and $\mathcal{L}_1$-equivalence similarly. A Thom–Levine theorem for these equivalence relations can be proved in the same way. ”

12.9 Constant tangent spaces

In this final section of this chapter, we restrict attention to those equivalence relations where the tangent spaces are finitely generated $\mathcal{E}_n$-modules, namely to $\mathcal{G} = \mathcal{R}, \mathcal{C}$ and $\mathcal{K}$. Except for a brief use in the proof of the finite determinacy theorem for contact equivalence, we will not make use of this material until Part III of the book.

There is one very useful case where the 'smoothness in s' assumption in the Thom–Levine theorem above is automatically satisfied, and that is where the tangent space to f_s is constant (i.e., independent of s), or indeed where it contains a 'constant' submodule. We make this precise by using the idea of relative tangent space as follows. First recall from Chapter 5 that we defined $\mathcal{E}_{n,I}$ to be the germs along $\{0\} \times [0, 1]$ of smooth functions in $x \in \mathbb{R}^n$ and $s \in [0, 1]$; we now wish to allow greater flexibility in s, so let $I \subset \mathbb{R}$ be any interval, and denote by $\mathcal{E}_{n,I}$ the germs along $\{0\} \times I$ of smooth functions in (x, s). Similarly, $\mathfrak{m}_{n,I}$ consists of those germs vanishing identically on $\{0\} \times I$, and $\theta_{n,I} = \mathcal{E}_{n,I}\theta_n$. Thus (see Definition D.14) any $\mathbf{v} \in \theta_{n,I}$ can be written as derivations as

$$\mathbf{v}(x, s) = \sum_{j=1}^{n} a_j(x, s)\frac{\partial}{\partial x_j},$$

(with no $\frac{\partial}{\partial s}$ component), where the a_j are smooth functions defined in a neighbourhood of $\{0\} \times I$ in $\mathbb{R}^n \times I$.

Given a family of germs $f_s \in \mathcal{E}^p_{n,I}$, we define the ***relative tangent space*** for $\mathcal{G}$-equivalence to be the $\mathcal{E}_{n,I}$-module,

$$T_{\mathrm{rel}}\mathcal{G} \cdot f_s = \mathcal{E}_{n,I}\, T\mathcal{G}{\cdot}f_s, \tag{12.7}$$

(for $\mathcal{G} = \mathcal{R}, \mathcal{C}$ or $\mathcal{K}$).

For example, let $f_s(x, y) = x^2 + y^3 + s(x^3 + y^2) \in \mathcal{E}_{2,I}$ and I any interval. Then

$$T_{\mathrm{rel}}\mathcal{R}{\cdot}f_s = \left\langle x(2 + 3sx),\ 2sy + 3y^2\right\rangle = \langle x,\ y(2s + 3y)\rangle \lhd \mathcal{E}_{2,I}.$$

The second equality follows because $(2 + 3sx)$ is a unit in $\mathcal{E}_{2,I}$ for any interval I (even on $I = \mathbb{R}$). On the other hand, if $0 \in I$ then $(2s + 3y)$ is not a unit.

Given $T_{\mathrm{rel}}\mathcal{G} \cdot f_s$ one can reclaim the $\mathcal{E}_n$-module $T\mathcal{G}{\cdot}f_{s_0}$ (for fixed $s_0 \in I$) by applying the 'evaluation homomorphism',

$$\begin{aligned} \mathrm{ev}\colon \mathcal{E}_{n,I} &\to \mathcal{E}_n, \\ h &\mapsto h|_{s\,=\,s_0}. \end{aligned} \tag{12.8}$$

That is, $\mathrm{ev}(h)(x) = h(x, s_0)$. This is clearly a homomorphism of $\mathcal{E}_n$-modules, and extends to $\mathcal{E}_{n,I}$-modules, and moreover

$$\mathrm{ev}(T_{\mathrm{rel}}\,\mathcal{G}\cdot f_s) = T\mathcal{G}\cdot f_{s_0}.$$

Definition 12.19. Consider the equivalence relations $\mathcal{G} = \mathcal{R}, \mathcal{C}, \mathcal{K}$. Let $f_s \in \mathcal{E}^p_{n,I}$ be any smooth family of germs. We say the ***tangent space is constant*** if there are vector fields $v_1, \dots, v_r \in \theta(f) = \mathcal{E}^p_n$ (for some $r > 0$), such that

$$T_{\mathrm{rel}}\,\mathcal{G}\cdot f_s = \mathcal{E}_{n,I}\,\{v_1, v_2, \dots, v_r\}. \tag{12.9}$$

We emphasize that the v_j must be independent of s. ✯

Continuing the example of $T_{\mathrm{rel}}\,\mathcal{R}\cdot f_s$ above, suppose $0 \notin I$. Then we can write $v_1 = x$ and $v_2 = y$, and because $(2s + 3y)$ is a unit in $\mathcal{E}_{n,I}$, we have

$$T_{\mathrm{rel}}\,\mathcal{R}\cdot f_s = \mathcal{E}_{n,I}\,\{v_1, v_2\},$$

showing it is constant. On the other hand, if $0 \in I$ then $T_{\mathrm{rel}}\,\mathcal{R}\cdot f_s$ is not constant.

Lemma 12.20. *If $T_{\mathrm{rel}}\,\mathcal{G}\cdot f_s$ is constant (for $s \in I$), then for any fixed $s_0 \in I$ the v_j in (12.9) are generators of the $\mathcal{E}_n$-module $T\mathcal{G}\cdot f_{s_0}$.*

Proof: This follows from the evaluation homomorphism introduced above. Indeed, if

$$T_{\mathrm{rel}}\,\mathcal{G}\cdot f_s = \mathcal{E}_{n,I}\,\{v_1, v_2, \dots, v_r\},$$

then applying ev to each side gives

$$T\mathcal{G}\cdot f_{s_0} = \mathcal{E}_n\,\{v_1, v_2, \dots, v_r\}.$$

Thus $v_1, \dots, v_r$ are generators of $T\mathcal{G}\cdot f_{s_0}$ for any $s_0 \in I$. ✔

We continue to suppose $\mathcal{G} = \mathcal{R}, \mathcal{C}$ or $\mathcal{K}$, and for simplicity, we now assume $0 \in I$. The following final result of this chapter is useful for classifying singularities, as we will see particularly in Chapter 21.

Theorem 12.21. *Suppose $f_s = f + sh$ is a family of germs, with s in an interval I containing 0, and let M be a finitely generated $\mathcal{E}_n$-submodule of $\theta(f)$. Suppose*

(i). $h \in M$, and

(ii). $\mathcal{E}_{n,I}M \lhd T_{\text{rel}}\mathcal{G}\cdot f_s$.

Then the family f_s is $\mathcal{G}$-trivial for $s \in I$.

PROOF: This is a consequence of the Thom–Levine principle (Theorem 12.17). With $f_s = f + sh$ we have $\dot{f}_s = h$, and hence the hypotheses imply $\dot{f}_s \in T_{\text{rel}}\mathcal{G}\cdot f_s$. If we can establish that this is 'smooth in s', then the result follows from the Thom–Levine theorem.

But this is simple: since $h \in M$ we can write

$$h(x) = \sum_j \beta_j(x)v_j(x),$$

with $\beta_j \in \mathcal{E}_n$ and where $v_1, \ldots, v_r$ are generators of M. This expression for $h = \dot{f}_s$ is independent of s, and hence trivially smooth! ✔

Thus far, we have only defined relative tangent spaces for f_s with s in an interval in $\mathbb{R}$ (i.e. for 1–parameter families). However, the definition is valid for any finite number of parameters. For example, consider the family $f_{a,b}(x,y) = ax^2 + by^2$ (a, b are parameters). Then $T_{\text{rel}}\mathcal{R}\cdot f = \langle 2ax,\ 2by\rangle$. If we put $U = \{(a,b) \in \mathbb{R}^2 \mid ab \neq 0\}$ then, over U, $T_{\text{rel}}\mathcal{R}\cdot f_u = \mathfrak{m}_{2,U}$ where $u = (a,b) \in U$.

Corollary 12.22. *Let $f\colon (\mathbb{R}^n, 0) \to (\mathbb{R}^p, 0)$ be a smooth map germ and M be a finitely generated submodule of $\theta(f)$. Suppose U is a connected open subset of a finite–dimensional linear subspace of M. For $u \in U$ write $f_u = f + u$ and suppose that, over U,*

$$\mathcal{E}_{n,U}M \subset T_{\text{rel}}\mathcal{G}\cdot f_u.$$

Then f_u is $\mathcal{G}$-equivalent to f_v for all $u, v \in U$.

PROOF: Fix $u, v \in U$. Since U is open and connected, there is a path P in U leading from u to v. For each point $p \in P$ consider the open ball, centre p radius ε_p (for some $\varepsilon_p > 0$ for which this ball is contained in U). These balls form an open cover of P. Since P is compact, we can extract a finite subcover, $U_1, \ldots, U_m$, where U_1 is the ball centre u and U_m the ball centre v, and suppose they are numbered so that $U_i \cap U_{i+1} \neq \emptyset$. Let $u_0 = u$, $u_m = v$, and for $i = 1, \ldots, m-1$ let $u_i \in U_i \cap U_{i+1}$. We will show that each f_{u_i} is $\mathcal{G}$-equivalent to $f_{u_{i+1}}$, from which the result follows.

Now, open balls in a finite–dimensional vector space are convex, and hence each segment $(1-s)u_i + su_{i+1}$ (for $s \in [0,1]$) is contained in U_i, and hence in U.

Let $h_i = u_{i+1} - u_i$, and put $f_s^i = f + u_i + sh_i$ for $s \in [0,1]$. Then $u_{i+1} = f_1^i = f_0^{i+1}$ and $\dot{f}_s^i = h_i \in M$. By the theorem above, each family f_s^i is therefore $\mathcal{G}$-trivial and hence $u_{i+1} = f_1^i$ is $\mathcal{G}$-equivalent to $u_i = f_0^i$, as required ✔

In Chapter 20 we meet another equivalence relation called $\mathcal{K}_V$-equivalence, whose tangent space is also an $\mathcal{E}_n$-module, and these results continue to hold for $\mathcal{G} = \mathcal{K}_V$. See Theorem 20.4 for the statement.

Remarks 12.23. (i). These results on constant tangent spaces do not require f to be of finite codimension.

(ii). The constant tangent space approach described here has in fact already been used for showing that $x^3 + xy^3$ is 4-determined (for $\mathcal{R}$-equivalence) in Problem 5.8.

”

Problems

12.1 Let $\mathcal{R}_1$ be the equivalence relation introduced in Section 5.4. Show that for $f \in \mathcal{E}_n^p$, the natural definition for the tangent space is

$$T\mathcal{R}_1 \cdot f = \mathrm{t}f(\mathfrak{m}_n^2\theta(n)) = \mathfrak{m}_n^2\, Jf.$$

12.2 Use Hadamard's lemma to show that the $\mathcal{E}_n$-module $\mathfrak{m}_n\theta_n$ has n^2 generators.

12.3 Let $f\colon (\mathbb{R}^n,0) \to (\mathbb{R}^p,0)$. Show $T_e\mathcal{R}{\cdot}f = \theta(f)$ if and only if f is a submersion. [*Hint*: *use linearly adapted coordinates*: *see Section B.8.*]

12.4 Show $T_e\mathcal{L}{\cdot}f = \theta(f)$ if and only if f is an immersion.

12.5 Show that in the above two problems, the map is 1-determined (for right equivalence in the case of a submersion and for left equivalence in the case of an immersion).

12.6 Let $f \in \mathcal{E}_n^p$ and suppose $\mathbf{v} \in \mathfrak{m}_n\theta(n)$ satisfies $\mathrm{t}f(\mathbf{v}) = 0$. Let ϕ_s be the integral flow of $\mathbf{v}$ (in particular, $\phi_0 = \mathsf{Id}$). Show that $f \circ \phi_s = f$ (for all $s \in \mathbb{R}$). Why is it not enough to assume $\mathbf{v} \in \theta(n)$? (†)

12.7 Let $f\colon (\mathbb{R}^2,0) \to (\mathbb{R}^3,0)$ be the germ $f(x,y) = (x^2, xy, y^2)$. Show $\operatorname{codim}(f, \mathcal{K}) = 7$.

12.8 Justify the expression for $T_e\mathcal{K}_{\mathrm{un}}{\cdot}f$ in (12.5).

12.9 Recall the notion of $\mathcal{R}^e$-equivalence for even functions from Chapter 9.

(i). Determine expressions for the tangent spaces $T\mathcal{R}^e\cdot f$ and $T_e\mathcal{R}^e\cdot f$. [You should find the two tangent spaces to be the same for this equivalence.]

(ii). Write down a statement and proof for the Thom–Levine theorem for $\mathcal{R}^e$-equivalence for germs of even functions.

12.10 Use the results of Section 12.9 to show that $f_s = xy + s(x^2 - y^2)$ is $\mathcal{R}$-trivial (for $s \in \mathbb{R}$). (†)

12.11 Let $f_s(x, y) = x^4 + 2sx^2y^2 + y^4$ be a family of germs in $\mathcal{E}_2$. Let I be any interval not containing $s = \pm 1$, and show that over I, $T_{\text{rel}}\mathcal{R}\cdot f_s$ contains $\mathcal{E}_{2,I}M$ where $M = \mathfrak{m}_2^6$.

12.12 Under the hypotheses of Theorem 5.10, show that $T_{\text{rel}}\mathcal{R}\cdot f_s$ is constant.

12.13 Find expressions for $T_e\mathcal{L}_{\text{un}} \cdot F$ and $T_e\mathcal{A}_{\text{un}} \cdot F$, adapting the expressions in (12.4) and (12.5). (†)

12.14 Analogous to the $\mathcal{R}_1$-equivalence of Section 5.4, one can refine contact equivalence: two maps f, g are ***$\mathcal{K}_1$-equivalent*** if they are $\mathcal{K}$-equivalent as in Equation (11.1), but with $\mathsf{d}\Psi(0,0) = \mathsf{Id}_{n+p}$.

(i). Show that this is equivalent to modifying Definition 11.1, by requiring ϕ and M to satisfy,
- $\mathsf{d}\phi_0 = \mathsf{Id}_n$
- $M(0) = \mathsf{Id}_p$,

where Id_n is the $n \times n$ identity matrix, and Id_p similarly.

(ii). Deduce that

$$T\mathcal{K}_1 \cdot f = \mathsf{t}f(\mathfrak{m}_n^2\theta_n) + \mathfrak{m}_n I_f \theta(f) = \mathfrak{m}_n T\mathcal{K} \cdot f.$$

(iii). Show that the Thom–Levine principle (Theorem 12.17), and the results of Section 12.9 hold for $\mathcal{K}_1$-equivalence.

13

Classification for contact equivalence

IN THIS CHAPTER, WE BEGIN with the finite determinacy theorem for contact equivalence, and then proceed by giving a classification of some germs of low codimension. For the classification, we consider the most important case of equidimensional maps $f\colon (\mathbb{R}^n,0) \to (\mathbb{R}^n,0)$. The approach to finite determinacy is, *mutatis mutandis*, identical to the approach taken in Chapter 5 for $\mathcal{R}$-equivalence.

13.1 Finite determinacy

Analogous to the definition in Chapter 5, one says a smooth family of map germs $f_s\colon (\mathbb{R}^n,0) \to (\mathbb{R}^p,0)$ (with $s \in S$, a neighbourhood of 0 in $\mathbb{R}$) is a $\mathcal{K}$-trivial family if there is a smooth family of diffeomorphisms ϕ_s fixing 0 and a smooth family of matrices $M_s \in \mathsf{GL}_p(\mathcal{E}_n)$, such that

$$f_s \circ \phi_s = M_s f_0.$$

One can assume, without loss of generality, that ϕ_0 is the identity on $\mathbb{R}^n$ and $M_0 = \mathsf{Id}_p$ (the $p \times p$ identity matrix).

Discussions in the previous chapter motived the following definition for contact tangent spaces:

Definition 13.1. Let $f\colon (\mathbb{R}^n,0) \to (\mathbb{R}^p,0)$ be a smooth germ. The tangent space and extended tangent spaces for contact equivalence are

$$T\mathcal{K}{\cdot}f = \mathsf{t}f(\mathfrak{m}_n\theta_n) + I_f\,\theta(f)$$

and

$$T_e\mathcal{K}{\cdot}f = \mathsf{t}f(\theta_n) + I_f\,\theta(f),$$

respectively. ☆

Recall from Chapter 5 that a germ $f \in \mathcal{E}_n$ is k-determined with respect to $\mathcal{R}$-equivalence if $\mathfrak{m}_n^{k+1} \subset \mathfrak{m}_n^2 Jf$. In terms of tangent spaces, $\mathfrak{m}_n Jf = T\mathcal{R} \cdot f$ so that f is k-determined if $\mathfrak{m}_n^{k+1} \subset \mathfrak{m}_n T\mathcal{R} \cdot f$. The finite determinacy theorem for contact equivalence is very similar, as follows.

Theorem 13.2. *Let $f\colon (\mathbb{R}^n, 0) \to (\mathbb{R}^p, 0)$ be a smooth map germ. If*

$$\mathfrak{m}_n^{k+1}\,\theta(f) \subset \mathfrak{m}_n T\mathcal{K}\cdot f$$

then f is k-determined with respect to $\mathcal{K}$-equivalence.

PROOF: In essence this uses the homotopy method, via the Thom–Levine principle and is very similar to the proof of finite determinacy for right equivalence. We give a slightly different presentation of the proof based on the 'constant tangent space' results of Section 12.9.

Recall the notation introduced before the proof of Lemma 5.19: $\mathcal{E}_{n,I}$ is the ring of germs along $\{0\} \times I \subset \mathbb{R}^n \times I$, where I is the interval $[0, 1]$, of smooth functions (representatives are smooth functions defined for all $s \in [0, 1]$ and for x in a neighbourhood of the origin in $\mathbb{R}^n$). The ideal $\mathfrak{m}_{n,I}$ consists of those germs vanishing identically on the s-axis, and let $\theta_{n,I} = \mathcal{E}_{n,I}\theta_n$.

Let $h \in \mathfrak{m}_n^{k+1}\mathcal{E}_n^p$ and put $f_s = f + sh$. Define the *relative* $\mathcal{K}$-tangent space for $s \in I$ to be

$$T_{\mathrm{rel}}\mathcal{K}\cdot f_s = \mathrm{t}f_s(\mathfrak{m}_{n,I}\theta_n) + I_{f_s}\theta(f),$$

where I_{f_s} is now the ideal of f_s in $\mathcal{E}_{n,I}$. (Note that the modules $\mathcal{E}_{n,I}\theta(f_s)$ and $\mathcal{E}_{n,I}\theta(f)$ can be identified, and identified with $\mathcal{E}^p_{n,I}$.) In the case that f is independent of s, this definition says that $T_{\mathrm{rel}}\mathcal{K}\cdot f = \mathcal{E}_{n,I}T\mathcal{K}\cdot f$. The proof relies on the following lemma.

Lemma 13.3. *Suppose f is such that $\mathfrak{m}_n^k\theta(f) \subset T\mathcal{K}\cdot f$ and $h \in \mathfrak{m}_n^{k+1}\mathcal{E}_n^p$. Then, the $\mathcal{E}_{n,I}$-module $T_{\mathrm{rel}}\mathcal{K}\cdot f_s$ is constant (for $s \in I = \mathbb{R}$).*

PROOF: The proof is similar to the proof of Lemma 5.19 for $\mathcal{R}$-finite determinacy. Since $\mathfrak{m}_n T_{\mathrm{rel}}\mathcal{K}\cdot f$ is by definition constant, we need only show that $\mathfrak{m}_{n,I}T_{\mathrm{rel}}\mathcal{K}\cdot f_s = \mathfrak{m}_{n,I}T_{\mathrm{rel}}\mathcal{K}\cdot f$.

We have, firstly,

$$\begin{aligned}
\mathfrak{m}_{n,I}T_{\mathrm{rel}}\mathcal{K}\cdot f_s &= \mathfrak{m}_{n,I}T_{\mathrm{rel}}\mathcal{K}\cdot(f + sh)\\
&\subset \mathfrak{m}_{n,I}T_{\mathrm{rel}}\mathcal{K}\cdot f + s\mathfrak{m}_{n,I}T_{\mathrm{rel}}\mathcal{K}\cdot h\\
&\subset \mathfrak{m}_{n,I}T_{\mathrm{rel}}\mathcal{K}\cdot f + s\mathfrak{m}_{n,I}^{k+1}\mathcal{E}_n^p\\
&= \mathfrak{m}_{n,I}T_{\mathrm{rel}}\mathcal{K}\cdot f.
\end{aligned}$$

Note that since $h \in \mathfrak{m}_n^{k+1}\mathcal{E}_n^p$ it follows from the definition that $T\mathcal{K}{\cdot}h \subset \mathfrak{m}_n^k\mathcal{E}_n^p$. Conversely, writing $f = f_s - sh$,

$$\begin{aligned} \mathfrak{m}_{n,I}T_{\mathrm{rel}}\mathcal{K}{\cdot}f &= \mathfrak{m}_{n,I}T_{\mathrm{rel}}\mathcal{K}{\cdot}(f_s - sh) \\ &\subset \mathfrak{m}_{n,I}T_{\mathrm{rel}}\mathcal{K}{\cdot}f_s + s\mathfrak{m}_{n,I}T_{\mathrm{rel}}\mathcal{K}{\cdot}h \\ &\subset \mathfrak{m}_{n,I}T_{\mathrm{rel}}\mathcal{K}{\cdot}f_s + s\mathfrak{m}_{n,I}^{k+1}\mathcal{E}_n^p \\ &\subset \mathfrak{m}_{n,I}T_{\mathrm{rel}}\mathcal{K}{\cdot}f_s + \mathfrak{m}_{n,I}\,T_{\mathrm{rel}}\mathcal{K}{\cdot}f. \end{aligned}$$

Hence, by Nakayama's lemma, $\mathfrak{m}_{n,I}T_{\mathrm{rel}}\mathcal{K}{\cdot}f \subset \mathfrak{m}_{n,I}T_{\mathrm{rel}}\mathcal{K}{\cdot}f_s$, and consequently the two modules coincide. ✔

The theorem now follows: the lemma above shows that the $\mathfrak{m}_{n,I}T_{\mathrm{rel}}\mathcal{K}{\cdot}f_s$ is constant, and contains h, which is the hypothesis of Theorem 12.21. Hence $f \sim_{\mathcal{K}} f + h$ as required. ✔

Example 13.4. Continuing Example 12.12 (the folded handkerchief) from the last chapter, we saw that the map germ $f(x, y) = (x^2, y^2)$ satisfies $T\mathcal{K}{\cdot}f = \mathfrak{m}_2^2\theta(f)$. Consequently, f is 2-determined. ✏

Remark 13.5. There is a refinement of contact equivalence, namely $\mathcal{K}_1$-equivalence, where the linear part of the diffeomorphisms at the origin are equal to the identity (cf. $\mathcal{R}_1$-equivalence in Section 5.4), and the tangent space for this equivalence relation is

$$T\mathcal{K}_1{\cdot}f = \mathfrak{m}_n\,T\mathcal{K}{\cdot}f.$$

See Problem 12.14 for details.

The hypothesis of the theorem above implies

$$\mathfrak{m}_n^{k+1}\,\theta(f) \;\subset\; T\mathcal{K}_1{\cdot}f.$$

This condition is equivalent to f being k-determined w.r.t. $\mathcal{K}_1$-equivalence. One implication involves the same argument as above, while the other goes beyond this text (cf. Theorem 5.17 and the discussion following). ❞

Definition 13.6. A smooth map germ $f\colon (\mathbb{R}^n, 0) \to (\mathbb{R}^p, 0)$ which is finitely determined with respect to contact equivalence is said to be of ***finite singularity type***. ✩

Finite singularity type is often abbreviated FST. Mather showed that, in a sense we won't make precise, almost all map germs are FST (see [115, Theorem 5.1]).

13.2 Classification

A map germ $f\colon (\mathbb{R}^n,0) \to (\mathbb{R}^n,0)$ is said to be of[1] ***corank*** k if

$$\operatorname{rk} \mathsf{d}f_0 = n - k,$$

or equivalently if the kernel of $\mathsf{d}f_0$ has dimension k. If f is of corank k then we can choose linearly adapted coordinates (see Appendix B), for which f takes the form

$$f(x,y) = (g(x,y), y), \tag{13.1}$$

with $x \in \mathbb{R}^k$ and $y \in \mathbb{R}^{n-k}$, and $\mathsf{d}g(0) = 0$. More explicitly,

$$f(x_1,\dots,x_k,y_1,\dots,y_{n-k}) = (g_1(x,y),\ g_2(x,y),\dots,g_k(x,y),y_1\dots,y_{n-k}).$$

The first step in the classification for contact equivalence is as follows.

Lemma 13.7. *The map given in (13.1) is contact equivalent to the map*

$$(x,y) \longmapsto (g(x,0),y).$$

PROOF: Now for each i, $g_i(x,y) - g_i(x,0)$ is zero when $y = 0$, so by Hadamard's lemma is an element of $\langle y_1,\dots,y_{n-k}\rangle$. We can therefore write

$$g_i(x,y) - g_i(x,0) = \sum_{j=1}^{n-k} y_j h_{ij}(x,y).$$

Write $H = (h_{ij}(x,y))$ (a $k \times (n-k)$ matrix), and then we have

$$\begin{pmatrix} g(x,y) \\ y \end{pmatrix} = \begin{pmatrix} \mathsf{Id}_k & H \\ 0 & \mathsf{Id}_{n-k} \end{pmatrix} \begin{pmatrix} g(x,0) \\ y \end{pmatrix}.$$

The matrix is upper triangular with 1s down the diagonal so is invertible, showing the two maps are indeed contact equivalent (and in fact $\phi = \mathsf{Id}$ so they are $\mathcal{C}$-equivalent). ✔

Thus if we are classifying maps up to contact equivalence, we can assume our map of corank k is of the form given in the lemma. The map $x \mapsto g(x,0)$ is called the ***core*** of the map f in (13.1).

[1] If $g \in \mathcal{E}_n$ then its corank as defined on p. 81 coincides with the corank of $f = \nabla g$ as defined here.

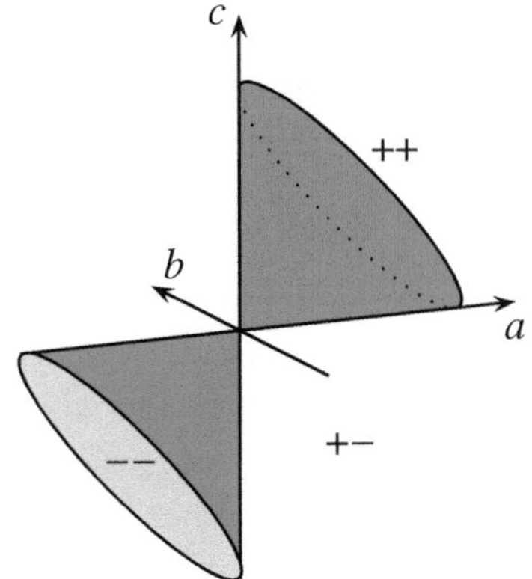

FIGURE 13.1 Sym(2) and the cone of degenerate quadratic forms.

13.2A Corank 1

By the lemma above, a map germ $f\colon (\mathbb{R}^n,0) \to (\mathbb{R}^n,0)$ of corank 1 is contact equivalent to one of the form

$$f(x,y) = (g(x),y)$$

where $x \in \mathbb{R}$ and $y \in \mathbb{R}^{n-1}$. There remains therefore to classify $g \in \mathcal{E}_1$.

Theorem 13.8. *If $f\colon (\mathbb{R}^n,0) \to (\mathbb{R}^n,0)$ is of corank 1 and $\mathcal{K}$-codimension k, then it is $\mathcal{K}$-equivalent to the map germ*

$$f(x,\mathbf{y}) = (x^{k+1},\mathbf{y}),$$

where $x \in \mathbb{R}$ and $\mathbf{y} \in \mathbb{R}^{n-1}$.

This singularity is called an A_k singularity – compare with Chapter 6.

The proof of this theorem is very similar to that of the analogous result for right equivalence (Theorem 6.2) and is left as an exercise.

13.2B Corank 2

In the last of his foundational series of six papers, Mather [76, VI] gives a classification up to contact equivalence of germs of corank 2. We construct the first ones.

Since $f\colon (\mathbb{R}^n,0) \to (\mathbb{R}^n,0)$ is of corank 2, it is $\mathcal{K}$-equivalent to a map of the form

$$f(x,y,z_1,\ldots,z_{n-2}) = (g_1(x,y),g_2(x,y),z_1,\ldots,z_{n-2}),$$

Pencils of Binary Quadratic Forms Consider the space of all binary quadratic forms $ax^2 + 2bxy + cy^2$. Since this form can be identified with the symmetric matrix $S = \begin{pmatrix} a & b \\ b & c \end{pmatrix}$, we denote the space of binary quadratic forms by $\mathsf{Sym}(2)$. It is a 3–dimensional vector space with coordinates a, b, c (the basis vectors are the quadratic forms x^2, $2xy$ and y^2).

It is useful to divide the space $\mathsf{Sym}(2)$ into six parts according to the eigenvalues of the symmetric matrix, or equivalently in terms of its trace and determinant. Let S be as above. Then S is either positive definite (eg $x^2 + y^2$), negative definite (eg $-x^2 - y^2$), indefinite (eg $x^2 - y^2$ or xy), positive semidefinite (eg x^2), negative semidefinite (eg $-x^2$), both of which are degenerate, or finally zero. These six types are best illustrated by their position in the space $\mathsf{Sym}(2)$ in relation to the cone $\det S = 0$, see Figure 13.1 ($\det S = 0$ if and only if the quadratic form is degenerate or zero).

Now suppose we are given two binary quadratic forms $q_1(x, y)$ and $q_2(x, y)$ in $\mathsf{Sym}(2)$, then they either span a plane (if they are linearly independent as vectors in $\mathsf{Sym}(2)$) or a line or they are both zero. In the first case, the plane they span is called a ***pencil of quadratic forms***. See Figure 13.2, illustrating two such pencils.

There are three types of pencil, according to how it meets the cone:

- ***Elliptic pencil***: in this case the pencil meets the cone at the origin only;
- ***Hyperbolic pencil***: in this case the pencil intersects the cone in two distinct lines;
- ***Parabolic pencil***: in this case the pencil is tangent to the cone, and the intersection is just one line.

Note that every non-zero quadratic form in an elliptic pencil is indefinite, while the hyperbolic pencils are the only ones containing definite quadratic forms (both positive and negative definite).

If we are given a pair of linearly independent quadratic forms S_1, S_2, it is useful to determine whether the corresponding pencil P is elliptic, hyperbolic or parabolic. The algebraic method is as follows: the plane P consists of the symmetric matrices $S_\lambda := \lambda_1 S_1 + \lambda_2 S_2$. Now form the quadratic form $Q(\lambda_1, \lambda_2) = \det S_\lambda$. If this form is indefinite, then P meets the cone in 2 lines (the two roots of Q) and P is hyperbolic, while if it is definite then P does not meet the cone (except at the origin) and it is elliptic. If Q has a double root, then P is parabolic. See Problem 13.11.

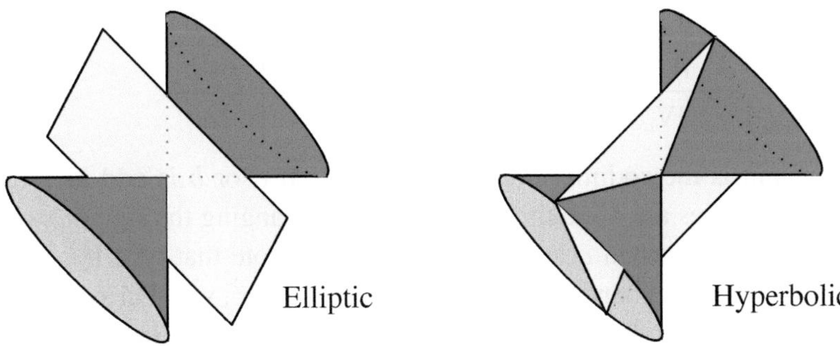

FIGURE 13.2 An elliptic and a hyperbolic pencil of quadratic forms

where $g_1, g_2 \in \mathfrak{m}_2^2$. We therefore need to classify map germs $g\colon (\mathbb{R}^2, 0) \to (\mathbb{R}^2, 0)$ with $g(0) = 0$, $\mathsf{d}g(0) = 0$. The 2-jet of g is given by a pair of quadratic forms:

$$q_1 = a_1x^2 + 2b_1xy + c_1y^2, \quad q_2 = a_2x^2 + 2b_2xy + c_2y^2.$$

That is $\mathrm{j}^2 g = (q_1, q_2)$. If q_1, q_2 are linearly independent (which will 'generically' be the case) then they span a pencil of quadratic forms; see the box on p. 168. Moreover, if g is replaced by a map germ $\mathcal{C}$-equivalent to it, then the 2-jet of the new map will span the same pencil. In this way the $\mathcal{C}$-class of the map is determined by (and determines) the pencil of quadratic forms. There remains to consider the effect of changes of coordinates. The proof of the following proposition is the subject of Problem 13.9; the nomenclature is explained in the box on p. 168.

Proposition 13.9 (Normal forms for pencils of quadratic forms). *Let P be a pencil of quadratic forms. If it is elliptic, coordinates in the x-y-plane can be chosen so that P is spanned by $\{x^2 - y^2, xy\}$. If it is hyperbolic, coordinates can be chosen so that P is spanned by $\{x^2 + y^2, xy\}$. Finally if it is parabolic, coordinates can be chosen so that P is spanned by $\{x^2, xy\}$.*

Lemma 13.10. *The germs $g_\pm\colon (x, y) \mapsto (x^2 \pm y^2, xy)$ are 2-determined.*

To prove this one shows $T\mathcal{K}{\cdot}g_\pm = \mathfrak{m}_2^2\,\theta(g_\pm)$ – see Example 13.4 for a similar case. Thus, the list of contact classes of corank-2 maps begins with the two maps in the lemma; which are denoted $\mathrm{I}_{2,2}$ (for g_+) and $\mathrm{II}_{2,2}$ (for g_-). A fuller list was provided by Mather [76, VI]. The simplest cases are,

$\mathrm{I}_{a,b}$	$(xy,\, x^a + y^b)$	$a \geq b \geq 2$	$a+b$
$\mathrm{II}_{a,b}$	$(xy,\, x^a - y^b)$	$a \geq b \geq 2$	$a+b$
IV_a	$(x^2 + y^2,\, x^a)$	$a \geq 3$	$2a$

(13.2)

The last column is the codimension $\operatorname{codim}(f, \mathcal{K})$. If a or b is odd then the $\mathrm{I}_{a,b}$ and $\mathrm{II}_{a,b}$ singularities are $\mathcal{K}$-equivalent (as seen by changing the sign of x or y), so the $\mathrm{II}_{a,b}$ is only used when a and b are both even. Note that type $\mathrm{I}_{2,2}$ is contact equivalent to the 'folded handkerchief' germ $(x, y) \mapsto (x^2, y^2)$. Further discussion of this classification can be found in [41, Chapter V] as well as in [94, Chapter 8], where they also consider the cases $\mathbb{R}^n \to \mathbb{R}^p$ with $p > n$.

The corank-2 singularity of least codimension not appearing on this list is

$$f(x, y) = (x^2,\, y^3),$$

which is of codimension 7. A more complete list can be found in [6, §9.8].

Boardman symbol Singular map germs can be partially classified by their ***Boardman symbol***, which is a symbol of the form $\Sigma^{i_1, i_2, \dots}$, where $i_1 \geq i_2 \geq \dots$. If $f \in \mathcal{E}_n^p$ then the indices $i_1, i_2, \dots$ are defined recursively by:

- $i_1 = \dim \ker \mathrm{d}f$,
- $i_2 = \dim \ker \mathrm{d}(f|_{S_{i_1}})$, where S_{i_1} is the locus of points where f is of type Σ^{i_1},
- $i_3 = \dim \ker \mathrm{d}(f|_{S_{i_1, i_2}})$, where S_{i_1, i_2} is the locus of points where f is of type Σ^{i_1, i_2}, and so on, and
- $i_k = \dim \ker (f|_{S_I})$, where $I = (i_1, \dots, i_{k-1})$ and S_I is the locus of points where f has type Σ^I.

This definition assumes each S_I is a submanifold, but there is also an algebraic approach to defining the Boardman symbol which is more general; see for example [6].

A singularity of type A_k has Boardman symbol $\Sigma^{1,1,\dots,1,0}$, with k 1s. On the other hand the Σ^i for $i \geq 2$ is more complex, and the classification in the table above is the complete list of germs of type $\Sigma^{2,0}$. Mather [76, VI] also gives a partial classification of germs of type $\Sigma^{2,1}$ for map germs $f \in \mathcal{E}_n^p$ with $p \geq n$, the first of which is the map germ $(x, y) \mapsto (x^2, y^3)$ mentioned above. See [94] for futher details.

13.3 Geometric criterion

Recall from Section 5.7 that there is a geometric criterion for ensuring a given analytic function germ is of finite $\mathcal{R}$-codimension. Here we describe the analogous criterion for finite $\mathcal{K}$-determinacy (finite singularity type) for an analytic map germ.

Let $f\colon (\mathbb{R}^n, 0) \to (\mathbb{R}^p, 0)$ be an analytic map germ (e.g. a polynomial map). The complexification of f is obtained by replacing the real variables $x_1, \ldots, x_n$ in the infinite Taylor series of f with complex variables $z_1, \ldots, z_n$. This defines a complex analytic map germ $f^{\mathbb{C}}\colon (\mathbb{C}^n, 0) \to (\mathbb{C}^p, 0)$.

Theorem 13.11. *A complex analytic map germ is FST if and only if it satisfies*

$$S(f) \cap f^{-1}(0) = \{0\},$$

where $S(f)$ is the germ at 0 of the set of points in $\mathbb{C}^n$ where the rank of f is less than p. A real analytic germ is FST iff its complexification is.

In particular, if $n < p$ then every point is singular and f is FST if and only if $f^{-1}(0) = \{0\}$.

Like the corresponding criterion for $\mathcal{R}$-finite determinacy (Theorem 5.20), this is proved using the Nullstellensatz. One can use these geometric criteria to prove that for $p = 1$ the function-germ f is of finite $\mathcal{R}$-codimension if and only if it is FST. For details, see [115].

While we are on the subject of complex analytic germs, there is an important geometric property of the local algebra $Q(f)$ which is only valid in the complex case. Recall (Definition 11.9) that the algebraic multiplicity of a germ f is $m_A(f) = \dim Q_f$, where now this is a complex vector space and dim is the complex dimension.

Theorem 13.12. *Let $f\colon \mathbb{C}^n \rightarrowtail \mathbb{C}^n$ be an analytic map of FST, defined on a neighbourhood U of 0. Then there is a neighbourhood V of 0 in $\mathbb{C}^n$ such that for almost every $v \in V$, $f^{-1}(v)$ has cardinality $m_A(f)$.*

That is, in the holomorphic setting, geometric multiplicity is equal to algebraic multiplicity. The proof uses techniques from commutative algebra that go beyond this text: see for example [34, Chapter 3], [53, Chapter 1] and [90], or from topology [78]. For a real germ $f\colon (\mathbb{R}^n, 0) \to (\mathbb{R}^n, 0)$ this theorem implies that the geometric multiplicity cannot be greater than the algebraic multiplicity, as stated in Chapter 11.

Corollary 13.13. *Let $h\colon (\mathbb{C}^n, 0) \to \mathbb{C}$ be the germ of a holomorphic function of finite $\mathcal{R}$-codimension. Then there is a deformation $H\colon \mathbb{C}^n \times \mathbb{C} \rightarrowtail \mathbb{C}$ of h such that for arbitrarily small $\varepsilon \in \mathbb{C}$ the function h_ε has μ critical points, where $\mu = \dim_{\mathbb{C}} \mathcal{O}_n / Jh$.*

Here $\mathcal{O}_n$ is the ring of germs at 0 of holomorphic functions. The finite determinacy and unfolding theorems for holomorphic germs have identical expressions to those for smooth germs (and the proofs are the same, using everywhere holomorphic vector fields and holomorphic diffeomorphisms).

PROOF: Consider the analytic map germ

$$f(z_1, \dots, z_n) = \nabla h = \left(\frac{\partial h}{\partial z_1}, \dots, \frac{\partial h}{\partial z_n}\right).$$

Since h is of finite codimension, by the geometric criterion for $\mathcal{R}$-equivalence (Theorem 5.20), the origin is an isolated critical point of h, and hence $f^{-1}(0) = \{0\}$ which implies f is of finite singularity type. Now let U, V be as in the theorem (after taking a representative of the germ f), and consider the unfolding

$$\begin{aligned} H\colon \mathbb{C}^n \times \mathbb{C}^n &\longrightarrow \mathbb{C} \\ (z, v) &\longmapsto h(z) - \sum_j v_j z_j. \end{aligned}$$

Then $\nabla h_v(z) = f(z) - v$. Now let v be such that $f^{-1}(v)$ has precisely $m_A(f)$ points in U. It follows that for such v, h_v has $m_A(f)$ critical points in U. Finally, observe that the MIlnor number $\mu(h) = m_A(f)$. ✔

Problems

13.1 Show that the germ $g \in \mathcal{E}_2^2$ given by $g(x, y) = (x^2 + y^2, xy)$ is 2-determined with respect to $\mathcal{K}$-equivalence. (†)

13.2 Show that the germ $g \in \mathcal{E}_n^n$ given by $g(x, y) = (x^r, y)$ is r-determined with respect to $\mathcal{K}$-equivalence, where $x \in \mathbb{R}$ and $y = (y_1, \dots, y_{n-1}) \in \mathbb{R}^{n-1}$.

13.3 Find the $\mathcal{K}$-codimension of each of the map germs g in the previous two problems. (†)

13.4 Suppose $f \in \mathcal{E}_n$ is weighted homogeneous. Show that $T\mathcal{R}\cdot f = T\mathcal{K}\cdot f$. [*Hint: see Problem* 4.11.]

13.5 Let $f_1, f_2 \in \mathfrak{m}_2$ be non-singular (so not in $\mathfrak{m}_2^2$). Show that the germ $f: (\mathbb{R}^2, 0) \to (\mathbb{R}^2, 0)$ with $f = (f_1, f_2)$ is of type A_1 if and only if the smooth curves $f_1^{-1}(0)$ and $f_2^{-1}(0)$ are transverse.

13.6 Find the $\mathcal{K}$-codimension and degree of determinacy of the germ at 0 of $g(x, y) = (x^3, y^3)$. (†)

13.7 Show that the map germ $(x, y) \mapsto (x^a, y^b)$ has $\mathcal{K}$-codimension equal to $2ab - (a + b)$.

13.8 Provide the details proving Theorem 13.8. (†)

13.9 Prove Proposition 13.9 as follows. (1) Suppose P is hyperbolic, and let $q_1, q_2 \in P$ be independent forms in the cone, then $q_1 = (ax + by)^2$ and $q_1 = (cx + dy)^2$ for some a, b, c, d. Now change basis replacing $ax + by$ by x, and $cx + dy$ by y (this is invertible as (a, b) and (c, d) are linearly independent by the independence of q_1 and q_2). Thus, any hyperbolic pencil can be generated by the forms $q_1 = x^2$ and $q_2 = y^2$. By replacing these by their sum and difference, we can obtain forms $q_1 = x^2 + y^2$ and $q_2 = x^2 - y^2$. Finally by rotating by $\pi/4$ in the plane, q_1 remains the same while q_2 changes to $2xy$. (2) Suppose now P is elliptic. Firstly let q_1 be any indefinite form, which therefore factors as $q_1 = (ax + by)(cx + dy)$; again change basis replacing $ax + by$ by x and $cx + dy$ by y, so making $q_1 = xy$. By subtracting multiples of q_1 from q_2 and rescaling the x, y coordinates one arrives at the given normal form. (3) The parabolic case is left to the reader.

13.10 (i) Show that $f(x, y) = (x^2, y^2)$ is of type $\mathrm{I}_{2,2}$.
(ii) Show that $\mathrm{I}_{a,b}$ and $\mathrm{II}_{a,b}$ are $\mathcal{K}$-equivalent if a or b is odd.

13.11 Let $S_1 = \begin{pmatrix} a_1 & b_1 \\ b_1 & c_1 \end{pmatrix}$ and $S_2 = \begin{pmatrix} a_2 & b_2 \\ b_2 & c_2 \end{pmatrix}$, and consider the pencil of quadratic forms they determine. Let

$$D = (a_1c_2 - a_2c_1)^2 + 4(b_1c_2 - b_2c_1)(b_1a_2 - b_2a_1).$$

Show that the pencil is elliptic if $D < 0$, hyperbolic if $D > 0$ and parabolic if $D = 0$. (The expression D is called the ***discriminant*** of the pencil.)

13.12 Let C be a binary cubic form (as in the box on p. 85), and define two binary quadrics $q_1 = \partial C/\partial x$ and $q_2 = \partial C/\partial y$. Show that the pencil defined by q_1, q_2 is of the same type (elliptic/hyperbolic/parabolic) as C.

13.13 Use the geometric criterion to show that the algebraic map $f(x, y, z) = (xy, yz, zx)$ (real or complex) is not of finite singularity type.

13.14 Show that the simplest map germ of corank n, namely $f \in \mathcal{E}_n^n$ given by

$$f(x_1, \ldots, x_n) = (x_1^2, \ldots, x_n^2),$$

has $\mathcal{K}$-codimension equal to $n2^{n-1}$.

13.15 Mather's classification in [76, VI] of germs with Boardman symbol $\Sigma^{2,0}$ includes the following two families for $p = n + 1$,

$\mathrm{III}_{a,b}$	(x^a, y^b, xy)	$a \geq b \geq 2$
V_a	$(x^2 + y^2, x^a, x^{a-1}y)$	$a \geq 3$

Find the $\mathcal{K}$-codimension of each of these.

14

Contact equivalence and unfoldings

Families of maps and equations arise in many settings, in particular in bifurcation theory, and here we study such families using contact equivalence. This chapter is mathematically very close to Chapter 7 on unfoldings of (critical points of) functions, allowing for a more streamlined treatment.

Before proceeding, recall the definitions of the zero-set Z_G, the singular set Σ_G and the discriminant Δ_G of a family of equations $G(x,u) = 0$, given in Chapter 10.

14.1 Equivalence of unfoldings and versality

Here we adapt the definitions from Chapter 7, passing from right equivalence to contact equivalence. The study of unfoldings or deformations of maps up to contact equivalence was first introduced by Jean Martinet [73], who called it *V-equivalence*.

Let $f\colon (\mathbb{R}^n,0) \to (\mathbb{R}^p,0)$ be a given map germ and let

$$F\colon (\mathbb{R}^n \times \mathbb{R}^a,\ (0,0)) \to (\mathbb{R}^p,0)$$

be a given unfolding of f with base space $\mathbb{R}^a$, meaning that $F(x,0) = f(x)$. Let $\phi\colon (\mathbb{R}^b,0) \to (\mathbb{R}^a,0)$ be a map germ. As for functions, one defines the ***unfolding of f induced from F by ϕ***, written ϕ^*F, to be the family,

$$(\phi^* F)\,(x;w) = F(x;\phi(w)).$$

Two families $F,\ G\colon (\mathbb{R}^n \times \mathbb{R}^a,\ (0,0)) \to (\mathbb{R}^p,\ 0)$ are ***equivalent*** (or more precisely $\mathcal{K}_{\rm un}$-equivalent, or in words, *unfolding-contact equivalent*) if there is a diffeomorphism-germ $\Phi\colon (\mathbb{R}^n \times \mathbb{R}^a, (0,0)) \to (\mathbb{R}^n \times \mathbb{R}^a, (0,0))$ of the form

$$\Phi(x,u) = (\phi(x,u), \psi(u)),$$

and invertible matrices $M(x,u)$ depending smoothly on (x,u), such that,

$$F(x,u) = M(x,u)\, G(\phi(x,u), \psi(u)).$$

In particular, this means that, for each $u \in \mathbb{R}^a$, $(\psi^* F)_u$ and G_u are $\mathcal{K}$-equivalent. Note that, because ψ and Φ are diffeomorphisms and $M(x,u)$ is invertible, this equivalence is indeed an equivalence relation.

Finally, an unfolding $F\colon (\mathbb{R}^n \times \mathbb{R}^a,\ (0,0)) \to (\mathbb{R}^p, 0)$ of f is ***$\mathcal{K}$-versal*** if given any other unfolding $G\colon (\mathbb{R}^n \times \mathbb{R}^b,\ (0,0)) \to (\mathbb{R}^p, 0)$ there is a map germ $\psi\colon (\mathbb{R}^b, 0) \to (\mathbb{R}^a,\ 0)$ such that the unfoldings $\psi^* F$ and G are $\mathcal{K}_{\rm un}$-equivalent.

The versal unfolding theorem is almost identical to the versal unfolding theorem for right equivalence (Theorem 7.8, p. 97). Recall that

$$T_e\mathcal{K}{\cdot}f = {\rm t}f(\theta_n) + I_f\,\theta(f),$$

and recall the notation for the initial speeds of an unfolding:

$$\dot{F}_j = \frac{\partial F}{\partial u_j}(x,0), \qquad \dot{F} := \mathbb{R}\left\{\dot{F}_1, \dots, \dot{F}_a\right\} \subset \theta(f).$$

Theorem 14.1 (Versality theorem for $\mathcal{K}$-equivalence). *Let $f \in \mathcal{E}_n^p$ be of finite $\mathcal{K}$-codimension. An unfolding $F\colon (\mathbb{R}^n \times \mathbb{R}^a,\ (0,0)) \to (\mathbb{R}^p, 0)$ of f is versal if and only if*

$$T_e\mathcal{K}{\cdot}f + \dot{F} = \theta(f).$$

In particular, if $h_1, \dots, h_a$ form a cobasis for $T_e\mathcal{K}{\cdot}f$ in $\theta(f)$, then

$$F(x;u) = f(x) + \sum_{j=1}^{a} u_j h_j(x)$$

is a miniversal unfolding.

This theorem is due to Martinet; see [75], where the proof may be found. The proof relies on the preparation theorem, and is very similar to the proof of the versality theorem for right equivalence given in Chapter 16.

Definition 14.2. An unfolding $F\colon (\mathbb{R}^n \times \mathbb{R}^a,\ (0,0)) \to (\mathbb{R}^p, 0)$ is said to be ***regular*** if, as a map (germ), F is a submersion. ✮

It follows from the regular value theorem (Theorem B.23) that for a regular unfolding, the zero-set Z_F is a submanifold of $\mathbb{R}^n \times \mathbb{R}^a$ of dimension $n - p + a$.

Proposition 14.3. *Every versal unfolding is regular.*

Proof: Suppose F is a versal unfolding of f. We can use linearly adapted coordinates to write f as $f(x,y) = (f_1(x), y)$, with $\mathsf{d}(f_1)_0 = 0$ and $x \in \mathbb{R}^k$. Then F can be taken to be of the form $F(x,y,u) = (F_1(x,u), y)$. Now, since $\mathsf{d}(f_1)_0 = 0$ it follows that $T\mathcal{K}{\cdot}f_1 \in \mathfrak{m}_k\mathcal{E}_k^p$. A simple choice of cobasis of $T\mathcal{K}{\cdot}f_1$ in $\mathcal{E}_k^p$ therefore includes the constant basis vectors $e_1, \dots, e_k$. Writing out the resulting form of F shows it to be a submersion. ✔

The interpretation of the discriminant Δ_F depends on the relative dimensions n and p, as the following example illustrates.

Example 14.4. We are usually interested in equidimensional maps (i.e., $n = p$) but here we consider a case with $n < p$. Consider the map $f\colon \mathbb{R} \to \mathbb{R}^2$, with $f(t) = (t^3, 0)$. Now $\theta(f) \simeq \mathcal{E}_1^2 = \left\{ \begin{pmatrix} h_1(t) \\ h_2(t) \end{pmatrix} \mid h_1, h_2 \in \mathcal{E}_1 \right\}$, and

$$T_e\mathcal{K}{\cdot}f = \left\langle \begin{pmatrix} t^2 \\ 0 \end{pmatrix} \right\rangle + \langle t^3 \rangle\, \theta(f) = \left\langle \begin{pmatrix} t^2 \\ 0 \end{pmatrix}, \begin{pmatrix} 0 \\ t^3 \end{pmatrix} \right\rangle .$$

As a cobasis for $T_e\mathcal{K}{\cdot}f$ one can take

$$\left\{ \begin{pmatrix} 1 \\ 0 \end{pmatrix}, \begin{pmatrix} t \\ 0 \end{pmatrix}, \begin{pmatrix} 0 \\ 1 \end{pmatrix}, \begin{pmatrix} 0 \\ t \end{pmatrix}, \begin{pmatrix} 0 \\ t^2 \end{pmatrix} \right\},$$

and thus $\operatorname{codim}(f, \mathcal{K}) = 5$, and

$$F(t;\, a,\, b,\, c,\, d,\, e) = \begin{pmatrix} t^3 + at + b \\ ct^2 + dt + e \end{pmatrix}$$

is a versal unfolding of f. Then

$$Z_F = \left\{ (t,a,b,c,d,e) \in \mathbb{R} \times \mathbb{R}^5 \mid b = -(t^3 + at),\ e = -(ct^2 + dt) \right\}.$$

This is a submanifold of $\mathbb{R}^6$ of dimension 4, and may be parametrized by (t,a,c,d). Since here $p = 2 > n = 1$, every point of Z_F is also a point of Σ_F; so $\Sigma_F = Z_F$. The projection Δ_F of Σ_F to $\mathbb{R}^5$ is then also of dimension 4, and one finds that $u \in \Delta_F$ if and only if $\exists t \in \mathbb{R}$ such that $F(t,u) = 0$. Thus, in this case, Δ_F discriminates between there being or not being a solution to the equation $f_u(t) = 0$.

If instead $p > n+1$ then the image of Z_F would be of codimension greater than 1, so would no longer be a hypersurface. ✎

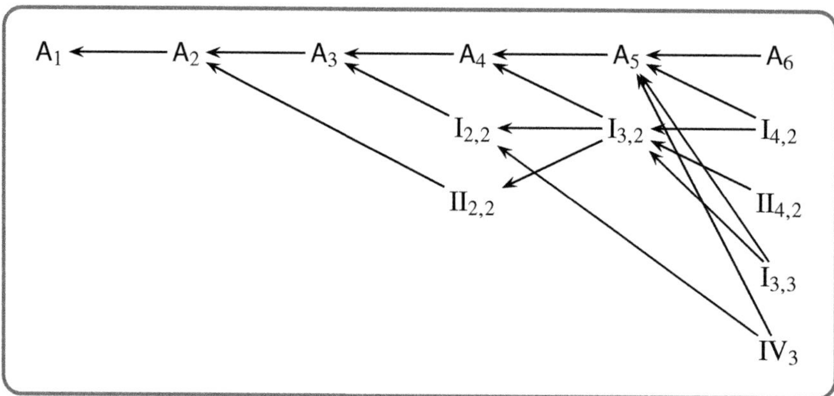

FIGURE 14.1 Adjacencies between the different $\mathcal{K}$-classes: an arrow from A to B means that B occurs in the versal unfolding of A; see p. 170 for the notation. The columns consist of singularities of the same $\mathcal{K}$-codimension (recall $\operatorname{codim}(\mathsf{A}_k, \mathcal{K}) = k$). These and similar adjacencies can be found in [68]; see also Problems 14.13–14.18.

Example 14.5. Consider the cusp family

$$F(x; u, v) = x^3 + ux + v$$

as an unfolding of $f(x) = x^3$. Now, $T_e\mathcal{K}{\cdot}f = \langle x^2 \rangle$, and the unfolding terms $\{x, 1\}$ form a cobasis of this ideal, which shows that F is a $\mathcal{K}$-versal unfolding of f. This is of course the same as (the gradient of) the cusp catastrophe family discussed in Section 2.4. For example, the discriminant is the cusp curve (semicubical parabola) given by the equation $4u^3 + 27v^2 = 0$. ✐

Adjacencies One says a singularity type A is ***adjacent to*** a singularity type B if type B appears in the versal unfolding of type A. In Figure 14.1 we show the adjacencies between the different $\mathcal{K}$-classes up to codimension 6.

14.2 Versal unfoldings of corank 1 maps

Consider the A_k singularity in one variable $f(x) = x^{k+1}$. Then $T_e\mathcal{K}{\cdot}f = \langle x^k \rangle$, whence $\operatorname{codim}(f, \mathcal{K}) = k$, and a miniversal unfolding of this A_k singularity is

$$F(x; u_1, \dots, u_k) = x^{k+1} + u_1 x^{k-1} + \cdots + u_{k-1}x + u_k.$$

More generally, in n variables the A_k singularity is $f(x,y) = (x^{k+1}, y)$ (with $y \in \mathbb{R}^{n-1}$), see Section 13.2A. The versal unfolding is then

$$F(x,y;u_1,\dots,u_k) = \left(x^{k+1} + u_1 x^{k-1} + \cdots + u_{k-1}x + u_k,\ y\right).$$

The geometry of these is identical to the geometry of the $\mathcal{R}^+$-versal unfolding of the A_{k+1} singularity described in Chapter 7. For example, the swallowtail singularity arises from the function $f(x) = x^5$ under $\mathcal{R}^+$-equivalence, and from the equation $x^4 = 0$ under $\mathcal{K}$-equivalence. These are called ***generalized swallowtail*** singularities.

14.3 Versal unfoldings of corank 2 maps

Recall the table on p. 170 for the beginning of the classification of corank 2 map germs. Here we just consider one example, the singularity $\mathrm{I}_{2,2}$, given by

$$f(x,y) = (xy, x^2 + y^2).$$

Then

$$T_e\mathcal{K}\cdot f = \left\langle \begin{pmatrix} y \\ 2x \end{pmatrix}, \begin{pmatrix} x \\ 2y \end{pmatrix}, \begin{pmatrix} xy \\ 0 \end{pmatrix}, \begin{pmatrix} 0 \\ xy \end{pmatrix}, \begin{pmatrix} x^2+y^2 \\ 0 \end{pmatrix}, \begin{pmatrix} 0 \\ x^2+y^2 \end{pmatrix} \right\rangle.$$

The Newton diagram for this is

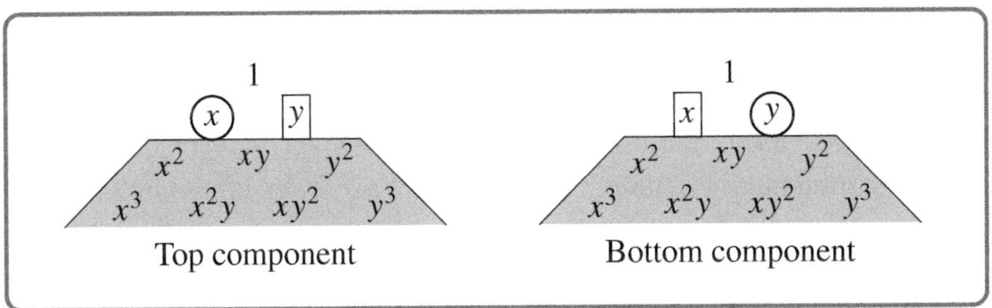

The shaded regions account for the last four (quadratic) generators in the expression above for $T_e\mathcal{K}\cdot f$, but to get a cobasis one need also to take into account the first two, illustrated by the circles and boxes in the diagram. It follows that the singularity has codimension 4, and as a miniversal unfolding we can take

$$F(x;\ a,b,c,d) = (xy + a,\ x^2 + y^2 + bx + cy + d). \tag{14.1}$$

See Problems 14.5 and 14.6 for more such examples.

Problems

14.1 Let $f \in \mathcal{E}_2^3$ be given by $f(x, y) = (x^2, xy, y^2)$. Show that $\operatorname{codim}(f, \mathcal{K}) = 7$ and write down a versal unfolding F of f. Determine Z_F and the projection $\pi_F : Z_F \to \mathbb{R}^7$. Interpret what Δ_F means in this case. [*This is type* $\mathrm{III}_{2,2}$ *in the notation of [76, VI].*] (†)

14.2 Show that $\operatorname{codim}(f, \mathcal{K}) = 0$ if and only if $f : (\mathbb{R}^n, 0) \to (\mathbb{R}^p, 0)$ is the germ of a submersion.

14.3 Find a $\mathcal{K}$-versal unfolding of the germ $f : \mathbb{R} \to \mathbb{R}$, $f(x) = x^4$. Relate this to the swallowtail catastrophe of Chapter 7.

14.4 Show that equivalence of unfoldings defines an equivalence relation.

14.5 Find miniversal unfoldings of each of the corank 2 singularities given on p. 170.

14.6 For the unfolding given in Equation (14.1), show Z_F is a submanifold and find Σ_F. Show that, in the parametrization of Z_F given by (x, y, b, c), for each fixed value of b, c, Σ_F is in general a hyperbola through the origin (in (x, y) coordinates). Sketch this, being careful to distinguish the cases where $b = \pm c$. (†)

14.7 Repeat the previous question for the $\mathrm{II}_{2,2}$ singularity, $f(x, y) = (xy, x^2 - y^2)$. In particular there is a difference in Σ_F – the curves are no longer hyperbolae – what are they?

14.8 Find the $\mathcal{K}$-versal unfolding F of the map germ $f(t) = (t^2, 0)$. Show that the discriminant (here the image of Z_F) is the Cayley cross-cap (see Figure B.6 in Appendix B).

14.9 Let $f \in \mathcal{E}_n^p$ be a map germ of finite singularity type, and $F(x; u)$ be an a–parameter unfolding of f. Now let $k > 0$, and let g be the k–dimensional suspension of f; that is, $g \in \mathcal{E}_{n+k}^{p+k}$ is given by $g(x, y) = (f(x), y)$. Similarly, let

$$G(x, y; u) = (F(x; u), y).$$

Show that F is a versal unfolding of f if and only if G is a versal unfolding of g. In particular $\operatorname{codim}(f, \mathcal{K}) = \operatorname{codim}(g, \mathcal{K})$.

14.10 Prove Proposition 14.3. [*Hint: see the proof of Proposition* 7.14.]

14.11 Suppose $F\colon \mathbb{R}^n \times \mathbb{R}^a \to \mathbb{R}^n$ is a family of map germs (with $n = p$), and let $(x,u) \in Z_F$. Show that if $(x,u) \notin \Sigma_F$ then there is a neighbourhood V of (x,u) in Z_F such that $V \cap Z_F$ is a submanifold of V and the projection $\pi_F\colon (V \cap Z_F) \to \mathbb{R}^a$ is a local diffeomorphism. (†)

14.12 Let F be a regular k–parameter unfolding of a germ $f \in \mathcal{E}_n^p$ (recall 'regular' means that F itself is a submersion). Let $\pi_F\colon Z_F \to \mathbb{R}^k$ be the usual projection. Show that π_F is contact equivalent to a suspension of the core f_0 of f. (†)

In the remaining problems, we discuss some of the adjacencies appearing in Figure 14.1.

14.13 Consider the IV_3 singularity $f(x,y) = (x^2 + y^2,\ x^3)$. Show that for $\varepsilon \neq 0$, the perturbation $f_\varepsilon(x,y) = (x^2 + y^2 + \varepsilon x,\ x^3)$ has an A_5 singularity at the origin. (†)

14.14 Let $a > 2$ and consider the 1–parameter unfolding of the $\mathrm{I}_{a,b}$ singularity given by

$$F(x,y;\varepsilon) = (xy,\ x^a + \varepsilon x^{a-1} + y^b).$$

Show that for $\varepsilon > 0$, this has two zeros, one being a singularity of type $\mathrm{I}_{a-1,b}$ and the other a simple zero. Deduce that the versal unfolding of the $\mathrm{I}_{a,b}$ singularity contains $\mathrm{I}_{a-1,b}$ singularities and $\mathrm{I}_{a,b-1}$ (the latter if $b > 2$). If a is odd, show it also has a $\mathrm{II}_{a-1,b}$ in its versal unfolding. Deduce the same adjacencies for the versal unfoldings of $\mathrm{II}_{a,b}$.

14.15 Consider the $\mathrm{I}_{2,2}$ singularity in the form $f(x,y) = (x^2, y^2)$. Show that the perturbation $f_\varepsilon(x,y) = (x^2 + \varepsilon y,\ y^2)$ has a singularity at the origin of type A_3.

14.16 Show that the germ $f(x,y) = (xy, (x+y)^3)$ is of type $\mathrm{I}_{3,3}$. Consider the unfolding of f given by

$$f_\varepsilon(x,y) = (xy + \varepsilon(x+y),\ (x+y)^3).$$

Show that for $\varepsilon \neq 0$ this has a singularity of type A_5 at the origin.

14.17 Show that, for $\varepsilon \neq 0$, the unfolding of the $\mathrm{I}_{a,2}$ singularity given by

$$f_\varepsilon(x,y) = (xy,\ x^a + y^2 + \varepsilon y)$$

has two zeros, one being a singularity of type A_a and the other a simple zero.

14.18 Consider the $\mathrm{I}_{a,b}$ singularity $f(x,y) = (xy, x^a + y^b)$ and its versal unfolding

$$F(x,y;u,v,r) = \left(xy - r,\ x^a + y^b + \sum_{i=0}^{a-1} u_i x^i + \sum_{j=1}^{b-1} v_j y^j \right).$$

Show that, for $t \neq 0$, the unfolding of f given by

$$f_t(x,y) = F(x,y;\ u(t),\ v(t),\ t^{a+b})$$

has a zero at $\mathbf{x}_t = (x,y) = (-t^b, -t^a)$ of type A_{a+b-1}, where

$$u_i(t) = \binom{a+b}{b+i} t^{b(a-i)}, \quad \text{and} \quad v_j(t) = \binom{a+b}{b-j} t^{a(b-j)}.$$

[*Hint*: *in a neighbourhood of* $\mathbf{x}_t$ *solve the first component for* y *and substitute into the second component of* f.] (†)

15

Geometric applications

René Thom suggested many applications of catastrophe theory, to the Natural Sciences, to Biological and Human Sciences, as well as within Mathematics. Here we discuss two applications to Geometry which began with the suggestions of Thom in [111]: the first to envelopes and caustics and the second to the geometry of surfaces.

15.1 Envelopes and caustics

The first application we consider is to the study of envelopes of families of lines or curves (or planes in space ...). Thom's suggestion was continued in greater depth by Klaus Jänich [63], who in particular considered the envelope of the family of normals to a 'wave front'. These normals are geometric light rays, and their envelope forms the caustic for that wave front. See also the work of the physicist Michael Berry [9].

Consider a 1–parameter family of lines in the plane, with parameter s, which in Cartesian form is,

$$L_s : \qquad a(s)x + b(s)y + c(s) = 0, \tag{15.1}$$

where a, b, c are smooth functions of s. We will assume that a and b never vanish together (otherwise the equation does not define a line). The classical definition of the envelope was as the locus of points where 'infinitely close' lines meet, and we make this more precise with the following lemma.

Lemma 15.1. *Given a family L_s of lines as in (15.1), let L'_s be the line with equation*

$$L'_s : \qquad a'(s)x + b'(s)y + c'(s) = 0,$$

where primes denote differentiation with respect to s. Suppose $a'(s)$ and $b'(s)$ do not have a common zero. Suppose $L_s \neq L'_s$. Then the limiting intersection point is given by

$$\lim_{\delta \to 0,\, \delta \neq 0} (L_s \cap L_{s+\delta}) = L_s \cap L'_s.$$

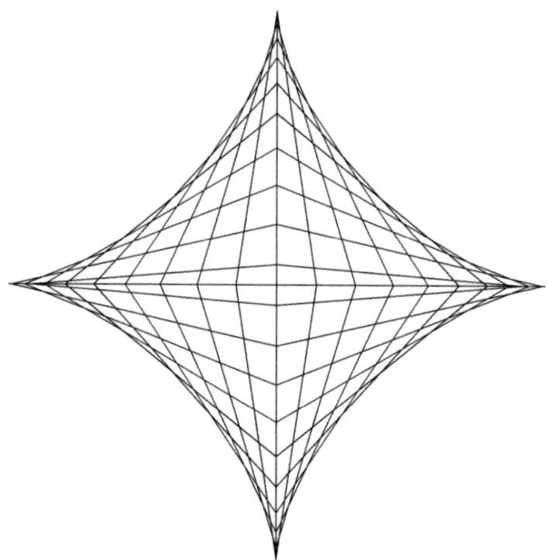

FIGURE 15.1 The astroid as the envelope of a family of lines.

If L_s and L_s' are parallel but distinct, then the right-hand side is empty, and the limit tends to infinity in the direction parallel to L_s.

The proof is left to the reader. In this way one defines the envelope of the family of lines to be the locus of intersections $L_s \cap L_s'$, as s varies, at least for those values of s for which $L_s \neq L_s'$. This enables one to compute the envelope of a family of lines. However, it does not answer the question of what forms one expects an envelope to assume, or what singularities it may have. Using contact equivalence and unfoldings, we answer that question after the following example.

Example 15.2. The lines shown in Figure 15.1 join the points $(2\sin s,\, 0)$ and $(0, 2\cos s)$; these have equation

$$L_s: \quad x\cos s + y\sin s = \sin 2s \quad (s \in [0, 2\pi]).$$

The envelope is clearly visible as the outer curve formed by the lines. This curve is called an ***astroid***. It follows from the lemma that the astroid is parametrized by $(x, y) = (2\sin^3 s,\ 2\cos^3 s)$. ✐

Using a vector notation that is readily extended to higher dimensions, we let $\mathbf{n}(s) = (a(s), b(s))$. Then a point $\mathbf{x} \in \mathbb{R}^2$ lies on L_s if and only if,

$$\mathbf{n}(s) \cdot \mathbf{x} + c(s) = 0. \tag{15.2}$$

Define $G\colon \mathbb{R} \times \mathbb{R}^2 \rightarrowtail \mathbb{R}$ by

$$G(s; \mathbf{x}) = \mathbf{n}(s) \cdot \mathbf{x} + c(s). \tag{15.3}$$

We will think of G as a family of functions of s, parametrized by $\mathbf{x}$. Then the line L_s contains $\mathbf{x}$ if and only if $G(s;\mathbf{x}) = 0$; that is, if and only if $(s,\mathbf{x}) \in Z_G$. By the lemma above, $\mathbf{x}$ belongs to the envelope if and only if

$$G(s,\mathbf{x}) = 0, \quad \text{and} \quad \frac{\partial}{\partial s}G(s,\mathbf{x}) = 0,$$

at least when $L_s \neq L'_s$. Thus the envelope arises from the bifurcation set Σ_G of G, and one can make the following alternative definition of envelope.

Definition 15.3. The envelope of the family of lines is the discriminant Δ_G of the family $G(s;\mathbf{x})$. ✭

Given that $G(s,\mathbf{x})$ is a 2–parameter family of functions of s, it is natural to ask whether the singularities that arise are versally unfolded in the family. This depends on the family of lines; here we derive conditions on $a(s)$ and $b(s)$ which guarantee the unfoldings are versal. Since it is a 2–parameter family, we expect to see only fold (saddle-node) and cusp bifurcations – these are both illustrated in the astroid in Figure 15.1.

Theorem 15.4. *Let $(s_0,x_0,y_0) \in Z_G$, and let*

$$g(t) = G(s_0+t,x_0,y_0) = x_0a(s_0+t) + y_0b(s_0+t) + c(s_0+t),$$

(in particular $g(0) = 0$). Denote by $\Sigma^{(j)}$ the subset of Z_G for which the j-jet of g vanishes.

(i). If $(s_0,x_0,y_0) \in Z_G \setminus \Sigma^{(1)}$ then the family G has no bifurcation at that point,

(ii). At any point $(s_0,x_0,y_0) \in \Sigma^{(1)} \setminus \Sigma^{(2)}$, the family G has a fold bifurcation at that point.

(iii). Suppose $(s_0,x_0,y_0) \in \Sigma^{(2)} \setminus \Sigma^{(3)}$. The family G has a cusp bifurcation at that point if and only if the the lines L_{s_0} and L'_{s_0} are not parallel.

PROOF: For simplicity, we suppose $s_0 = 0$. Write the Taylor series of g as a function of s at the point $(0,x_0,y_0)$ as

$$g = g_0 + g_1s + \tfrac{1}{2}g_2s^2 + \cdots,$$

with $g_j \in \mathbb{R}$. By hypothesis, $g_0 = 0$. Now,

$$T_e\mathcal{K}{\cdot}G = \mathcal{E}_1\left\{g_1 + g_2s + \ldots,\ g_1s + \tfrac{1}{2}g_2s^2 + \cdots\right\}.$$

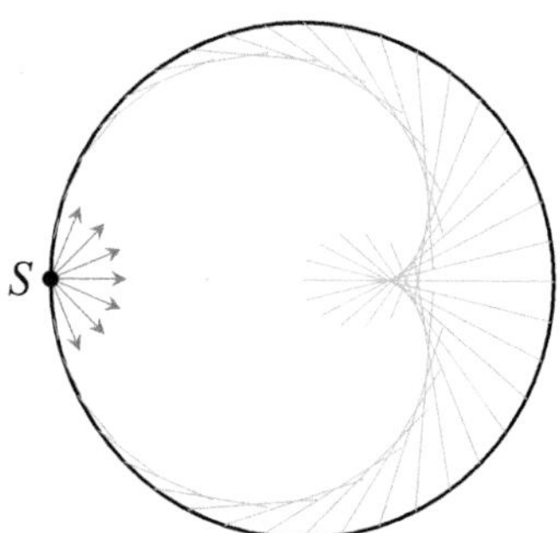

FIGURE 15.2 Caustic obtained by reflecting light rays from a source S on the rim of a coffee cup; see Example 15.6.

Moreover, $\dot{G} = \mathbb{R}\{a(s), b(s)\}$. Proceeding case by case, the result follows easily. For example in (ii), one shows that the condition for versality is $(a(s_0), b(s_0)) \neq (0,0)$, but this is required from the start (otherwise L_{s_0} fails to be a line). For (iii) the condition that the lines are not parallel is that the determinant $D := \begin{vmatrix} a(s_0) & b(s_0) \\ a'(s_0) & b'(s_0) \end{vmatrix} \neq 0$. ✔

Now consider the projection to the plane and the discriminant Δ_G of G, which is the envelope of the family of lines. Since we are no longer just considering the germ of the family of lines, but a global family, with $s \in \mathbb{R}$ (or in some bounded interval), one should include the possibility that a given $(x_0, y_0) \in \Delta_G$ is the image of more than one point (x_0, y_0, s_0) and (x_0, y_0, s_0') in Σ_G. In the meantime, we can conclude the following.

Corollary 15.5. *Let L_s $(s \in \mathbb{R})$ be a family of lines in the plane, and let G be the associated family of maps (15.3). Suppose the projection $\Sigma_G \to \mathbb{R}^2$ is injective and proper. Then the envelope of the family is a smooth curve possibly with isolated cusps.*

Example 15.6. Refer to Figure 15.2 – we show the caustic indeed has a single cusp. Consider the ray emanating from $(-1, 0)$ in a direction subtending an angle $\theta \in (-\pi/2, \pi/2)$ with the positive x-axis. It meets the circle (the rim of the coffee cup) at the point $(\cos 2\theta, \sin 2\theta)$. The reflected line L_θ has equation

$$G(\theta, x, y) := x(\sin 2\theta + \sin 4\theta) - y(\cos 2\theta + \cos 4\theta) - \sin 2\theta = 0.$$

Comparing notation with that above, we have (with s replaced by θ)

$$a(\theta) = \sin 2\theta + \sin 4\theta, \quad b(\theta) = \cos 2\theta + \cos 4\theta, \quad \text{and} \quad c(\theta) = -\sin 2\theta.$$

Solving $G = G_\theta = G_{\theta\theta} = 0$ one finds only one solution, namely $\theta = 0$ and $(x, y) = (1/3, 0)$. At this point, L_θ is the x-axis, while L'_θ is the line $x = 1/3$ which is distinct from L_θ, and hence the bifurcation is a cusp bifurcation, and the discriminant (envelope) has a genuine cusp at the point $(x, y) = (1/3, 0)$.

Note that at the point $\theta = \pi/2$ the two lines L_θ and L'_θ coincide, so the theory is not applicable there. ✐

There are many other interesting envelopes of families of lines, apart from caustics: two such examples are the Steiner deltoid associated to a triangle, and the solutions of Clairaut equations which we now discuss briefly.

Example 15.7. A Clairaut equation is any differential equation of the form

$$x\frac{\mathrm{d}y}{\mathrm{d}x} + f\left(\frac{\mathrm{d}y}{\mathrm{d}x}\right) = y, \tag{15.4}$$

where $f\colon \mathbb{R} \to \mathbb{R}$ is a smooth function. Differentiating this leads to the equation,

$$\frac{\mathrm{d}^2y}{\mathrm{d}x^2}\left(x + f'\left(\frac{\mathrm{d}y}{\mathrm{d}x}\right)\right) = 0. \tag{15.5}$$

There are therefore two possibilities for solutions of the differential equation, depending on which factor vanishes.

The first possibility is where $\frac{\mathrm{d}^2y}{\mathrm{d}x^2} = 0$ which is a straight line $y = px + q$. On substituting into (15.4) this requires $q = f(p)$. The *general solution* to the Clairaut equation is therefore the line

$$L_p: \quad y - px = f(p) \quad (p \in \mathbb{R}).$$

The envelope of this family of straight lines is given by intersecting L_p and L'_p. An easy calculation shows that L'_p is given by the second factor in (15.5), and together with (15.4) gives the so-called *singular solution*, which in parametric form is

$$x = -f'(p), \quad y = f(p) - pf'(p).$$

The singular solution is therefore the envelope of the general solution(s), as Clairaut himself discovered. See Problem 15.7, and for more details on the classification of singularities of Clairaut equations, see [30]. ✐

15.2 Curvature and flatness

Following the inital suggestion by Thom, this geometric application of singularity theory was developed in detail by Ian Porteous, whose work culminated in the book [97], and then by others, see for example [8] and the more recent [61] and references therein. In fact Porteous started by looking at the envelope of the family of normals to a surface in $\mathbb{R}^3$ and singularities of the exponential map of the normal bundle, but later this developed into the discussion below, partly thanks to the author's PhD thesis [82].

The aim is to measure the geometry of a surface or more general submanifold by comparing it to a class of archetypal geometric subsets. For example, to study the flatness of a surface, one would consider its contact with planes or lines. Or to discuss curvature (roundness), one might consider the contact with spheres or circles.

The general set-up is the following. Suppose the geometric archetypes we have chosen in $\mathbb{R}^n$ are submanifolds of codimension p (e.g. each line in $\mathbb{R}^3$ is of codimension 2). Then each one can be expressed as the zero-set of a smooth map $\mathbb{R}^n \to \mathbb{R}^p$, where 0 is a regular value (cf. the regular value theorem, Theorem B.23). Now suppose that the set of all such archetypal objects forms a family with k parameters. In this way we are led to a smooth family of maps

$$F\colon \mathbb{R}^n \times M \longrightarrow \mathbb{R}^p, \tag{15.6}$$

with the property that for each $m \in M$, the set $\{x \in \mathbb{R}^n \mid F(x, m) = 0\}$ is one of our archetypal geometric objects. Here M is assumed to be a manifold of dimension k, parametrizing our chosen collection of archetypes (it may be a submanifold of some larger $\mathbb{R}^K$).

For example, suppose we choose our archetypes to be the planes in $\mathbb{R}^3$. A plane is determined by an equation of the form $\mathbf{n} \cdot \mathbf{x} = r$, where $\mathbf{n}$ is a unit vector; that is, $\mathbf{n} \in S^2 \subset \mathbb{R}^3$, and $r \in \mathbb{R}$. Let us write $\Pi_{(\mathbf{n},r)}$ for this plane. (Note that $\Pi_{(\mathbf{n},r)} = \Pi_{(-\mathbf{n},-r)}$, but that is not a problem.) Let $M = S^2 \times \mathbb{R}$. Then our required family of maps is

$$\begin{array}{rcl} F\colon \mathbb{R}^3 \times M & \longrightarrow & \mathbb{R} \\ (\mathbf{x}, (\mathbf{n}, r)) & \longmapsto & \mathbf{n} \cdot \mathbf{x} - r. \end{array} \tag{15.7}$$

And indeed, $\Pi_{(\mathbf{n},r)} = \{\mathbf{x} \in \mathbb{R}^3 \mid F(\mathbf{x}, (\mathbf{n}, r)) = 0\}$, as required. It is easy to see how this is extended to hyperplanes in $\mathbb{R}^n$, for any $n \geq 2$.

Alternatively suppose our archetypes are lines in $\mathbb{R}^3$. A line may be determined by a plane through the origin, perpendicular to the line, and the point of intersection of the line with the plane. To write the cartesian equations for the line, we choose an orthonmal basis $\{\mathbf{u}, \mathbf{v}\}$ for the plane, and the coordinates (a, b) for the point of intersection. Then the line associated to this data is the zero-set of the equation

$$\begin{aligned} F\colon \mathbb{R}^3 \times M &\longrightarrow \mathbb{R}^2 \\ (\mathbf{x}, (\mathbf{u}, \mathbf{v}, a, b)) &\longmapsto (\mathbf{u}\cdot\mathbf{x} - a,\ \mathbf{v}\cdot\mathbf{x} - b). \end{aligned}$$

We should note that, with this set-up, there is a 1–dimensional redundancy in M corresponding to the fact that for any given plane, there is a 1–parameter set of pairs of orthonormal bases. But that can be accounted for in any analysis.

The procedure for studying the geometry via singularity theory is as follows:

(i). fix a parametrization of our set of archetypes by a manifold M, with a corresponding family of maps F as in (15.6);

(ii). let S be a given manifold and $g\colon S \to \mathbb{R}^n$ a smooth immersion, whose image is the submanifold of geometric interest;

(iii). define the family

$$\Phi\colon S \times M \to \mathbb{R}^p, \quad \Phi(x; m) = F(g(x); m),$$

and write $\phi_m\colon S \to \mathbb{R}^p$ for $\phi_m(x) = \Phi(x, m)$, that is, $\phi_m = f_m \circ g$;

(iv). finally, the ***contact type*** of $g(S)$ with an archetype $m \in M$ at a point $x \in S$ is defined to be the $\mathcal{K}$-class of the germ at x of the map ϕ_m.

Remark 15.8. Suppose that S_1, S_2 are submanifolds of $\mathbb{R}^n$ of dimensions k_j, defined (i) by embeddings $g_j\colon S_j \to \mathbb{R}^n$, and (ii) as level sets of a submersion $f_j\colon \mathbb{R}^n \to \mathbb{R}^{p_j}$ $(p_j = n - k_j)$. It can be shown [83, 61] that the three local rings are isomorphic:

$$\frac{\mathcal{E}_{k_1}}{I_{f_2 \circ g_1}} \simeq \frac{\mathcal{E}_{k_2}}{I_{f_1 \circ g_2}} \simeq \frac{\mathcal{E}_n}{I_{f_1} + I_{f_2}},$$

(where I_f denotes the ideal generated by the components of the map f) so that any of these can be used to define contact type. ❞

15.2A Flatness

Let us illustrate these ideas by measuring the flatness of a smooth surface S in $\mathbb{R}^3$ in terms of its contact with planes. The geometric terms we use can be found in, for example, do Carmo's book [23]. Let F be the family of maps given in (15.7). Let

(u, v) be local coordinates on the surface, then $g(u, v) \in \mathbb{R}^3$. Fix $m = (\mathbf{n}, r) \in M$. Then, for $x \in S$,

$$\phi_m(x) = \mathbf{n} \cdot g(x) - r. \tag{15.8}$$

In this case, $\phi_m\colon \mathbb{R}^2 \rightarrowtail \mathbb{R}$. Clearly $\phi_m(x) = 0$ if and only if $g(x) \in \Pi_{(\mathbf{n},r)}$. Thus,

$$Z_\Phi = \{(x, m) \in S \times M \mid x \in \Pi_m\}.$$

Now consider the singularity type. First,

$$\mathsf{d}\phi_m(x) = \mathbf{n} \cdot \mathsf{d}g_x.$$

The image of $\mathsf{d}g_x$ is the tangent plane T_xS (or more precisely $T_{g(x)}g(S)$, but we allow ourselves to be slack with the notation), and hence ϕ_m has a critical point at x if and only if the plane represented by m is the tangent plane to the surface at $g(x)$. In this case ϕ_m has a singularity of type at least A_1 at x. Thus,

$$\Sigma_\Phi = \{(x, m) \in S \times M \mid \Pi_m = T_xS\}.$$

Now we pass to higher singularity types. The second derivative of ϕ_m is $\mathsf{d}^2\phi_m = \mathbf{n} \cdot \mathsf{d}^2 g_x$. This quadratic form is called the *second fundamental form*[1] of the surface at x, and the sign of its determinant agrees with the sign of the curvature. Moreover, it is degenerate if and only if x is a point of zero curvature (a *parabolic point* of the surface). Thus the contact at x of the surface with its tangent plane is of type at least A_2 if and only if x is a parabolic point. If it is of type precisely A_2 then the surface intersects its tangent plane in a cuspidal curve (the zero-set of an A_2-singularity), whose limiting tangent direction at the point x is the null direction of the Hessian, $\mathbf{n} \cdot \mathsf{d}^2 g_x$.

That is as far as classical differential geometry goes. However, since $\dim M = 3$, the family Φ has three parameters, so one would expect there to be isolated A_3 points (swallowtail singularities). And indeed, the singularity theory allows us to determine such points, using the following conditions,

(i). At least A_1: $\mathsf{d}\phi_m = 0$;

(ii). At least A_2: there is a non-zero vector $\mathbf{a} \in \mathbb{R}^2$ for which $\mathsf{d}^2\phi_m\mathbf{a} = 0$;

(iii). At least A_3: there is a vector $\mathbf{b} \in \mathbb{R}^2$ such that

$$\mathsf{d}^3\phi_m\mathbf{a}^2 + \mathsf{d}^2\phi_m\mathbf{b} = 0.$$

This condition is equivalent to $\mathsf{d}^3\phi_m\mathbf{a}^3 = 0$.

[1]the second fundamental form is often defined as $-\mathsf{d}\mathbf{n} \cdot \mathsf{d}g$, but differentiating the identity $\mathbf{n} \cdot \mathsf{d}g = 0$ shows the two expressions to be equal

For an A_3 singularity of ϕ_m one requires first that $d^2\phi_m\mathbf{a} = 0$ (for A_2, $\mathbf{a}$ is the limiting tangent vector mentioned above) and in addition the existence of a vector $\mathbf{b} \in \mathbb{R}^2 = T_x S$ for which

$$d^3\phi_m\mathbf{a}^2 + d^2\phi_m\mathbf{b} = 0.$$

Equivalently, A_3 contact occurs at points of the parabolic locus where $d^3\phi_m\mathbf{a}^3 = 0$. These points are geometrically interesting, even though they were not studied in classical differential geometry. We call them ***biplanar points***, for reasons explained below. They were first studied in detail in [8], where they are shown to be cusps of the Gauss map, based on a different but related approach also using singularity theory (see remark below). A biplanar point is said to be ***generic*** if ϕ_m has precisely an A_3 singularity there, and the family Φ presents a versal unfolding of the singularity.

A simple geometric consequence of this analysis is the following.

Theorem 15.9. *In any neighbourhood of a generic biplanar point, there are pairs of points of the surface that share a common tangent plane.*

Such planes are called ***bitangent planes***; thus the proposition states that there are bitangent planes in any neighbourhood of the tangent plane at a generic biplanar point.

PROOF: Let Π_0 be the plane having A_3 contact with the surface. Since, by hypothesis, this singularity is versally unfolded, the discriminant in M is a swallowtail. This swallowtail surface represents the set of planes near Π_0 that are tangent to the surface (singularity of type at least A_1). The swallowtail surface has a curve (or half-curve) of self-intersection points (see Figure 7.1). These points represent precisely the bitangent planes, since they have two A_1 singularities at distinct points on the surface (see Problem 7.14(ii)). ✔

This proof uses the fact that the singularity in question is the swallowtail, and it is natural to ask what the cuspidal edge of the swallowtail surface corresponds to. In fact we already have an answer to that: they correspond to the tangent planes at the parabolic points, for that is where the contact is of type A_2.

Remark 15.10. An alternative approach to this flat geometry was taken by Banchoff, Gaffney and McCrory [8], where they consider singularities of the Gauss map $G: S \to S^2$, where $G(x)$ is the unit normal vector to the surface at x.

Let us modify our approach and consider right equivalence of the family

$$\Psi: S \times S^2 \to \mathbb{R}, \quad \Psi(x; \mathbf{n}) = \mathbf{n} \cdot g(x).$$

(This is a family of 'height functions' on the surface S, and as before $\mathbf{n} \in S^2$ is a unit vector.) Differentiating with respect x shows that the catastrophe set of this family is,

$$C_\Psi = \{(x, \mathbf{n}) \mid \mathbf{n} \text{ is normal to the surface at } x\}.$$

If we identify C_Ψ with S (in fact there are two unit normals to the surface at each point, but we select one of them), then the catastrophe map $\pi_\Psi \colon C_\Psi \to S^2$ is precisely the Gauss map. A local analysis then shows that Σ_Ψ consists of the parabolic points on the surface, and the biplanar points exhibit cusp singularities of the family. Details can be found in [8]. ❞

15.2B Roundness

The original geometric application suggested by Thom [111, Section 5.4] is to measure the contact of a surface in $\mathbb{R}^3$ with spheres. A high degree of contact with a sphere would mean the surface being very 'round', and it should not be a surprise that the radius of a sphere with high contact is closely related to the radius of curvature of the surface at the point of contact.

Since a sphere is determined by its centre $\mathbf{c}$ and radius $r > 0$, this can be made precise by introducing the family of ***distance squared functions***,

$$F \colon \mathbb{R}^3 \times M \to \mathbb{R}, \quad (x, (\mathbf{c}, r)) \mapsto \|\mathbf{c} - \mathbf{x}\|^2 - r^2, \tag{15.9}$$

where $M = \mathbb{R}^3 \times \mathbb{R}^+$ represents the family of all spheres in $\mathbb{R}^3$.

Now, let $g \colon S \hookrightarrow \mathbb{R}^3$ be an immersion of a surface. For $m \in M$, let $\phi_m = f_m \circ g$ (usual notation as described above). Without going into details, one finds, for $m \in M$:

- ϕ_m has a critical point at $x \in S$ if and only if $\mathbf{c}$ lies on the normal to the surface at $g(x)$ (i.e. the sphere represented by m is tangent to the surface at $g(x)$),
- ϕ_m has a critical point of type at least A_2 iff $\mathbf{c}$ lies at a focal point (or principal centre of curvature) of the surface at $g(x)$,
- ϕ_m has a critical point of type at least A_3 if in addition x lies on a so-called ridge point of the surface (a notion due to Porteous [97]), and
- ϕ_m has a critical point of type at least D_4 iff $g(x)$ is an umbilic of the surface (i.e. the two principal curvatures coincide) and $\mathbf{c}$ lies at the unique centre of curvature for the surface at $g(x)$.

This last point gives rise to the name 'umbilic' for critical points of type D_k. Many further details can be found in [61, 82] and references therein.

There has been considerable interest in how the types of phenomena described above evolve in a 1–parameter family of surfaces. See for example [19].

Problems

15.1 Consider the family of lines L_θ with equation

$$x\cos\theta + y\sin\theta = 1.$$

Show that the envelope is the unit circle.

15.2 Prove Lemma 15.1.

15.3 Consider the ellipse $\gamma(t) = (3\cos t, 2\sin t)$, and let N_t be the normal line through the point $\gamma(t)$. Show that the envelope of this family of lines is the evolute of the ellipse given in Example 1.1. (†)

15.4 Prove more generally that the two definitions of the evolute of a smooth plane curve coincide: the one given in Section 1.4 as the locus of centres of curvature, and the definition as the envelope of the set of normal lines to the curve (see [97] for further details).

15.5 Extend Example 15.6 as follows. Let $S(u, v)$ be a point on the plane (the source of light), and let C be the unit circle with centre at $(0, 0)$. Consider the family of lines (light rays) emitted from S and reflected by the usual law of reflection from the circle C. Find the caustic formed by (or the envelope of) the reflected rays. (If S lies on the circle C the envelope is a cardioid.)

15.6 In the definition of envelope above, we used the Cartesian equation for the lines L_s. Suppose instead the lines are parametrized by $\mathbf{x} = \gamma_s(t)$. Consider the map

$$\begin{aligned} F\colon \mathbb{R}^2 &\longrightarrow \mathbb{R}^2, \\ (s,t) &\longmapsto \gamma_s(t). \end{aligned}$$

Show that the envelope is the set of singular values of F. (†)

15.7 Show that the envelope of the general solutions to a Clairaut equation (see Example 15.7) is the singular solution. Show this singular solution has a singular point if $f''(p) = 0$, and find the condition on f for this singular point to be an ordinary cusp.

Analyse the Clairaut equations for $f(p) = p^2$ and $f(p) = p^3$.

15.8 For a surface in $\mathbb{R}^3$, at each point of the parabolic curve there is the associated *parabolic direction*, which is the direction in which the curvature vanishes. Show the following:

(i). this is equal to the kernel of the Hessian matrix of ϕ_m in (15.8), and

(ii). at biplanar points this direction is tangent to the parabolic curve.

16

Preparation theorem

ONE OF THE CENTRAL TECHNICAL theorems underlying singularity theory is the preparation theorem. It has an interesting history, and is based on the Weierstrass preparation theorem, which is about complex analytic functions of several variables (see for example [34] or [53]). The theorem is used to prove all the versality theorems. It is also crucial for computations for left–right ($\mathcal{A}$)-equivalence.

In this chapter we discuss the theorem and its applications, but we do not give the proof. Some standard sources for the proof are the books by Bröcker [12], by Golubitsky and Guillemin [45] and by Martinet [75].

The history, very briefly, is that the statement of the theorem was conceived by Thom, who needed it in order to prove the versality theorem of Chapter 7. However, lacking the expertise in analysis, he attempted to convince the French analyst, Bernard Malgrange to prove it. Malgrange himself says[1] that at first he did not believe the statement to be true, in spite of Thom's insistence, but eventually was persuaded by J.-P. Serre that it was worth pursuing, and did then succeed in finding a proof. The version we give is a more algebraic statement due to John Mather [76, III].

Outside of singularity theory, the Malgrange preparation theorem also has applications in invariant theory (see Section 16.4 below) and in the study of PDEs, in particular Fourier integral operators; see [58], which also contains a proof of the preparation theorem, and [112].

16.1 Statement

Recall that a map germ $\phi\colon (\mathbb{R}^n, 0) \to (\mathbb{R}^p, 0)$ defines a homomorphism $\phi^*\colon \mathcal{E}_p \to \mathcal{E}_n$ by $\phi^* h = h \circ \phi$. Let A be an $\mathcal{E}_n$-module, then it becomes an $\mathcal{E}_p$-module via ϕ^* as follows: let $w \in \mathcal{E}_p$ and $a \in A$ then $w.a := (\phi^* w)a \in A$. Recall also that $I_\phi \lhd \mathcal{E}_n$ is the ideal generated by the components of ϕ.

[1] Personal communication; see also [72].

The main question is whether, if A is finitely generated as an $\mathcal{E}_n$-module, is it also finitely generated as an $\mathcal{E}_p$-module? Answer: not always! A simple example where it fails is provided by the map $\phi(x, y) = x$, and $A = \mathcal{E}_2$ which is finitely generated as an $\mathcal{E}_2$-module of course, but not as an $\mathcal{E}_1$-module.

Theorem 16.1 (Malgrange–Mather preparation theorem). *Let $\phi\colon (\mathbb{R}^n, 0) \to (\mathbb{R}^p, 0)$ be the germ of a smooth map, and let A be a finitely generated $\mathcal{E}_n$-module for which $A/I_\phi A$ is finite–dimensional. Then A is finitely generated as an $\mathcal{E}_p$-module. More precisely, let $\{u_1 \dots, u_r\} \subset A$ be a cobasis for $I_\phi A$ in A. Then A is generated by $\{u_1 \dots, u_r\}$ as an $\mathcal{E}_p$-module.*

Explicitly, to say A is generated by $\{u_1 \dots, u_r\}$ as an $\mathcal{E}_p$-module means that for each $a \in A$ there are $h_1, \dots, h_r \in \mathcal{E}_p$ for which

$$a = (h_1 \circ \phi)u_1 + \cdots + (h_r \circ \phi)u_r.$$

In general, the h_j are not uniquely determined.

Proofs can be found in the references cited above. Note that the converse of the theorem is also true: if A is finitely generated as an $\mathcal{E}_p$-module then $A/I_\phi A$ is of finite dimension, but this is elementary.

The theorem is stated in the real C^∞ setting, but is also valid (and useful) in the real or complex analytic setting. See for example [65] for a proof, explaining also its relation to the Weierstrass preparation theorem.

16.2 Simple applications

Here we give some more or less simple applications of the preparation theorem; further examples are given in the problems. In the following section, we show how the theorem is used to prove the versality theorems.

Example 16.2. Consider the simplest singular map $\phi\colon \mathbb{R} \to \mathbb{R}$ defined by $\phi(x) = x^2$ (or rather its germ at 0), and let $A = \mathcal{E}_1$. Then $I_\phi = \langle x^2 \rangle$ and so $A/I_\phi A = \mathcal{E}_1/\langle x^2 \rangle \simeq \mathbb{R}\{1, x\}$; that is, $\{1, x\}$ forms the required cobasis. It follows from the preparation theorem that given any element $f \in \mathcal{E}_1$ there are functions h_1, h_2 in $\mathcal{E}_1$ such that

$$f(x) = h_1(x^2) + h_2(x^2)x.$$

In particular, if f is *even*, that is $f(-x) = f(x)$, then $h_2 = 0$ (since $0 = f(x) - f(-x) = 2xh_2(x^2)$), and hence $f(x) = h_1(x^2)$. This fact, that any smooth even function of x can be written as a smooth function of x^2, was used in Chapter 9, and was first proved by Whitney (without the benefit of the preparation theorem). See Problems 16.4 and 16.5 for an extension to more variables. ✐

Example 16.3. Some applications are a little more subtle. In Example 12.6 we used the fact that any smooth germ $f \in \mathfrak{m}_1^2$ can be written as a smooth function of the two arguments t^2 and t^3. Using the preparation theorem this can be proved as follows. Let $f \in \mathfrak{m}_1^2$ and write f as $f(t) = t^2 f_1(t)$.

Consider the map germ $\phi\colon (\mathbb{R}, 0) \to (\mathbb{R}^2, 0)$ defined by $\phi(t) = (t^2, t^3)$. Clearly $\{1, t\}$ is a cobasis for I_ϕ in $A = \mathcal{E}_1$. It follows from the Preparation Theorem that, given our $f_1 \in \mathcal{E}_1$ there are functions h_1, h_2 such that

$$f_1(t) = h_1(t^2, t^3) + t\, h_2(t^2, t^3).$$

Now substitute for f_1 and we have

$$f(t) = t^2 \left(h_1(t^2, t^3) + t\, h_2(t^2, t^3)\right) = t^2 h_1(t^2, t^3) + t^3 h_2(t^2, t^3),$$

and this is a smooth function of t^2 and t^3 as required. See Problem 16.7 for a similar example. ✐

Example 16.4. A similar analysis can be applied to maps: let

$$\phi(x, y) = (x, y^3 + xy).$$

(Whitney's 'cusp' map). A cobasis for I_ϕ in $\mathcal{E}_2$ is $\{1, y, y^2\}$. Let $A = \theta_2 = \mathcal{E}_2^2$. It follows that a cobasis for $I_\phi A$ in A is

$$\left\{\begin{pmatrix}1\\0\end{pmatrix}, \begin{pmatrix}y\\0\end{pmatrix}, \begin{pmatrix}y^2\\0\end{pmatrix}, \begin{pmatrix}0\\1\end{pmatrix}, \begin{pmatrix}0\\y\end{pmatrix}, \begin{pmatrix}0\\y^2\end{pmatrix}\right\}.$$

Denote these vector fields (germs) by $u_j \in \theta_2$ (for $j = 1, \ldots, 6$). It follows from the Preparation Theorem that, given any vector field (or map) $v \in \theta_2$ there are germs $h_j \in \mathcal{E}_2$ such that

$$v(x, y) = \sum_j h_j(x, y^3 + xy)\, u_j.$$

For example, one can write $v = \begin{pmatrix}y^3\\xy^2\end{pmatrix}$ as

$$\begin{pmatrix}y^3\\xy^2\end{pmatrix} = (y^3 + xy)\begin{pmatrix}1\\0\end{pmatrix} - x\begin{pmatrix}y\\0\end{pmatrix} + x\begin{pmatrix}0\\y^2\end{pmatrix}.$$

✐

16.3 Application to versality theorems

We show how the preparation theorem is used to prove the versality theorem for right equivalence. A similar argument can be used for contact equivalence, as well as the other equivalence relations of singularity theory. The approach taken here is based on the proof for contact equivalence given by Martinet [75].

Since we use several different rings of germs in this section, we use a self evident notation of $\mathcal{E}_x$ or $\mathcal{E}_{x,\lambda}$ etc. (notation previously used in Chapter 12).

Let $f \in \mathcal{E}_x$ be a germ of finite $\mathcal{R}^+$-codimension, and let

$$F(x;u) = f(x) + \sum_{j=1}^{\ell} u_j \psi_j(x)$$

be an $\mathcal{R}^+$ *infinitesimally versal* unfolding of f: i.e., one satisfying

$$Jf + \mathbb{R}\{\psi_0, \psi_1, \ldots, \psi_\ell\} = \mathcal{E}_n, \tag{16.1}$$

where $\psi_0 = 1$, arising from the constant term in $\mathcal{R}^+$-equivalence.

We wish to prove Theorem 7.8, which states that every infinitesimally versal unfolding is versal, or, more briefly, *infinitesimal versality implies versality.*

Recall the definition of $\mathcal{R}^+_{\mathrm{un}}$-equivalence of unfoldings (Definition 7.6) and the corresponding tangent space defined in Chapter 12: for any unfolding $G(x;v)$ of f

$$T\mathcal{R}^+_{\mathrm{un}} \cdot G = J_x G + \mathcal{E}_v \left\{1, \frac{\partial G}{\partial v_j}\right\},$$

where

$$J_x G = \left\langle \frac{\partial G}{\partial x_j} \,\middle|\, j = 1, \ldots, n \right\rangle \lhd \mathcal{E}_{x,v},$$

which is the Jacobian ideal of the unfolding G relative to the x-variables (also denoted $T_{\mathrm{rel}}\mathcal{R}{\cdot}G$). The following property of infinitesimally versal unfoldings follows from the preparation theorem.

Proposition 16.5. *(i). Suppose $F(x;u)$ is an infinitesimally $\mathcal{R}^+$-versal unfolding of a germ f, with ℓ parameters (as above). Then*

$$T\mathcal{R}^+_{\mathrm{un}} \cdot F = \mathcal{E}_{x,u}.$$

(ii). More generally, suppose $F(x;u,v)$ is an unfolding of f for which F_1 is infinitesimally $\mathcal{R}^+$-versal, where $F_1(x;u) := F(x;u,0)$, then

$$J_x F + \mathcal{E}_{u,v}\left\{1, \frac{\partial F}{\partial u_1}, \ldots, \frac{\partial F}{\partial u_\ell}\right\} = \mathcal{E}_{x,u,v}.$$

PROOF: Since (i) is a special case of (ii), we just prove (ii). Let A be the quotient $\mathcal{E}_{x,u,v}$-module $A = \mathcal{E}_{x,u,v}/J_xF$, and let ϕ be the projection $\mathbb{R}^{n+\ell+k} \to \mathbb{R}^{\ell+k}$, $\phi(x,u,v) = (u,v)$. Then $A/I_\phi A$ is (isomorphic to) the $\mathcal{E}_x$-module $\mathcal{E}_x/Jf$. The fact that F_1 is infinitesimally versal means

$$Jf + \mathbb{R}\{\psi_0, \dots, \psi_\ell\} = \mathcal{E}_x,$$

where $\psi_0 = 1$ and for $j > 0$, $\psi_j = \dot{F}_j = \frac{\partial F}{\partial u_j}\big|_{u=0,\,v=0}$. That is, $\{\psi_0, \dots, \psi_\ell\}$ forms a cobasis for $I_\phi A$ in A. It follows from the preparation theorem that

$$S := \left\{1, \frac{\partial F}{\partial u_1}, \dots, \frac{\partial F}{\partial u_\ell}\right\},$$

is a set of generators for A as an $\mathcal{E}_{u,v}$-module. That is,

$$\mathcal{E}_{x,u,v} = J_xF + \mathcal{E}_{u,v}S,$$

as required. ✔

Now let $G(x;\lambda)$ (for $\lambda \in \mathbb{R}^k$) be any unfolding of f. To prove the versality theorem, we need to show that G is equivalent to an unfolding induced from F. To compare F and G, we form the 'direct sum' $H := F \oplus G$ of the two unfoldings:

$$H(x;u,\lambda) = F(x;u) + G(x;\lambda) - f(x).$$

Note that $H(x,0,0) = f(x)$, so H is also an unfolding of f, with $k+\ell$ parameters. By the following lemma, to show F is versal it is sufficient to show that, for any given unfolding G, the direct sum $H = F \oplus G$ is $\mathcal{R}^+_{\mathrm{un}}$-equivalent to $\widetilde{F}$, where

$$\widetilde{F}(x;u,\mu) = F(x;u).$$

Lemma 16.6. *Suppose the unfoldings H and $\widetilde{F}$ defined above are $\mathcal{R}^+_{\mathrm{un}}$-equivalent. Then G is equivalent to an unfolding induced from F.*

PROOF: Since H and $\widetilde{F}$ are equivalent, there is a diffeomorphism Φ of the form

$$\Phi(x,u,\lambda) = (\phi(x,u,\lambda), \psi(u,\lambda), \chi(u,\lambda)) \tag{16.2}$$

and a smooth function $C(u,\lambda)$ for which

$$\begin{aligned} H(x,u,\lambda) &= \widetilde{F}(\phi(x,u,\lambda), \psi(u,\lambda), \chi(u,\lambda)) + C(u,\lambda) \\ &= F(\phi(x,u,\lambda), \psi(u,\lambda)) + C(u,\lambda). \end{aligned}$$

Putting $u = 0$ shows that

$$G(x,\lambda) = H(x,0,\lambda) = F(\phi(x,0,\lambda), \psi(0,\lambda)),$$

as required. ✔

There remains to show that indeed H and $\widetilde{F}$ are equivalent unfoldings.

For $j = 0, \dots, k$ consider the unfoldings,

$$H_j(x, u, \lambda_1, \dots, \lambda_k) = H(x, u, \lambda_1, \dots, \lambda_j, 0, \dots, 0).$$

In particular $H_0 = \widetilde{F}$ and $H_k = H$. Since F is infinitesimally versal, so are all the H_j, and we show that for $j > 0$, H_j is $\mathcal{R}^+_{\text{un}}$-equivalent to H_{j-1}. Stringing these together shows that $H_k = H$ is equivalent to $H_0 = \widetilde{F}$, as required.

We present the argument for $j = k$ (in order not to introduce extraneous notation) and write $\lambda = (\lambda', \lambda_k)$.

Consider $\partial H_k / \partial \lambda_k \in \mathcal{E}_{x,u,\lambda}$. By Proposition 16.5(ii) there are functions $a_i(x, u, \lambda)$ for $i = 1, \dots, n$, and $b_j(u, \lambda)$ for $j = 0, \dots, \ell$, such that

$$\frac{\partial H_k}{\partial \lambda_k} = \sum_i a_i(x, u, \lambda) \frac{\partial H_k}{\partial x_i} + b_0(u, \lambda) + \sum_{j \geq 1} b_j(u, \lambda) \frac{\partial H_k}{\partial u_j}. \tag{16.3}$$

Consider the resulting vector field,

$$X = -\frac{\partial}{\partial \lambda_k} + \sum_i a_i(x, u, \lambda) \frac{\partial}{\partial x_i} + \sum_{j \geq 1} b_i(u, \lambda) \frac{\partial}{\partial u_j}.$$

Integrating this vector field gives a diffeomorphism of the form

$$\Phi_t(x, u, \lambda) = (\phi_t(x, u, \lambda), \psi_t(u, \lambda), \lambda', \lambda_k - t). \tag{16.4}$$

Let us also write $\Psi_t(u, \lambda) = (\psi_t(u, \lambda), \lambda', \lambda_k - t)$. Both Φ_t and Ψ_t are diffeomorphism germs. Now, (16.3) is equivalent to

$$\frac{\mathrm{d}}{\mathrm{d}t}(H_k \circ \Phi_t(x, u, \lambda)) = -b_0(\Psi_t(u, \lambda)).$$

Integrating from $t = 0$ shows

$$H_k \circ \Phi_t(x, u, \lambda) - H_k(x, u, \lambda) = C_t(u, \lambda)$$

for some smooth function $C_t(u, \lambda)$.

Putting $t = \lambda_k$ shows that

$$H_k \circ \Phi_{\lambda_k}(x, u, \lambda) - H_k(x, u, \lambda) = C_{\lambda_k}(u, \lambda).$$

The final component of Φ_{λ_k} is zero, so that

$$H_k \circ \Phi_{\lambda_k}(x, u, \lambda) = H_{k-1} \circ \Phi_{\lambda_k}(x, u, \lambda).$$

Now Φ_{λ_k} is not a diffeomorphism, so we adjust it by changing the final component, so defining the germ $\widetilde{\Phi}$ by

$$\widetilde{\Phi}(x,u,\lambda) = (\phi_{\lambda_k}(x,u,\lambda), \psi_{\lambda_k}(u,\lambda), \lambda', \lambda_k)$$

(compare with (16.4)). It follows from the inverse function theorem that this is the germ of a diffeomorphism.

Since Φ_{λ_k} and $\widetilde{\Phi}$ differ only in the last component it follows that $H_{k-1} \circ \widetilde{\Phi} = H_{k-1} \circ \Phi_{\lambda_k}$. Consequently,

$$H_{k-1} \circ \widetilde{\Phi} = H_k + C(u,\lambda)$$

for some smooth function C derived from b_0. Note that $\widetilde{\Phi}$ is a diffeomorphism of the type required for equivalence of unfoldings, and hence H_k and H_{k-1} are indeed equivalent.

Repeating this for descending values of $j = k, k-1, \ldots, 1$ completes the proof that infinitesimal versality implies versality for $\mathcal{R}^+$-equivalence. ✔

16.4 Application to invariant theory

We end this chapter with a classical application of the preparation theorem, generalizing the result of Whitney described in Example 16.2 for even functions. It requires readers to have some knowledge of invariant theory, and can safely be skipped as it is not used elsewhere. The result is due to E. Bierstone [10].

Let G be a finite group and let the real vector space V be a representation of G of dimension n. Denote by $R = \mathbb{R}[x_1, \ldots, x_n]$ the ring of polynomials on V and by R^G the subring consisting of invariant polynomials. We need two classical results from invariant theory, proofs of which can be found for example in the book by Neusel and Smith [89].

Firstly, Hilbert showed in a remarkable theorem that the ring of invariant polynomials is a finitely generated ring: that is, there are polynomials $p_1, \ldots, p_k \in R^G$ such that any invariant polyomial can be written as a polynomial in $p_1, \ldots, p_k$. (This result led to the notion of Noetherian rings, and arguably to the whole of commutative algebra.)

Let $p_1, \ldots, p_k \in R^G$ be generators of the ring R^G: it is easy to see these can be chosen to be homogeneous of positive degree. Define the polynomial map

$$\begin{aligned} \phi\colon V &\longrightarrow \mathbb{R}^k \\ x &\longmapsto (p_1(x), \ldots, p_k(x)). \end{aligned}$$

Let I_ϕ be the ideal in R generated by $p_1, \ldots, p_k$. The second result we require, due to Emmy Noether, is that the ideal I_ϕ is of finite codimension in R.

Hilbert's theorem states therefore that if f is an invariant polynomial, then there is a polynomial h on $\mathbb{R}^k$ such that $f = h \circ \phi$. The extension of Hilbert's theorem to the smooth setting is the following.

Theorem 16.7. *Let $f\colon (V, 0) \to \mathbb{R}$ be the germ of a smooth invariant function. Then there is a smooth germ $h \in \mathcal{E}_k$ such that $f = h \circ \phi$.*

PROOF: First let $J_\phi = \langle p_1, \ldots, p_k \rangle \lhd \mathcal{E}_n$. Since the original algebraic I_ϕ is of finite codimension in R, it follows that J_ϕ is of finite codimension in $\mathcal{E}_n$. Indeed, let $u_0 = 1$ and let $\{u_0, \ldots, u_r\} \subset R$ be a cobasis for I_ϕ in R. Then this also forms a cobasis for J_ϕ in $\mathcal{E}_n$. We will moreover choose the u_j to have average 0 for $j > 0$, as described in the lemma below.

By the preparation theorem applied to the map ϕ and module $A = \mathcal{E}_n$, for any $f \in \mathcal{E}_n$, there are smooth functions $h_j \in \mathcal{E}_k$ $(j = 0, \ldots, \ell)$ such that

$$f = \sum_{j=0}^{\ell} u_j h_j \circ \phi.$$

This means $f(x) = \sum_j u_j(x) h_j(\phi(x))$. Now suppose f is invariant: $f(gx) = f(x)$ for all $g \in G$. Then, for each g,

$$f(x) = \sum_j u_j(gx) h_j(\phi(x)),$$

since ϕ is also invariant: $\phi(gx) = \phi(x)$. Summing over $g \in G$ (the group is finite), we obtain

$$|G| f(x) = \sum_j \left(\sum_g u_j(gx) \right) h_j(\phi(x)).$$

Since for $j > 0$, the sum $\sum_g u_j(gx) = 0$, it follows that $f(x) = h_0(\phi(x))$ so that h_0 is the required h in the statement of the theorem. ✔

Lemma 16.8. *A cobasis $\{u_0, u_1, \ldots, u_\ell\}$ for I_ϕ in $R = \mathbb{R}[V]$ can be chosen so that $u_0 = 1$, and for $j > 0$, u_j has average 0; that is, for each $j = 1, \ldots, \ell$,*

$$\frac{1}{|G|} \sum_{g \in G} u_j(gx) = 0, \quad \forall x \in \mathbb{R}^n.$$

PROOF: First let $\{v_0, \ldots, v_\ell\}$ be any cobasis, with $v_0 = 1$ and $v_j \in \mathfrak{m}_n$. For each $j > 0$, the function $\rho_j(x) := \frac{1}{|G|}\sum_{g\in G} v_j(gx)$ is invariant. This is because, for $h \in G$,

$$\begin{aligned}\rho_j(hx) &= \frac{1}{|G|}\sum_{g\in G} v_j(ghx)\\ &= \frac{1}{|G|}\sum_{k\in G} v_j(kx)\\ &= \rho_j(x),\end{aligned}$$

as required. The second equality follows because, for any $h \in G$ the set $\{gh \mid g \in G\} = G$. Since ρ_j is invariant, it follows that $\rho_j \in I_\phi$ (by Hilbert's theorem).

Now set $u_0 := v_0 = 1$ and for $j > 0$ set $u_j := v_j - \rho_j$. Then the average of each u_j is zero (for $j > 0$) and $\{u_0, \ldots, u_\ell\}$ also forms a cobasis for I_ϕ (since they differ from the v_j by elements of I_ϕ). ✔

The analogous theorem for representations of compact Lie groups is due to G. Schwarz [103]. In that case, although the ring of polynomial invariants is still finitely generated, the argument above fails because the ideal I_ϕ is no longer of finite codimension in R.

Problems

16.1 Show directly that the two germs h_1 and h_2 in Example 16.2 are uniquely determined by f.

16.2 By considering the germ at 0 of the function $\phi(x) = x^3$, use the preparation theorem to show that given any function germ $f \in \mathcal{E}_1$ there are germs $h_1, h_2, h_3 \in \mathcal{E}_1$ such that

$$f(x) = h_1(x^3) + h_2(x^3)x + h_3(x^3)x^2.$$

16.3 Use Theorem 12.7 to show that the space curve singularity $t \mapsto (t^3, t^4, t^5)$ is 5-determined for left equivalence.

16.4 Extend Example 16.2 to 2 variables in the following two ways. Let $f \in \mathcal{E}_2$.

(i) Show there are smooth germs $h_j \in \mathcal{E}_3$ ($j = 0, 1, 2$) such that

$$f(x, y) = h_0(x^2, xy, y^2) + xh_1(x^2, xy, y^2) + yh_2(x^2, xy, y^2).$$

(ii) Secondly, show there are smooth germs $g_j \in \mathcal{E}_2$ $(j = 0, \dots, 3)$ such that

$$f(x, y) = g_0(x^2, y^2) + xg_1(x^2, y^2) + yg_2(x^2, y^2) + xyg_3(x^2, y^2).$$

(iii) Deduce that if f is even, i.e., $f(-x, -y) = f(x, y)$, then there are smooth functions h_0, g_0 and g_3 such that

$$f(x, y) = h_0(x^2, xy, y^2) = g_0(x^2, y^2) + xyg_3(x^2, y^2).$$

What is the analogous conclusion if f is odd: $f(-x, -y) = -f(x, y)$? (†)

16.5 Extend the previous problem to n variables.

16.6 Referring to Example 12.16, let $R = \mathcal{E}_{x,u}$ and let A be the quotient module $A = \mathcal{E}_{x,u}/\mathcal{E}_{x,u}\{3x^2 + u\}$. Let $\phi\colon \mathbb{R}^2 \to \mathbb{R}$ be the projection $\phi(x, u) = u$. Use the preparation theorem to show that

$$\mathcal{E}_{x,u}\left\{3x^2 + u\right\} + \mathcal{E}_u\{x, 1\} = \mathcal{E}_{x,u}.$$

16.7 Let $f \in \mathcal{E}_1$. Show there exists $h \in \mathcal{E}_2$ and constants $a, b, c \in \mathbb{R}$ such that

$$f(t) = h(t^3, t^4) + at + bt^2 + ct^5.$$

[*Hint*: *see Example* 16.3.] (†)

16.8 Let $F\colon \mathbb{R}^n \times \mathbb{R}^\ell \to \mathbb{R}^p$ be a $\mathcal{K}$-versal unfolding of a map germ $f \in \mathcal{E}_n^p$. Prove the analogue of Proposition 16.5(i), namely that $T\mathcal{K}_{\text{un}}{\cdot}F = \theta(F)$.

16.9 Adapt the argument in Section 16.3 to write down the proof of the versality theorem for contact equivalence (Theorem 14.1). [*See [75] for the solution.*]

16.10 Let p be a polynomial which is invariant under a linear transformation. Show that each homogeneous part of p is also invariant under that transformation.

17

Left–right equivalence

N THIS CHAPTER WE OUTLINE a topic we have not discussed much previously but which plays an important role in singularity theory. We do not go into any great detail, but to omit it entirely would be an oversight.

Recall from Chapter 12 that two map germs (or indeed, maps) are $\mathcal{A}$-equivalent, or left–right equivalent, if they coincide after applying diffeomorphisms of source and target. Often what one wants to understand about the (local) structure of a map between two manifolds, especially a non-linear one, would not depend on the particular coordinate systems chosen to describe it, and as such it is the $\mathcal{A}$-class that contains all the important structure.

We begin the chapter by describing what it means for a map (germ) to be stable, and then discuss Martinet's work relating $\mathcal{A}$-equivalence to unfoldings and contact equivalence. We also illustrate this equivalence with the beginnings of two classifications up to $\mathcal{A}$-equivalence.

17.1 Stable maps and germs

Usually, in mathematics, a property of an object is said to be ***stable*** (or structurally stable) if in some suitable topology, the object has a neighbourhood all of whose elements share that property. Some simple examples are the invertibility of a square matrix, the non-degeneracy of a critical point, and another is the hyperbolicity of an equilibrium point of a vector field, which requires that the linear part of the vector field at the equilibrium point has no purely imaginary eigenvalues.

In [76, II], Mather considers stability of smooth maps between manifolds $f\colon N \to P$, in the sense that a map f is stable if there is a neighbourhood U of f in $C^\infty(N,P)$ in a suitable topology (the 'Whitney topology') such that for every $g \in U$ there are diffeomorphisms ϕ of N and ψ of P such that $g = \psi \circ f \circ \phi$ (i.e. f and g are left–right equivalent). Mather defines the notion of infinitesimal stability of a map, involving vector fields along f, and proves that infinitesimal stability implies stability of f, provided f is a proper map.

On the other hand, since sets of germs have no useful topology, one says a map germ $f\colon (\mathbb{R}^n,0) \to (\mathbb{R}^p,0)$ is ***stable*** if every unfolding of f is $\mathcal{A}$-trivial in the following sense.

Definition 17.1. A smooth germ $f\colon (\mathbb{R}^n,0) \to (\mathbb{R}^p,0)$ is said to be ***stable*** if every unfolding $F(x;u)$ of f is $\mathcal{A}_{\mathrm{un}}$-equivalent to the trivial unfolding $(x,u) \mapsto f(x)$. More explicitly, there are unfoldings $\Phi(x,u)$ of the identity on $\mathbb{R}^n$ and $\Psi(y,u)$ of the identity on $\mathbb{R}^p$ such that

$$\Psi(F(\Phi(x,u),u),u) = f(x).$$

☆

To be an unfolding of the identity means $\Phi(x,0) = x$ (and similarly for Ψ). The infinitesimal version of this notion of stability is that $T_e\mathcal{A}\cdot f = \theta(f)$ (see Section 12.4 for definitions), and one shows, using the preparation theorem, that infinitesimal stability implies stability [75, Chapter XIV].

We have essentially shown in Part I that the only stable germs $(\mathbb{R}^n,0) \to (\mathbb{R},0)$ are the nondegenerate critical points (where $\mathcal{R}^+$-equivalence suffices, rather than the full strength of $\mathcal{A}$-equivalence). For map germs between surfaces, that is $(\mathbb{R}^2,0) \to (\mathbb{R}^2,0)$, Whitney [118] showed that the only two are the fold and the cusp, with local normal forms

fold $(x,y) \mapsto (x,y^2)$

cusp $(x,y) \mapsto (x,y^3+xy)$.

See Section 17.3 for a higher–dimensional case, and Section 17.4 for the start of a classification of germs of parametric curves in the plane.

Although we concentrate on germs throughout this book, one cannot discuss stability without a brief digression into more global questions. Here one does use the topological notion of stability in terms of open sets. Details about the discussion below can be found for example in the book of Golubitsky and Guillemin [45].

Let N,P be two compact[1] manifolds, and $C^\infty(N,P)$ the set of smooth maps from the first to the second. This set can be endowed with a topology, called the Whitney topology, where the open sets are defined in terms of the jet spaces.

A map $f\colon N \to P$ is said to be ***stable*** if there is a neighbourhood U of f in $C^\infty(N,P)$ in the Whitney topology such that every map in U is $\mathcal{A}$-equivalent to f. The principal question is whether the stable maps are dense; that is, whether any map can be approximated by a stable one.

[1]One can relax the compactness assumption by restricting attention to proper maps, provided N,P allow such maps to exist, but for simplicity we restrict our discussion to compact manifolds.

In 1955, Whitney [118] published the first work in this direction, where he shows that if $\dim N = \dim P = 2$, with both compact, then there is an open dense subset U of $C^\infty(N, P)$ in the Whitney topology, for which every map in U has only fold or cusp singularities.

The original hope (in the 1960s) was that this density of the set of stable maps was universally true, but Mather showed this not to be the case. Indeed, whether or not the set of stable maps is dense depends only on the dimensions of the source and target, and Mather determined precisely for which dimensions the stable maps are dense: what he calls the *nice dimensions* [76, VI] . For example, if $\dim N = \dim P \leq 8$ then stable maps are dense, but for $\dim N = \dim P \geq 9$ they are not. The obstruction to the density of stable maps arises when singularities with a modulus cannot be perturbed away, for then an arbitrarily small perturbation of the given map may change the value of the modulus, so producing an inequivalent map.

A solution to this lack of density of stable maps is to introduce the notion of *topological stability*, where one asks that the nearby maps are left–right equivalent, but by homeomorphisms rather than diffeomorphisms. The conclusion is that the set of topologically stable smooth (and proper) maps between two manifolds is dense. This was first conjectured by Thom in the 1960s, and proved by Mather in the 1970s. The first full account of the proof appeared in [42]. An in-depth study of topological stability can be found in the monograph [94].

17.2 Versal unfoldings and stable maps

Recall that if $F\colon (\mathbb{R}^n \times \mathbb{R}^a, (0,0)) \to (\mathbb{R}^p, 0)$ is an unfolding of $f \in \mathcal{E}_n^p$ then we write $Z_F = F^{-1}(0)$ and $\pi_F\colon Z_F \to \mathbb{R}^a$ is the restriction to Z_F of the natural projection, and the unfolding is *regular* if the map F is itself a submersion (germ). For a regular unfolding, Z_F is therefore a submanifold of $\mathbb{R}^n \times \mathbb{R}^a$ of dimension $n - p + a$.

Let $F, G\colon (\mathbb{R}^n \times \mathbb{R}^a, (0,0)) \to (\mathbb{R}^p, 0)$ be two a–parameter unfoldings of a map germ $f\colon (\mathbb{R}^n, 0) \to (\mathbb{R}^p, 0)$. The following fundamental property of unfoldings was noticed by Martinet, and which we refer to as ***Martinet's first theorem***.

Theorem 17.2 (Martinet [73]). *Suppose F and G are regular unfoldings of a germ f. Then F and G are $\mathcal{K}_{\mathrm{un}}$-equivalent if and only if the map germs π_F and π_G are left–right equivalent.*

PROOF: Suppose first that F and G are $\mathcal{K}_{\mathrm{un}}$-equivalent. The resulting equivalence maps Z_F to Z_G and takes π_F to π_G. Indeed, suppose $F(\phi(x,u), \psi(u)) =$

$M(x,u)\,G(x,u)$. Then $(x,u) \in Z_G$ if and only if $\Phi(x,u) \in Z_F$, where $\Phi(x,u) = (\phi(x,u), \psi(u))$. Moreover, the following diagram commutes:

$$\begin{array}{ccc} Z_F & \xrightarrow{\Phi} & Z_G \\ \pi_F \downarrow & & \downarrow \pi_G \\ (\mathbb{R}^a, 0) & \xrightarrow{\psi} & (\mathbb{R}^a, 0) \end{array}$$

which shows that, provided Z_F and Z_G are submanifolds, π_F and π_G are $\mathcal{A}$-equivalent.

Conversely, suppose unfoldings F and G are such that π_F and π_G are $\mathcal{A}$-equivalent, as in the commutative diagram. Then, for $(x,u) \in Z_F$, $\Phi(x,u) = (\phi(x,u), \psi(u))$, for some smooth germ ϕ defined on Z_F.

Let $\tilde{\phi}\colon \mathbb{R}^n \times \mathbb{R}^a \to \mathbb{R}^n$ be any extension of ϕ, and define $\widetilde{\Phi}\colon \mathbb{R}^n \times \mathbb{R}^a \to \mathbb{R}^n \times \mathbb{R}^a$ in the obvious way,

$$\widetilde{\Phi}(x,u) = (\tilde{\phi}(x,u), \psi(u)).$$

We claim $\tilde{\phi}$ can be chosen so that $\widetilde{\Phi}$ is the germ of a diffeomorphism. It then follows that $\widetilde{\Phi}^* I_G = I_F$ whence (by the arguments used in Theorem 11.5) there is a matrix-valued function $M(x,u)$ such that $G(x,u) = M(x,u)\,F \circ \widetilde{\Phi}(x,u)$, and hence F and G are $\mathcal{K}_{\mathrm{un}}$ equivalent as required.

To prove the claim, it suffices to show that $\tilde{\phi}$ can be chosen so that $\mathsf{d}_x\tilde{\phi}_{(0,0)} : \mathbb{R}^n \to \mathbb{R}^n$ is invertible. Let $K = \ker(\mathsf{d}f_0) \subset \mathbb{R}^n$. Then the intersection of $\mathbb{R}^n$ with the tangent spaces to both Z_F and Z_G at the origin is equal to K; that is, $(\mathbb{R}^n \times \{0\}) \cap T_0 Z_F = K$, and likewise for Z_G.

Let L be a complementary subspace to K in $\mathbb{R}^n$, let $R_F, R_G \subset \mathbb{R}^a$ be the respective images of $\mathsf{d}\pi_F$ and $\mathsf{d}\pi_G$, and let S_F, S_G be respective complementary subspaces to R_F, R_G in $\mathbb{R}^a$. By the implicit function theorem, we can express each of V_F and V_G as the graph of a map $(\alpha_F, \beta_F)\colon K \times R_F \to L \times S_F$ and $(\alpha_G, \beta_G)\colon K \times R_G \to L \times S_G$. Let $(x,y,u,v) \in K \times L \times R_F \times S_F$. Then the map $(x,y,u,v) \mapsto (x, y - \alpha_F(x,u), u, v - \beta_F(x,u))$ and analogously for G, are straightening maps for V_F and V_G respectively. After straightening, the given map $\Phi\colon V_F \to V_G$ satisfies $\phi(x,0,u,0) = (\phi_1(x,u), 0) \in \mathbb{R}^n$. Define $\tilde{\phi}\colon \mathbb{R}^n \times \mathbb{R}^a$ by, in the simplest fashion:

$$\tilde{\phi}(x,y,u,v) = (\phi(x,u), y).$$

This is clearly an extension of ϕ and its restriction on $\mathbb{R}^n$ is a diffeomorphism, as required. ✔

This result can be extended to cases where Z_F and Z_G are not submanifolds, if one extends appropriately the definition of $\mathcal{A}$-equivalence, see [80].

Furthermore, Martinet shows the following central result, which we refer to as ***Martinet's second theorem***.

Theorem 17.3 (Martinet [73]). *Let F be a regular unfolding of $f \in \mathcal{E}_n^p$. Then F is $\mathcal{K}$-versal if and only if π_F is stable.*

Note that the hypothesis that F is regular is not really necessary: if it is not regular then Z_F is not a submanifold, and so π_F cannot be stable since a non-trivial unfolding of π_F can be obtained by deforming Z_F to be smooth.

PROOF: Martinet proves this using the infinitesimal criteria; here we give a direct proof using the definitions of stability and versality, and the theorem above.

Firstly, suppose π_F is a stable germ, and let H be any c–parameter extension of F; that is, $H\colon (\mathbb{R}^n \times \mathbb{R}^a \times \mathbb{R}^c, (0,0,0)) \to (\mathbb{R}^p, 0)$ is such that $H(x,u,0) = F(x,u)$. We wish to show that H is equivalent to $\widehat{F}(x,u,v) = F(x,u)$ (independent of v), which would imply that F is versal. Note that since F is regular, so is H. Consider the associated map $\pi_H : Z_H \to \mathbb{R}^a \times \mathbb{R}^c$. Restriciting to $v = 0$ reduces H to F, and hence $\pi_H(x,u,0) = (\pi_F(x,u), 0)$. Thus π_H is an extension of π_F and since we are assuming that π_F is stable, it follows that π_H is $\mathcal{A}$-equivalent to the trivial extension $\pi_{\widehat{F}}$. It follows from the theorem above that H is $\mathcal{K}_{\mathrm{un}}$-equivalent to $\widehat{F}$, as required.

Conversely, suppose F is versal. For simplicity, we assume f is of rank 0 at the origin, and F is of the form $F(x;u,v) = F_0(x;u) - v$ with $F_0(0;u) = 0$. If this is not the case, we can write f in the form $f(x,y) = (f_0(x), y)$ (after a contact equivalence), and then carry the variable y throughout the argument in a trivial manner. Moreover, since all versal unfoldings are $\mathcal{K}_{\mathrm{un}}$-equivalent, by Martinet's first theorem (Theorem 17.2) above we have not lost any generality.

Thus Z_F can be parametrized by (x,u), and $\pi_F(x,u) = (u, F_0(x,u))$. Now let $\phi\colon Z_F \times \mathbb{R}^c \to \mathbb{R}^a \times \mathbb{R}^c$ be an extension of π_F. If we can show that there is an unfolding H of f extending F for which $\phi = \pi_H$ then we are done, for by the versality of F, H is equivalent to the trivial extension $\widehat{F}$ of F, and hence $\phi = \pi_H$ is a trivial extension (suspension) of π_F. Since this would hold for an arbitrary extension ϕ, it would follow that π_F is indeed a stable germ.

To this end write,

$$\phi(x,u,w) = (u + w\phi_1(x,u,w),\ F_0(x,u) + w\phi_2(x,u,w),\ w).$$

Since the first component here is a submersion, changing u to $\bar{u} = u + w\phi_1(x, u, w)$ is a right equivalence that reduces ϕ to the form (still writing u rather than $\bar{u}$) $\phi(x, u, w) = (u, H_0(x, u, w), w)$. Now let $H(x, u, v, w) = H_0(x, u, w) - v$. Then H is an extension of F, and by the versality of F it is $\mathcal{K}_{\text{un}}$-equivalent to $H'(x, u, v, w) = F(x, u) - v$ (independent of w). It follows by the previous theorem that $\phi = \pi_H$ is $\mathcal{A}$-equivalent to a suspension of π_F as required. ✔

For example, the fold map arises from the versal unfolding of the germ $y \mapsto y^2$ while the cusp arises from the versal unfolding of $y \mapsto y^3$.

We return to these ideas briefly in Chapter 22.

Lemma 17.4. *Let F be a regular unfolding of f, and π_F the associated projection. The maps $f, \widetilde{F}$ and π_F all have the same core, where*

$$\widetilde{F}(x, u) = (F(x, u), u).$$

PROOF: That f and $\widetilde{F}$ have the same core follows from Proposition B.32, using the fact that the second component of $\widetilde{F}$ is a submersion. On the other hand, π_F and $\widetilde{F}$ also have the same core, as follows from the same proposition using the fact that the first component of $\widetilde{F}$ is a submersion. ✔

Example 17.5. Let us show briefly how the above results due to Martinet can be used to find stable map germs. Let us find all stable map germs $(\mathbb{R}^3, 0) \to (\mathbb{R}^2, 0)$.

Begin with a germ $f_0\colon (\mathbb{R}^n, 0) \to (\mathbb{R}^p, 0)$ with $\operatorname{codim}(f_0, \mathcal{K}) = \ell$, and without loss of generality we can assume f_0 is of rank 0 at the origin. Let $F\colon \mathbb{R}^n \times \mathbb{R}^k \to \mathbb{R}^p$ be a $\mathcal{K}$-versal unfolding of f_0 (necessarily, $k \geq \ell$). Then $\dim Z_F = (n - p + k)$, and so $\pi_F\colon Z_F \simeq \mathbb{R}^{n-p+k} \to \mathbb{R}^k$.

Now, with F versal, π_F will be stable. Therefore we require $n - p + k = 3$ and $k = 2$, whence $p = n - 1$, and $\operatorname{codim}(f_0, \mathcal{K}) \leq 2$. We go through case by case.

- First take $\ell = 0$. This requires f_0 to be a submersion. Therefore by the lemma above, π_F will also be a submersion; that is $\pi_F(x, y, z) = (x, y)$ say.
- Now consider $\ell = 1$, and $f_0(x, y) = x^2 \pm y^2$ (if $n > 2$ and $\operatorname{rk} f_0 = 0$ then $\operatorname{codim}(f_0, \mathcal{K}) > 2$). This gives

  $$F(x, y, u, v) = x^2 \pm y^2 - v,$$

  say, whence $\pi_F(x, y, u) = (x^2 \pm y^2, u)$.

- Now consider $\ell = 2$. Up to $\mathcal{K}$-equivalence, the only germ of rank 0 and $\mathcal{K}$-codimension 2 is $f_0(x, y) = x^3 + y^2$. A versal unfolding is,

$$F(x, y; u, v) = x^3 + y^2 + ux - v,$$

in which case $\pi_F(x, y, u) = (x^3 + ux + y^2, u)$.

Summarizing, we obtain

Theorem 17.6. *Any stable map germ* $(\mathbb{R}^3, 0) \to (\mathbb{R}^2, 0)$ *is* $\mathcal{A}$*-equivalent to one of*

- $(x, y, z) \mapsto (x, y)$,
- $(x, y, z) \mapsto (x^2 \pm y^2, z)$,
- $(x, y, z) \mapsto (x^3 + xz + y^2, z)$.

These are known as a submersion, fold and cusp respectively. ✐

17.3 Example: maps of the plane to 3-space

David Mond [79] has given a classification of the simple map germs $(\mathbb{R}^2, 0) \to (\mathbb{R}^3, 0)$, up to $\mathcal{A}$-equivalence. The only singular stable map germ in these dimensions is known as the ***pinch-point***, and given by

$$(x, y) \longmapsto (x,\ xy,\ y^2). \tag{17.1}$$

This singularity was already known to Whitney [116, 117] (where he considers maps $\mathbb{R}^n \to \mathbb{R}^{2n-1}$; see Problem 17.6). The image is the famous *Cayley cross-cap*, which is the subset of $\mathbb{R}^3$ with equation $Y^2 = X^2 Z$ and $Z \geq 0$. A curved version of this is shown in Figure 17.1 (a more 'geometric' form is shown in Figure 19.1).

The simplest method to show this germ is stable is to use Martinet's theorem above, see Problem 17.1.

Mond shows there are just two inequivalent map germs of $\mathcal{A}$-codimension 1, differing by a sign, which are

$$f_{\pm}(x, y) = (x, y^2, y^3 \pm x^2 y). \tag{17.2}$$

Their versal unfoldings are,

$$F_{\pm}(x, y; u) = (x, y^2, y^3 \pm (x^2 + u) y). \tag{17.3}$$

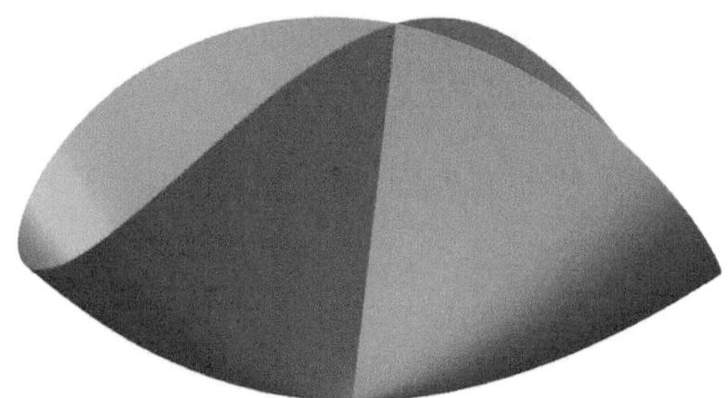

FIGURE 17.1 A pinch-point singularity.

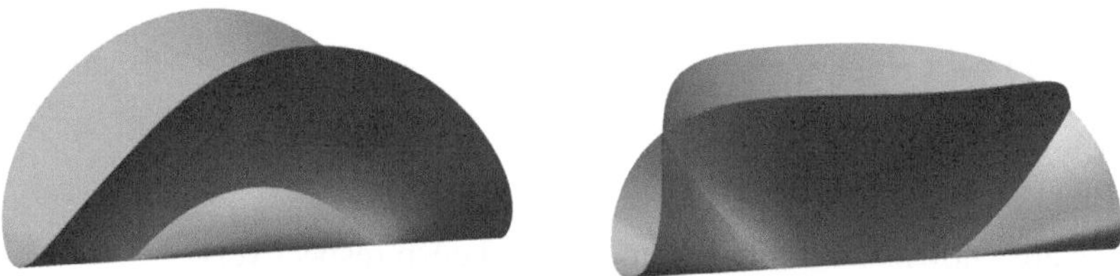

FIGURE 17.2 Deformations of the degenerate pinch-point singularities: f_- on the left and f_+ on the right; they show the images of $F_{\pm,u}$ with $u < 0$. Both deformations have two non-degenerate pinch-points.

For $u < 0$, the maps $F_{\pm,u}$ have two pinch-points, while for $u > 0$ they have none. For $u < 0$, the image of the map $F_{+,u}$ has a double locus consisting of a single curve joining the two pinch-points, while in the image of $F_{-,u}$ the double locus forms two separate curves (Figure 17.2 and Problem 17.5). See [79] for further details, and singularities of higher codimension.

17.4 Plane curves revisited

In Chapter 8 we considered plane curves as the zero-set of a smooth map $\mathbb{R}^2 \to \mathbb{R}$. The alternative approach, giving rise to a different theory, is to consider them as parametric curves, $\gamma : \mathbb{R} \to \mathbb{R}^2$. The principal difference between the two is that perturbations can be very different: for example, there are perturbations of the cusp curve described in Chapter 8 with two connected components, which is evidently not possible as the image of a map $\mathbb{R} \to \mathbb{R}^2$; see Figure 8.3.

For the classification, we consider smooth map germs $\gamma\colon (\mathbb{R}, 0) \to (\mathbb{R}^2, 0)$ up to left–right equivalence. For this we need the left tangent space $T_e\mathcal{L}{\cdot}\gamma$ described in Section 12.3; recall that L_γ is the $\mathcal{E}_2$-module

$$L_\gamma = \{h \circ \gamma \mid h \in \mathcal{E}_2\} \subset \mathcal{E}_1,$$

and then $T_e\mathcal{L}{\cdot}\gamma = L_\gamma \times L_\gamma \subset \theta(\gamma)$. The classification is based on the theorem of Bruce, Gaffney and du Plessis [17] which states that two map germs γ, δ are left-equivalent if and only if $L_\gamma = L_\delta$. The corresponding finite determinacy theorem is stated as Theorem 12.7.

Preliminary preparation We can write the Taylor series of γ as

$$\gamma(t) = \mathbf{a}_1 t + \mathbf{a}_2 t^2 + \mathbf{a}_3 t^3 + \cdots,$$

where $\mathbf{a}_j = (a_j, b_j) \in \mathbb{R}^2$. We say γ is of ***order*** k if

$$\mathbf{a}_j = \mathbf{0} \text{ for } j < k, \quad \text{but} \quad \mathbf{a}_k \neq \mathbf{0}.$$

If γ is of order k then we can choose a basis in $\mathbb{R}^2$ so that $\mathbf{a}_k = (1, 0)$. Moreover, if γ is of finite codimension, then there is an $\ell > k$ and a $b \neq 0$ such that

$$\gamma(t) = (t^k + \cdots, \; bt^\ell + \cdots).$$

If there were no such ℓ then one can show that, for every $n > 0$, $L_\gamma \not\supset \mathfrak{m}_2^n$. By using a left equivalence, we may assume that ℓ is coprime to k, and $b = 1$. Using right equivalence we may reduce the first component to t^k (Theorem 6.2). Thus our pre-normal form for a finitely determined curve singularity is

$$\gamma(t) = (t^k, \; t^\ell + \cdots), \tag{17.4}$$

with $\ell > k$.

Non-singular curves Suppose $\gamma'(0) \neq 0$; that is, γ is of order 1. Then the local immersion theorem tells us coordinates can be chosen on $\mathbb{R}^2$ such that $\gamma(t) = (t, 0)$. These are the only stable germs.

Cusps Now consider curves of order 2. These are called (generalized) cusps. The curve $\gamma(t) = (t^2, t^3)$ is the standard 'ordinary' cusp, while $\gamma(t) = (t^2, t^5)$ is called the ramphoid cusp.[2]

[2] *Ramphoid* derives from the Greek, meaning beak-like; originally it meant a cusp where both branches lie on the same side of the limiting tangent line: this would hold for the left-equivalent curve $t \mapsto (t^2, t^4 + t^5)$; see Figure 17.3.

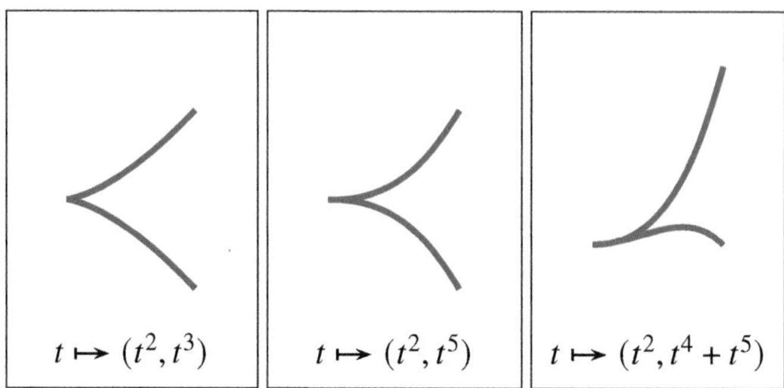

FIGURE 17.3 An ordinary cusp, a simple ramphoid cusp and a left-equivalent curved ramphoid cusp.

Proposition 17.7. *Let $\gamma\colon (\mathbb{R}, 0) \to (\mathbb{R}^2, 0)$ be a finitely $\mathcal{A}$-determined curve germ of order 2. Then there is an integer $r > 0$ such that γ is $\mathcal{A}$-equivalent to*

$$\gamma_r(t) = (t^2,\ t^{2r+1}).$$

PROOF: Using the argument above for the pre-normal form shows that we can choose coordinates so that $\gamma(t) = (t^2, h(t))$ for some $h \in \mathfrak{m}_1^3$. Write $h = h_0 + h_1$ where h_0 is even and h_1 is odd. Now by the preparation theorem (see Example 16.2), we can write $h_0(t) = h_2(t^2)$, and $h_1(t) = t^{2r+1}h_3(t^2)$ for some $r \geq 1$, with $h_3(0) \neq 0$ (one can show without difficulty that if the odd part of h vanishes to all orders then γ is not finitely determined). Now consider the diffeomorphism germ of the plane

$$\psi(x, y) = (x, y + h_2(x) + yh_3(x)).$$

Then $\gamma = \psi \circ \gamma_r$, showing the two are left-equivalent. ✔

Further details of the classification can be found in the paper of Bruce and Gaffney [16], where they also discuss the relation between parametric and Cartesian forms of a curve singularity. A classification of space curves up to $\mathcal{A}$-equivalence can be found in [43], though they restrict attention to complex curves.

Applications of singularities of plane curves to fluid mechanics are described in the review [35].

17.5 Multigerm singularities

To pass from stability of germs to stability of maps one needs to take into account the possibility that two or more singular points of a map have the same value.

If all the germs of a (proper) map are stable, and distinct singular points have distinct values, then the map is stable (the argument involves gluing together local diffeomorphisms). On the other hand, if two singular points have the same value then the local discriminants need to be in some sense in 'general position'. Making sense of these ideas involves introducing multigerms as follows.

Let S be a finite subset of $\mathbb{R}^n$, say $S = \{s_1, \ldots, s_r\}$. A neighbourhood of S is a union of neighbourhoods of the points s_i. Consider the set of smooth functions defined on neighbourhoods of S, and we say two such functions f, g are ***multigerm equivalent*** at S if there is a neighbourhood U of S on which $f(x) = g(x)$ $(\forall x \in U)$.

The definition of $\mathcal{A}$-equivalence of two multigerms is as one would expect: $f, g\colon (\mathbb{R}^n, S) \to (\mathbb{R}^p, 0)$ are $\mathcal{A}$-equivalent if there is a diffeomorphism germ ϕ of $\mathbb{R}^n$ at S and a diffeomorphism germ at 0 of $\mathbb{R}^p$ such that $f = \psi \circ g \circ \phi$. We give one very simple example, but not treat this interesting topic any further. Some details and classifications can be found in [119, 24, 59].

Example 17.8. Let $S = \{x_1, x_2\} \subset \mathbb{R}$ consist of two points. Consider the multigerm $f\colon (\mathbb{R}, S) \to (\mathbb{R}^2, 0)$ given by $(f(s) = (s, 0),\ f(t) = (t, t^2))$ where s is a local coordinate about x_1 and t about x_2. The two local images are tangent curves, and f is therefore not stable. A versal unfolding is given simply by deforming one of the components, for example, $(F(s; u) = (s, 0),\ F(t, u) = (t, t^2 + u))$. We leave it to the reader to visualize this elementary family. ✐

17.6 Geometric Criterion

We have seen several geometric criteria for the finite determinacy of holomorphic map germs, for $\mathcal{R}, \mathcal{L}$ and $\mathcal{K}$-equivalence. There is a similar criterion for $\mathcal{A}$-equivalence, but it is more difficult to work with. Some details can be found in the survey paper [115].

Theorem 17.9 (Geometric criterion for $\mathcal{A}$). *The holomorphic map germ $f\colon (\mathbb{C}^n, 0) \to (\mathbb{C}^p, 0)$ is of finite $\mathcal{A}$-codimension if and only if there is a neighbourhood U of the origin in $\mathbb{C}^n$ on which there is defined a representative (which we also denote f) such that for every finite subset $S \subset U \setminus \{0\}$ the multigerm of f at S is stable.*

Problems

17.1 Let $f\colon (\mathbb{R},0) \to (\mathbb{R}^2,0)$ be given by $f(t) = (t^2,0)$. Find a $\mathcal{K}$-versal unfolding F of this map germ, and show that π_F is equivalent to the pinch-point map (17.1), and use this to prove that this map germ is stable.

17.2 Suppose $f \in \mathcal{E}_n^p$ is a stable germ. Show that the so-called 'graph unfolding' $F\colon \mathbb{R}^n \times \mathbb{R}^p \to \mathbb{R}^p$ of f given by

$$F(x;u) = f(x) - u$$

is a $\mathcal{K}$-versal unfolding of f.

17.3 Deduce from the previous problem a theorem due to Mather: if f and g are stable map germs, then they are $\mathcal{A}$-equivalent if and only if they are $\mathcal{K}$-equivalent. (†)

17.4 Extend Example 17.5 to classify stable map germs $(\mathbb{R}^n,0) \to (\mathbb{R}^2,0)$. (†)

17.5 Consider the codimension-1 maps $f_\pm$ given in (17.2).

(i). Show, using the first 2 components, that if $f(x,y) = f(x',y')$ then $x = x'$ and $y = \pm y'$. Deduce that f_+ has no double points in its image, whereas f_- has a line of double points.

(ii). Now consider the double point locus in the image for the versal unfoldings (17.3) and justify the description following that equation (cf. Figure 17.2).

17.6 In [116, 117], Whitney considers the following map $W\colon \mathbb{R}^n \to \mathbb{R}^{2n-1}$, which extends the pinch-point to higher dimensions:

$$W(x_1,\dots,x_n) = (x_1^2, x_2,\dots,x_n,\ x_1x_2,\dots,x_1x_n).$$

Show that (i) W has an isolated singular point, (ii) the double locus is the image of the x_1-axis, and (iii) the germ at the origin of W is stable (using Martinet's second theorem). (†)

17.7 Consider the map germ $f\colon (\mathbb{R}^4,0) \to (\mathbb{R}^4,0)$ given by

$$f(x,y,u,v) = (x^2 + uy,\ y^2 + vx,\ u,\ v).$$

Show that f is stable. (This gives an example of a stable corank-2 map germ with the least possible dimensions for source and target.)

Part III

Bifurcation theory

18

Bifurcation problems and paths

BIFURCATION THEORY, IN THE FORM we consider, is the study of solutions of equations, and how they change as the equations are perturbed. The methods of finite determinacy and versal unfoldings lend themselves to such questions, and were more or less motivated by them. The uses of singularity theory in the study of bifurcations was to a large extent pioneered by Martin Golubitsky in the late 1970s and early 1980s; in particular there are two seminal papers by Golubitsky and David Schaeffer published in 1979 [48] (and the subsequent book [49]) where they use singularity theory to study so-called *imperfect* bifurcations; see for example the discussion of the pitchfork bifurcation in Chapter 1. There are many applications of these ideas, and for example the book by Ikeda and Murota [60] illustrates them as they arise in engineering problems.

As described briefly in Chapter 10, a k–parameter ***bifurcation problem*** is an equation of the form

$$g(x;\,\lambda) = 0, \tag{18.1}$$

with $\lambda \in \mathbb{R}^k$ the parameter and $x \in \mathbb{R}^n$ often called the state variable. One assumes the map

$$g\colon \mathbb{R}^n \times \mathbb{R}^k \rightarrowtail \mathbb{R}^p.$$

to be smooth. As usual we write $g_\lambda(x) = g(x,\,\lambda)$. For each value of $\lambda \in \mathbb{R}^k$, (18.1) is a system of p equations in n unknowns and we are interested in the solution set $g_\lambda^{-1}(0)$. This suggests that contact equivalence is the appropriate equivalence relation to use on g_λ, and then contact equivalence of families (or unfoldings) for g, as described in Chapter 14.

Definition 18.1. Two bifurcation problems $F, G\colon (\mathbb{R}^n \times \mathbb{R}^a,\ (0,0)) \to (\mathbb{R}^p, 0)$ are ***equivalent*** if firstly the germs F_0 and G_0 are $\mathcal{K}$-equivalent, and secondly the unfoldings F and G are $\mathcal{K}_{\mathrm{un}}$-equivalent. ✯

It is customary in bifurcation theory to consider primarily 1–parameter families, so $\lambda \in \mathbb{R}$. However, there is no intrinsic mathematical reason for doing this, and one can consider multi–parameter bifurcation theory using the same ideas.

It is also mathematically reasonable to allow the source and target dimensions to differ (as we allow in Chapter 10), although in applications it is not so common.

If $n = p$, the most common type of problem, there will be a finite number of isolated solutions x for each value of λ, and this number can change as λ crosses the discriminant.

If $n > p$ then the solution sets $g_\lambda^{-1}(0)$ will not consist of isolated points, but will be sets of higher dimension. For example, if $n = p + 1$, the solution sets will be curves in $\mathbb{R}^n$. We made a study of these in the case $n = 2$ and $p = 1$, including finite determinacy and unfoldings, in Chapter 8 (although we used right equivalence rather than contact equivalence there).

If $n < p$, then the system of equations is overdetermined, and generically one would expect there to be no solutions. In this case the discriminant Δ would correspond to those parameter values for which a solution exists. See Problem 18.3.

In practice, a bifurcation problem usually arises in some manner from a family of differential equations. The first step is to distinguish bifurcation points (x_0, λ_0) in the family where the equations are most degenerate in some sense. One can then use such points as *organizing centres* and apply the methods of bifurcation theory we describe below to find which phenomena should occur nearby and how the different phenomena (types of solution or bifurcation) fit together.

In the path approach to bifurcation theory, which was introduced in a heursitic manner in [48, Chapter 3, §12], and which we adopt, one begins by finding the versal unfolding of the organizing centre, the germ of g_{λ_0} at x_0. We write $F(x; u)$ for this versal unfolding. Then by the definition of versality, the given family $g(x, \lambda)$ is equivalent to an unfolding induced by a map h taking λ to $u = h(\lambda)$. In the notation we've been using before, g is equivalent to h^*F.

In this way we can represent any bifurcation problem $g(x, \lambda)$ whose core g_0 is of finite $\mathcal{K}$-codimension, as a map (germ) $h(\lambda)$ into the base space of a versal unfolding of the core. And recall that bifurcations within the family g occur at points where $h(\lambda) \in \Delta$, the discriminant of the versal deformation. One is therefore interested in the set of parameter values mapped to Δ; that is, in $h^{-1}(\Delta)$.

In this chapter we consider some examples of 1–parameter bifurcations and the corresponding paths and how they deform. We begin with a detailed study of the pitchfork bifurcation, and then proceed with a less detailed look at some other well-known bifurcations. In the following chapters we introduce the notion of $\mathcal{K}_\Delta$-equivalence, which measures the required contact of h with Δ.

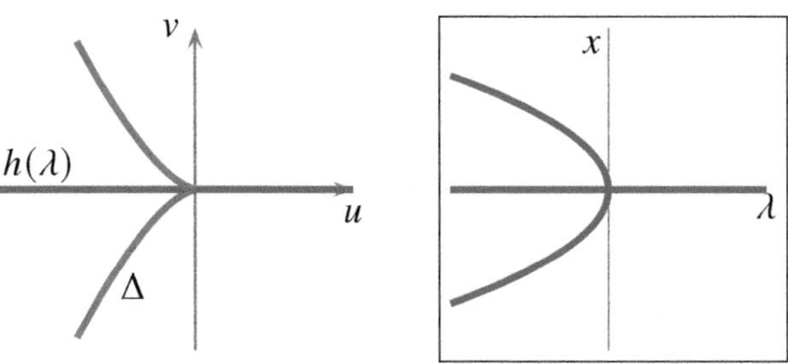

FIGURE 18.1 On the left, (the image of) the path h defining the pitchfork bifurcation, with resulting bifurcation diagram on the right.

18.1 The pitchfork bifurcation

Recall from Chapter 1 that the pitchfork bifurcation is given by

$$g(x;\lambda) = x^3 + \lambda x = 0.$$

(Changing λ to $-\lambda$ makes no essential difference.) The core is therefore $g_0(x) = x^3$ and we saw in Section 14.2 that the $\mathcal{K}$-versal deformation of x^3 is given by

$$F(x;u,v) = x^3 + ux + v.$$

The pitchfork is therefore defined by the path $(u,v) = h(\lambda) = (\lambda, 0)$; see Figure 18.1. The important feature of the path is how it meets the discriminant Δ of the family F, which is given by the algebraic equation $4u^3+27v^2 = 0$, as seen in Section 14.1.

Any perturbation of the pitchfork bifurcation can be obtained by perturbing the path h. This suggests we consider the following 2–parameter deformation of the path h (which we show in Chapter 21 to be a versal deformation in the appropriate sense):

$$H(\lambda;a,b) = (\lambda, a\lambda + b). \tag{18.2}$$

Different choices of $(a,b) \in \mathbb{R}^2$ will determine different perturbations of the path h. The coefficient a corresponds to a rotation of the path, while b translates it off the origin. Non-zero values of a, b result in so-called imperfect versions of the pitchfork.

As a bifurcation, the family H of paths becomes

$$G(x,\lambda;a,b) = x^3 + \lambda x + a\lambda + b = 0. \tag{18.3}$$

Diagrams of these perturbations for different values of (a,b) are shown in Figure 18.2 on p. 223, with a discussion below. Note that (18.3) is equivalent to the versal unfolding of the pitchfork bifurcation found by different methods in [49] (see Problem 18.2).

Discussion of Figure 18.2: This shows the different types of deformation of the pitchfork bifurcation, each obtained by deforming the path $h(\lambda) = (\lambda, 0)$. The diagram at the foot is the base space of the unfolding parametrized by (a, b). We distinguish 4 regions, labelled A–D. Note that changing (a, b) to $(-a, -b)$ corresponds to changing x to $-x$ in the equation $G(x, \lambda; a, b) = 0$, whence the bifurcation diagrams for regions C and D are the same as for A and B, but reflected in the λ-axis.

The red curve (discriminant $\mathcal{D}$ of this family) in this lower diagram corresponds to degenerate bifurcations: there are two components, one ($b = 0$) where the perturbed path passes through the cusp of Δ, while the other (given by $b = a^3$) is where the path is tangent to Δ. These correspond to the 2 types of codimension 1 bifurcation: transcritical and hysteresis respectively, as we see in Chapter 21.

Region A For (a, b) in this region, the path only meets Δ in a single point, corresponding to the single saddle-node bifurcation;

Transition A/B Here the path is tangent to Δ, corresponding to the transcritical bifurcation, and intersects it in one other point, corresponding to the saddle-node;

Region B In this region the path meets the discriminant Δ in three distinct points, corresponding to the three saddle-node points in the bifurcation diagram;

Transition B/C the path passes through the cusp point (the hysteresis bifurcation point; see Section 18.2D below) and one other point of Δ (a saddle-node).

Regions C and D and the corresponding transitions are similar to A and B as explained above.

18.2 Some standard bifurcations

We consider briefly several well-known 1–parameter bifurcations from the path point of view, but with fewer details than for the pitchfork case above. In the following chapters we proceed to find a general classification of bifurcations, with up to 3 parameters.

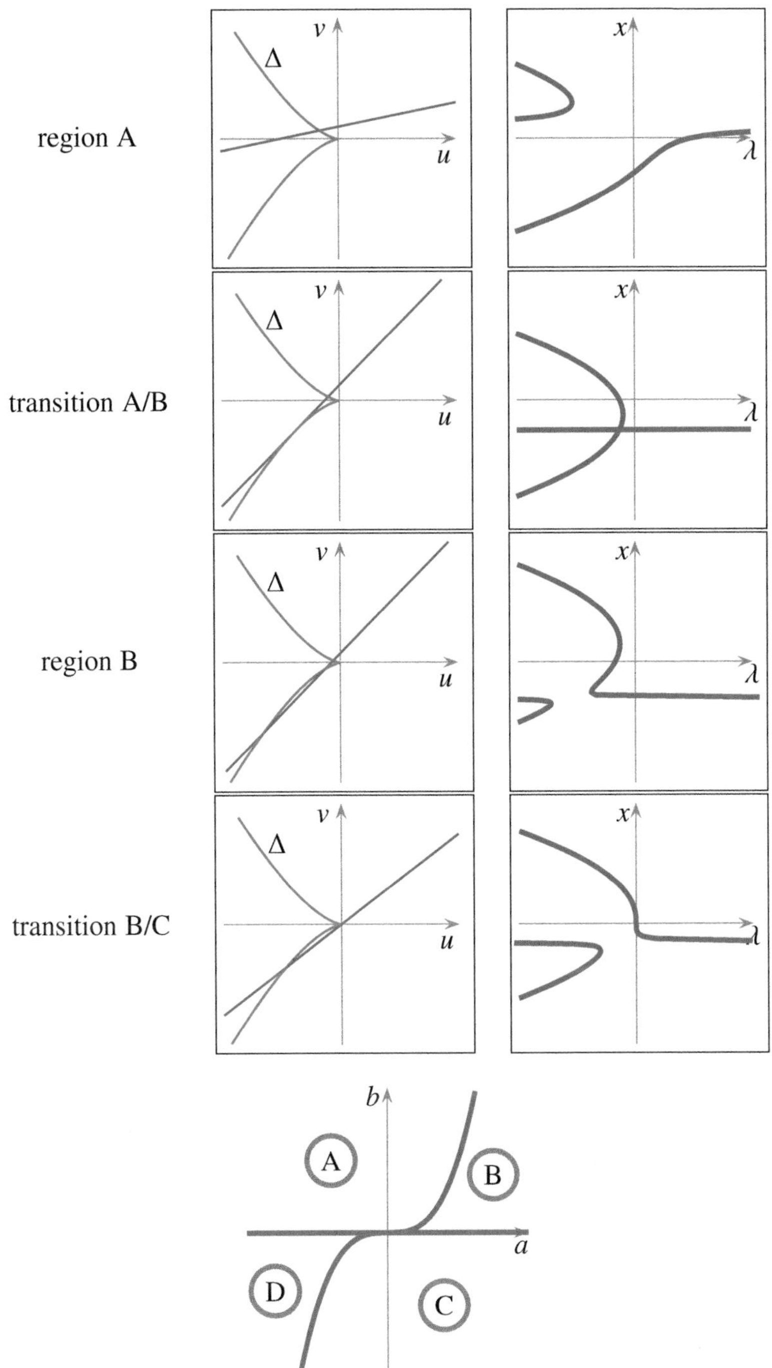

FIGURE 18.2 Deformations of the pitchfork bifurcation: see the discussion in the text for explanations, and compare with the undeformed pitchfork in Figure 18.1.

18.2A Saddle-node bifurcation

Also called the fold bifurcation, this was discussed in Section 1.1. It is the simplest bifurcation and occurs when two simple solutions to the equations (equilibria of a differential equation, perhaps) coalesce and disappear as the parameter is varied. Its normal form is

$$g(x, \lambda) = x^2 + \lambda.$$

This a versal unfolding of the core $g_0(x) = x^2$, so the path is the identity map $h(\lambda) = \lambda$ and any deformation of the path is equivalent to it. (By changing the sign of λ, one can also use $x^2 - \lambda$ as an equivalent normal form.)

18.2B Transcritical bifurcation

Consider first the transcritical bifurcation in the form it is best known:

$$g(x, \lambda) = x^2 - \lambda x = 0,$$

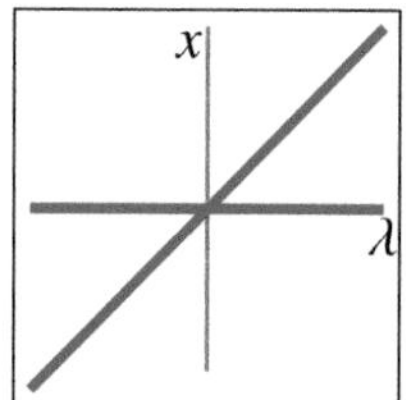

where $x, \lambda \in \mathbb{R}$. The bifurcation diagram is shown on the right. The core of this bifurcation is $g_0(x) = x^2$, and the $\mathcal{K}$-versal unfolding of the core is

$$F(x, u) = x^2 + u. \tag{18.4}$$

By the versality theorem the given unfolding g of g_0 is equivalent to one induced from the versal unfolding by some map $u = h(\lambda)$. Indeed, if we complete the square, we obtain

$$g(x, \lambda) = (x - \tfrac{1}{2}\lambda)^2 - \tfrac{1}{4}\lambda^2.$$

Thus g is equivalent to the unfolding

$$g'(X, \mu) = X^2 - \mu^2$$

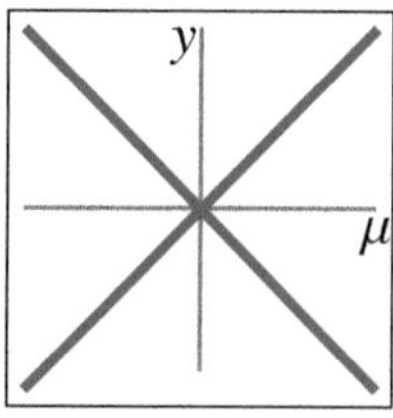

whose zero-set (bifurcation diagram) is shown on the left. Here $X = x - \frac{1}{2}\lambda$ and $\mu = \frac{1}{2}\lambda$, an example of equivalence of unfoldings ($\mathcal{K}_{\mathrm{un}}$-equivalence to be precise; see Section 14.1).

We can therefore consider the alternative form for the transcritical bifurcation

$$g(x, \lambda) = x^2 - \lambda^2 = 0.$$

In this form, the transcritical bifurcation is induced by the path

$$u = h(\lambda) = -\lambda^2.$$

That is, $g(x, \lambda) = F(x, h(\lambda))$, where F is given in (18.4). As with the pitchfork bifurcation, the main idea is to study perturbations of the bifurcation problem by perturbing the path h. In this example, we can perturb h by taking

$$H(\lambda; a) = -\lambda^2 + a,$$

where $a \in \mathbb{R}$ is the deformation parameter. This corresponds to a perturbed bifurcation problem,

$$G(x, \lambda; a) = x^2 - \lambda^2 + a = 0.$$

See Figure 18.3A for the perturbations of the figure above. Notice that for $a > 0$ there are two saddle-node bifurcations (occurring on the x-axis) and these correspond to the values of (x, λ) where the perturbed path meets the discriminant $\Delta = \{0\}$ in the base of the versal unfolding of g_0. For $a < 0$ the perturbed path does not meet Δ, and indeed there are no bifurcation points in the perturbed problem for $a > 0$.

18.2c Isola bifurcation

This is similar to the transcritical bifurcation above, but with the opposite sign:

$$g(x, \lambda) = x^2 + \lambda^2 = 0.$$

The corresponding path is $h(\lambda) = \lambda^2$. An unfolding of this is $H(x, \lambda; a) = \lambda^2 + a$, whose associated family of bifurcations is

$$G(x, \lambda; a) = x^2 + \lambda^2 + a.$$

For $a < 0$ the solution is a circle of radius $\sqrt{|a|}$, manifesting two saddle-node bifurcations. As $a \to 0$, the circle shrinks to a point and disappears. See Figure 18.3B.

18.2D Hysteresis bifurcation

The hysteresis bifurcation is given by

$$g(x, \lambda) = x^3 + \lambda = 0.$$

The core $g_0 = x^3$ has versal unfolding

$$F(x, u, v) = x^3 + ux + v,$$

as for the pitchfork studied above. The discriminant of this unfolding is the semicubical parabola $4u^3 + 27v^2 = 0$. The bifurcation equation g is induced from F

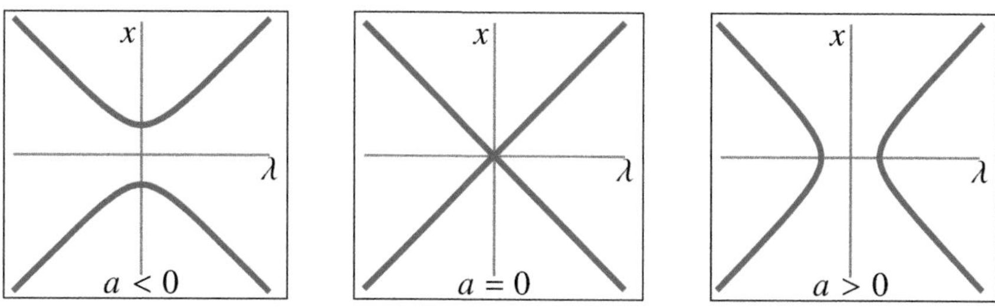

(A) The transcritical bifurcation (at $a = 0$) and its unfolding

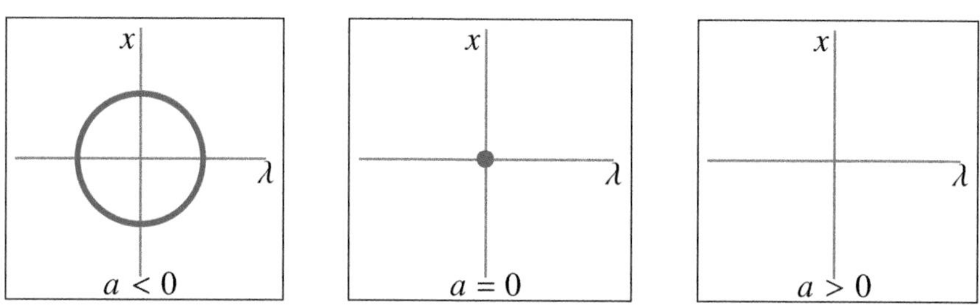

(B) The isola bifurcation and its unfolding

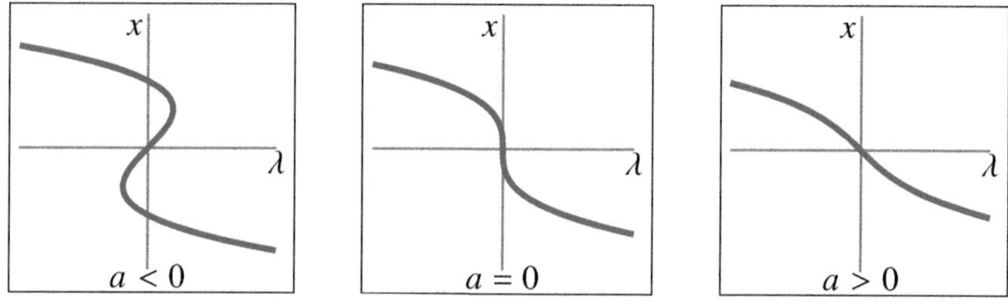

(C) The hysteresis bifurcation and its unfolding

FIGURE 18.3 The three codimension-1 bifurcations.

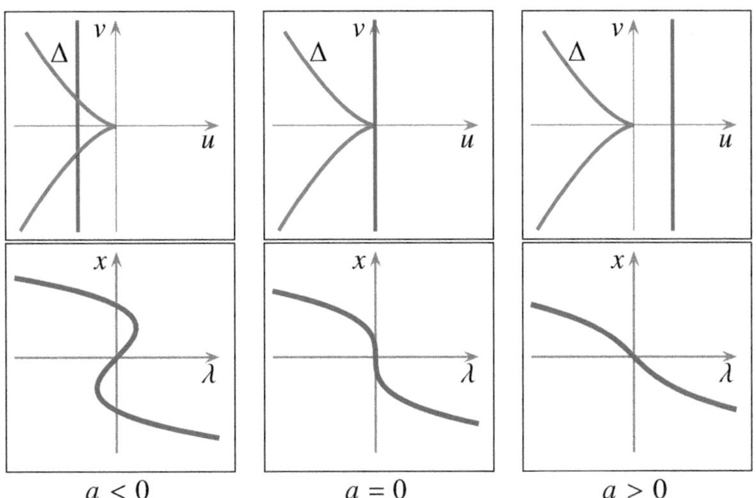

FIGURE 18.4 The hysteresis bifurcation and its unfolding: ABOVE, the perturbations of the path, and BELOW, the corresponding perturbations of the bifurcation diagram.

by putting $(u, v) = h(\lambda) = (0, \lambda)$; the image of this path is the v-axis, which is transverse to the limiting tangent to Δ at the origin (which is along the u-axis), so meets Δ in a less singular fashion than the path for the pitchfork.

Deforming the hysteresis bifurcation is simple: we merely translate the path to the left or the right:

$$H(\lambda; a) = (a, \lambda),$$

with corresponding bifurcation problem

$$G(x, \lambda; a) = x^3 + \lambda + ax = 0.$$

See Figure 18.3C for the bifurcation diagrams.

In Figure 18.4, the upper diagrams correspond to paths in the base space $B = \mathbb{R}^2$, and the lower ones to the resulting bifurcation diagrams; in the upper ones, the red curve is the discriminant Δ and the green line is the path: the original one where $a = 0$ and its perturbations obtained by shifting it to the left or right.

Notice that for $a < 0$, the perturbed path meets the discriminant in two points, and the lower–left bifurcation diagram has correspondingly two bifurcation points, both saddle-node bifurcations. For $a > 0$ the perturbed path does not meet the discriminant, and correspondingly there are no bifurcations in the right-hand diagram.

18.3 Conclusion

As we will prove in Chapter 21, the three 1–parameter bifurcations discussed above are the only ones of codimension 1 (i.e. have 1–parameter versal unfoldings). The first example with codimension 2 is the pitchfork bifurcation, which we also saw above.

From these examples, one can see that the crux for the bifurcations is how the perturbed path meets the discriminant Δ. This suggests a variation of contact equivalence: contact equivalence measures the 'form' of $f^{-1}(0)$ (or how f meets the origin), and in so-called $\mathcal{K}_V$-equivalence, the origin is replaced by a subset V of the target, and one is interested in how f meets V. The resulting equivalence relation is known as $\mathcal{K}_V$-equivalence. In particular, for these bifurcation problems, the subset V is the discriminant Δ giving $\mathcal{K}_\Delta$-equivalence. This equivalence is the subject of Chapter 20.

Before defining that equivalence, and in order to talk about the associated tangent space $T\mathcal{K}_V{\cdot}h$, we need to introduce the notion of vector fields tangent to the subset V: this is the subject of the next chapter.

Problems

18.1 Consider the *quartic fold bifurcation*, $x^4 - \lambda = 0$, and the 1–parameter deformation

$$G(x, \lambda; a) = x^4 - \lambda + ax.$$

Sketch the bifurcation diagram for $a < 0$, $a = 0$ and $a > 0$. Find the associated path in the base space of the $\mathcal{K}$-versal unfolding of the core $g_0(x) = x^4$, whose discriminant is the swallowtail, and show that the number of bifurcations occurring in g_a is equal to the number of intersections of the perturbed path with the swallowtail surface (see Figure 7.1).

18.2 Show that the unfolding of the pitchfork given by Golubitsky and Schaeffer [49], namely

$$G_1(y, \mu; \alpha, \beta) = y^3 - \mu y + \alpha + \beta y^2,$$

is equivalent to the unfolding in (18.3). [*Hint*: *start with* $x = y + \beta/3$.]

18.3 Consider the overdetermined system $g(x, y; \lambda, \mu) = (x^2 - \lambda, xy + \lambda - \mu, y^2 - \mu) = \mathbf{0}$. Find the values of (λ, μ) for which a solution exists: this curve is the discriminant of this problem.

19

Vector fields tangent to a variety

IN THE FOLLOWING CHAPTERS, we discuss an equivalence relation based on contact equivalence, known as $\mathcal{K}_V$-equivalence, where V is a subset of $\mathbb{R}^n$. We will restrict attention to algebraic subsets, or varieties, although much of the theory holds more generally for analytic subsets. In our application to bifurcation theory, the algebraic subset V is the discriminant of a versal unfolding, but in this and the next chapter we consider an arbitrary algebraic set V.

The theory of vector fields tangent to a variety defined over $\mathbb{C}$ is relatively straightforward, but over $\mathbb{R}$, where it is natural to use smooth vector fields, it is more subtle, as described in Section 19.5. For this reason, we limit our calculations to algebraic (i.e., polynomial) vector fields.

Definition 19.1. A subset $V \subset \mathbb{R}^n$ is an ***algebraic variety*** if there are polynomial functions $p_1, \ldots, p_r$ on $\mathbb{R}^n$ such that

$$V = \{x \in \mathbb{R}^n \mid p_1(x) = p_2(x) = \cdots = p_r(x) = 0\}.$$

A subset $W \subset \mathbb{R}^n$ is a ***semialgebraic variety*** if it is a finite union, intersection or difference of sets of the form $S_p = \{x \in \mathbb{R}^n \mid p(x) \geq 0\}$, where p is a polynomial. See also Section D.3 in Appendix D. ✯

Any semialgebraic variety S is contained in a unique smallest algebraic variety, called its ***algebraic closure***. It is most simply defined as the algebraic variety given by the zeros of all polynomials that vanish identically on S.

Semialgebraic varieties arise naturally, and in particular the discriminants of versal unfoldings are often merely semialgebraic rather than algebraic. This follows from the famous Tarski–Seidenberg theorem, which states that the image of an algebraic (or semialgebraic) variety under an algebraic map is semialgebraic. The singular set of any algebraic map is an algebraic variety, so its image is semialgebraic.

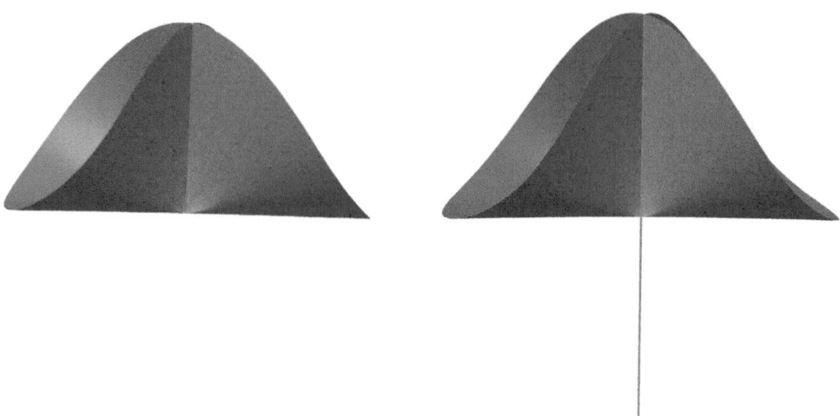

FIGURE 19.1 (LEFT) Cayley cross-cap and (RIGHT) Whitney umbrella. The former is a semialgebraic set, while the latter is its algebraic closure; see Example 19.2.

Example 19.2. A simple example of a semialgebraic variety is the ***Cayley cross-cap***: the image of the map $f(u,v) = (u, uv, v^2)$ from $\mathbb{R}^2$ to $\mathbb{R}^3$ (see left-hand image in Figure 19.1). This cross-cap is only semialgebraic, given by

$$x^2z - y^2 = 0, \quad \text{and} \quad z \geq 0.$$

The algebraic set satisfying $x^2z - y^2 = 0$ is the algebraic closure of the cross-cap and includes the negative z-axis. It is called the ***Whitney umbrella***; the negative z-axis being the handle of the umbrella. This is illustrated in the right-hand image in Figure 19.1. ✐

In this chapter it is important that we assume V to be algebraic and not merely semialgebraic.

In fact everything we state in this chapter also holds for analytic varieties, or germs of analytic varieties; these are defined in the same way as algebraic varieties above, but where the p_j are allowed to (germs of) analytic functions.

19.1 Ideals and vector fields

In this section, we describe the module of polynomial vector fields tangent to an algebraic variety in $\mathbb{R}^n$, and use this to define the module of smooth (germs) of vector fields tangent to an algebraic set that we will use in later chapters.

19.1A Algebraic vector fields

We assume throughout that V is an algebraic variety in $\mathbb{R}^n$, as in Definition 19.1. Note that since polynomials are globally defined, a germ of an algebraic variety determines the global variety. Denote the ring of polynomials in n variables by $\mathcal{P}_n$, and the ideal of polynomials vanishing on V by

$$\mathcal{I}(V) = \{p \in \mathcal{P}_n \mid \forall x \in V,\ p(x) = 0\}.$$

Since the ring of polynomials $\mathcal{P}_n$ is Noetherian, the ideal $\mathcal{I}(V)$ is finitely generated (see Theorem D.16).

To calculate finite determinacy and versal unfoldings with respect to the $\mathcal{K}_V$-equivalence mentioned above, we need to consider the vector fields on $\mathbb{R}^n$ tangent to V. These can be defined algebraically as follows.

Given a vector field η on $\mathbb{R}^n$, and a function f, we write $\eta(f)$ for the derivative of the function f in the direction of η (equivalently, $\eta(f) = \mathrm{t}f(\eta)$). This notation reflects the interpretation of a vector field as a derivation (see Section D.4 in Appendix D). Write $\mathcal{VF}_n$ for the set of all polynomial vector fields on $\mathbb{R}^n$, which is a module over $\mathcal{P}_n$.

Definition 19.3. The $\mathcal{P}_n$-module of ***polynomial vector fields tangent to*** V is

$$\mathcal{VF}(V) = \{\eta \in \mathcal{VF}_n \mid \eta(\mathcal{I}(V)) \subset \mathcal{I}(V)\}.$$ ☆

The condition $\eta(\mathcal{I}(V)) \subset \mathcal{I}(V)$ in this definition means $p \in \mathcal{I}(V) \Rightarrow \eta(p) \in \mathcal{I}(V)$. It is easy to check that[1] $\mathcal{VF}(V)$ is indeed a module over $\mathcal{P}_n$. Of course, since $\mathcal{I}(V)$ is finitely generated, then to guarantee $\eta(\mathcal{I}(V)) \subset \mathcal{I}(V)$, it suffices to check that $\eta(p) \in \mathcal{I}(V)$ for each of the generators p of $\mathcal{I}(V)$.

Calling such vector fields 'tangent' to V is justified by the proposition below, but first we illustrate the definition with a simple example.

Example 19.4. Let V be the algebraic variety in the plane ($n = 2$),

$$V = \left\{(u, v) \in \mathbb{R}^2 \mid u^3 - v^2 = 0\right\},$$

a semicubical parabola, or cusp curve. Let η be the polynomial vector field

$$\eta(u, v) = \begin{pmatrix} a(u, v) \\ b(u, v) \end{pmatrix} = a(u, v)\frac{\partial}{\partial u} + b(u, v)\frac{\partial}{\partial v}$$

[1]One often finds $\mathcal{VF}(V)$, or rather its analytic counterpart, called Derlog(V) or Der(log V), for 'logarithmic derivations', for reasons that go beyond the scope of this text; see [101].

in the plane, with $a, b \in \mathcal{P}_2$. Then by definition, $\eta \in \mathcal{VF}(V)$ if and only if $\eta(h) \in \langle h \rangle \lhd \mathcal{P}_2$, for $h = u^3 - v^2$. Explicitly, we require

$$\eta(h) = 3u^2\, a(u, v) - 2v\, b(u, v) \in \left\langle u^3 - v^2 \right\rangle .$$

This module has in fact just two generators:

$$\mathcal{VF}(V) = \mathcal{P}_2 \left\{ \begin{pmatrix} 2u \\ 3v \end{pmatrix}, \begin{pmatrix} 2v \\ 3u^2 \end{pmatrix} \right\}. \tag{19.1}$$

We prove that this is indeed equal to $\mathcal{VF}(V)$ in Example 19.9; for now notice that the two generators satisfy the equation above, and are linearly independent if $u^3 - v^2 \neq 0$. ✐

Proposition 19.5. *Let ξ be a polynomial vector field and $V \subset \mathbb{R}^n$ an algebraic variety. Then $\xi \in \mathcal{VF}(V)$ if and only if the flow ϕ_t resulting from ξ preserves V.*

PROOF: First suppose $\phi_t(x) \in V$ for all t in the domain of the flow. Now, under the hypothesis that V is algebraic, $x \in V$ iff $h(x) = 0$ for all $h \in \mathcal{I}(V)$. Then $h \circ \phi_t(x) = 0$ for all $h \in \mathcal{I}(V)$. Differentiating this at $t = 0$ shows that $\mathsf{d}h_x(\xi(x)) = 0$ which is equivalent to $\xi(h)(x) = 0$. This holds for all $x \in V$, and hence $\xi(h) \in \mathcal{I}(V)$. Since this holds for all $h \in \mathcal{I}(V)$, it follows that indeed $\xi \in \mathcal{VF}(V)$.

Conversely, suppose $\xi(\mathcal{I}(V)) \subset \mathcal{I}(V)$ and let $x \in V$. We wish to show $\phi_t(x) \in V$ for all t. Write $\mathcal{I}(V) = \langle p_1, \dots, p_r \rangle$. Since $\xi(\mathcal{I}(V)) \subset \mathcal{I}(V)$ it follows that for each i there are polynomials $u_{ij} \in \mathcal{P}_n$ $(j = 1, \dots, r)$ such that

$$\xi(p_i) = \sum_{j=1}^{r} u_{ij} p_j .$$

Write $x_t = \phi_t(x)$, and let $G(t) \in \mathbb{R}^r$ be the vector $G(t) = (p_1(x_t), \dots, p_r(x_t))^T$ (column vector). Similarly let $U(t)$ be the matrix whose entries are $u_{ij}(x_t)$. Now, $\frac{\mathsf{d}}{\mathsf{d}t} p_i(t) = \xi(p_i)(x_t)$, so it follows that

$$\frac{\mathsf{d}}{\mathsf{d}t} G(t) = U(t) G(t).$$

This is a smooth (indeed, polynomial) system of linear ODEs in $\mathbb{R}^r$ and so has a unique solution for each initial condition (see Theorem C.9). Now $G(0) = 0$, and it is clear that $G(t) \equiv 0$ is a solution to the differential equation, and hence is the only solution with initial condition $G(0) = 0$.

This proves the result, since $G(t) = 0$ means $p_i(x_t) = 0$ for all i, which in turn means that $\phi_t(x) \in V$. ✔

19.1B Smooth vector fields

The difficulty with this theory is that even if V is an algebraic variety, the ideal in $\mathcal{E}_n$ of smooth germs vanishing on V may not be finitely generated, and in particular may not be generated by the polynomials that vanish on V. See Section 19.5 for a discussion, and Problem 19.8 for an example. We therefore make the following 'half way house' definitions, which will be important for the chapters that follow.

Definition 19.6. Let $V \subset \mathbb{R}^n$ be an algebraic variety, and recall the definitions of $\mathcal{I}(V)$ and $\mathcal{VF}(V)$ given above. Define $I_V = \mathcal{E}_n \mathcal{I}(V)$ and $\theta_V = \mathcal{E}_n \mathcal{VF}(V)$. In contrast, we denote by $\mathcal{VF}^\infty(V)$ the $\mathcal{E}_n$-module of *all* germs of smooth vector fields tangent to V. ✯

By construction, both I_V and θ_V are finitely generated. On the other hand, $\mathcal{VF}^\infty(V)$ is not always finitely generated.

For example, if V is the semicubical parabola of Example 19.4, then

$$\theta_V = \mathcal{E}_2 \left\{ \begin{pmatrix} 2u \\ 3v \end{pmatrix}, \begin{pmatrix} 2v \\ 3u^2 \end{pmatrix} \right\}. \tag{19.2}$$

Remark 19.7. Clearly $\theta_V \subset \mathcal{VF}^\infty(V)$, and as we see below there are some cases where the two coincide. The analogue of Proposition 19.5 holds for $\mathcal{VF}^\infty(V)$ (with the same proof), but consequently not in general for θ_V. However, the proof of the second part is the same, namely that *if* $\xi \in \theta_V$ *then the associated flow preserves* V.

In Corollary 19.11 we describe some varieties where $\theta_V = \mathcal{VF}^\infty(V)$, but general conditions for equality are not known. Further discussion of these modules in the smooth and real-analytic settings can be found in Section 19.5. ❞

19.2 Calculation of θ_V

Recall that for a given algebraic variety $V \subset \mathbb{R}^n$, we write $\mathcal{I}(V)$ for the ideal consisting of all polynomial functions vanishing on V, and $\mathcal{VF}(V)$ the $\mathcal{P}_n$-module of polynomial vector fields tangent to V as defined above. Given a polynomial $h \in \mathcal{P}_n$ we write

$$\mathcal{VF}_h := \{\xi \in \mathcal{VF}_n \mid \xi(h) = 0\},$$

which is the $\mathcal{P}_n$-module of polynomial vector fields annihilating h, and we write θ_h for the $\mathcal{E}_n$-module generated by $\mathcal{VF}_h$.

19.2A Weighted homogeneous hypersurfaces

See Section D.8 in Appendix D for details on weighted homogeneity, and especially the Euler vector field $\mathfrak{e}$ defined in Definition D.24.

One says V is an ***algebraic hypersurface*** if $\mathcal{I}(V)$ is generated by a single polynomial $\mathcal{I}(V) = \langle h \rangle$, for some $h \in \mathcal{P}_n$. If $V \subset \mathbb{R}^n$ is an algebraic hypersurface, with defining equation $h(x) = 0$, then by definition $\eta \in \mathcal{VF}(V)$ if and only if $\eta(h) = ah$ for some $a \in \mathcal{P}_n$.

We assume from now on that $V \subset \mathbb{R}^n$ is defined by an equation $h(x) = 0$ where h is a weighted homogeneous polynomial, with weight vector $\mathbf{w} = (w_1, \ldots, w_n)$ say. We have that $\mathfrak{e}(h) = rh$, where $r > 0$ is the (weighted) degree of h and $\mathfrak{e}$ is the Euler vector field for the given weights.

Lemma 19.8. *Suppose V is a weighted homogeneous algebraic hypersurface, and denote by $\mathfrak{e}$ the Euler vector field for the given weights. Then*

$$\mathcal{VF}(V) = \mathcal{P}_n\{\mathfrak{e}\} \oplus \mathcal{VF}_h,$$

and similarly $\theta_V = \mathcal{E}_n\{\mathfrak{e}\} \oplus \theta_h$.

PROOF: Let $\xi \in \mathcal{VF}(V)$, and suppose $\xi(h) = ah$, with $a \in \mathcal{P}_n$, Then put $\bar{\xi} = \xi - (a/r)\mathfrak{e}$, where r is the weighted degree of h. It follows that

$$\bar{\xi}(h) = \xi(h) - \tfrac{1}{r} a\mathfrak{e}(h) = 0,$$

whence $\bar{\xi} \in \mathcal{VF}_h$. Thus we can write any $\xi \in \mathcal{VF}(V)$ uniquely as $\xi = \frac{1}{r}a\mathfrak{e} + \bar{\xi}$, where $\xi(h) = ah$, showing that $\mathcal{VF}(V) = \mathcal{P}_n\mathfrak{e} \oplus \mathcal{VF}_h$, as claimed. The statement about smooth vector fields follows immediately. ✔

The purpose of the above lemma is that $\mathcal{VF}_h$ (hence also θ_h) can be calculated algebraically as the following example shows (see also Corollary 19.11).

Example 19.9. As in Example 19.4, let $V = \{(u, v) \in \mathbb{R}^2 \mid u^3 - v^2 = 0\}$, and $h(u, v) = u^3 - v^2$. This is weighted homogeneous with weight vector $\mathbf{w} = (2, 3)$ and h has weighted degree $r = 6$. The corresponding Euler vector field is

$$\mathfrak{e} = 2u\frac{\partial}{\partial u} + 3v\frac{\partial}{\partial v}.$$

To find $\mathcal{VF}_h$, note that a vector field $\xi = a\frac{\partial}{\partial u} + b\frac{\partial}{\partial v}$ in $\mathcal{VF}_h$ satisfies the identity $3u^2a - 2vb = 0$. It follows that if $v = 0$ then $a = 0$ so $a = 2vc$ for some $c \in \mathcal{P}_2$.

Then $2v(3u^2c - b) = 0$ whence $b = 3cu^2$ and $\xi = c(2v, 3u^2)^T$. The module $\theta_h = \mathcal{E}_2\,\mathcal{VF}_h$ is therefore

$$\theta_h = \mathcal{E}_2\left\{2v\frac{\partial}{\partial u} + 3u^2\frac{\partial}{\partial v}\right\}.$$

This shows that θ_V is generated by $\mathfrak{e}$ and the vector field ξ above, as claimed in Example 19.4. Note that this argument relies solely on Hadamard's lemma, and so shows that θ_V coincides with the module $\mathcal{VF}^\infty(V)$ of *all* germs of smooth vector fields tangent to V, as we see more generally below. ✐

In the example above, the vector field generating θ_h is equal to $\frac{\partial h}{\partial v}\frac{\partial}{\partial u} - \frac{\partial h}{\partial u}\frac{\partial}{\partial v}$. One can show more, as follows.

Proposition 19.10. *Suppose V is an algebraic hypersurface, defined by $V = h^{-1}(0)$ and that $h \in \mathcal{P}_n$ is of finite $\mathcal{R}$-codimension. Then $\mathcal{VF}_h$ is generated by the* ***Koszul vector fields***

$$\xi_{ij} = \frac{\partial h}{\partial x_i}\frac{\partial}{\partial x_j} - \frac{\partial h}{\partial x_j}\frac{\partial}{\partial x_i}, \qquad (i \neq j).$$

Moreover, $\mathcal{VF}_h^\infty = \mathcal{E}_n\,\mathcal{VF}_h$.

Proof: It is clear that (for any h), $\xi_{ij} \in \mathcal{VF}_h$, but the proof that these generate $\mathcal{VF}_h$ or $\mathcal{VF}_h^\infty$ relies on the fact that if h is of finite codimension then the partial derivatives $\frac{\partial h}{\partial x_j}$ form a so-called *regular sequence* in $\mathcal{P}_n$ and in $\mathcal{E}_n$ [36]. A vector field $\xi(x) = \sum_i a_i(x)\partial_x$ belongs to $\mathcal{VF}_h$ if and only if,

$$a_1\frac{\partial h}{\partial x_1} + \cdots + a_n\frac{\partial h}{\partial x_n} = 0.$$

That is, in the language of commutative algebra, the vector $(a_1, \ldots, a_n)$ is a relation between the partial derivatives. And for any regular sequence the module of relations is generated by the so-called Koszul relations, which are those of the form $a_ie_j = a_je_i$, where e_i is the ith basis vector of $\mathbb{R}^n$. ✔

Corollary 19.11. *If V is a hypersurface defined as $h^{-1}(0)$ with h weighted homogenous of finite codimension, then*

$$\begin{aligned} \mathcal{VF}(V) &= \mathcal{P}_n\{\mathfrak{e}\} \oplus \mathcal{VF}_h, \quad \textit{and} \\ \mathcal{VF}^\infty(V) &= \mathcal{E}_n\{\mathfrak{e}\} \oplus \mathcal{VF}_h^\infty, \end{aligned}$$

where $\mathcal{VF}_h^\infty$ is the $\mathcal{E}_n$ module generated by the Koszul vector fields of Proposition 19.10.

PROOF: This is simply a combination of Propositions 19.8 and 19.10. ✔

Example 19.12. Let $V \subset \mathbb{R}^3$ be the algebraic cone

$$V = \left\{(x, y, z)) \in \mathbb{R}^3 \mid h(x, y, z) = 0\right\}, \tag{19.3}$$

where $h(x, y, z) = x^2 + y^2 - z^2$. This is the right circular cone with axis of symmetry along the z-axis. The function h is homogeneous, so the Euler vector field tangent to V is

$$\mathfrak{e} = x\frac{\partial}{\partial x} + y\frac{\partial}{\partial y} + z\frac{\partial}{\partial z}.$$

By the results above, the modules $\mathcal{VF}(V)$ and $\mathcal{VF}^\infty(V)$ are generated by the Euler field $\mathfrak{e}$ and the Koszul vector fields, which here are (removing the common factor of 2)

$$\xi_{12} = x\partial_y - y\partial_x, \quad \xi_{23} = y\partial_z + z\partial_y, \quad \xi_{13} = x\partial_z + z\partial_x,$$

where we have written ∂_x for $\frac{\partial}{\partial x}$ etc. Thus

$$\mathcal{VF}(V) = \mathcal{P}_3 \left\{ \begin{pmatrix} x \\ y \\ z \end{pmatrix}, \begin{pmatrix} -y \\ x \\ 0 \end{pmatrix}, \begin{pmatrix} 0 \\ z \\ y \end{pmatrix}, \begin{pmatrix} z \\ 0 \\ x \end{pmatrix} \right\}. \tag{19.4}$$

And $\mathcal{VF}^\infty(V)$ is the $\mathcal{E}_3$-module generated by the same four vector fields. ✐

19.2B Vector fields tangent to discriminants

In the approach to studying bifurcation problems described in the previous chapter, we are interested in the way the path meets the discriminant of a versal unfolding. The resulting singularity theoretic equivalence on paths is called $\mathcal{K}_V$-equivalence (defined in the next chapter), and requires us to know the module of (polynomial) vector fields tangent to such a discriminant. Note that (except for codimension 1 and 2 versal unfoldings) the discriminant does not have an isolated singularity, so Corollary 19.11 does not apply.

To find the vector fields tangent to the discriminant, one can proceed by calculating the equation of the discriminant and then applying the approach described above. This is rather lengthy, but is not unreasonable using computer algebra. The results of the computer algebra computations are shown in Table 19.1 (taken from [85], with a correction) and were computed using MACAULAY2 [52]. It shows generators for the module of vector fields tangent to the discriminant for each singularity up to $\mathcal{K}$-codimension 4 in the equidimensional setting (i.e. where $n = p$). The six

TABLE 19.1 Generators for $\mathcal{VF}_\Delta$ for equidimensional cores of codimension up to 4.

Type	$G(\mathbf{x},\mathbf{u})$	Generators of θ_Δ
A_1	x^2+u	u
A_2	x^3+ux+v	$\begin{pmatrix}2u\\3v\end{pmatrix};\begin{pmatrix}9v\\-2u^2\end{pmatrix}$
A_3	x^4+ux^2+vx+s	$\begin{pmatrix}2u\\3v\\4s\end{pmatrix};\begin{pmatrix}6v\\8s-2u^2\\-uv\end{pmatrix},\begin{pmatrix}16s-4u^2\\-8uv\\-3v^2\end{pmatrix}$
A_4	$x^5+ux^3+vx^2+$ $+wx+s$	$\begin{pmatrix}2u\\3v\\4w\\5s\end{pmatrix};\begin{pmatrix}15v\\20w-6u^2\\25s-4uv\\-2uw\end{pmatrix},\begin{pmatrix}20w\\25s-4uv\\10uw-6v^2\\15us-3vw\end{pmatrix},\begin{pmatrix}25s\\-2uw\\15us-3vw\\10vs-4w^2\end{pmatrix}$
$I_{2,2}$	$\begin{pmatrix}x^2+uy+s\\y^2+vx+t\end{pmatrix}$	$\begin{pmatrix}u\\v\\2s\\2t\end{pmatrix};\begin{pmatrix}3u\\-3v\\2s\\-2t\end{pmatrix},\begin{pmatrix}0\\8t\\3u^2v\\-8vs\end{pmatrix},\begin{pmatrix}8s\\0\\-8ut\\3uv^2\end{pmatrix}$
$II_{2,2}$	$\begin{pmatrix}x^2-y^2+u(x+y)+s\\2xy+v(x+y)+t\end{pmatrix}$	$\begin{pmatrix}u\\v\\2s\\2t\end{pmatrix};\begin{pmatrix}-3v\\3u\\u^2-2t\\2s+uv\end{pmatrix},\begin{pmatrix}2s\\2t-u^2-v^2\\vs\\-us\end{pmatrix},\begin{pmatrix}2t\\-2s\\-us\\2ut-3vs\end{pmatrix}$

Note: the semicolon in the third column separates the Euler vector field from the submodule $\mathcal{VF}_h$.

lines of MACAULAY2 code for this computation are shown in the box at the end of this chapter, on p. 246.

An alternative, more elegant approach is due to K. Saito [101] for discriminants of versal unfoldings for $\mathcal{R}$-equivalence for scalar functions (i.e. $f\colon \mathbb{C}^n \to \mathbb{C}$) and is based on the properties of versal unfoldings. Here we present Saito's method without proof. The proof is given in [101], and an independent proof can be found in [13].

Let $f \in \mathfrak{m}_n^2$ be a germ of $\mathcal{R}$-codimension ℓ, and let $\{g_1, \ldots, g_\ell\}$ be a cobasis for Jf in $\mathcal{E}_n$ (with $g_1 = 1$):

$$\mathcal{E}_x = Jf + \mathbb{R}\{g_1, \ldots, g_\ell\}.$$

Let $F\colon \mathbb{R}^n \times \mathbb{R}^\ell \to \mathbb{R}$ be defined as usual by

$$F(x,u) = f(x) + \sum_j u_j g_j(x).$$

It follows from the Preparation Theorem (Theorem 16.1, with $\phi\colon \mathbb{R}^n \times \mathbb{R}^\ell \to \mathbb{R}^\ell$ being the natural projection), that

$$\mathcal{E}_{x,u} = J_x F + \mathcal{E}_u \{g_1, \ldots, g_\ell\}.$$

Here $J_x F = \mathrm{t}f(\theta_{x/u})$ is the Jacobian ideal in $\mathcal{E}_{x,u}$ generated by the $\partial F/\partial x_i$. In other words, given any $\psi \in \mathcal{E}_{x,u}$ we can find a vector field η on $\mathbb{R}^\ell$ such that

$$\psi = \eta(F) \bmod J_x F.$$

Here $\eta(F) = \sum_{j=1}^{\ell} a_j(u)\partial F/\partial u_j = \sum_{j=1}^{\ell} a_j(u) g_j(x)$ for some smooth coefficients a_j.

For $i = 1, \ldots, \ell$, let $\eta_i = \sum_j a_{ij}(u)\partial/\partial u_j$ satisfy

$$g_i F = \eta_i(F) \bmod J_x F,$$

with notation as before. Saito proves that the vector fields $\eta_1, \ldots \eta_\ell$ generate $\mathcal{VF}_\Delta$.

Example 19.13. Consider the A_3 singularity $f(x) = x^4$ and its $\mathcal{R}$-versal unfolding $F(x,u,v,s) = x^4 + ux^2 + vx + s$. Then $JF = \mathcal{E}_{x,u,v,s}(4x^3 + 2ux + v)$, and let $g_3 = x^2$, $g_2 = x$, $g_1 = 1$. Easy calculations show

$$\begin{aligned}
g_1 F &= x^4 + ux^2 + vx + s &&= \tfrac{1}{2}ug_3 + \tfrac{3}{4}vg_2 + sg_1 + \tfrac{1}{4}x[4x^3 + 2ux + v],\\
g_2 F &= x(x^4 + ux^2 + vx + s) &&= \tfrac{3}{4}vg_3 + (s - \tfrac{1}{4}u^2)g_2 - \tfrac{1}{8}uvg_1 + \\
&&&\quad \tfrac{1}{8}(2x^2 + u)[4x^3 + 2ux + v],\\
g_3 F &= x^2(x^4 + ux^2 + vx + s) &&= \tfrac{1}{4}(4s - u^2)g_3 - \tfrac{1}{2}uvg_2 - \tfrac{3}{16}v^2 g_1 + \\
&&&\quad \tfrac{1}{16}(4x^3 + 2ux + 3v)[4x^3 + 2ux + v].
\end{aligned}$$

In each case, the coefficients of g_j form the coefficients of the corresponding vector field η. Thus,

$$\eta_1 = \tfrac{1}{4}\left(2u\partial_u + 3v\partial_v + 4s\partial_s\right), \quad \eta_2 = \tfrac{1}{8}\left(6v\partial_u + (8s - 2u^2)\partial_v - uv\partial_s\right),$$

and

$$\eta_3 = \tfrac{1}{16}\left(4(4s - u^2)\partial_u - 8uv\partial_v - 3v^2\partial_s\right).$$

Up to scalar multiples, these agree with the generators of θ_Δ for the A_3 singularity given in Table 19.1. ✐

Remark 19.14. One consequence of Saito's method is that if $\Delta \subset \mathbb{C}^n$ is the discriminant of a versal unfolding, then the module $\mathcal{VF}(\Delta)$ has precisely n generators. In contrast, if $V \subset \mathbb{C}^n$ is a hypersurface with isolated singularity then $\mathcal{VF}(V)$ has $\frac{1}{2}n(n-1)+1$ generators (the Koszul vector fields and the Euler vector field). These numbers only coincide if $n \leq 2$. ❞

19.3 An example: the Whitney Umbrella

Shown in Figure 19.1, the ***Whitney umbrella*** is the algebraic surface W in $\mathbb{R}^3$ with equation $h(x,y,z) = 0$, where

$$h(x,y,z) = x^2z - y^2.$$

This function (germ) h is not finitely determined: the entire z-axis is contained in this set as the (non-isolated) singular set. However, if $(x,y) \neq (0,0)$ then $z \geq 0$. The part of the 'surface' which is the negative z-axis is called the *handle* of the umbrella. This is similar to the so-called whisker or thread in the swallowtail surface described in Problem 7.14(ii). This equation is weighted homogeneous with weights $(1,2,2)$ and degree 4 (alternatively, $(2,3,2)$ and degree 6).

Proposition 19.15. *The module $\mathcal{VF}(W)$ of polynomial vector fields tangent to the Whitney umbrella W has 4 generators:*

$$\mathcal{VF}(W) = \mathcal{P}_3\left\{\begin{pmatrix}x\\2y\\2z\end{pmatrix};\ \begin{pmatrix}0\\x^2\\2y\end{pmatrix},\ \begin{pmatrix}x\\0\\-2z\end{pmatrix},\ \begin{pmatrix}y\\xz\\0\end{pmatrix}\right\},$$

and furthermore $\mathcal{VF}^\infty(W) = \mathcal{E}_n\,\mathcal{VF}(W)$.

The first is the Euler vector field which exists because h is weighted homogeneous; the other 3 annihilate h (and a combination of the first and third is the Euler field for the alternative weights). Two of the three generators of $\mathcal{VF}_h$ are Koszul vector fields, the other is not.

PROOF: Since h is weighted homogeneous, we can apply Lemma 19.8, with

$$\mathfrak{e} = x\partial_x + y\partial_y + 2z\partial_z.$$

We now need to find $\mathcal{VF}_h$ and $\mathcal{VF}_h^\infty$. Let R denote either ring $\mathcal{P}_3$ or $\mathcal{E}_3$ (the arguments for both are the same). Let $\xi = a\partial_x + b\partial_y + 2c\partial_z$, with $a, b, c \in R$. We require $\xi(h)$=0, or

$$axz - yb + x^2c = 0.$$

If we put $x = 0$ we find $b(0, y, z) = 0$, whence $b(x, y, z) = xb_1(x, y, z)$ for some $b_1 \in R$ (by Hadamard's lemma in the smooth case). Then $\xi(h) = 2x(az - 2yb_1 + xc) = 0$, or

$$az - yb_1 + xc = 0. \tag{19.5}$$

Putting, say, $x = y = 0$ shows that there are functions $g_1, g_2 \in R$ for which $a = xg_1 + yg_2$ (again, by Hadamard's lemma). Then

$$x(g_1z + c) + y(g_2z - b_1) = 0.$$

Finally, putting $x = 0$ again shows that there is a $g_3 \in R$ such that $g_2z - b_1 = g_3x$, or $b_1 = g_2z - g_3x$. Substituting into (19.5) shows

$$(xg_1 + yg_2)z - y(g_2z - g_3x) + xc = 0,$$

whence $c = -g_1z - g_3y$. Substituting back into ξ shows that any $\xi \in \mathcal{VF}_h$ (or $\mathcal{VF}_h^\infty$) has the form

$$\xi = g_1(x\partial_x - 2z\partial_z) + g_2(y\partial_x + xz\partial_y) - g_3(x^2\partial_y + 2y\partial_z)$$

for some $g_1, g_2, g_3 \in R$, as required. ✔

19.4 Functions on singular varieties

A relatively straightforward use of the module of vector fields tangent to a variety is the study of critical points of functions relative to the variety.

Let $V \subset \mathbb{R}^n$ be a variety (either algebraic or analytic). We say $f\colon V \to \mathbb{R}$ is ***smooth*** if there is a neighbourhood U of V and a smooth function $\bar{f}\colon U \to \mathbb{R}$ for which $\bar{f}|_V = f$. Such an $\bar{f}$ is called a smooth extension of f. Furthermore, a *diffeomorphism* of V is by definition a diffeomorphism from one neighbourhood U_1 of V to another neighbourhood U_2 that preserves V (i.e. if $\phi\colon U_1 \to U_2$ is such a diffeomorphism, then one requires $v \in V \Leftrightarrow \phi(v) \in V$). We denote the group of germs at 0 of such diffeomorphisms by $\mathcal{R}(V)$.

Given two smooth function germs $f, g \in \mathcal{E}_n$ one can refine the notion of right equivalence, to one of $\mathcal{R}(V)$-equivalence in the obvious way. The $\mathcal{R}(V)$-tangent spaces for a germ $f \in \mathcal{E}_n$ are

$$T\mathcal{R}(V)\cdot f = \mathrm{t}f(\theta_{V,0}), \quad \text{and} \quad T_e\mathcal{R}(V)\cdot f = \mathrm{t}f(\theta_V),$$

where $\theta_{V,0} = \theta_V \cap \mathfrak{m}_n\theta_n$ (in fact the two tangent spaces are often the same). This equivalence was introduced by Damon [27]. With these definitions of tangent spaces, the finite determinacy and versal unfolding theorems hold for $\mathcal{R}(V)$-equivalence, as follows.

Theorem 19.16. *Let $f \in \mathcal{E}_n$. Then*

(i). f is k-determined with respect to $\mathcal{R}(V)$-equivalence if

$$\mathfrak{m}_n^{k+1} \subset \mathfrak{m}_n T\mathcal{R}(V)\cdot f, \tag{19.6}$$

(ii). $F\colon (\mathbb{R}^n \times \mathbb{R}^a, (0,0)) \to \mathbb{R}$ is an $\mathcal{R}(V)^+$-versal unfolding of f if its initial speeds satisfy

$$\dot{F} + T_e\mathcal{R}(V)\cdot f + \mathbb{R} = \mathcal{E}_n.$$

The proofs are very similar to the corresponding theorems for $\mathcal{R}$-equivalence and are left to the reader.

Example 19.17. Let V be the cone with equation $x^2 + y^2 - z^2 = 0$, and let f be the linear function $f(x, y, z) = ax + by + cz$. Then, using Proposition 19.15,

$$T\mathcal{R}(V)\cdot f = T_e\mathcal{R}(V)\cdot f = \langle ax + by + cz, bz + cy, az + cx, ay - bx\rangle \lhd \mathcal{E}_3.$$

Some simple linear algebra shows that this ideal is equal to $\mathfrak{m}_3$ if and only if $a^2 + b^2 - c^2 \neq 0$, and in that case the germ f is 1-determined relative to $\mathcal{R}(V)$. See Problem 19.7 for a geometric interpretation of this condition. ✐

Remark 19.18. In the same way that one defines the multiplicity μ of a critical point ($\mu = \dim(\mathcal{E}_n/Jf)$ see Remark 5.21), Bruce and Roberts [22] define the algebraic multiplicity μ_V of a critical point on a variety in the complex setting to be $\dim(\mathcal{O}_n/J_V(f))$, where $J_V(f) = \mathrm{t}f(\theta_V)$; however, the algebraic multiplicity is not always equal to a geometric multiplicity (numbers of critical points in a deformation, counting local multiplicitites), even in the complex analytic setting. Whether they are equal depends on the detailed geometry of the variety in question and in particular of the so-called logarithmic characteristic variety. (For example, one finds that for the cone they are equal.)

An important class of singular variety is the orbit space by a finite (or more generally reductive) group, and functions on such a variety can be identified with functions invariant under the action. The case of finite group actions is well-understood, see [99] and [86]. However, the corresponding results for reductive groups is still open, although multiplicities of critical points of functions invariant under $\mathbb{C}^*$-actions is addressed in [88]. ❞

19.5 Some technical considerations

This final section of this chapter aims to elucidate some of the finer points related to this chapter, but is not required for the further development of the subject in this book.

Given an algebraic variety V in $\mathbb{R}^n$ or $\mathbb{C}^n$, there are several ideals associated with it, depending on the 'category' of function germs we are interested in. Firstly, there is the ideal $\mathcal{I}(V) \lhd \mathcal{P}_n$ of polynomials vanishing on V, the ideal we have based this chapter on. Secondly, there is the ideal $\mathcal{I}^\omega(V) \lhd \mathcal{O}_n$ of germs at 0 of analytic functions that vanish on V (in both the real and complex setting), and thirdly, in the real case, the ideal $\mathcal{I}^\infty(V) \lhd \mathcal{E}_n$ of germs of smooth functions that vanish on V.

Denote similarly $\mathcal{VF}^\omega(V)$ and $\mathcal{VF}^\infty(V)$ the modules of all germs at 0 of analytic (real or complex) and smooth vector fields tangent to V, respectively.

The ideals $\mathcal{I}(V)$ and $\mathcal{I}^\omega(V)$ are finitely generated, since the ambient rings are Noetherian (see Section D.6 in Appendix D), but the smooth version $\mathcal{I}^\infty(V)$ may not be in general. Likewise the modules $\mathcal{VF}(V)$ and $\mathcal{VF}^\omega(V)$ are finitely generated for the same reason, but not in general $\mathcal{VF}^\infty(V)$.

For the complex case, Serre [104] shows that the ideal $\mathcal{I}^\omega(V) \lhd \mathcal{O}_n$ is generated over $\mathcal{O}_n$ by $\mathcal{I}(V)$:

$$\mathcal{I}^\omega(V) = \mathcal{O}_n\,\mathcal{I}(V).$$

The proof of this, as well as of statements below, is based on formal power series, and a similar argument shows that for the vector fields, $\mathcal{VF}^\omega(V) = \mathcal{O}_n\,\mathcal{VF}(V)$.

In the real case, the analogous results follow by complexifying $V \subset \mathbb{R}^n$ to $V^{\mathbb{C}} \subset \mathbb{C}^n$ (whose ideal is generated by $\mathcal{I}(V)$). Then

$$\mathcal{I}^\omega(V) = \mathcal{A}_n \mathcal{I}(V) \quad \text{and} \quad \mathcal{VF}^\omega(V) = \mathcal{A}_n \mathcal{VF}(V),$$

where $\mathcal{A}_n$ is the ring of germs at 0 of real analytic functions in $\mathbb{R}^n$. This implies in particular that the constructions in this chapter would be unchanged if we replaced $\mathcal{VF}(V)$ by $\mathcal{VF}^\omega(V)$.

The smooth case is more subtle, even for the ideal $\mathcal{I}^\infty(V)$. For this ideal, when V is an algebraic variety, Malgrange [72] points out that the crucial notion is that of *coherent* real analytic or algebraic variety, namely, (the germ of) a variety $V \subset \mathbb{R}^n$ is said to be ***coherent*** at the origin if there is a set of analytic functions $\{g_1, \dots, g_n\}$ defined on a neighbourhood U of the origin, such that, for every point $x \in U$ the germs of the g_j at x generate the ideal $\mathcal{I}^\omega(V)_x \lhd \mathcal{A}_x$ (the ring of germs at x of analytic functions).

Two examples of non-coherent algebraic surfaces in $\mathbb{R}^3$ are the Whitney umbrella and the swallowtail surface, while the cone $z^2 = x^2 + y^2$ in $\mathbb{R}^3$ is coherent. On the other hand, if a polynomial $h \in \mathfrak{m}_n$ is finitely determined then the variety $h^{-1}(0)$ is coherent, because, except at $\{0\}$, it is everywhere a smooth hypersurface defined by $h = 0$. It turns out that in the complex analytic world, every analytic variety is coherent.

The important theorem of Malgrange in this context [72, Theorem 3.10] states that an analytic or algebraic $V \subset \mathbb{R}^n$ is coherent at 0 if and only if

$$\mathcal{I}^\infty(V) = \mathcal{E}_n \mathcal{I}(V).$$

The right-hand side is the ideal I_V that we use in this chapter, which is therefore equal to $\mathcal{I}^\infty(V)$ if and only if the variety V is coherent.

Turning to vector fields, the results are less clear-cut. Damon [28] proves that, if V is coherent, then

$$\mathcal{VF}^\infty(V) = \mathcal{E}_n \mathcal{VF}^\omega(V) \bmod \mathfrak{m}_n^\infty \theta_n.$$

In the case that V is a weighted homogenous hypersurface with isolated singularity (isolated over $\mathbb{C}$), we have shown that the two modules are in fact equal: $\mathcal{VF}^\infty(V) = \mathcal{E}_n \mathcal{VF}(V)$. In general, more needs to be understood.

Problems

19.1 Let $X \subset \mathbb{R}^2$ be the subset satisfying $xy = 1$. Let $\pi\colon \mathbb{R}^2 \to \mathbb{R}$ be $\pi(x, y) = x$. Show that the image $\pi(X)$ is semialgebraic. (†)

19.2 Let $V = \{(u, v) \in \mathbb{R}^2 \mid uv = 0\}$. Show that

$$\theta_V = \mathcal{E}_2 \left\{ \begin{pmatrix} u \\ 0 \end{pmatrix}, \begin{pmatrix} 0 \\ v \end{pmatrix} \right\}.$$

19.3 Extend the previous exercise to find generators for the module of vector fields on $\mathbb{R}^n$ tangent to the hypersurface equal to the union of all the coordinate hyperplanes,

$$V = \{(u_1, \dots, u_n) \in \mathbb{R}^n \mid u_1 u_2 \dots u_n = 0\}.$$ (†)

19.4 Let $h \in \mathcal{P}_n$ and let $\xi \in \mathcal{VF}_h$. Show that the flow ϕ_t associated to ξ preserves every level set $h^{-1}(c)$ of h. [*Hint: this is easier than Proposition* 19.5.]

19.5 Find the generators of the module $\mathcal{VF}_h$ for $h \in \mathcal{P}_3$ given by $h(x, y, z) = x^2 + y^2 + z^2$. There should be three generators. For all (x, y, z) outside the origin, these vector fields span a 2–dimensional subspace of $\mathbb{R}^3$. Find a geometric interpretation of that space in terms of the function h. (†)

19.6 Consider the cone $x^2 + y^2 - z^2$ (see Example. 19.12).

(i). Let $f(x, y, z) = ax + by + cz$. Show that f is 1-determined with respect to $\mathcal{R}(V)$-equivalence if and only if $a^2 + b^2 - c^2 \neq 0$. Show that in this case, $\operatorname{codim}(f, \mathcal{R}(V)^+) = 0$.

(ii). Show there are at most three $\mathcal{R}(V)$-equivalence classes of such functions f, and then by applying the diffeomorphism $(x, y, z) \to (x, y, -z)$ which preserves V, show that there are in fact only 2 such classes.

(iii). Now consider the quadratic function $g(x, y, z) = 2x^2 + y^2 + z^2$. Show it is 2-determined, and find its codimension and a versal unfolding.

(iv). Show that, on the other hand, the function $g_0(x, y, z) = x^2 + y^2 + z^2$ is of infinite codimension.

19.7 Given $\alpha, \beta, \gamma > 0$, let V be the cone $\alpha x^2 + \beta y^2 - \gamma z^2 = 0$. Find the $\mathcal{E}_3$-module θ_V of germs at the origin of smooth vector fields tangent to V (cf. Example. 19.12). Let $f = ax + by + cz \in \mathfrak{m}_3$. Show that f is 1-determined with respect to $\mathcal{R}(V)$-equivalence if and only if the plane $f = 0$ is not tangent to the cone.

19.8 Consider the Whitney Umbrella W, given by the equation $h = 0$, where $h(x, y, z) = x^2 z - y^2$. Consider the smooth germ

$$f(x, y, z) = \begin{cases} x e^{1/z} & \text{if } z < 0 \\ 0 & \text{otherwise.} \end{cases}$$

Show f vanishes on W but is not in the ideal I_W generated by h, showing that W is not coherent. (†)

19.9 As a simple example of a codimension-2 algebraic variety, let V be the union of the three axes in $\mathbb{R}^3$ (whose germ at the origin is sometimes known as the ***triod*** singularity[2]). Show that $\mathcal{I}(V) = \mathcal{P}_3\{xy, yz, zx\}$ and $\mathcal{I}^\infty(V) = \mathcal{E}_3\mathcal{I}(V)$.

19.10 For readers who know about Lie algebras: show that $\mathcal{VF}(V)$ and $\mathcal{VF}_h$ are Lie algebras (with the usual Lie bracket of vector fields). Deduce that the same is true for θ_V and θ_h. Show moreover that the three linear vector fields generating $\mathcal{VF}_h$ in Problem 19.5 form the Lie algrebra $\mathfrak{so}(3)$, while those in Example, 19.12 form the Lie algebra $\mathfrak{so}(2,1) = \mathfrak{sl}(2,\mathbb{R})$.

[2]CTC. Wall, private communication.

MACAULAY2 CODE for finding the module $\mathcal{VF}$ of polynomial vector fields tangent to the discriminant of the A_3 singularity. The code seems simple enough not to require commenting. See [52] for the language.

```
R = QQ[x,u,v,s]
F = x∧4 + u*x∧2 + v*x + s
Sigma = ideal { F, diff(x,F) }
Delta = eliminate ( {x}, Sigma )
load "VectorFields.m2"
VF = der ( Delta, Delta )
```

The output of the final command is

```
image | -1   0     0       0     |
      |  0  -2u    16s     6v    |
      |  0  -3v   -2uv   -2u2+8s |
      |  0  -4s -3v2+8us  -uv    |
```

The first row and column correspond to the x-variable, which should be ignored. The columns of the remaining 3×3 matrix coincide (after some simplification) with the vector fields appearing in Table 19.1 (note that, for example, `-2u2+8s` represents $-2u^2 + 8s$).

20

$\mathcal{K}_V$-equivalence

THIS IS A GEOMETRIC VARIANT of contact equivalence, introduced in the 1980s by James Damon. In contact equivalence, the crucial feature is how a map f 'meets' the origin in the target, or in other words the form of $f^{-1}(0)$. In this modified version, the origin is replaced by a subset V of the target and the equivalence measures how the map meets V, or the form of $f^{-1}(V)$. As we have seen, this is important in the path approach to bifurcation theory, which we expand on in Chapter 21.

The earliest applications of $\mathcal{K}_V$-equivalence are probably to the study of caustics in [84] and [62]. Moreover, there are applications of $\mathcal{K}_V$-equivalence where V is not the discriminant of a versal unfolding; one (by the author) in Hamiltonian systems is [87], where V is the algebraic subset of the space of real 2×3 matrices, consisting of those of rank at most 1.

20.1 The definition and tangent space

Let $f, g\colon (\mathbb{R}^n, 0) \to (\mathbb{R}^p, 0)$ be two map germs, and let $V \subset \mathbb{R}^p$ be an algebraic variety. The definition is analogous to the 'alternative' definition of contact equivalence given in Section 11.3.

Definition 20.1. The germs f and g as above are ***$\mathcal{K}_V$-equivalent*** if there is a diffeomorphism germ Ψ of $(\mathbb{R}^n \times \mathbb{R}^p, (0,0))$ of the form

$$\Psi(x, y) = (\phi(x),\ \psi(x, y)), \tag{20.1}$$

for $x \in \mathbb{R}^n$ and $y \in \mathbb{R}^p$, such that

(i). $\Psi(\mathbb{R}^n \times V) = \mathbb{R}^n \times V$, and

(ii). $\Psi(\Gamma_f) = \Gamma_g$. ✯

Recall that Γ_f and Γ_g are the graphs of f and g respectively. More explicitly, these two conditions are equivalent to, (i)′. $y \in V \Rightarrow \psi(x, y) \in V$,

(ii)′. $g \circ \phi(x) = \psi(x, f(x))$.

In particular $x \in f^{-1}(V) \Leftrightarrow \phi(x) \in g^{-1}(V)$. Note that it follows from (20.1) that ϕ is a diffeomorphism of $(\mathbb{R}^n, 0)$, whence it follows that the germs $f^{-1}(V)$ and $g^{-1}(V)$ are diffeomorphic.

To define the tangent space for this equivalence relation, one needs to consider the infinitesimal version of diffeomorphisms preserving V. This is provided by the set of vector fields that are everywhere tangent to V (discussed in Section 19.1). We start with an algebraic definition.

Let V be an algebraic variety in $\mathbb{R}^p$. Then θ_V is the $\mathcal{E}_p$-module generated by $\mathcal{VF}_V$ (the polynomial vector fields on $\mathbb{R}^p$ tangent to V), and let

$$\theta_{V,0} = \theta_V \cap \mathfrak{m}_p\theta_p, \tag{20.2}$$

which is the set of those vector fields in θ_V that vanish at the origin. Note that for many varieties, such as that of the cone in Example 19.4, $\theta_{V,0} = \theta_V$. A similar analysis to the one in Section 12.5 suggests the following definition.

Definition 20.2. Let $f\colon (\mathbb{R}^n, 0) \to (\mathbb{R}^p, 0)$ be a smooth map germ, and $V \subset \mathbb{R}^p$ an algebraic variety. The ***$\mathcal{K}_V$-tangent space*** of f is defined to be

$$T\mathcal{K}_V{\cdot}f = \mathrm{t}f(\mathfrak{m}_n\theta_n) + f^*\theta_{V,0}.$$

The ***extended*** tangent space required for unfoldings is

$$T_e\mathcal{K}_V{\cdot}f = \mathrm{t}f(\theta_n) + f^*\theta_V. \qquad ✯$$

Here $f^*\theta_V$ is the $\mathcal{E}_n$-module of vector fields along f that are tangent to V; that is, it is the $\mathcal{E}_n$-module generated by $\omega f(\theta_V) := \{\mathbf{w} \circ f \mid \mathbf{w} \in \theta_V\}$ (see Section 12.3). And similarly for $f^*\theta_{V,0}$. See the example below.

Example 20.3. With V the semicubical parabola from Example 19.4, let $f \in \mathcal{E}_2^2$ be given by $f(x, y) = (x^2, y^2)$. Then $\mathrm{t}f = \begin{pmatrix} 2x & 0 \\ 0 & 2y \end{pmatrix}$, so that

$$\mathrm{t}f(\mathfrak{m}_n\theta_n) = \mathcal{E}_2\left\{\begin{pmatrix} x^2 \\ 0 \end{pmatrix}, \begin{pmatrix} xy \\ 0 \end{pmatrix}, \begin{pmatrix} 0 \\ xy \end{pmatrix} \begin{pmatrix} 0 \\ y^2 \end{pmatrix}\right\}$$

and, substituting $u = x^2$, $v = y^2$ into (19.2),

$$f^*\theta_V = \mathcal{E}_2\left\{\begin{pmatrix} 2x^2 \\ 3y^2 \end{pmatrix}, \begin{pmatrix} 2y^2 \\ 3x^4 \end{pmatrix}\right\}.$$

Note that for this variety, $\theta_{V,0} = \theta_V$. Simplifying $T\mathcal{K}_V \cdot f$ gives

$$T\mathcal{K}_V \cdot f = \mathcal{E}_2 \left\{ \begin{pmatrix} x^2 \\ 0 \end{pmatrix}, \begin{pmatrix} xy \\ 0 \end{pmatrix}, \begin{pmatrix} 0 \\ xy \end{pmatrix}, \begin{pmatrix} 0 \\ y^2 \end{pmatrix}, \begin{pmatrix} 2y^2 \\ 3x^4 \end{pmatrix} \right\}.$$

This includes the monomial vectors $(0, x^5)^T$ and $(y^3, 0)^T$, and hence $T\mathcal{K}_V \cdot f$ is of finite codimension.

For the extended tangent space,

$$\begin{aligned} T_e\mathcal{K}_V \cdot f &= \mathcal{E}_2 \left\{ \begin{pmatrix} x \\ 0 \end{pmatrix}, \begin{pmatrix} 0 \\ y \end{pmatrix} \right\} + \mathcal{E}_2 \left\{ \begin{pmatrix} 2x^2 \\ 3y^2 \end{pmatrix}, \begin{pmatrix} 2y^2 \\ 3x^4 \end{pmatrix} \right\} \\ &= \mathcal{E}_2 \left\{ \begin{pmatrix} x \\ 0 \end{pmatrix}, \begin{pmatrix} 0 \\ y \end{pmatrix}, \begin{pmatrix} 2y^2 \\ 3x^4 \end{pmatrix} \right\}. \end{aligned}$$

One cobasis for this in $\theta(f)$ is

$$\left\{ \begin{pmatrix} 1 \\ 0 \end{pmatrix}, \begin{pmatrix} y \\ 0 \end{pmatrix}, \begin{pmatrix} y^2 \\ 0 \end{pmatrix}, \begin{pmatrix} 0 \\ 1 \end{pmatrix}, \begin{pmatrix} 0 \\ x \end{pmatrix}, \begin{pmatrix} 0 \\ x^2 \end{pmatrix}, \begin{pmatrix} 0 \\ x^3 \end{pmatrix} \right\}.$$

The germ f therefore has $\mathcal{K}_V$-codimension $\operatorname{codim}(f, \mathcal{K}_V) = 7$. ✐

Now $\mathcal{K}_V$ is a new equivalence relation on germs and it is natural to ask whether the Thom–Levine principle and the constant tangent space results of Secs 12.8 and 12.9 hold for $\mathcal{G} = \mathcal{K}_V$. Indeed they do:

Theorem 20.4. *Theorems 12.17 and 12.21 and Corollary 12.22 hold for* $\mathcal{G} = \mathcal{K}_V$.

The proofs are very similar to those for $\mathcal{K}$-equivalence and details are left to the reader.

20.2 Finite determinacy

We saw in Chapters 5 and 13 that the optimal finite determinacy theorems are obtained for $\mathcal{R}_1$- and $\mathcal{K}_1$-equivalence, rather than $\mathcal{R}$ and $\mathcal{K}$. A similar definition for $\mathcal{K}_{V,1}$-equivalence can be made, but it is rather involved and we leave it for Section 20.5. Here we state the basic theorem using $\mathcal{K}_V$-equivalence.

Recall that $T\mathcal{K}_V \cdot f = tf(\mathfrak{m}_n\theta_n) + f^*\theta_{V,0}$, where $\theta_{V,0} = \theta_V \cap \mathfrak{m}_p\theta_p$, which is the set of those vector fields tangent to V that vanish at the origin. The finite determinacy theorem we state here is very similar to the one for contact equivalence.

Theorem 20.5. *f is k-determined with respect to $\mathcal{K}_V$-equivalence if*

$$\mathfrak{m}_n^{k+1}\theta(f) \subset \mathfrak{m}_n T\mathcal{K}_V \cdot f.$$

A weaker, but often useful, statement is that f is k-determined if

$$\mathfrak{m}_n^k\theta(f) \subset T\mathcal{K}_V \cdot f. \tag{20.3}$$

Thus every map of finite $\mathcal{K}_V$-codimension is finitely $\mathcal{K}_V$-determined. The proof of this theorem is very similar to that for finite determinacy for $\mathcal{K}$-equivalence (Theorem 13.2), and relies on Theorem 20.4 above. There is a stronger version of this theorem below (Theorem 20.11). The details of both proofs are left to the reader.

For example, the map f in Example 20.3 is 4 determined for the given V, since we have shown $\mathfrak{m}_2^5\theta(f) \subset \mathfrak{m}_n T\mathcal{K}_V \cdot f$.

Example 20.6. Let $V \subset \mathbb{R}^2$ be the semicubical parabola $u^3 - v^2 = 0$, and let $f(t) = (t, 0)$. We show f is 1-determined with respect to $\mathcal{K}_V$-equivalence. (This example is related to the pitchfork bifurcation as described on p. 263.)

In Example 19.4 we found that the module of vector fields tangent to V is,

$$\theta_V = \mathcal{E}_2\left\{\begin{pmatrix}2u\\3v\end{pmatrix}, \begin{pmatrix}2v\\3u^2\end{pmatrix}\right\}.$$

Consider first the path

$$t \longmapsto (t + O(t^2),\ at^2 + O(t^3)).$$

This is a general path with the same 1-jet as f. Now we can change the coordinate t (a right equivalence) so that $t' = t + O(t^2)$. Then (renaming t' by t), the given path is equivalent to the path

$$g(t) = (t, at^2 + O(t^3)).$$

Now let

$$M = \mathcal{E}_1\left\{\begin{pmatrix}t\\0\end{pmatrix}, \begin{pmatrix}0\\t^2\end{pmatrix}\right\}.$$

Since both generating vector fields of θ_V vanish at the origin, the $\mathcal{K}_V$ tangent space of g satisfies

$$T\mathcal{K}_V \cdot g + \mathfrak{m}_1 M = \mathcal{E}_1\left\{\begin{pmatrix}t^2\\0\end{pmatrix}, \begin{pmatrix}0\\t^3\end{pmatrix}\right\} + \mathcal{E}_1\left\{\begin{pmatrix}t\\2at^2\end{pmatrix}, \begin{pmatrix}2t\\3at^2\end{pmatrix}, \begin{pmatrix}0\\2t^2\end{pmatrix}\right\}.$$

(The first summand is $\mathfrak{m}_1 M$.) This is equal to M, and so, by Nakayama's lemma, we deduce that

$$T\mathcal{K}_V \cdot g = M.$$

In particular $\mathfrak{m}_1^2\theta(g) \subset T\mathcal{K}_V \cdot g$ and hence g is 2-determined (by the theorem above), and is $\mathcal{K}_V$-equivalent to

$$f_s(t) = (t, st^2),$$

for some $s \in \mathbb{R}$. To show this is $\mathcal{K}_V$-equivalent to $f_0 = f$, we note that $\dot{f}_s \in M$, and

$$T_{\text{rel}}\mathcal{K}_V \cdot f_s = \mathcal{E}_{1,I} M$$

(for $I = \mathbb{R}$), and hence it follows from Theorem 12.21 that the family f_s is $\mathcal{K}_V$-trivial, as required. ✐

20.3 An example: the cone

Let $V \subset \mathbb{R}^3$ be the algebraic cone with equation $x^2 + y^2 - z^2 = 0$. The module θ_V of vector fields tangent to V is given in Proposition 19.15.

Proposition 20.7. *Let V be the cone above and let $\gamma\colon (\mathbb{R}, 0) \to (\mathbb{R}^3, 0)$ be the germ of a non-singular curve. Then γ is 1-determined with respect to $\mathcal{K}_V$-equivalence iff the tangent vector $\dot{\gamma}(0)$ does not lie in the cone V.*

PROOF: Write $\gamma(t) = t(a, b, c) + O(t^2)$. Since γ is non-singular, a, b, c cannot all be zero. Let $\gamma_0(t) = t(a, b, c)$ (the 1-jet of γ). Then using the generators for θ_V (which in this case is the same as $\theta_{V,0}$) given in Proposition 19.15, we find

$$\begin{aligned} T\mathcal{K}_V \cdot \gamma_0 &= \mathfrak{m}_1 \left\{ \begin{pmatrix} a \\ b \\ c \end{pmatrix} \right\} + \mathcal{E}_1 \left\{ \begin{pmatrix} at \\ bt \\ ct \end{pmatrix}, \begin{pmatrix} 0 \\ ct \\ bt \end{pmatrix}, \begin{pmatrix} ct \\ 0 \\ at \end{pmatrix}, \begin{pmatrix} bt \\ -at \\ 0 \end{pmatrix} \right\} \\ &= \mathfrak{m}_1 \left\{ \begin{pmatrix} a \\ b \\ c \end{pmatrix}, \begin{pmatrix} 0 \\ c \\ b \end{pmatrix}, \begin{pmatrix} c \\ 0 \\ a \end{pmatrix}, \begin{pmatrix} b \\ -a \\ 0 \end{pmatrix} \right\}. \end{aligned}$$

We claim that if $(a, b, c) \notin V$, that is, $a^2 + b^2 - c^2 \neq 0$, then

$$T\mathcal{K}_V \cdot \gamma_0 = \mathfrak{m}_1\theta(\gamma_0), \tag{20.4}$$

in which case γ_0 is 1-determined, and hence so is γ.

To prove the claim, consider the determinant of any three of the four column vectors in the brackets. The last three have determinant 0. The first three have determinant $a(a^2 + b^2 - c^2)$. Since we are assuming $a^2 + b^2 - c^2 \neq 0$, if $a \neq 0$ then by linear combinations of them, we can obtain any vector in $\mathbb{R}^3$ and hence any vector field in $\theta(\gamma_0)$ as required. If instead $a = 0$, then the other determinants give $b(a^2 + b^2 - c^2)$ and $c(a^2 + b^2 - c^2)$, so at least one of these will give the required result.

For the converse, assume $(a, b, c) \in V$. Then $\gamma_0^{-1}(V) = \mathbb{R}$, while if γ has higher–order terms ensuring the curve does not lie in the cone, then $\gamma^{-1}(V) = \{0\}$ showing that γ and γ_0 cannot be $\mathcal{K}_V$-equivalent. ✔

Remark 20.8. Using 'constant tangent space' arguments based on (20.4) one can go further and show that under the condition that $\dot{\gamma}(0) \notin V$, then γ is $\mathcal{K}_V$-equivalent to one of two curves (lines): to $t \mapsto (t, 0, 0)$ or to $t \mapsto (0, 0, t)$. The first arises if $a^2 + b^2 > c^2$, the second if $a^2 + b^2 < c^2$. ❞

20.4 Versal unfoldings

The definition of induced and versal deformations is entirely analogous to those of the corresponding notions for $\mathcal{K}$-equivalence (see Section 14.1), and it is unnecessary to spell it all out again: just replace $\mathcal{K}$-equivalence by $\mathcal{K}_V$-equivalence in those definitions. Note that the theorem here is only that infinitesimal versality implies versality not in general the converse. This is because the vector fields θ_V we use do not (in general) contain all vector fields tangent to V, If V is such that θ_V does contain all smooth germs of vector fields tangent to V then the theorem becomes 'if and only if'.

Theorem 20.9. *Let V be an algebraic subset of $\mathbb{R}^p$ and $f \in \mathcal{E}_n^p$. A deformation $F\colon (\mathbb{R}^n \times \mathbb{R}^a, (0,0)) \to \mathbb{R}^p$ of f is $\mathcal{K}_V$-versal if*

$$T_e\mathcal{K}_V \cdot f + \dot{F} = \theta(f).$$

In particular, if $h_1, \ldots, h_c$ form a cobasis for $T_e\mathcal{K}_V \cdot f$ in $\theta(f)$, then

$$F(x, u) = f(x) + \sum_{j=1}^{c} u_j h_j(x)$$

is a miniversal unfolding.

Example 20.10. Continuing the cone example above (Section 20.3), one has

$$T_e\mathcal{K}_V\cdot\gamma = \mathcal{E}_1\left\{\begin{pmatrix}a\\b\\c\end{pmatrix}, \begin{pmatrix}0\\ct\\bt\end{pmatrix}, \begin{pmatrix}ct\\0\\at\end{pmatrix}, \begin{pmatrix}bt\\-at\\0\end{pmatrix}\right\}.$$

This contains $\mathfrak{m}_1\theta(\gamma)$ and has codimension 2 in $\theta(\gamma)$, with cobasis $\{\mathbf{v}_1, \mathbf{v}_2\}$, where $\mathbf{v}_1, \mathbf{v}_2$ are any pair of constant vectors which together with $(a, b, c)^T$ form a basis for $\mathbb{R}^3$. Thus

$$G(t; \lambda, \mu) = \gamma(t) + \lambda\mathbf{v}_1 + \mu\mathbf{v}_2$$

is a $\mathcal{K}_V$-versal unfolding of γ. Varying the parameters λ, μ corresponds simply to shifting the line $\gamma(t)$ in the appropriate direction. ✐

20.5 Refined finite determinacy theorem

There is a refinement of the finite determinacy theorem above which involves defining $\mathcal{K}_{V,1}$-equivalence as follows. Analogous to $\mathcal{K}_1$-equivalence (see Problem 12.14), we define $\mathcal{K}_{V,1}$-equivalence by requiring $\mathsf{d}\Psi(0,0) = \mathsf{Id}_{n+p}$ in Definition 20.1. Then

$$\Psi(x, y) = (x + \overline{\phi}(x),\ y + \overline{\psi}(x, y)),$$

where the Taylor series of both $\overline{\phi}$ and $\overline{\psi}$ have zero linear part. This leads to

$$T\mathcal{K}_{V,1}\cdot f = \mathrm{t}f(\mathfrak{m}_n^2\theta_n) + \mathfrak{m}_n^2 f^*\theta_V + \mathfrak{m}_n f^*\theta_{V,0} + f^*\theta_{V,1}. \tag{20.5}$$

Here $\theta_{V,1} = \theta_V \cap \mathfrak{m}_p^2\theta_p$. Clearly, $\mathfrak{m}_n T\mathcal{K}_V\cdot f \subset T\mathcal{K}_{V,1}\cdot f$, and for many algebraic varieties V the two modules are in fact equal.

The resulting finite determinacy theorem, which in general is stronger than Theorem 20.5 because of the $f^*\theta_{V,1}$ term which is absent from $\mathfrak{m}_n T\mathcal{K}_V\cdot f$, states the following.

Theorem 20.11. *The germ f is k-determined with respect to $\mathcal{K}_{V,1}$-equivalence if*

$$\mathfrak{m}_n^{k+1}\theta(f) \subset T\mathcal{K}_{V,1}\cdot f.$$

See Problem 20.12 for an example.

20.6 Geometric criterion

Analogous to the geometric criteria for finite right and contact codimension, there is a geometric criterion for when a holomorphic map germ is of finite $\mathcal{K}_V$ codimension. For the statement, we need to define a notion of transversality to an algebraic variety: in this context, rather than using curves in V, one uses vector fields to define the tangent space. Define the ***logarithmic tangent space*** at $y \in V$ to be

$$\mathrm{LT}_y V = \{\xi(y) \mid \xi \in \theta_V\}. \tag{20.6}$$

Here θ_V is the $\mathcal{O}_n$-module of (germs of) holomorphic vector fields tangent to V. If V is a submanifold, then the logarithmic tangent space coincides with the usual one: $\mathrm{LT}_y V = T_y V$. Transversality to V is then defined in the usual way (see Section B.7), but using this notion of tangent space: f is ***transverse*** to V at y if

$$\mathrm{im}(\mathrm{d}f_x) + \mathrm{LT}_y V = \mathbb{C}^p \quad (\forall x \in f^{-1}(y)),$$

and is transverse to V if it is transverse for all $y \in V$.

Theorem 20.12. *Let $V \subset \mathbb{C}^p$ be an algebraic variety containing 0, and $f\colon (\mathbb{C}^n, 0) \to (\mathbb{C}^p, 0)$ be a holomorphic germ. Then f is of finite $\mathcal{K}_V$ codimension if and only if there is a punctured neighbourhood U of the origin in $\mathbb{C}^n$ such that f is transverse to V at $f(x)$ for all $x \in U$.*

Like the earlier geometric criteria, the proof is based on the Nullstellensatz, and also holds if V is the germ of an analytic variety, where the statement of the theorem needs to use representatives of the germs.

Problems

20.1 Write out in detail the equivalence of (i) and (ii) with (i') and (ii') in Definition 20.1. Show also that $x \in g^{-1}(V) \iff \phi(x) \in f^{-1}(V)$.

20.2 Let $V = \{(u, v) \in \mathbb{R}^2 \mid uv = 0\}$. Find the $\mathcal{K}_V$-codimension and a $\mathcal{K}_V$-versal unfolding of the map germ $f\colon (\mathbb{R}^2, 0) \to (\mathbb{R}^2, 0)$, with $f(x, y) = (xy,\ x^2 + y^2)$. See Problem 19.2 for the module of vector fields tangent to V.

In the problems below, we assume that V is an algebraic subset of $\mathbb{R}^p$

20.3 Suppose $V \subset \mathbb{R}^p$ is such that $\theta_V \subset \mathfrak{m}_p\theta_p$. Show that if f is of $\mathcal{K}_V$-finite codimension then it is of finite $\mathcal{K}$-codimension.

20.4 Suppose $V \subset \mathbb{R}^p$ is such that $\theta_V \subset \mathfrak{m}_p\theta_p$, and let $f\colon (\mathbb{R}^n,0) \to (\mathbb{R}^p,0)$ with $n < p$. Show that the codimension satisfies $\operatorname{codim}(f,\mathcal{K}_V) \geq p-n$. [*Hint*: *consider* $T_e\mathcal{K}_V \cdot f + \mathfrak{m}_p\theta(f)$.]

20.5 Let $V = \{0\} \times \mathbb{R}^{p-k} \subset \mathbb{R}^p$, and let $f\colon (\mathbb{R}^n,0) \to (\mathbb{R}^p,0)$ be a smooth map germ. Show that in this case the definition of $\mathcal{K}_V$-equivalence of f reduces to the definition of $\mathcal{K}$-equivalence of f_1 where $f(x) = (f_1(x), f_2(x)) \in \mathbb{R}^k \times \mathbb{R}^{p-k}$. Show moreover that

$$T\mathcal{K}_V \cdot f = (T\mathcal{K} \cdot f_1) \times \mathfrak{m}_p\,\theta(f_2). \qquad (\dagger)$$

20.6 Extend the previous exercise as follows. Let $V' \subset \mathbb{R}^k$, and let $V = V' \times \mathbb{R}^{p-k} \subset \mathbb{R}^p$ be algebraic varieties. Show that $f, g\colon (\mathbb{R}^n,0) \to (\mathbb{R}^p,0)$ are $\mathcal{K}_V$-equivalent if and only if f' and g' are $\mathcal{K}_{V'}$-equivalent, where $f' = \pi \circ f$ where $\pi\colon \mathbb{R}^p \to \mathbb{R}^k$, $(u,v) \mapsto u$, and $g' = \pi \circ g$. (In other words, f' and g' are the first k components of f and g respectively.) Show moreover that if F is a $\mathcal{K}_V$-versal unfolding of f, then F' is a $\mathcal{K}_V$-versal unfolding of f' (with F' defined from F in a way to be made explicit).

20.7 Let V be the algebraic variety from Problem 20.2. Classify map germs $f\colon (\mathbb{R},0) \to (\mathbb{R}^2,0)$ up to $\mathcal{K}_V$-equivalence, up to codimension 3. $(\dagger)$

20.8 Let $f\colon (\mathbb{R}^n,0) \to (\mathbb{R}^p,0)$, be the germ of a submersion, and let $V \subset \mathbb{R}^p$ be any algebraic variety. Show that $\operatorname{codim}(f,\mathcal{K}_V) = 0$. $(\dagger)$

20.9 Let $f \in \mathfrak{m}_n\mathcal{E}_n^p$ and $V \subset \mathbb{R}^p$ be any algebraic variety. Show that the $\operatorname{codim}(f,\mathcal{K}_V) = 0$ if and only if f is transverse to V at $0 \in \mathbb{R}^n$: that is, if and only if $\operatorname{im}(\mathrm{d}f_0) + T_0V = \mathbb{R}^p$, where T_0V is the logarithmic tangent space.

20.10 Consider the algebraic variety V in $\mathbb{R}^3$ given by the union of the 3 coordinate axes (the *triod* singularity; see Problem 19.9 for its ideal). Show that $\mathcal{VF}(V)$ has the six generators,

$$\mathcal{VF}(V) = \mathcal{P}_3\left\{x\partial_x,\ y\partial_y,\ z\partial_z,\ yz\partial_x,\ xz\partial_y,\ xy\partial_z\right\},$$

and moreover that $\mathcal{VF}^\infty(V) = \mathcal{E}_3\mathcal{VF}(V)$, where ∂_x means $\frac{\partial}{\partial x}$ etc. (This suggests a variation of Corollary 19.11 may hold for varieties of higher codimension.) Show that the map germ $f\colon (\mathbb{R}^2,0) \to (\mathbb{R}^3,0)$ given by

$$f(u,v) = (u, v, u+v)$$

is 1-determined with respect to $\mathcal{K}_V$-equivalence, and find $\operatorname{codim}(f,\ \mathcal{K}_V)$.

20.11 Write out the proof of Theorem 20.5.

20.12 Let $V \subset \mathbb{R}^2$ be the union of the three lines in the plane given by $u^3 - 3uv^2 = 0$, and let $f(t) = (t, 0)$. Show that Theorem 20.5 predicts that f is 2-determined with respect to $\mathcal{K}_V$-equivalence, while the refinement in Section 20.5 shows that f is in fact 1-determined.

20.13 Adapt the argument in Section 16.3 to prove the versality theorem for $\mathcal{K}_V$-equivalence (Theorem 20.9).

20.14 Prove the claim in Remark 20.8.

21

Classification of paths relative to discriminants

THIS CHAPTER ILLUSTRATES how the path approach can be used to classify bifurcation problems, using $\mathcal{K}_V$-equivalence. For 1–parameter problems, the lists we find and their codimensions are the same as those found in, for example, the book of Golubisky and Schaeffer [49] since the approaches are equivalent. However, for more than 1 parameter some of the classification appears to be new. We should emphasize here that the bifurcations here are just bifurcations of zero-sets of a map – they do not involve bifurcations of dynamics; see for example the books of Kuznetsov [67] or Guckenheimer and Holmes [55] for associated dynamics.

Two–parameter bifurcation problems were classified, using the distinguished parameter approach, in the thesis of Peters [91, 92], but only for corank-1 problems (i.e. for bifurcation problems that can be reduced to a single state variable).

In Chapter 18 we analyzed some of the fundamental and well-known 1–parameter bifurcations from the path point of view. In this chapter we revisit these as part of a classification, and prove that the deformations we described in that chapter are indeed versal deformations of the (degenerate) bifurcation problem.

Let us summarize the path approach. Let $g(x;\lambda) = 0$ be a k–parameter bifurcation problem (germ); the bifurcation of interest takes place at $\lambda = 0$. We may assume that $g_0(x) := g(x, 0)$ is of rank 0 at the origin: if not then we apply Lyapunov–Schmidt reduction (or the implicit function theorem) to reduce the dimensions. Let $F(x, u)$ be a $\mathcal{K}$-versal unfolding of g_0, with $u \in \mathbb{R}^\ell$. By the definition of versality, the bifurcation problem is then equivalent as an unfolding to one induced by a map germ $u = h(\lambda)$ giving $g'(x, \lambda) = F(x, h(\lambda))$, where $h\colon \mathbb{R}^k \to \mathbb{R}^\ell$ is called the path (even if $k > 1$). We saw in Chapter 18 that the important feature of a path is the intersection of its image in the base $\mathbb{R}^\ell$ with the discriminant Δ of the versal unfolding. This motivates the following definition.

Definition 21.1. Let $g_0 \in \mathcal{E}_n^p$ be a germ, with $\mathcal{K}$-codimension $\operatorname{codim}(g_0, \mathcal{K}) = \ell$, and let $F\colon (\mathbb{R}^n \times \mathbb{R}^\ell, (0,0)) \to (\mathbb{R}^p, 0)$ be a versal deformation of g_0. We say two k–parameter bifurcation problems g_1, g_2 with core g_0 are ***path-equivalent*** if the resulting maps $h_1, h_2\colon (\mathbb{R}^k, 0) \to (\mathbb{R}^\ell, 0)$ inducing g_1, g_2 are $\mathcal{K}_\Delta$-equivalent, where Δ is the discriminant of F. ☆

Given the path $h\colon (\mathbb{R}^k, 0) \to (\mathbb{R}^\ell, 0)$, let $r = \operatorname{codim}(h, \mathcal{K}_\Delta)$, and let

$$H\colon (\mathbb{R}^k \times \mathbb{R}^r, (0,0)) \to (\mathbb{R}^\ell, 0)$$

be the $\mathcal{K}_\Delta$-versal deformation of h. Then the corresponding versal unfolding of the bifurcation problem g is

$$\begin{aligned} G\colon (\mathbb{R}^n \times \mathbb{R}^k \times \mathbb{R}^r, (0,0,0)) &\longrightarrow (\mathbb{R}^p, 0), \\ (x; \lambda; a) &\longmapsto F(x, H(\lambda; a)). \end{aligned}$$

The relationship between path equivalence and bifurcation equivalence (or distinguished parameter equivalence, or unfolding equivalence) is discussed in the next chapter (Section 22.1), where we show that a $\mathcal{K}_\Delta$-versal unfolding of a path h induces a $\mathcal{K}_{\mathrm{un}}$-versal unfolding of the corresponding bifurcation problem.

We now turn to a general classification of paths relative to the discriminant of each core, proceeding core by core. We limit ourselves to equidimensional bifurcation problems ($n = p$). We also limit ourselves to 1–parameter paths with $\mathcal{K}_\Delta$-codimension at most 3, to 2–parameter paths of codimension at most 2, and to 3–parameter paths of codimension at most 1. Under these limitations, the core is of $\mathcal{K}$-codimension at most 4 (see Problem 20.4) and hence of corank at most 2 (see the classification of such germs in Chapter 13). The results are again summarised in three tables at the end of the chapter, but written there in terms of bifurcation problems.

Organization This chapter is divided into sections, one for each of the six cores that arise (namely, A_1–A_4, $\mathrm{I}_{2,2}$ and $\mathrm{II}_{2,2}$).

The procedure, within each section, is as follows. Firstly the section is divided into three parts according to the number of parameters (1, 2 or 3). Subsections have titles such as

(A_2.2) Pitchfork: $h(\lambda) = (\lambda, 0)$ (codim=2)

The (A_2.2) is simply an enumeration of the cases we consider with the given core (here A_2); this is followed by the standard name (if one exists); then a normal form for the path, and finally the $\mathcal{K}_\Delta$-codimension of that path, which, as explained in the next chapter, is equal to the codimension of the corresponding bifurcation problem.

Following this heading is the calculation justifying the given normal form together with a calculation of its $\mathcal{K}_\Delta$-codimension and versal unfolding. The normal form is justified in two steps: firstly finite determinacy, and secondly using constant tangent space arguments from Theorem 20.4 (and in particular the $\mathcal{K}_V$-version of Corollary 12.22) to simplify the lower–order terms.

Some of the calculations are presented in a very minimalist fashion, although I hope in sufficient detail for anyone to check them without too much further effort.

Notation In the following calculations, if h is the germ of a path $h\colon (\mathbb{R}^k, 0) \to (\mathbb{R}^\ell, 0)$ we will denote the coefficients of the Jacobian matrix by α_i, β_i etc. (for $i = 1, \ldots, k$). For example, if $k = 2$ and $\ell = 3$ we write

$$\mathsf{d}h(0) = \begin{pmatrix} \alpha_1 & \alpha_2 \\ \beta_1 & \beta_2 \\ \gamma_1 & \gamma_2 \end{pmatrix}. \tag{21.1}$$

21.1 A_1 core

The first singular core (with $n = p$) is $g_0(x) = x^2$, whose $\mathcal{K}$-versal unfolding is of course

$$F(x; u) = x^2 + u. \qquad (21.A_1.1)$$

The discriminant here is just $\Delta = \{0\}$, and it follows that $\mathcal{K}_\Delta$-equivalence is simply ordinary $\mathcal{K}$-equivalence. As mentioned previously, for low codimension this is the same as the classification up to $\mathcal{R}$-equivalence, although the $\mathcal{K}$-codimension differs from the $\mathcal{R}^+$-codimension by 1 because of the constant term in the $\mathcal{K}$-versal unfolding.

Remark 21.2. If there are more state variables and correspondingly more equations (we are assuming $n = p$), the unfolding becomes

$$F(x_1, \ldots, x_n; u) = \left(x_1^2 + u,\ x_2,\ \ldots, x_n\right).$$

This 'suspension' to more variables can be applied to all the bifurcation problems in this chapter, and makes no difference to the conclusions. ❞

One parameter

If h is finitely determined, then $u = h(\lambda)$ is $\mathcal{K}$-equivalent to $\pm\lambda^{r+1}$, which is of $\mathcal{K}$-codimension r. Thus, every 1–parameter bifurcation problem of finite codimension, in one state variable with A_1 core is equivalent to

$$g(x;\lambda) = x^2 \pm \lambda^{r+1}. \tag{21.A$_1$.2}$$

The signs $\pm$ are equivalent if $r + 1$ is odd.

The versal unfolding of the bifurcation problem is given by the $\mathcal{K}$-versal unfolding of the path $u = \pm\lambda^{r+1}$, which is

$$H(\lambda; a_1, \dots, a_r) = \pm\lambda^{r+1} + \sum_{j=0}^{r-1} a_j\lambda^j.$$

Thus, the versal unfolding of the bifurcation problem (21.A_1.2) is given by

$$G(x;\lambda; a_0, \dots, a_{r-1}) = x^2 \pm \lambda^{r+1} + \sum_{j=0}^{r-1} a_j\lambda^j = 0.$$

The bifurcations of this type with 1 parameter are listed in Table 21.1.

TABLE 21.1 The path and corresponding 1–parameter bifurcation problem with A_1 core.

codim	$u = h(\lambda)$	bifurcation problem	name	unfolding
0	λ	$x^2 + \lambda = 0$	saddle-node	–
1	$-\lambda^2$	$x^2 - \lambda^2$	transcritical	$+a$
1	λ^2	$x^2 + \lambda^2$	isola	$+a$
2	λ^3	$x^2 + \lambda^3 = 0$	asymmetric cusp	$+a\lambda + b$
3	$\pm\lambda^4$	$x^2 \pm \lambda^4 = 0$		$+a\lambda^2 + b\lambda + c$

Two parameters

Suppose now there are 2 parameters λ and μ. The 'path' is now a smooth map germ $(\mathbb{R}^2, 0) \to (\mathbb{R}, 0)$, namely $u = h(\lambda, \mu)$. The discriminant is the origin, so $\mathcal{K}_\Delta$-equivalence is just $\mathcal{K}$-equivalence. For low–codimension problems, the classification is the same as $\mathcal{R}$-classification of functions of 2 variables. See Table 21.2.

TABLE 21.2 The path and corresponding 2–parameter bifurcation problem with A_1 core.

codim	$u = h(\lambda, \mu)$	bifurcation problem	unfolding terms
0	λ	$x^2 + \lambda = 0$	–
1	$\pm\lambda^2 \pm \mu^2$	$x^2 \pm \lambda^2 \pm \mu^2 = 0$	$+a$
2	$\lambda^3 \pm \mu^2$	$x^2 - \lambda^3 \pm \mu^2 = 0$	$+a\lambda + b$
3	$\pm\lambda^4 \pm \mu^2$	$x^2 \pm \lambda^4 \pm \mu^2 = 0$	$+a\lambda^2 + b\lambda + c$
4	$\lambda^5 \pm \mu^2$	$x^2 + \lambda^5 \pm \mu^2 = 0$	$+a\lambda^3 + b\lambda^2 + c\lambda + d$
4	$\lambda^3 \pm \lambda\mu^2$	$x^2 \pm \lambda^3 \pm \lambda\mu^2 = 0$	$+a(\lambda^2 \mp \mu^2) + b\lambda + c\mu + d$

More parameters

These are similar. We write $g(x, \lambda) = x^2 + h(\lambda)$, where $\lambda = (\lambda_1, \ldots, \lambda_k)$, and the path is classified by the $\mathcal{K}$-classification of h.

21.2 A_2 core

Now consider the core $g_0(x) = x^3$. The $\mathcal{K}$-versal unfolding is

$$F(x; u, v) = x^3 + ux + v, \qquad (21.A_2.1)$$

(see Section 14.2). Throughout this section, Δ refers to the discriminant of this versal unfolding, namely $\Delta = \{(u, v) \in \mathbb{R}^2 \mid 4u^3 + 27v^2 = 0\}$. The vector fields on $\mathbb{R}^2$ tangent to this discriminant are given in Table 19.1; in this case we have,

$$\theta_\Delta = \mathcal{E}_2 \left\{ \begin{pmatrix} 2u \\ 3v \end{pmatrix}, \begin{pmatrix} 9v \\ -2u^2 \end{pmatrix} \right\}. \qquad (21.A_2.2)$$

Remark 21.3. The cusp has a reflection symmetry given by $(u, v) \mapsto (u, -v)$. It is useful to note that this symmetry arises from (or lifts to) the symmetry of $F^{-1}(0)$ given by $(x, u, v) \mapsto (-x, u, -v)$. It follows that two paths of the form

$$h(\lambda) = (h_1(\lambda), h_2(\lambda)), \quad \text{and} \quad h'(\lambda) = (h_1(\lambda), -h_2(\lambda))$$

are not only path equivalent but give rise to equivalent bifurcation problems. ❞

One parameter

Any 1–parameter bifurcation problem $g(x, \lambda) = 0$ with core g_0 is equivalent to one of the form $x^3 + h_1(\lambda)x + h_2(\lambda)$. We need therefore to classify germs $h\colon (\mathbb{R}, 0) \to (\mathbb{R}^2, 0)$ up to $\mathcal{K}_\Delta$-equivalence, where Δ is the semicubical parabola above. We first consider the possibility that the 1-jet of h is non-zero. Since $j^1h(\lambda) = (r\lambda,\ s\lambda)$, we first suppose $s \neq 0$ and then $s = 0, r \neq 0$. Afterwards, we proceed with $r = s = 0$.

(A$_2$.1) Hysteresis: $h(\lambda) = (0, \lambda)$ **(codim=1)** Consider the path

$$\lambda \longmapsto (r\lambda,\ s\lambda) + O(\lambda^2), \tag{21.A$_2$.3}$$

with $r, s \in \mathbb{R}$ and $s \neq 0$. Since $s \neq 0$ we can apply a right equivalence by rescaling λ to make $s = 1$, and we assume that has been done, and write

$$h_r(\lambda) = (r\lambda, \lambda).$$

If we can show this to be 1-determined with respect to $\mathcal{K}_\Delta$-equivalence then we can ignore any higher–order terms. Now, for fixed $r \in \mathbb{R}$,

$$T\mathcal{K}_\Delta{\cdot}h_r = \mathcal{E}_1\left\{\begin{pmatrix} r\lambda \\ \lambda \end{pmatrix}, \begin{pmatrix} 2r\lambda \\ 3\lambda \end{pmatrix}, \begin{pmatrix} 9\lambda \\ -2r^2\lambda^2 \end{pmatrix}\right\} = \mathfrak{m}_1\theta(h_r).$$

It follows that h_r is indeed 1-determined (for all $r \in \mathbb{R}$).

Now, with interval $I = \mathbb{R}$, the calculation above shows that

$$T_{\rm rel}\,\mathcal{K}_\Delta \cdot h_r = \mathfrak{m}_{1,I}\,\theta(h),$$

which is constant, and contains $\dot{h}_r = (\lambda, 0)$. Hence the h_r for different values of r are all $\mathcal{K}_\Delta$-equivalent to h_0. The calculation involves noting that

$$\begin{pmatrix} \lambda \\ 0 \end{pmatrix} = \tfrac{1}{9}\left(\gamma + 2r^2\lambda(\beta - 2\alpha)\right), \quad \text{and} \quad \begin{pmatrix} 0 \\ \lambda \end{pmatrix} = \beta - 2\alpha$$

where α, β, γ are the 3 generators as given for $T\mathcal{K}_\Delta{\cdot}h_r$ given above. Finally, with $r = 0$,

$$T_e\mathcal{K}_\Delta{\cdot}h = \mathcal{E}_1\left\{\begin{pmatrix} 0 \\ 1 \end{pmatrix}, \begin{pmatrix} \lambda \\ 0 \end{pmatrix}\right\}.$$

The path h is therefore of $\mathcal{K}_\Delta$-codimension 1. A versal unfolding is $H(\lambda; a) = (a, \lambda)$ and the corresponding bifurcation equation is

$$G(x; \lambda; a) = x^3 + ax + \lambda,$$

which is the unfolding of the hysteresis bifurcation discussed in Section 18.2D.

(A_2.2) Pitchfork: $h(\lambda) = (\lambda, 0)$ **(codim=2)** The previous case considered the map (21.A_2.3) with $s \neq 0$. Now put $s = 0$ but assume $r \neq 0$. By rescaling λ we can assume

$$h_s(\lambda) = (\lambda,\ s\lambda^2 + O(\lambda^3)),$$

for some $s \in \mathbb{R}$ (a different meaning for s; we will often use r, s as coefficients in calculations, and in different cases they may represent different terms).

We first show h_s is 1-determined, proceeding as follows. Let $M = \mathfrak{m}_1 \times \mathfrak{m}_1^2 \lhd \mathcal{E}_1^2$. Since both generating vector fields of θ_Δ vanish at the origin, the $\mathcal{K}_\Delta$ tangent space of h_s satisfies

$$T\mathcal{K}_\Delta{\cdot}h_s + \mathfrak{m}_1 M = \mathcal{E}_1 \left\{ \begin{pmatrix} \lambda^2 \\ 0 \end{pmatrix}, \begin{pmatrix} 0 \\ \lambda^3 \end{pmatrix} \right\} + \mathcal{E}_1 \left\{ \begin{pmatrix} \lambda \\ 2s\lambda^2 \end{pmatrix}, \begin{pmatrix} 2\lambda \\ 3s\lambda^2 \end{pmatrix}, \begin{pmatrix} 0 \\ 2\lambda^2 \end{pmatrix} \right\}.$$

(The first summand is $\mathfrak{m}_1 M$.) This is equal to M, and hence, by Nakayama's lemma, we deduce that, for all $s \in \mathbb{R}$,

$$T\mathcal{K}_\Delta{\cdot}h_s = M.$$

In particular $\mathfrak{m}_1^2\theta(h) \subset T\mathcal{K}_\Delta{\cdot}h$ and hence h_s is 2-determined, and is equivalent to

$$h_s = (\lambda, s\lambda^2)$$

(renaming h_s). To show this is $\mathcal{K}_\Delta$-equivalent to h_0, we note that $\dot{h}_s \in M$, and

$$T_{\mathrm{rel}}\,\mathcal{K}_\Delta \cdot h_s = \mathcal{E}_{1,I} M$$

(for $I = \mathbb{R}$), and hence it follows from Theorem 20.4 that, for all s, h_s is equivalent to h_0.

For the versal unfolding of the pitchfork bifurcation, we require the appropriate tangent space of the path $h = h_0$:

$$T_e\mathcal{K}_\Delta{\cdot}h = \mathcal{E}_1 \left\{ \begin{pmatrix} 1 \\ 0 \end{pmatrix}, \begin{pmatrix} 3\lambda \\ 0 \end{pmatrix}, \begin{pmatrix} 0 \\ -6\lambda^2 \end{pmatrix} \right\} = \mathcal{E}_1 \left\{ \begin{pmatrix} 1 \\ 0 \end{pmatrix}, \begin{pmatrix} 0 \\ \lambda^2 \end{pmatrix} \right\}.$$

The terms $(0, 1)$ and $(0, \lambda)$ therefore form a cobasis, and a versal unfolding is given by

$$H(\lambda; a, b) = (\lambda,\ a\lambda + b). \qquad (21.A_2.4)$$

with corresponding bifurcation

$$G(x;\lambda;a,b,c) = x^3 + \lambda x + a\lambda + b = 0.$$

See Section 18.1 for a full discussion of this bifurcation. Note that the paths $\lambda \mapsto (\lambda, 0)$ and $\lambda \mapsto (-\lambda, 0)$ are right equivalent so from the classification point of view the difference is immaterial.

(A$_2$.3) Winged cusp: $h(\lambda) = (0, \lambda^2)$ **(codim=3)** Now assume that the 1-jet of the path vanishes, and the 2-jet is $(r\lambda^2, s\lambda^2)$ with $s \neq 0$. By rescaling λ, we can make $s = \pm 1$, and by the simple $\mathcal{K}_\Delta$-equivalence $v \mapsto -v$ we can choose $s = 1$ (see Remark 21.3). So let

$$h_r(\lambda) = (r\lambda^2, \lambda^2).$$

Then

$$T\mathcal{K}_\Delta{\cdot}h_r = \mathcal{E}_1\left\{\begin{pmatrix} r\lambda^2 \\ \lambda^2 \end{pmatrix}, \begin{pmatrix} 2r\lambda^2 \\ 3\lambda^2 \end{pmatrix}, \begin{pmatrix} 9\lambda^2 \\ -2r^2\lambda^4 \end{pmatrix}\right\} = \mathfrak{m}_1^2\theta(h_r).$$

Thus h_r is 2-determined, and moreover $T_{\mathrm{rel}}\mathcal{K}_\Delta \cdot h_r$ is independent of r and contains $\dot{h}_r$. The family h_r is therefore $\mathcal{K}_\Delta$-trivial and equivalent to h_0.

We have, with $h = h_0$,

$$T_e\mathcal{K}_\Delta{\cdot}h = \mathcal{E}_1\left\{\begin{pmatrix} 0 \\ \lambda \end{pmatrix}, \begin{pmatrix} \lambda^2 \\ 0 \end{pmatrix}\right\},$$

which is of codimension 3 in $\theta(h)$. The resulting bifurcation is known as the ***winged cusp*** bifurcation. A versal deformation is given by

$$H(\lambda;a,b,c) = (a\lambda + b,\ \lambda^2 + c), \qquad (21.\mathrm{A}_2.5)$$

with corresponding bifurcation

$$G(x;\lambda;a,b,c) = x^3 + (a\lambda + b)x + \lambda^2 + c = 0.$$

The next case is beyond our stated aim of codimension 3, but is included as an illustration of a unimodal path.

(A$_2$.4): $h(\lambda) = (\pm\lambda^2, r\lambda^3)$ **(unimodal, codim=5)** We continue to assume the 1-jet of h is zero, the second component has zero 2-jet (i.e., $s = 0$ in the previous case), but the 2-jet of the first component does not vanish. Applying a right equivalence by rescaling λ, we can assume the first component is $\pm\lambda^2$, and hence we assume

$$h^{\pm}_{r,s}(\lambda) = (\pm\lambda^2, r\lambda^3 + s\lambda^4),$$

which we show is 3-determined provided

$$\pm 4 + 27r^2 \neq 0, \qquad (21.A_2.6)$$

First, using Nakayama's lemma, and assuming (21.A_2.6), it is easy to show that $\mathfrak{m}_1^3 \times \mathfrak{m}_1^4 \subset T\mathcal{K}_\Delta \cdot h^{\pm}_{r,s}$, and hence

$$T\mathcal{K}_\Delta \cdot h^{\pm}_{r,s} = \mathbb{R}\left\{\begin{pmatrix}\pm 2\lambda^2\\ 3r\lambda^3\end{pmatrix}\right\} + \left(\mathfrak{m}_1^3 \times \mathfrak{m}_1^4\right).$$

This contains $\mathfrak{m}_1^4\theta(h)$ showing $h^{\pm}_{r,s}$ is 4-determined under condition (21.A_2.6).

Now, the generators of the tangent space $T\mathcal{K}_\Delta \cdot h^{\pm}_{r,s}$ are independent of s, and $\frac{\partial}{\partial s}h^{\pm}_{r,s} = (0, \lambda^4) \in T\mathcal{K}_\Delta \cdot h^{\pm}_{r,s}$ (smoothly in $s \in \mathbb{R}$) and hence the s-family is trivial, again assuming (21.A_2.6). Thus $h^{\pm}_{r,s}$ is equivalent to $h^{\pm}_r = h^{\pm}_{r,0}$ (with $s = 0$).

Now,

$$T_e\mathcal{K}_\Delta \cdot h^{\pm}_r = \mathcal{E}_1\left\{\begin{pmatrix}\pm 2\lambda\\ 3r\lambda^2\end{pmatrix}, \begin{pmatrix}9r\lambda^3\\ -2\lambda^4\end{pmatrix}\right\}.$$

Assuming $\pm 4 + 27r^2 \neq 0$ this is of codimension 5. Morever, for all r satisfying that condition, $\dot{h}_r = (0, \lambda^3)^T$ which is an element of a cobasis for h_r, showing that the members of the family are all inequivalent (at least, locally in r); that is, h_r is a unimodular family. As a family, $\{h_r\}$ is of codimension 4, so could generically appear in a family with 4 parameters, but not with 3 parameters.

A versal unfolding of $h^{\pm}_r$ is

$$H^{\pm}(\lambda; a, b, c, d, e) = (\pm\lambda^2 + b\lambda + c, (r + a)\lambda^3 + d\lambda + e),$$

with associated bifurcation problems

$$G^{\pm}(x, \lambda; a, b, c, d, e) = x^3 + (\pm\lambda^2 + b\lambda + c)x + (r + a)\lambda^3 + d\lambda + e.$$

It is clear from a diagram showing the image of the path and the discriminant that the $\pm$ cases are not equivalent.

Note that the condition (21.A_2.6) fails if the path has, in some sense, higher–order contact with Δ.

The set of paths $h: \mathbb{R} \rightarrowtail \mathbb{R}^2$ with vanishing 2-jet at some point is of codimension 5 (the value, first and second derivatives of each component is zero at some $t \in \mathbb{R}$, making 6 independent conditions at t, but t is arbitrary so leaving 5 conditions). If we limit to 3–parameter families then such paths can safely be avoided.

Two parameters

Now we consider $h\colon (\mathbb{R}^2,0) \to (\mathbb{R}^2,0)$ and classify relative to $\mathcal{K}_\Delta$-equivalence. The first case (the 'cusp') is where h is the germ of a diffeomorphism. After that we consider the cases where h is of rank 1 at the origin: firstly where h_2 (the second component of h) is a submersion, and then where it is not. The set of map germs where the 1-jet vanishes is of of codimension 4, which is too high for our classification.

(A$_2$.5) Cusp: $h(\lambda,\mu) = (\lambda,\mu)$ **(codim=0)** If h is a local diffeomorphism (that is $\mathsf{d}h_0$ is invertible), then h is right equivalent to the identity map. And regardless of Δ, one has

$$T_e\mathcal{K}_\Delta{\cdot}h = \theta(h),$$

showing of course that h is of codimension 0 (see Problem, 20.8). The corresponding bifurcation problem is the cusp,

$$G(x;\lambda,\mu) = x^3 + \lambda x + \mu = 0. \tag{21.A$_2$.7}$$

This 2–parameter family arises in, for example, the Zeeman Catastrophe Machine, described in Section 1.3.

We now consider the path to have rank 1 at the origin, beginning with 3 cases where h_2 is a submersion (where $h = (h_1,h_2)$).

(A$_2$.6) Lips/beaks: $h(\mu,\lambda) = (\pm\lambda^2,\mu)$ **(codim=1)** Suppose h is of rank 1, and h_2 is a submersion. Then we can choose coordinates λ,μ in the source so that $h_2(\lambda,\mu) = \mu$. Then

$$h(\mu,\lambda) = (\alpha\mu + p(\lambda,\mu),\ \mu), \tag{21.A$_2$.8}$$

with $\alpha \in \mathbb{R}$ and $p \in \mathfrak{m}_2^2$. We write p as Taylor series to some order, so $p(\lambda,\mu) = \sum_{i+j=2}^N p_{i,j}\mu^i\lambda^j + O(N+1)$, where N is a sufficiently large integer (for finite determinacy).

Let $M = \langle \mu, \lambda^2\rangle\,\theta(h)$. Using Nakayama's lemma one shows that

$$T\mathcal{K}_\Delta{\cdot}h = M \oplus \mathbb{R}\left\{\begin{pmatrix}\alpha\lambda\\ \lambda\end{pmatrix}\right\}$$

precisely when

$$p_{0,2} \neq 0, \tag{21.A$_2$.9}$$

and hence under this condition, h is 2-determined.

Let us now assume (21.A$_2$.9) holds: we wish to show that h is $\mathcal{K}_\Delta$-equivalent to $(\pm\lambda^2, \mu)$. Let U be the subset of 2-jets as above, parametrized by (α, p), and with $p_{0,2} \neq 0$. This has two connected components $U_\pm$, distinguished by the sign of $p_{0,2}$. These two components satisfy the conditions of the $\mathcal{K}_V$ version of Corollary 12.22, with $U_\pm$ contained in a linear subspace of M. Thus $(\alpha, p_{0,2})$ with $p_{0,2} \neq 0$ is equivalent to $(0, \pm 1)$. This shows that h is $\mathcal{K}_\Delta$-equivalent to $(\pm\lambda^2, \mu)$.

Now, with $h(\lambda, \mu) = (\pm\lambda^2, \mu)$,

$$T_e\mathcal{K}_\Delta{\cdot}h = \mathfrak{m}_2 \times \mathcal{E}_2,$$

showing that h is of codimension 1, with versal unfolding

$$H_\pm(\lambda, \mu; a) = (\pm(\lambda^2 + a),\ \mu). \qquad (21.\mathsf{A}_2.10)$$

See Figure 21.1. The resulting bifurcation problem is

$$G_\pm(x, \lambda, \mu; a) = x^3 \pm (\lambda^2 + a)x + \mu = 0.$$

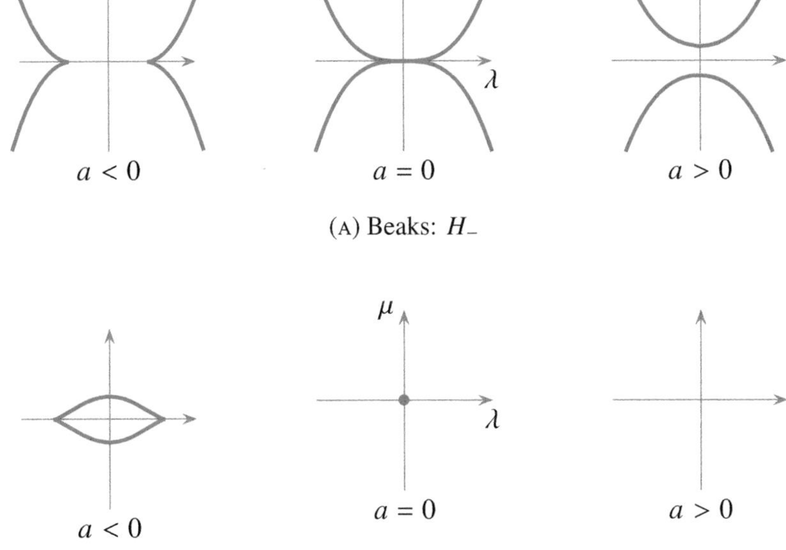

FIGURE 21.1 The curves show $h_a^{-1}(\Delta)$ for each of the beaks and lips bifurcations (see (A$_2$.6)); this is equal to the discriminant of the associated bifurcation problem G_a.

(A_2.7): $h(\lambda,\mu) = (\lambda^3,\mu)$ (codim=2) We continue with the assumption that the path is of rank 1 at the origin and h_2 is a submersion. It is simpler to first choose coordinates in the source so that $h_2(\lambda,\mu) = \mu$. Thus

$$h = (p(\lambda,\mu),\ \mu)$$

with now $p \in \mathfrak{m}_2^2$.

With h in the form (21.A_2.8), we now continue to suppose $\beta = 1$, but now with $\delta = 0$, and in that case $p_{0,2} = \alpha q_{0,2}$. (We treat the case $\beta = 0$ below.)

Let $M = \langle \mu,\ \lambda^3 \rangle \theta(h)$. Then one shows, as usual using Nakayama's lemma, that $M \subset T\mathcal{K}_\Delta{\cdot}h$ provided $p_{0,3} \neq \alpha q_{0,3}$. Hence firstly, under this condition, h is 3-determined, and by the constant tangent space theorems, h is equivalent to $(\pm\lambda^3,\mu)$. But of course the sign can be changed by changing the sign of λ.

For $h = (\lambda^3,\mu)$, one finds $T_e\mathcal{K}_\Delta{\cdot}h$ to be of codimension 2, with versal unfolding

$$H(\lambda,\mu;a,b) = (\lambda^3 + a\lambda + b,\ \mu) \qquad (21.A_2.11)$$

and associated bifurcation problem,

$$G(x;\lambda,\mu;a) = x^3 + (\lambda^3 + a\lambda + b)x + \mu = 0.$$

The set of paths with h_2 being a submersion but more degenerate than those above has codimension at least 4 so we ignore such maps. We now turn to those paths whose 1-jet is non-zero, but for which $\mathrm{d}h_2(0) = 0$.

(A_2.8): $h(\lambda,\mu) = (\lambda,\mu^2)$ (codim=2) We continue with the assumption that the path is of rank 1 at the origin. With h in the form (21.A_2.8), we now suppose $\beta = 0$, and in that case $\alpha \neq 0$. We can rescale λ so that $\alpha = 1$. Thus $h(\lambda,\mu) = (\lambda + p(\lambda,\mu),\ q(\lambda,\mu))$. In fact we can simplify further, by changing the λ-coordinate to $\lambda' = \lambda + p(\lambda,\mu)$. Relabelling λ' as λ, we now write

$$h(\lambda,\mu) = (\lambda,\ q(\lambda,\mu))\,. \qquad (21.A_2.12)$$

We write, as before, $q(\lambda,\mu) = q_{2,0}\lambda^2 + q_{1,1}\lambda\mu + q_{0,2}\mu^2 + O(3)$, and using Nakayama's lemma, one obtains

$$T\mathcal{K}_\Delta{\cdot}h = \mathfrak{m}_2 \times \mathfrak{m}_2^2 \quad \text{if and only if} \quad q_{0,2} \neq 0.$$

Using a similar constant tangent space argument to previous cases, one can then reduce h to the form $h(\lambda,\mu) = (\lambda,\pm\mu^2)$. Note that the reflection $(u,v) \mapsto (u,-v)$ preserves Δ and changes $(\lambda,-\mu^2)$ to (λ,μ^2); see Remark 21.3.

Finally, one finds, with $h(\lambda, \mu) = (\lambda, \mu^2)$,

$$T_e\mathcal{K}_\Delta{\cdot}h = \mathcal{E}_2\left\{\begin{pmatrix}1\\0\end{pmatrix}, \begin{pmatrix}0\\\mu\end{pmatrix}, \begin{pmatrix}0\\\lambda^2\end{pmatrix}\right\}.$$

Thus h is of codimension 2, with versal unfolding

$$H(\lambda, \mu; a, b) = (\lambda, \mu^2 + a\lambda + b), \tag{21.A$_2$.13}$$

and consequent bifurcation problem

$$G(x; \lambda, \mu; a, b) = x^3 + \lambda x + \mu^2 + a\lambda + b = 0.$$

Remark 21.4. The curve $h^{-1}(\Delta)$ is given by $27\mu^4 - 4\lambda^3 = 0$, a curve with an E_6 singularity. Writing $h_{a,b}(\lambda, \mu) = (-\lambda, \mu^2 + a\lambda + b)$ then $h_{a,b}^{-1}(\Delta)$ is given by

$$27(\mu^2 + a\lambda + b)^2 = 4\lambda^3.$$

Examples of this are shown in Figure 21.2. Notice that the map $h_{a,b}$ is a fold; its singular locus is $\mu = 0$, and the fold line in the image is $v = au + b$. This is the same line as the deformed path for the unfolding of the pitchfork bifurcation described in Section 18.1 and Figs 18.1 and 18.2. Comparing with the latter figure, the image of $h_{a,b}$ lies above the fold line. The nature of the discriminant $h_{a,b}^{-1}(\Delta)$ can be determined by analysing the 4 regions A–D in Figure 18.2. See Figure 21.2 for diagrams. ❞

Three parameters

Now we consider germs $h\colon (\mathbb{R}^3, 0) \to (\mathbb{R}^2, 0)$, and classify these with respect to $\mathcal{K}_\Delta$-equivalence, up to codimension 1.

(A$_2$.9) Submersion/cusp: $h(\lambda, \mu, \nu) = (\lambda, \mu)$ **(codim=0)** Any submersion is right equivalent to this map, and is of $\mathcal{K}_\Delta$-codimension 0. The corresponding bifurcation is

$$G(x; \lambda, \mu, \nu) = x^3 + \lambda x + \mu$$

(A$_2$.10) Flying saucer/3D-beaks/Yo-yo: $h(\lambda, \mu, \nu) = (\pm\lambda^2 \pm \mu^2, \nu)$ **(codim=1)** Let $h = (h_1, h_2)$ be of rank 1 and in particular suppose that h_2 is a submersion. In that case, we can choose coordinates in the source so that $h_2 = \nu$.

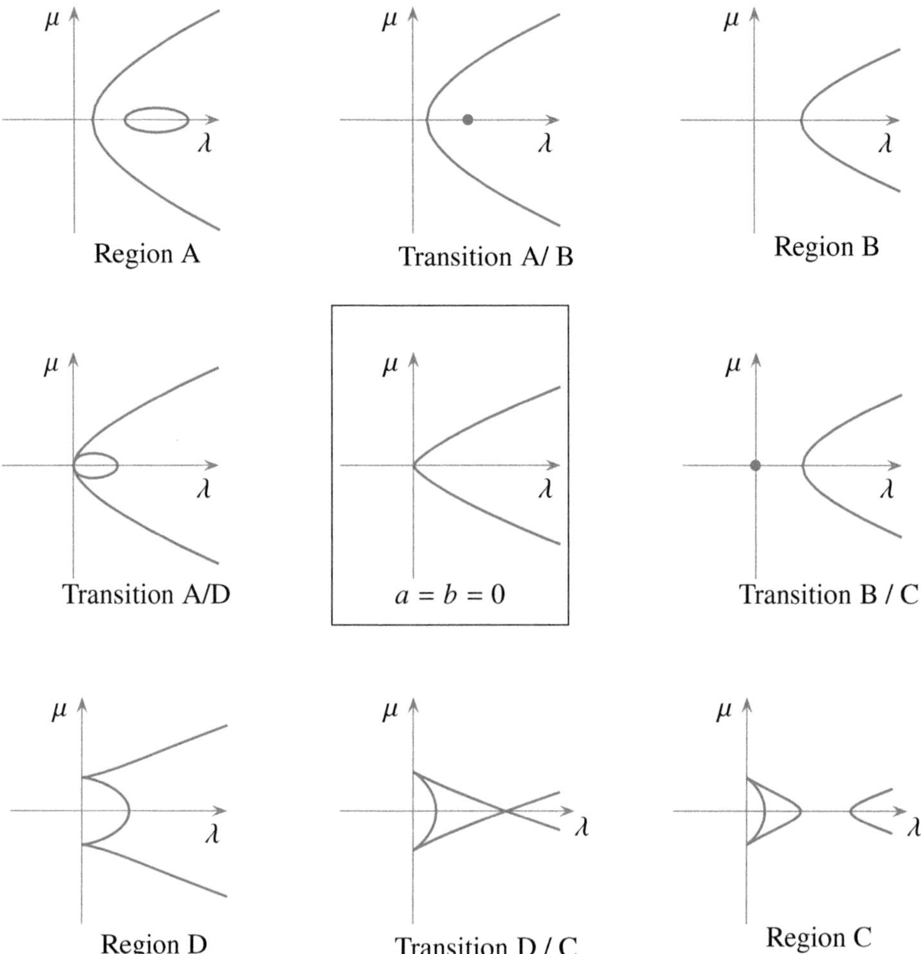

FIGURE 21.2 The discriminants for the bifurcation in (21.A_2.13); the regions A–D are those shown in Figure 18.2.

Consider $h(\lambda, \mu, \nu) = (\alpha\nu + q(\lambda, \mu, \nu),\ \nu)$, with $q \in \mathfrak{m}_3^2$. Then

$$T\mathcal{K}_\Delta{\cdot}h = \mathfrak{m}_3\left\{\begin{pmatrix}\alpha + q_\nu\\ 1\end{pmatrix}, \begin{pmatrix}q_\lambda\\ 0\end{pmatrix}, \begin{pmatrix}q_\mu\\ 0\end{pmatrix}\right\} + \mathcal{E}_3\left\{\begin{pmatrix}2(\alpha\nu + q)\\ 3\nu\end{pmatrix}, \begin{pmatrix}9\nu\\ -2(\alpha\nu + q)^2\end{pmatrix}\right\}.$$

Applying Nakayama's lemma with $M = \langle\lambda^2, \lambda\mu, \mu^2, \nu\rangle \times \mathfrak{m}_3$ one finds that $T\mathcal{K}_\Delta{\cdot}h = M$ provided the quadratic form $Q := \mathrm{j}^2 q(\lambda, \mu, 0)$ is nondegenerate.

Map germs h satisfying this nondegeneracy condition are all 2-determined, and can be divided into 3 classes according to the signature of Q, giving rise to three normal forms,

$$h_0 = (\lambda^2 + \mu^2, \nu), \quad h_1 = (\lambda^2 - \mu^2, \nu), \quad \text{and} \quad h_2 = (-\lambda^2 - \mu^2, \nu).$$

Here we use the constant tangent space argument to change α to 0.

For each of the three cases, a versal unfolding is given by

$$H(\lambda, \mu, \nu; a) = (\pm\lambda^2 \pm \mu^2 + a, \nu). \tag{21.A$_2$.14}$$

See Figure 21.3 for illustrations of $h^{-1}(\Delta)$ as a varies. The associated bifurcation problem is

$$G(x; \lambda, \mu, \nu; a) = x^3 + (\pm\lambda^2 \pm \mu^2 + a)x + \nu = 0.$$

If h_2 fails to be a submersion or the nondegeneracy condition above is not satisfied then the 3–parameter path is of codimension greater than 1.

21.3 A$_3$ core

Now consider the core $g_0(x) = x^4$. The versal unfolding is

$$F(x; u, v, s) = x^4 + ux^2 + vx + s, \tag{21.A$_3$.1}$$

whose discriminant is the swallowtail surface (the equation of the surface is given in Problem 7.14, although we don't need to make use of it). From Table 19.1 we have (slightly simplified),

$$\theta_\Delta = \mathcal{E}_3\left\{\begin{pmatrix}2u\\ 3v\\ 4s\end{pmatrix}, \begin{pmatrix}6v\\ 8s - 2u^2\\ -uv\end{pmatrix}, \begin{pmatrix}16s\\ -2uv\\ 8us - 3v^2\end{pmatrix}\right\}.$$

(A) Flying saucer: $h_a(\lambda, \mu, \nu) = (\lambda^2 + \mu^2 + a, \nu)$
(for $a > 0$ the discriminant is empty)

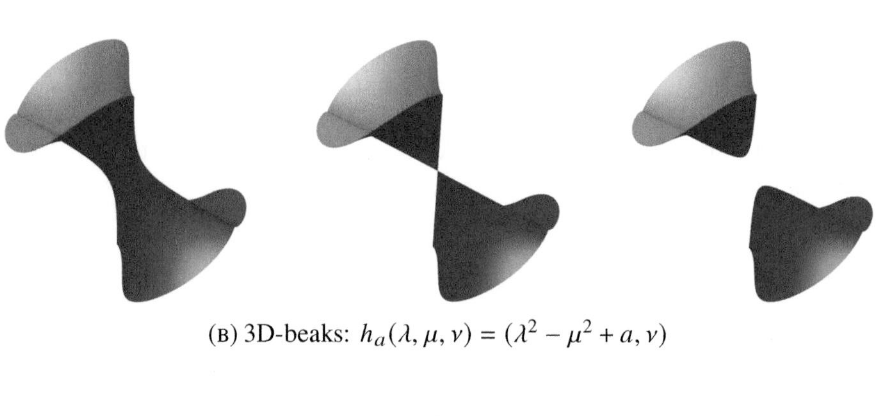

(B) 3D-beaks: $h_a(\lambda, \mu, \nu) = (\lambda^2 - \mu^2 + a, \nu)$

(C) Yo-yo: $h_a(\lambda, \mu, \nu) = (-\lambda^2 - \mu^2 + a, \nu)$

FIGURE 21.3 Discriminants for the codimension-1 3–parameter paths with A_2 core (A_2.10) (see (21.A_2.14)): in each figure the cuspidal edge is given by $h_a^{-1}(0,0)$. For each path, $a < 0$ is on the left, $a = 0$ in the middle and $a > 0$ on the right.

One parameter

(A$_3$.1) Quartic fold: $h(\lambda) = (0, 0, \lambda)$ (codim=2) We start with $h(\lambda) = (\alpha\lambda,\ \beta\lambda,\ \gamma\lambda) + O(\lambda^2)$, with $(\alpha, \beta, \gamma) \neq 0$. We have,

$$T\mathcal{K}_\Delta{\cdot}h + \mathfrak{m}_1^2\theta(h) = \mathcal{E}_1\left\{\begin{pmatrix}\alpha\lambda\\ \beta\lambda\\ \gamma\lambda\end{pmatrix}, \begin{pmatrix}2\alpha\lambda\\ 3\beta\lambda\\ 3\gamma\lambda\end{pmatrix}, \begin{pmatrix}6\beta\lambda\\ 8\gamma\lambda\\ 0\end{pmatrix}, \begin{pmatrix}16\gamma\lambda\\ 0\\ 0\end{pmatrix}\right\} + \mathfrak{m}_1^2\theta(h).$$

Thus $T\mathcal{K}_\Delta{\cdot}h = \mathfrak{m}_1\theta(h)$ if and only if $\gamma \neq 0$. This shows that for $\gamma \neq 0$, firstly h is 1-determined, and secondly the constant tangent space argument shows that h is $\mathcal{K}_\Delta$-equivalent to $\lambda \mapsto (0, 0, \lambda)$.

Finally, with $h(\lambda) = (0, 0, \lambda)$, we find

$$T_e\mathcal{K}_\Delta{\cdot}h = \mathbb{R}\left\{\begin{pmatrix}0\\ 0\\ 1\end{pmatrix}\right\} + \mathfrak{m}_1\theta(h),$$

showing that h is of codimension 2, and a versal unfolding is given by

$$H(\lambda; a, b) = (a, b, \lambda). \tag{21.A$_3$.2}$$

The corresponding family of bifurcation problems is

$$G(x; \lambda; a, b) = x^4 + ax^2 + bx + \lambda = 0.$$

The path h is the vertical axis in Figure 7.1 (or Figure **??**), and varying a, b translates the path horizontally.

(A$_3$.2) Quartic Pitchfork: $h(\lambda) = (0, \lambda, 0)$ (codim=3) The previous case assumed $\gamma \neq 0$. We now put $\gamma = 0$. Then with $h(\lambda) = (\alpha\lambda, \beta\lambda, \delta\lambda^2) + O(\lambda^2, \lambda^2, \lambda^3)$ one finds

$$T\mathcal{K}_\Delta{\cdot}h + \mathfrak{m}_1 M = \mathcal{E}_1\left\{\begin{pmatrix}\alpha\lambda\\ \beta\lambda\\ 2\delta\lambda^2\end{pmatrix}, \begin{pmatrix}2\alpha\lambda\\ 3\beta\lambda\\ 4\delta\lambda^2\end{pmatrix}, \begin{pmatrix}6\beta\lambda\\ 0\\ -\alpha\beta\lambda^2\end{pmatrix}, \begin{pmatrix}0\\ 0\\ -3\beta^2\lambda^2\end{pmatrix}\right\} + \mathfrak{m}_1 M,$$

where $M = \left(\mathfrak{m}_1 \times \mathfrak{m}_1 \times \mathfrak{m}_1^2\right)$. Thus, $T\mathcal{K}_\Delta{\cdot}h = M$ if and only if $\beta \neq 0$. By rescaling λ we can choose $\beta = 1$. Thus h is 2-determined, but moreover by the constant tangent space argument of Theorem 20.4, h is $\mathcal{K}_\Delta$-equivalent to $h(\lambda) = (0, \lambda, 0)$.

The extended $\mathcal{K}_\Delta$-tangent space for this path h is

$$T_e\mathcal{K}_\Delta\cdot = \left\{ \begin{pmatrix}0\\1\\0\end{pmatrix}, \begin{pmatrix}\lambda\\0\\0\end{pmatrix}, \begin{pmatrix}0\\0\\\lambda^2\end{pmatrix} \right\}.$$

Consequently, h is of codimension 3 with $\mathcal{K}_\Delta$-versal unfolding

$$H(\lambda; a, b, c) = (a, \lambda, b\lambda + c). \tag{21.A$_3$.3}$$

The associated family of bifurcation problems is

$$G(x; \lambda; a, b) = x^4 + ax^2 + \lambda x + b\lambda + c.$$

The next path in the classification is of codimension greater than 3 and is described in Problem 21.8.

Two parameters

We now classify map germs $h\colon (\mathbb{R}^2, 0) \to (\mathbb{R}^3, 0)$ relative to $\mathcal{K}_\Delta$-equivalence. The least possible codimension is 1.

(A$_3$.3): $h(\lambda, \mu) = (0, \lambda, \mu)$ **(codim=1)** One finds that a necessary and sufficient condition for $T\mathcal{K}_\Delta\cdot h = \mathfrak{m}_2\theta(h)$ is that (h_2, h_3) is a submersion at $(0, 0)$. In that case one can choose coordinates in $\mathbb{R}^2$ so that

$$h(\lambda, \mu) = (p(\lambda, \mu),\ \lambda,\ \mu).$$

The constant tangent space argument, shows these are all $\mathcal{K}_\Delta$-equivalent to the same map with $p = 0$. For $h(\lambda, \mu) = (0, \lambda, \mu)$,

$$T_e\mathcal{K}_\Delta\cdot h = \mathfrak{m}_2 \times \mathcal{E}_2 \times \mathcal{E}_2,$$

whence a versal unfolding is given by

$$H(\lambda, \mu; a) = (a, \lambda, \mu) \tag{21.A$_3$.4}$$

which corresponds to the bifurcation problem,

$$G(x; \lambda, \mu; a) = x^4 + ax^2 + \lambda x + \mu = 0.$$

(A_3.4): $h(\lambda,\mu) = (\lambda,\lambda^2,\mu)$ (codim=2) Now suppose (h_2, h_3) fails to be a submersion. If $\mathsf{d}h_3(0) = 0$ or if h fails to be an immersion then the codimension is greater than 2. We can therefore assume $\mathsf{d}h_2(0) = 0$, and h takes the form

$$h(\lambda,\mu) = (\lambda,\ p(\lambda,\mu),\ \mu),$$

with $p \in \mathfrak{m}_2^2$. In this case, provided $p_{2,0} \neq 0$, one finds $\mathfrak{m}_2^2\theta(h) \subset T\mathcal{K}_\Delta{\cdot}h$. The constant tangent space argument shows that such h are equivalent to $(\lambda,\mu) \mapsto (\lambda, \pm\lambda^2, \mu)$, and the symmetry $v \mapsto -v$ in (21.A_3.1) shows these are both equivalent to $h(\lambda,\mu) = (\lambda,\lambda^2,\mu)$.

A versal unfolding of this path is given by

$$H(\lambda,\mu;a,b) = (\lambda, \lambda^2 + a\lambda + b, \mu),$$

with associated bifurcation problem

$$G(x;\lambda,\mu;a,b) = x^4 + \lambda x^2 + (\lambda^2 + a\lambda + b)x + \mu = 0. \qquad (21.A_3.5)$$

Three parameters

Now we consider $h\colon (\mathbb{R}^3,0) \to (\mathbb{R}^3,0)$ and classify relative to $\mathcal{K}_\Delta$-equivalence. The first case (the 'swallowtail' itself) is where h is the germ of a diffeomorphism. After that we consider the cases where h is of rank 2 at the origin: firstly where (h_2, h_3) is a submersion, and then where it isn't. The set of maps of rank 1 at some point is of codimension 4, which is too high for our classification.

(A_3.5) Swallowtail: $h(\lambda,\mu,v) = (\lambda,\mu,v)$ (codim=0) If h is a local diffeomorphism (i.e. $\mathsf{d}h_0$ is invertible), then h is right equivalent to the identity map. And regardless of Δ one has

$$T_e\mathcal{K}_\Delta{\cdot}h = \theta(h),$$

showing of course that h is of codimension 0 (see Problem 20.8). The corresponding bifurcation problem is the swallowtail,

$$G(x;\lambda,\mu,v) = x^4 + \lambda x^2 + \mu x + v = 0. \qquad (21.A_3.6)$$

(A_3.6) Double swallowtail: $h(\lambda,\mu,\nu) = (\pm\lambda^2,\mu,\nu)$ **(codim=1)** Now assume h is of rank 2 at the origin. If (h_2,h_3) is not a submersion, then the codimension is higher than 1, and so we can choose coordinates in the source so that $h(\lambda,\mu,\nu) = (p(\lambda,\mu,\nu),\mu,\nu)$, with $p \in \mathfrak{m}_3^2$. A constant tangent space argument shows that provided $p_{2,0,0} \neq 0$ the germ is $\mathcal{K}_\Delta$-equivalent to $h(\lambda,\mu,\nu) = (\pm\lambda^2,\mu,\nu)$. These have codimension 1. A versal unfolding is given by,

$$H(\lambda,\mu,\nu;a) = (\pm(\lambda^2+a),\mu,\nu), \tag{21.A$_3$.7}$$

with associated bifurcation problem,

$$G(x;\lambda,\mu,\nu;a) = x^4 \pm (\lambda^2+a)x^2 + \mu x + \nu = 0.$$

21.4 A_4 core

Now consider the core $g_0(x) = x^5$ and its versal unfolding,

$$F(x,u,v,s,t) = x^5 + ux^3 + vx^2 + wx + s. \tag{21.A$_4$.1}$$

The discriminant of this family is the butterfly hypersurface in $\mathbb{R}^4$. See [98] for illustrations of 3–dimensional sections of this hypersurface. The module θ_Δ of vector fields tangent to this variety can be found in the table on p. 237.

One parameter

(A_4.1) Quintic fold: $h(\lambda) = (0,0,0,\lambda)$ **(codim=3)** We begin with

$$h(\lambda) = (\alpha\lambda,\ \beta\lambda,\ \gamma\lambda,\ \delta\lambda) + O(\lambda^2).$$

The calculation is very similar to that for the quartic fold (A_3.1) and details are left to the reader. The upshot is that h is 1-determined iff $\delta \neq 0$, and by a constant tangent space argument, is then equivalent to $h(\lambda) = (0,0,0,\lambda)$. The path has codimension 3, with versal unfolding $H(\lambda;a,b,c) = (a,b,c,\lambda)$ and resulting bifurcation equation

$$G(x;\lambda;a,b,c) = x^5 + ax^3 + bx^2 + cx + \lambda = 0. \tag{21.A$_4$.2}$$

If the fourth component of h is singular, then it is not hard to show the germ is of $\mathcal{K}_\Delta$-codimension at least 4. The first case would be $h(\lambda) = (0,0,\lambda,0)$ which is indeed of $\mathcal{K}_\Delta$-codimension 4.

Two parameters

We now classify map germs $h\colon (\mathbb{R}^2,0) \to (\mathbb{R}^4,0)$ relative to $\mathcal{K}_\Delta$-equivalence. The least possible codimension is 2.

(A$_4$.2): $h(\lambda,\mu) = (0,\pm\mu,\lambda,\mu)$ **(codim 2)** Begin by writing the general path as

$$h(\lambda,\mu) = \begin{pmatrix} \alpha_1 & \alpha_2 \\ \beta_1 & \beta_2 \\ \gamma_1 & \gamma_2 \\ \delta_1 & \delta_2 \end{pmatrix} \begin{pmatrix} \lambda \\ \mu \end{pmatrix} + O(2). \tag{21.A$_4$.3}$$

One finds that a necessary condition for $T\mathcal{K}_\Delta{\cdot}h = \mathfrak{m}_2\theta(h)$ is that $\gamma_1\delta_2 - \delta_1\gamma_2 \neq 0$; that is, that the map $(\lambda,\mu) \mapsto (h_3,h_4)$ is a submersion. With this assumption, we can choose coordinates in the source so that $(h_3,h_4) = (\lambda,\mu)$. Thus h takes the form,

$$h = (\alpha_1\lambda + \alpha_2\mu + p(\lambda,\mu),\ \beta_1\lambda + \beta_2\mu + q(\lambda,\mu),\ \lambda,\ \mu),$$

with $p,q \in \mathfrak{m}_2^2$. For such h, one finds $T\mathcal{K}_\Delta{\cdot}h = \mathfrak{m}_2\theta(h)$ if and only if $35\beta_1^2 - 40\alpha_1 + 32\beta_2 \neq 0$. This shows firstly that any such h satisfying this nondegeneracy condition is 2-determined, and secondly the constant tangent space argument shows that it is $\mathcal{K}_\Delta$-equivalent to

$$h(\lambda,\mu) = (\pm\lambda,0,\lambda,\mu).$$

(The $\pm$-sign depending on the sign of the nondegeneracy condition above – note that the relative sign of s and v is invariant under changes of signs allowed in expression (21.A$_4$.1).) A versal unfolding is

$$H(\lambda,\mu;a,b) = (\pm(\lambda+a),b,\lambda,\mu)$$

with consequent bifurcation problem

$$G(x;\lambda,\mu;a,b) = x^5 \pm (\lambda+a)x^3 + bx^2 + \lambda x + \mu = 0. \tag{21.A$_4$.4}$$

Three parameters

Now we consider $h\colon (\mathbb{R}^3,0) \to (\mathbb{R}^4,0)$ and classify relative to $\mathcal{K}_\Delta$-equivalence. On a dimension count, the minimal codimension is 1.

(A$_4$.3): $h(\lambda,\mu,\nu) = (0,\lambda,\mu,\nu)$ **(codim=1)** Writing h in the form (21.A$_4$.3) (but where the matrix has 3 columns), one finds that $T\mathcal{K}_\Delta{\cdot}h = \mathfrak{m}_3\theta(h)$ only if the map $(\lambda, \mu, \nu) \mapsto (h_2, h_3, h_4)$ is a diffeomorphism. Using this, we can choose coordinates in the source so that

$$h(\lambda, \mu, \nu) = (\alpha_1\lambda + \alpha_2\mu + \alpha_3\nu + p(\lambda, \mu, \nu), \lambda, \mu, \nu).$$

A computation shows that in this case $T\mathcal{K}_\Delta{\cdot}h = \mathfrak{m}_3\theta(h)$, independently of the α_j, p. Thus h is 1-determined and by a constant tangent space argument, is equivalent to

$$h(\lambda, \mu, \nu) = (0, \lambda, \mu, \nu).$$

Its versal unfolding is

$$H(\lambda, \mu, \nu; a) = (a, \lambda, \mu, \nu).$$

The corresponding bifurcation problem arises as a section of the butterfly singularity,

$$G(x; \lambda, \mu, \nu; a) = x^5 + ax^3 + \lambda x^2 + \mu x + \nu = 0.$$

Other 3–parameter paths will have codimension greater than 1.

21.5 $\mathrm{I}_{2,2}$ core

We now consider the core $g_0(x, y) = (x^2, y^2)$. This is the first case of a corank-2 bifurcation. The $\mathcal{K}$-versal unfolding of g_0 we use is

$$F(x, y; u, v, s, t) = \begin{pmatrix} x^2 + uy + s \\ y^2 + vx + t \end{pmatrix}. \qquad (21.\mathrm{I}_{2,2}.1)$$

The module θ_Δ of vector fields tangent to the discriminant can be found in the table on p. 237.

One parameter

($\mathrm{I}_{2,2}$.1): $h(\lambda) = (0, 0, \lambda, \pm\lambda)$ (codim=3) Let

$$h(\lambda) = (\alpha\lambda, \beta\lambda, \gamma\lambda, \delta\lambda) + O(\lambda^2).$$

A simple calculation similar to those above shows that

$$T\mathcal{K}_\Delta{\cdot}h = \mathfrak{m}_1\theta(h)$$

if and only if $\gamma\delta \neq 0$. Under this condition, the germ h is 1-determined.

The condition $\gamma\delta \neq 0$ partitions the space of 1-jets into 4 components. However, changing the sign of λ and exchanging x and y reduces this to two. A constant tangent space argument shows that h is equivalent to one of

$$h(\lambda) = (0, 0, \lambda, \pm\lambda).$$

This germ has $\mathcal{K}_\Delta$-codimension equal to 3, and a versal unfolding of h is

$$H(\lambda; a, b, c) = (a, b, \lambda + c, \pm\lambda),$$

and the associated bifurcation equation is

$$G(x, y, \lambda; a, b, c) = \begin{pmatrix} x^2 + ay + \lambda + c \\ y^2 + bx \pm \lambda \end{pmatrix} = 0.$$

The set of germs as above with $\gamma\delta = 0$ has codimension equal to 4.

Two parameters

Now consider $h\colon (\mathbb{R}^2, 0) \to (\mathbb{R}^4, 0)$, where $\mathbb{R}^4$ is the base space for the $\mathrm{I}_{2,2}$ versal unfolding. The minimal possible codimension is 2, and this can only occur if the map is an immersion. The set of such germs that fail to be an immersion is of codimension 3,

($\mathrm{I}_{2,2}$.2): $h(\lambda, \mu) = (\mu, \lambda, \lambda, \mu)$ (codim=2). We begin by assuming h has rank 2. To determine when $T\mathcal{K}_\Delta{\cdot}h = \mathfrak{m}_2\theta(h)$, we proceed as follows. Write

$$h(\lambda, \mu) = \begin{pmatrix} \alpha_1 & \alpha_2 \\ \beta_1 & \beta_2 \\ \gamma_1 & \gamma_2 \\ \delta_1 & \delta_2 \end{pmatrix} \begin{pmatrix} \lambda \\ \mu \end{pmatrix} + O(2). \qquad (21.\mathrm{I}_{2,2}.2)$$

The dimension of $\mathfrak{m}_2\theta(h)$ mod $\mathfrak{m}_2^2\theta(h)$ is equal to 8 (a basis is $\{\lambda e_i, \mu e_i\}_{i=1,\dots,4}$), and there are 8 generators of $T\mathcal{K}_\Delta{\cdot}h$. The determinant of the corresponding 8×8 matrix is (up to a constant)

$$(\delta_1\gamma_2 - \delta_2\gamma_1)^2(\beta_1\delta_2 - \beta_2\delta_1)(\alpha_1\gamma_2 - \alpha_2\gamma_1). \qquad (21.\mathrm{I}_{2,2}.3)$$

For this matrix to be invertible, the first factor $(\delta_1\gamma_2 - \delta_2\gamma_1)$ of the determinant must be nonzero, meaning that the map h must be a submersion onto the final two components s, t. We can therefore choose the coordinates λ, μ so that $s = \lambda$ and $t = \mu$. This simplifies h to

$$h(\lambda, \mu) = (\alpha_1\lambda + \alpha_2\mu,\ \beta_1\lambda + \beta_2\mu,\ \lambda,\ \mu) + O(2). \qquad (21.\mathrm{I}_{2,2}.4)$$

For this h,

$$T\mathcal{K}_\Delta \cdot h = \mathfrak{m}_2\, \theta(h)$$

if and only if $\beta_1\alpha_2 \neq 0$. By the now familiar constant tangent space argument, we can reduce h to one of 4 cases with $\beta_2 = \alpha_1 = 0$ and $\alpha_2 \pm 1, \beta_1 = \pm 1$:

$$(\pm\mu, \pm\lambda, \lambda, \mu).$$

Changing the signs of (u, y) and/or (v, x) in F shows that in fact these are all equivalent. We have therefore reduced this first case to

$$h(\lambda, \mu) = (\mu, \lambda, \lambda, \mu).$$

Calculation of $T_e\mathcal{K}_\Delta \cdot h$ shows this is indeed of codimension 2, and a versal unfolding is given by

$$H(\lambda, \mu; a, b) = (\mu + a,\ \lambda + b,\ \lambda,\ \mu).$$

The corresponding family of bifurcations is

$$G(x, y; \lambda, \mu; a, b) = \begin{pmatrix} x^2 + (\mu + a)y + \lambda \\ y^2 + (\lambda + b)x + \mu \end{pmatrix} = 0.$$

Three parameters

Any map germ $(\mathbb{R}^3, 0) \to (\mathbb{R}^4, 0)$ is of $\mathcal{K}_\Delta$-codimension at least 1. In order to equal one, the germ must be an immersion, for otherwise $T_e\mathcal{K}_\Delta \cdot h + \mathfrak{m}_3\theta(h)$ is of codimension greater than 1 in $\theta(h)$.

($\mathrm{I}_{2,2}$.3): $h(\lambda, \mu, \nu) = (\lambda, \lambda, \mu, \nu)$ (codim=1). We begin by noting that, due to the form of the third and fourth components of θ_Δ, if (h_3, h_4) fails to be a submersion then the codimension is greater than 1. We therefore consider immersions that are submersions onto the final 2 coordinates and chose μ, ν via a right equivalence so that,

$$h(\lambda, \mu, \nu) = (p(\lambda, \mu, \nu),\ q(\lambda, \mu, \nu),\ \mu,\ \nu), \qquad (21.\mathrm{I}_{2,2}.5)$$

with $p, q \in \mathfrak{m}_3$. A calculation with Nakayama's lemma shows that $T\mathcal{K}_\Delta \cdot h = \mathfrak{m}_3\,\theta(h)$ if and only if $p_{1,0,0} \neq 0$ and $q_{1,0,0} \neq 0$. Under these conditions therefore, h is 1-determined and by a constant tangent space argument can be reduced to

$$h(\lambda, \mu, \nu) = (\pm\lambda,\ \pm\lambda,\ \mu,\ \nu).$$

From the expression (21.$\mathrm{I}_{2,2}$.1), we see that the signs of u, v can be changed (with corresponding changes in y, x respectively), showing that the four cases above are all equivalent to

$$h(\lambda, \mu, \nu) = (\lambda,\ \lambda,\ \mu,\ \nu).$$

The $\mathcal{K}_\Delta$ codimension of this h is indeed equal to 1, with versal unfolding

$$H(\lambda, \mu, \nu; a) = (\lambda + a, \lambda - a, \mu, \nu).$$

The associated bifurcation problem is

$$G(x, y; \lambda, \mu, \nu; a) = \begin{pmatrix} x^2 + (\lambda + a)y + \mu \\ y^2 + (\lambda - a)x + \nu \end{pmatrix} = 0.$$

The set of more degenerate 3–parameter paths is of higher codimension.

21.6 $\mathrm{II}_{2,2}$ core

Now consider the core $g_0(x) = (x^2 - y^2,\ xy)$; the analysis is very similar to the $\mathrm{I}_{2,2}$ case. One versal unfolding is

$$F(x, y, u, v, s, t) = \begin{pmatrix} x^2 - y^2 + u(x + y) + s \\ xy + v(x + y) + t \end{pmatrix}. \qquad (21.\mathrm{II}_{2,2}.1)$$

Generators of the module θ_Δ of vector fields tangent to the discriminant can be found in the table on p. 237.

One parameter

($\mathrm{II}_{2,2}$.1): $h(\lambda) = (0, 0, 0, \lambda)$ (codim=3) We begin with

$$h(\lambda) = (\alpha\lambda,\ \beta\lambda,\ \gamma\lambda,\ \delta\lambda) + O(\lambda^2),$$

Then using Nakayama's lemma one checks that

$$T\mathcal{K}_\Delta{\cdot}h = \mathfrak{m}_1\theta(h)$$

if and only if $\gamma^2 + \delta^2 \neq 0$, and since we assume the map is real, this is equivalent to $(\gamma, \delta) \neq (0, 0)$. Using a constant tangent space argument, we deduce such h are equivalent to, say,

$$h(\lambda) = (0, 0, 0, \lambda).$$

This path is of codimension 3 with versal unfolding

$$H(\lambda; a, b, c) = (a, b, c, \lambda),$$

and associated bifurcation problem

$$G(x, y; \lambda; a, b, c) = \begin{pmatrix} x^2 - y^2 + a(x + y) + b \\ xy + c(x + y) + \lambda \end{pmatrix} = 0.$$

Two parameters

Now consider $h\colon (\mathbb{R}^2, 0) \to (\mathbb{R}^4, 0)$ relative to the $\mathrm{II}_{2,2}$ discriminant. The minimal possible codimension is 2, and this can only occur if the map is an immersion.

($\mathrm{II}_{2,2}$.2): $h(\lambda, \mu) = (0, \lambda, \lambda, \mu)$ (codim=2) Writing the 1-jet of h in the form of Equation (21.$\mathrm{I}_{2,2}$.2) we may follow an identical argument to the one of $\mathrm{I}_{2,2}$.2. Indeed, the appropriate nondegeneracy condition requires $(\delta_1\gamma_2 - \delta_2\gamma_1) \neq 0$, and we may choose coordinates in the source to reduce h to the expression in (21.$\mathrm{I}_{2,2}$.4). This turns out to be nondegenerate if and only if

$$(\alpha_1 - \beta_2)^2 + (\alpha_2 + \beta_1)^2 \neq 0.$$

That is, under this condition $T\mathcal{K}_\Delta{\cdot}h = \mathfrak{m}_2\theta(h)$, and h is 1-determined. Using the usual constant tangent space argument one sees that such h are $\mathcal{K}_\Delta$-equivalent to, for example,

$$h(\lambda, \mu) = (0, \lambda, \lambda, \mu).$$

This is of codimension 2, and has versal unfolding

$$H(\lambda, \mu; a, b) = (a,\ \lambda + b,\ \lambda,\ \mu).$$

The corresponding family of bifurcations is

$$G(x, y; \lambda, \mu; a, b) = \begin{pmatrix} x^2 - y^2 + a(x + y) + \lambda \\ xy + (\lambda + b)(x + y) + \mu \end{pmatrix} = 0.$$

In the previous case, there were two nondegeneracy conditions: firstly that (h_3, h_4) is a submersion, and the second that $(\alpha_1, \alpha_2) \neq (\beta_2, -\beta_1)$. If either of these conditions is violated then the germ is of codimension greater than 2.

Three parameters

As for the $I_{2,2}$ core considered further above, we note that in order for the codimension of $h\colon (\mathbb{R}^3, 0) \to (\mathbb{R}^4, 0)$ to be at most 1, h must be an immersion which is also a submersion on the last two coordinates. Thus, we can write h in the form $(21.I_{2,2}.5)$.

($II_{2,2}$.3): $h(\lambda, \mu, \nu) = (\lambda, \lambda, \mu, \nu)$ **(codim=1)** With h in the form $(21.I_{2,2}.5)$, one shows that $T\mathcal{K}_\Delta{\cdot}h = \mathfrak{m}_3\theta(h)$ if and only if $\partial h_j/\partial\lambda(0,0,0) \neq 0$, for $j = 1, 2$. Under this condition, h is 2-determined and the constant tangent space argument shows that such h are all $\mathcal{K}_\Delta$-equivalent to

$$h(\lambda, \mu, \nu) = (\pm\lambda,\ \pm\lambda,\ \mu,\ \nu).$$

Symmetries of F in $(21.II_{2,2}.1)$ include

$$(x, y, u, v, s, t) \mapsto (-x, -y, -u, -v, s, t) \quad \text{and} \quad (x, y, u, v, s, t) \mapsto (y, x, -u, v, s, t),$$

which show that all 4 choices of sign are $\mathcal{K}_\Delta$-equivalent. A $\mathcal{K}_\Delta$-versal unfolding of h is given by

$$H(\lambda, \mu, \nu; a) = (\lambda + a,\ \lambda,\ \mu,\ \nu).$$

The corresponding bifurcation problem is

$$G(x, y; \lambda, \mu, \nu; a) = \begin{pmatrix} x^2 - y^2 + (\lambda + a)(x + y) + \mu \\ xy + \lambda(x + y) + \nu \end{pmatrix} = 0.$$

21.7 Conclusion: classification of bifurcations

In this section, we collect into three tables the data from the calculations above for classification of paths and present them as bifurcations and their unfoldings.

The lists of corank-1 bifurcations with one parameter appear in the book of Golubitsky and Schaeffer [49], while the two parameter bifurcations appear in the thesis of Martin Peters [92]; see also [91], where he uses the methods developed by

Golubitsky and Schaeffer rather than the path approach. The bifurcation problems with 3 parameters and those of corank 2 are not covered by Golubitsky and Schaeffer or by Peters.

In Tables 21.3–21.5, we use x, y to denote state variables (we don't need more than 2), λ, μ, ν for bifurcation parameters (we consider up to 3–parameter bifurcation problems), and a, b, c for the unfolding coefficients (we stop at codimension 3). It turns out that under these restrictions, all the bifurcation problems are *simple*, in the sense that the versal unfoldings contain only finitely many distinct bifurcation types. The final column refers to the label given to each path detailed earlier in the chapter.

21.8 Bifurcations of hypersurfaces

The classification via path equivalence elaborated above also applies, with no further work, to other settings where the discriminants of versal unfoldings are the same as those above.

For example, hypersurfaces are given by an equation $f(x_1, \dots, x_n) = 0$, and the classification of their singularities is given by classifying the germ of f with respect to $\mathcal{K}$-equivalence, and for low codimension this coincides with the classification up to $\mathcal{R}$-equivalence described in Chapter 6.

Example 21.5. For example, consider the singularity of type A_3 in $\mathbb{R}^n$ as core, namely

$$f_0(x, y) = x^4 + Q(y) = 0,$$

where $Q(y)$ is a non-degenerate quadratic form in $y \in \mathbb{R}^{n-1}$ (see the splitting lemma). A $\mathcal{K}$-versal unfolding of f_0 is given by

$$F(x, y; u, v, s) = Q(y) + x^4 + ux^2 + vx + s = 0.$$

The discriminant Δ is the swallowtail surface in $\mathbb{R}^3$. The classification of paths $h\colon (\mathbb{R}, 0) \to (\mathbb{R}^3, 0)$ relative to Δ is given in Section 21.3. For example, the path of type ($\mathsf{A}_3.1$) would give rise to the 1–parameter family $g(x; \lambda) = Q(y) + x^4 + \lambda = 0$, which is of codimension 2, and this bifurcation problem has versal unfolding

$$G(x; \lambda; a, b) = Q(y) + x^4 + ax^2 + bx + \lambda = 0.$$

✐

The first case where the given classification fails to help is for umbilic hypersurface singularities of type $\mathsf{D}_4^{\pm}$. To classify the associated bifurcations, one needs first find the module of vector fields tangent to the discriminant in $\mathbb{R}^4$, for which see Problem 21.9.

TABLE 21.3 1–parameter bifurcation problems of codimension ≤ 3

codim	normal form	name	unfolding terms	path
codim 0	$g(x;\lambda) = x^2 + \lambda$	saddle–node	–	A_1
codim 1	$g(x;\lambda) = x^2 - \lambda^2$	transcritical	a	A_1
	$g(x;\lambda) = x^2 + \lambda^2$	isola	a	A_1
	$g(x;\lambda) = x^3 + \lambda$	hysteresis	ax	$(\mathsf{A}_2.1)$
codim 2	$g(x;\lambda) = x^2 + \lambda^3$	asymmetric cusp	$a\lambda + b$	A_1
	$g(x;\lambda) = x^3 + \lambda x$	pitchfork	$a\lambda + b$	$(\mathsf{A}_2.2)$
	$g(x;\lambda) = x^4 + \lambda$	quartic fold	$ax^2 + bx$	$(\mathsf{A}_3.1)$
codim 3	$g(x;\lambda) = x^2 \pm \lambda^4$	–	$a\lambda^2 + b\lambda + c$	A_1
	$g(x;\lambda) = x^3 + \lambda^2$	winged cusp	$a\lambda + bx + c$	$(\mathsf{A}_2.3)$
	$g(x;\lambda) = x^4 + \lambda x$	quartic pitchfork	$ax^2 + b\lambda + c$	$(\mathsf{A}_3.2)$
	$g(x;\lambda) = x^5 + \lambda$	quintic fold	$ax^3 + bx^2 + cx$	$(\mathsf{A}_4.1)$
	$g(x,y;\lambda) = \begin{pmatrix} x^2 + \lambda \\ y^2 \pm \lambda \end{pmatrix}$		$\begin{pmatrix} ay + c \\ bx \end{pmatrix}$	$(\mathrm{I}_{2,2}.1)$
	$g(x,y;\lambda) = \begin{pmatrix} x^2 - y^2 \\ xy + \lambda \end{pmatrix}$		$\begin{pmatrix} a(x+y) + b \\ c(x+y) \end{pmatrix}$	$(\mathrm{II}_{2,2}.1)$

TABLE 21.4 2–parameter bifurcation problems of codimension ≤ 2

codim	normal form	name	unfolding terms	path
codim 0	$g(x;\lambda,\mu) = x^2 + \lambda$	saddle–node	–	A_1
	$g(x;\lambda,\mu) = x^3 - \lambda x - \mu$	cusp	–	$(\mathsf{A}_2.5)$
codim 1	$g(x;\lambda,\mu) = x^2 \pm \lambda^2 \pm \mu^2$	–	a	A_1
	$g(x;\lambda,\mu) = x^3 + \lambda^2 x + \mu$	lips	ax	$(\mathsf{A}_2.6)$
	$g(x;\lambda,\mu) = x^3 - \lambda^2 x + \mu$	beaks	ax	$(\mathsf{A}_2.6)$
	$g(x;\lambda,\mu) = x^4 + \lambda x + \mu$	–	ax^2	$(\mathsf{A}_3.3)$
codim 2	$g(x;\lambda,\mu) = x^2 + \lambda^3 \pm \mu^2$		$a\lambda + b$	A_1
	$g(x;\lambda,\mu) = x^3 + \lambda^3 x + \mu$		$(a\lambda + b)x$	$(\mathsf{A}_2.7)$
	$g(x;\lambda,\mu) = x^3 + \lambda x + \mu^2$		$a\lambda + b$	$(\mathsf{A}_2.8)$
	$g(x;\lambda,\mu) = x^4 + \lambda x^2 + \lambda^2 x + \mu$		$(a\lambda + b)x$	$(\mathsf{A}_3.4)$
	$g(x;\lambda,\mu) = x^5 \pm \lambda x^3 + \lambda x + \mu$		$ax^3 + bx^2$	$(\mathsf{A}_4.2)$
	$g(x,y;\lambda,\mu) = \begin{pmatrix} x^2 + \mu y + \lambda \\ y^2 + \lambda x + \mu \end{pmatrix}$		$\begin{pmatrix} ay \\ bx \end{pmatrix}$	$(\mathrm{I}_{2,2}.2)$
	$g(x,y;\lambda,\mu) = \begin{pmatrix} x^2 - y^2 + \lambda \\ xy + \lambda(x+y) + \mu \end{pmatrix}$		$\begin{pmatrix} a(x+y) \\ b(x+y) \end{pmatrix}$	$(\mathrm{II}_{2,2}.2)$

TABLE 21.5 3–parameter bifurcation problems of codimension ≤ 1

codim	normal form	name	unfolding terms	path
codim 0	$x^2 + \lambda$	saddle node	–	A_1
	$x^3 + \lambda x + \mu$	cusp	–	$(\mathsf{A}_2.9)$
	$x^4 + \lambda x^2 + \mu x + \nu$	swallowtail	–	$(\mathsf{A}_3.5)$
codim 1	$x^2 \pm \lambda^2 \pm \mu^2 \pm \nu^2$	–	a	A_1
	$x^3 + (\lambda^2 + \mu^2)x + \nu$	flying saucer	ax	$(\mathsf{A}_2.10)$
	$x^3 + (\lambda^2 - \mu^2)x + \nu$	3D-beaks	ax	$(\mathsf{A}_2.10)$
	$x^3 - (\lambda^2 + \mu^2)x + \nu$	yo-yo	ax	$(\mathsf{A}_2.10)$
	$x^4 \pm \lambda^2 x^2 + \mu x + \nu$	double swallowtail	ax^2	$(\mathsf{A}_3.6)$
	$x^5 + \lambda x^2 + \mu x + \nu$		ax^3	$(\mathsf{A}_4.3)$
	$\begin{pmatrix} x^2 + \lambda y + \mu \\ y^2 + \lambda x + \nu \end{pmatrix}$		$\begin{pmatrix} ay \\ -ax \end{pmatrix}$	$(\mathrm{I}_{2,2}.3)$
	$\begin{pmatrix} x^2 - y^2 + \lambda(x+y) + \mu \\ xy + \lambda(x+y) + \nu \end{pmatrix}$		$\begin{pmatrix} a(x+y) \\ 0 \end{pmatrix}$	$(\mathrm{II}_{2,2}.3)$

Problems

21.1 For the hysteresis bifurcation and its versal unfolding, consider the ODE in 1 state variable,

$$\dot{x} = -x^3 + \lambda - ax.$$

Draw the phase diagrams for different values of λ, firstly for $a = 0$ and then for fixed values of $a > 0$ and $a < 0$, analogous to those of Figure 1.1B.

21.2 Use the path approach to show that the 2–parameter bifurcation problem

$$g(x;\lambda,\mu) = x^3 + \mu x + \lambda^3 = 0$$

is of codimension 4, and find a versal unfolding. (†)

21.3 Consider the A_2 core (Section 21.2), and the paths

$$h_s(\lambda,\mu) = (s\lambda\mu,\mu), \quad s \in \mathbb{R}.$$

Show these are all $\mathcal{K}_\Delta$-equivalent. [*Hint*: *in spite of not being finitely determined, Theorem* 12.21 *still applies.*] (†)

21.4 Consider the A_2 core (Section 21.2), and the path

$$h(\lambda,\mu) = (\lambda\mu,\ \lambda^2 \pm \mu^2).$$

Show that the $\mathcal{K}_\Delta$-codimension of h is equal to 4.

21.5 Consider again the A_2 core (Section 21.2), and the path

$$h(\lambda,\mu) = (\lambda, \mu^3 + s\lambda\mu)$$

(as a map, its discriminant is a cusp 'parallel' to Δ). Show that this is a unimodal family of germs of $\mathcal{K}_\Delta$-codimension 5, with exceptional case $s = 1$.

21.6 Let $h\colon (\mathbb{R}^2,0) \to (\mathbb{R}^2,0)$ be a smooth map germ of the form $h(\lambda,\mu) = (\alpha\lambda + p(\lambda,\mu),\lambda)$ with $p \in \mathfrak{m}_2^2$, and $\Delta \subset \mathbb{R}^2$ the discriminant of the A_2 core (as in Section 21.2). Show that

$$\lambda\,\mathfrak{m}_2\theta(h) \subset T\mathcal{K}_\Delta{\cdot}h,$$

and deduce that h is $\mathcal{K}_\Delta$-equivalent to the germ $(\lambda,\mu) \mapsto (p_0(\mu),\lambda)$, where $p_0(\mu) = p(0,\mu)$. [*Hint*: *Begin by writing* p *in the form*

$$p(\lambda,\mu) = p_0(\mu) + \lambda p_1(\mu) + \lambda^2 p_2(\mu) + r(\lambda,\mu),$$

where $r(0,\mu) = \frac{\partial r}{\partial\lambda}(0,\mu) = \frac{\partial^2 r}{\partial\lambda^2}(0,\mu) = 0$.] [*Note*: *we are not assuming any finite determinacy.*]

21.7 Consider the lips and beaks bifurcations $G_\pm(x;\lambda,\mu;a) = 0$ derived on p. 266; see also Figure 21.1. Find the numbers of solutions x of $G_\pm(x;\lambda,\mu;a) = 0$ for (λ,μ) in the different regions of $\mathbb{R}^2$, for $a < 0,\ a = 0,\ a > 0$.

21.8 Consider the 1–parameter bifurcation problems with A_3 core, continuing from (A_3.2). Show that the next case is equivalent to one of

$$h(\lambda) = (\lambda, 0, r\lambda^2), \quad r \notin \{0, \tfrac{1}{4}\}.$$

These are of codimension 5, and are unimodular with modal parameter r. Describe, in terms of the geometry of the swallowtail surface, why paths with $r = 0$ and $r = 1/4$ are exceptional.

21.9 Use Saito's method described in Section 19.2B to show that the module of vector fields tangent the discriminant of the versal unfolding

$$F(x,y;u,v,s,t) = x^3 + 3xy^2 + u(x^2 - y^2) + vx + sy + t$$

of the D_4^+ critical point is generated by the 4 vector fields,

$$\begin{pmatrix} u \\ 2v \\ 2s \\ 3t \end{pmatrix}, \begin{pmatrix} 9s \\ 9us \\ -4u^3 - 9uv - 27t \\ 2u^2s + 6vs \end{pmatrix}, \begin{pmatrix} 27t \\ 4u^2v - 6(v^2 + s^2) \\ 4u^2s + 12vs \\ 5u(v^2 - s^2) - 12u^2t \end{pmatrix}, \begin{pmatrix} 9v \\ 4u^3 - 9uv + 27t \\ 9us \\ 2u^2v - 3v^2 - 3s^2 \end{pmatrix}.$$

For the D_4^- versal unfolding

$$F(x,y;u,v,s,t) = x^3 - 3xy^2 + u(x^2 + y^2) + vx + sy + t$$

the 4 generators are,

$$\begin{pmatrix} u \\ 2v \\ 2s \\ 3t \end{pmatrix}, \begin{pmatrix} 9s \\ 9us \\ 4u^3 + 9uv + 27t \\ 2u^2s + 6vs \end{pmatrix}, \begin{pmatrix} 27t \\ 4u^2v - 6(v^2 + s^2) \\ 4u^2s + 12vs \\ 5u(v^2 + s^2) - 12u^2t \end{pmatrix}, \begin{pmatrix} 9v \\ 4u^3 - 9uv + 27t \\ 9us \\ 2u^2v - 3v^2 + 3s^2 \end{pmatrix}.$$

(Note that one singularity is changed to the other by replacing (y,s) by $(iy,-is)$.)

22

Loose ends

WE HAVE DISCUSSED THREE closely related equivalence relations: equivalence of bifurcation problems ($\mathcal{K}_{\text{un}}$-equivalence), path equivalence ($\mathcal{K}_\Delta$-equivalence) and left–right equivalence, each with their corresponding unfolding theories. In this chapter we discuss some results that tie these together; indeed, there are isomorphisms between all three unfolding theories. We do not give proofs as the required commutative algebra would require a lengthy detour.

The set-up is as follows. Let $f_0\colon (\mathbb{R}^n, 0) \to (\mathbb{R}^p, 0)$ be a germ of finite singularity type, and let $F\colon (\mathbb{R}^n \times \mathbb{R}^\ell, (0,0)) \to (\mathbb{R}^p, 0)$ be a $\mathcal{K}$-versal unfolding. Then we have the zero-set Z_F, the projection $\pi_F\colon Z_F \to (\mathbb{R}^\ell, 0)$, and the discriminant Δ. To simplify the exposition, we assume f_0 is of rank 0, and the versal unfolding is of the form

$$F(x; u, v) = f_0(x) + \sum_{j=1}^{\ell-p} u_j \beta_j(x) - v, \tag{22.1}$$

where $u \in \mathbb{R}^{\ell-p}$ and $v \in \mathbb{R}^p$, and $\{\beta_1, \ldots, \beta_{\ell-p}\}$ forms a cobasis for $T_e\mathcal{K}\cdot f_0$ in $\mathfrak{m}_n\theta(f_0)$. Then

$$Z_F = \{(x, u, v) \mid v = f_0(x) + \sum_{j=1}^{\ell-p} u_j \beta_j(x)\},$$

which can be parametrized by $(x, u) \in \mathbb{R}^n \times \mathbb{R}^{\ell-p}$, and then

$$\pi_F(x, u) = (u, f_0(x) + \sum u_j \beta_j(x)).$$

Given that data, our three constructions are as follows:

- an a–parameter unfolding $G\colon (\mathbb{R}^n \times \mathbb{R}^a, (0,0)) \to (\mathbb{R}^p, 0)$ of f_0
- a map germ $g\colon (\mathbb{R}^q, 0) \to (\mathbb{R}^a, 0)$ where $q = n + a - p$ (below we consider the possibility that the source of g is singular), and
- an a–parameter path $h\colon (\mathbb{R}^a, 0) \to (\mathbb{R}^\ell, 0)$.

Writing $\mathbb{R}^\ell = \mathbb{R}^{\ell-p} \times \mathbb{R}^p$, as above, and $h = (h_1, h_2)$ accordingly, then it is easy to see that h^*F is a regular unfolding if and only if h_2 is a submersion. We say a path h for which h_2 is a submersion, is a ***regular path***.

Any one of these determines the two others, as follows

- given the path h, we let $G = G_h = h^*F$ (the induced unfolding), and if h is regular then $\mathbb{R}^q = G^{-1}(0)$, and $g = \pi_G$;
- given the (regular) unfolding G, we set $Z = Z_G$ and $g = \pi_G$, and by the definition of versality, there is a path h such that G is $\mathcal{K}_{\mathrm{un}}$-equivalent to h^*F;
- given a map g with core f_0, we can define the graph unfolding by $G(x;\lambda) = g(x) - \lambda$, in which case π_G becomes g, and h is defined from G as in the previous case.

The natural question arises of how the deformations of the three objects, G, π_G and h are related, modulo their natural equivalence relations ($\mathcal{K}_{\mathrm{un}}$, $\mathcal{A}$ and $\mathcal{K}_\Delta$ respectively).

Notation: given any of the singularity theory equivalence relations $\mathcal{G}$ (such as $\mathcal{R}, \mathcal{A}, \mathcal{K}, \mathcal{K}_V, \mathcal{K}_{\mathrm{un}}$ etc) there is the associated tangent space $T_e\mathcal{G}{\cdot}f \subset \theta(f)$ for a germ f. These were mostly defined in Chapter 12. One defines the corresponding ***normal space*** to be the quotient module

$$N\mathcal{G}{\cdot}f \;=\; {}^{\theta(f)}\!\big/_{T_e\mathcal{G}\cdot f}\,.$$

This is a module over the same ring as $T_e\mathcal{G}{\cdot}f$. If f is of finite $\mathcal{G}$-codimension, then $\dim N\mathcal{G} \cdot f = \operatorname{codim}(f, \mathcal{G})$; indeed, if $\{\beta_1, \dots, \beta_k\}$ forms a cobasis for $T_e\mathcal{G}{\cdot}f$ in $\theta(f)$, then it projects to a basis of the normal space.

The theorems below were originally proved in the setting of complex analytic map germs. The proofs carry through for smooth real map germs provided one assumes they are finitely determined.

The property underlying all these theorems is that in the context of a versal unfolding, a vector field on $\mathbb{C}^\ell$ is tangent to Δ if and only if it lifts to a vector field on Z_G.

22.1 Path versus bifurcation equivalence

Let F be as in (22.1), and let $h\colon (\mathbb{R}^a, 0) \to (\mathbb{R}^\ell, 0)$. If one deforms the map germ h this gives rise to a deformation of G: more precisely the versal unfolding F defines a natural map of vector fields

$$\Psi\colon \theta(h) \longrightarrow \theta(G),$$

by $\Psi(\eta)(x,\lambda) = \sum_{j=1}^{\ell} \eta_j(\lambda)\,\beta_j(x)$, where $\eta = (\eta_1,\ldots,\eta_\ell) \in \theta(h)$. This map is a homomorphism of $\mathcal{E}_\lambda$-modules. (Here we do not need to assume h or G is regular.)

Now $N\mathcal{K}_\Delta \cdot h$ is a quotient of $\theta(h)$ and $N\mathcal{K}_{\mathrm{un}} \cdot G$ is a quotient of $\theta(G)$, and both are $\mathcal{E}_\lambda$-modules.

Theorem 22.1 ([80]). *If h is of finite $\mathcal{K}_\Delta$-codimension, then the homomorphism Ψ induces an isomorphism of $\mathcal{O}_\lambda$-modules*

$$N\mathcal{K}_\Delta \cdot h \stackrel{\psi}{\longrightarrow} N\mathcal{K}_{\mathrm{un}} \cdot G.$$

This tells us firstly that the $\mathcal{K}_\Delta$-codimension of h is equal to the bifurcation codimension of G, and moreover that the unfolding terms correspond under Ψ. In particular, a $\mathcal{K}_\Delta$-trivial deformation of h induces a $\mathcal{K}_{\mathrm{un}}$-trivial deformation of G, and a $\mathcal{K}_\Delta$-versal unfolding of h induces a $\mathcal{K}_{\mathrm{un}}$-versal unfolding of G.

This implies in particular that the unfoldings of the bifurcation problems given in the three tables at the end of Chapter 21 are indeed versal unfoldings in the bifurcation theoretic sense (i.e. for $\mathcal{K}_{\mathrm{un}}$-equivalence).

22.2 Path versus left–right equivalence

The unfolding theories were first related to one another by Damon [29] for regular paths and extended to the more general case in [80].

Let us assume h is a regular a–parameter path. Then we can choose coordinates on $\mathbb{R}^a = \mathbb{R}^{a-p} \times \mathbb{R}^p$ with $\lambda = (\sigma,\rho)$ so that $h(\lambda) = (\bar{h}(\sigma),\rho)$.

The regularity of h implies that the induced unfolding $G = h^*F$ is regular and

$$Z_G = \big\{(x,\sigma,\rho) \mid \rho = f_0(x) + \sum_j h_j(\sigma)\beta_j(x)\big\},$$

where $\bar{h} = (h_1,\ldots,h_{\ell-p})$. And then $\pi_G(x,\sigma) = \big(\sigma,\ f_0(x) + \sum_j h_j(\sigma)\beta_j(x)\big)$.

A deformation of h can be assumed to be simply a deformation of $\bar{h}$ (so the deformation depends only on σ and not on ρ). This gives rise to a deformation of π_G, and the resulting map on vector fields is

$$\begin{array}{rccl} \chi\colon \theta(h) & \longrightarrow & \theta(\pi_G) \\ \eta & \longmapsto & \big(0, \sum_j \eta_j(\sigma)\,\beta_j(x)\big). \end{array}$$

Theorem 22.2 (Damon [29]). *Let $G = h^*F$ be a regular unfolding. Then*

$$\operatorname{codim}(h, \mathcal{K}_\Delta) = \operatorname{codim}(\pi_G, \mathcal{A}),$$

and moreover, provided these are finite, χ induces an isomorphism

$$N\mathcal{K}_\Delta \cdot h \xrightarrow{\chi} N\mathcal{A} \cdot \pi_G.$$

There are similar consequences to this theorem as to the previous one. Referring to the classification of paths in Chapter 21, or indeed of the corresponding bifurcation problems, we see that not all correspond to regular unfoldings. Those that do determine low-codimension $\mathcal{A}$-equivalence classes of maps and their versal unfoldings. These are listed in Table 22.1 below.

22.3 Left-right versus bifurcation equivalence

The relation between the unfolding (bifurcation problem) G and the map π_G is given by Martinet's first theorem (Theorem 17.2), provided the unfolding is regular. We consider the non-regular case in the section below.

Given that theorem of Martinet, it should come as no surprise that the unfolding theories of G and of π_G are closely related, the first up to $\mathcal{K}_{\mathrm{un}}$-equivalence, the second up to $\mathcal{A}$-equivalence. This would be the extension of Martinet's second theorem to non-versal unfoldings, and non-stable map germs. Recall from Theorem 17.3 that G is a $\mathcal{K}$-versal unfolding if and only if π_G is stable. The following shows that the lack of versality of a regular unfolding G is directly related to the lack of stability of π_G; it is a straightforward consequence of the two theorems above.

Corollary 22.3. *Let $G\colon (\mathbb{R}^n \times \mathbb{R}^a, 0) \to (\mathbb{R}^p, 0)$ be a regular unfolding of finite $\mathcal{K}_{\mathrm{un}}$-codimension, of a germ $f_0 \in \mathcal{E}_n^p$ of finite singularity type. Then there is an isomorphism*

$$N\mathcal{K}_{\mathrm{un}} \cdot G \xrightarrow{\simeq} N\mathcal{A} \cdot \pi_G.$$

22.4 Non-regular unfoldings

Theorem 22.2 and Corollary 22.3 require the unfolding or path to be regular, in order that the source of π_G is a smooth submanifold. The possibility of extending

TABLE 22.1 Classification of low codimension maps up to $\mathcal{A}$-equivalence in the dimensions shown; these are derived from the classification of regular unfoldings of the different cores, found in Chapter 21. The 'codim' refers to $\operatorname{codim}(\pi_G, \mathcal{A})$.

codim	π_G	unfolding terms	path
	$\mathbb{R} \to \mathbb{R}$		
0	$x \mapsto x^2$	–	(A$_1$)
1	$x \mapsto x^3$	x	(A$_2$.1)
2	$x \mapsto x^4$	$x^2,\ x$	(A$_3$.1)
3	$x \mapsto x^5$	$x^3,\ x^2,\ x$	(A$_4$.1)
	$\mathbb{R}^2 \to \mathbb{R}^2$		
0	$(x,y) \mapsto (x^2, y)$	–	(A$_1$)
0	$(x,y) \mapsto (x^3 \pm xy, y)$	–	(A$_2$.5)
1	$(x,y) \mapsto (x^3 \pm xy^2, y)$	$(x,0)$	(A$_2$.6)
1	$(x,y) \mapsto (x^4 + xy, y)$	$(x^2,0)$	(A$_3$.3)
2	$(x,y) \mapsto (x^3 + xy^3, y)$	$(xy,0),\ (x,0)$	(A$_2$.7)
2	$(x,y) \mapsto (x^4 + x^2y + xy^2, y)$	$(xy,0),\ (x,0)$	(A$_3$.4)
2	$(x,y) \mapsto \left(x^5 \pm x^3y + xy, y\right)$	$(x^3,0),\ (x^2,0)$	(A$_4$.2)
2	$(x,y) \mapsto \frac{1}{1-xy}(x^2 - y^3,\ y^2 - x^3)$	$(y,0),\ (0,x)$	(I$_{2,2}$.2)
2	$(x,y) \mapsto \begin{pmatrix} x^2 - y^2 \\ xy + (x-y)(x+y)^2 \end{pmatrix}$	$\begin{pmatrix} x+y \\ 0 \end{pmatrix}, \begin{pmatrix} 0 \\ x+y \end{pmatrix}$	(II$_{2,2}$.2)
	$\mathbb{R}^3 \to \mathbb{R}^3$		
0	$(x,y,z) \mapsto (x^2, y, z)$	–	(A$_1$)
0	$(x,y,z) \mapsto (x^3 + xy, y, z)$	–	(A$_2$.9)
0	$(x,y,z) \mapsto (x^4 + yx^2 + zy, y, z)$	–	(A$_3$.5)
1	$(x,y,z) \mapsto (x^3 + x(y^2 \pm z^2), y, z)$	$(x,0)$	(A$_2$.10)
1	$(x,y,z) \mapsto (x^4 \pm y^2x^2 + zx)$	$(x^2,0)$	(A$_3$.6)
1	$(x,y,z) \mapsto (x^5 + yx^2 + zx, y, z)$	$(x^3,0,0)$	(A$_4$.3)
1	$(x,y,z) \mapsto (x^2 + yz, y^2 + zx, z)$	$(y,-x,0)$	(I$_{2,2}$.3)
1	$(x,y,z) \mapsto \begin{pmatrix} x^2 - y^2 + (x+y)z \\ xy + (x+y)z \\ z \end{pmatrix}$	$\begin{pmatrix} x+y \\ 0 \\ 0 \end{pmatrix}$	(II$_{2,2}$.3)

[*Note*: *two are written as column vectors merely to save width in the table.*]

these to non-regular unfoldings and paths was first considered in [80]. Here we outline the results of that paper, and for simplicity we consider the maps involved as complex analytic map germs.

The set up with f_0 and F is as before, although now we replace $\mathbb{R}$ by $\mathbb{C}$, and assume $n \geq p$. Let $h\colon (\mathbb{C}^a, 0) \to (\mathbb{C}^\ell, 0)$ be an arbitrary (not necessarily regular) a–parameter path, and let $G = h^*F$ be the induced unfolding of f_0. Now, $Z_G \subset \mathbb{C}^n \times \mathbb{C}^a$ may now be singular, and the projection $\pi_G\colon Z_G \to \mathbb{C}^a$ has a possibly singular source. In order to extend Theorem 22.2 to this setting we need to give a meaning to an unfolding of such a map (germ), and its $\mathcal{A}$-codimension.

Given an unfolding G, there are two ways to deform it: first as an unfolding up to $\mathcal{K}_{\mathrm{un}}$-equivalence) and secondly as a map germ itself $G\colon (\mathbb{C}^{n+a}, 0) \to (\mathbb{C}^p, 0)$ (up to $\mathcal{K}$-equivalence). If G is regular, then as a map germ it is a submersion and any deformation of it will be $\mathcal{K}$-trivial. However, if G is not regular then there are deformations of G where the singularity type of $Z_G = G^{-1}(0)$ remains unchanged, and then those deformations where Z_G does change.

For example, in the pitchfork bifurcation $G(x, \lambda) = x^3 + \lambda x$ described in Section 18.1, the zero-set Z_G has a simple node at the origin, and in its versal unfolding given in (18.3), for those deformations where $b = a^3$ the zero-set also has a node (see the A/B transition in Figure 18.2).

Let Z be the germ of an analytic (or algebraic) variety in $\mathbb{C}^N$. One says two map germs $f, f'\colon Z \to (\mathbb{C}^p, 0)$ are ***left–right equivalent*** (or $\mathcal{A}(Z)$-equivalent) if there are diffeomorphism germs ϕ of Z to itself, and ψ of $(\mathbb{C}^p, 0)$ such that $f' = \psi \circ f \circ \phi$. (A diffeomorphism of Z is an analytic diffeomorphism germ of the ambient space of Z that maps Z to itself.) This gives the $\mathcal{A}(Z)$-tangent space to f to be

$$T_e\mathcal{A}(Z) \cdot f \;=\; \mathrm{t}f(\theta_Z)\big|_Z + \omega f(\theta_p) \;\subset\; \theta(f).$$

Note that $\theta(f)$ consists of vector fields defined on Z with values in $\mathbb{C}^p$, and consequently the term $\mathrm{t}f(\theta_Z)$ must be restricted to Z. If Z is smooth, this reduces to the usual definition of $T_e\mathcal{A}{\cdot}f$ given in Chapter 12.

The first result of [80] (their Theorem 1.4) extends Theorem 22.2 and states the following.

Theorem 22.4. *With notation as above, assume the path h is of finite $\mathcal{K}_\Delta$-codimension. Then there is a short exact sequence of $\mathcal{O}_\lambda$-modules*

$$0 \longrightarrow N\mathcal{A}(Z_G) \cdot \pi_G \longrightarrow N\mathcal{K}_\Delta \cdot h \longrightarrow N\mathcal{K} \cdot G \longrightarrow 0.$$

In particular, $\operatorname{codim}(h, \mathcal{K}_\Delta) = \operatorname{codim}(\pi_G, \mathcal{A}(Z_G)) + \operatorname{codim}(G, \mathcal{K})$.

The last term in the exact sequence is the $\mathcal{K}$-normal space for G considered as a map; this corresponds to the space of deformations of the variety Z_G (which is zero if and only if G is a regular unfolding). We know from Theorem 22.1, which doesn't require regularity of the unfolding, that the middle term $N\mathcal{K}_\Delta \cdot h$ is isomorphic to $N\mathcal{K}_{\text{un}} \cdot G$.

Moving beyond this, one can define a left–right deformation theory of the pair (Z_G, π_G), where both Z_G and π_G are allowed to deform. See [80] for details, where the following is then proved.

Theorem 22.5 ([80]). *Let $n \geq p$ and $G: (\mathbb{C}^n \times \mathbb{C}^a, 0) \to (\mathbb{C}^p, 0)$ be an unfolding of a germ of finite singularity type $f_0 \in \mathcal{O}_n^p$. Then* $\operatorname{codim}(G, \mathcal{K}_{\text{un}}) = \operatorname{codim}((Z_G, \pi_G), \mathcal{A})$ *and, provided these are finite, there is an isomorphism of $\mathcal{O}_p$-modules*

$$N\mathcal{K}_{\text{un}} \cdot G \xrightarrow{\simeq} N\mathcal{A} \cdot (Z_G, \pi_G).$$

22.5 Multiple co-existing bifurcations

Consider a bifurcation problem $g(x, \lambda) = 0$, where x belongs to an open subset of $\mathbb{R}^n$ (e.g. for equilibria of a system of ODEs). It is possible that two distinct bifurcations occur at say x and x' in U for the same value of the parameter λ. This is equivalent to the projection π_g having multiple singularities (one at (x, λ) and the other at (x', λ)) with the same value λ).

The theorems of Martinet and their extensions discussed above continue to hold in the multigerm setting. This is made explicit in Damon's paper [29], but not in [80] although the arguments can be extended in a straightforward manner.

Problems

22.1 Find a regular unfolding G of $f_0: t \mapsto (t^2, 0)$ such that the associated projection π_G is of the form of (17.2). Find the map h which induces G from the versal deformation of f_0 and use its $\mathcal{K}_\Delta$-codimension to show that π_G is of $\mathcal{A}$-codimension 1.

22.2 Discuss the geometry of the corank-2 bifurcation problems given in Table 21.3.

23

Constrained bifurcation problems

THIS FINAL CHAPTER ILLUSTRATES the flexibility of the path approach to bifurcation theory. We consider three cases where there is some extra structure or constraint on the type of bifurcations under study: firstly bifurcation problems with symmetry, secondly bifurcations with a 'fixed solution', and finally variational (or gradient) bifurcation problems.

As we see below, one difference with such constrained bifurcation problems is that in the versal unfolding of the core, say $G(x,u) = 0$, the zero-set Z_G is no longer necessarily a submanifold of $\mathbb{R}^n \times \mathbb{R}^a$, and this implies that the projection $\pi\colon Z_G \to \mathbb{R}^a$ has two types of singular point: singularities of Z_G itself and singularities of the projection at points where Z_G is smooth. Correspondingly, the discriminant may have more than one component.

In his fundamental monograph, Damon [27] introduces so-called *geometric subgroups* of $\mathcal{A}$ and $\mathcal{K}$, and shows that the determinacy and versality theorems hold for these subgroups. The examples we consider here are examples of such geometric subgroups, as is the $\mathcal{K}_V$-equivalence we have considered above.

23.1 Symmetric bifurcations

Many mathematical models involve some degree of symmetry, and the effects of this on bifurcations are important. For example, as we have seen, the pitchfork bifurcation has codimension 2, but in a symmetric context it can have codimension 0 as we show below.

Here we illustrate the ideas with the simplest case, namely bifurcation problems with a reflection symmetry. We consider $\mathbb{R}^n$ with the action of the 2-element group $\mathbb{Z}_2$, as described in Chapter 9; that is the symmetry group consists of the two matrices $\pm\mathsf{Id}_n$. Recall the notation of $\mathcal{E}_n^+$, $\mathcal{E}_n^-$ and $\mathfrak{m}_n^+$ from that chapter.

Let $g\colon \mathbb{R}^n \rightarrowtail \mathbb{R}^n$. We say the equation $g(x) = 0$, is ***symmetric*** if

$$g(-x) = -g(x). \tag{23.1}$$

(The fact that the symmetry is $\mathbb{Z}_2$ is implicitly understood.)

As before, two equations $g_1(x) = 0$ and $g_2(x) = 0$ are equivalent if the two maps are $\mathcal{K}$-equivalent, $g_1 \circ \phi = Mg_2$, but now we require the contact equivalence to preserve the symmetry, which means that

$$\phi(-x) = -\phi(x), \quad M(-x) = M(x). \tag{23.2}$$

In other words, ϕ is odd and $M \in \mathsf{GL}_n(\mathcal{E}_n^+)$. It is straightforward to show that if g is symmetric then so is $M^{-1}g \circ \phi$, provided M and ϕ have these symmetry properties. Let us call the resulting equivalence relation $\mathcal{K}^{\mathbb{Z}_2}$-equivalence.

In contrast to the usual tangent space for $\mathcal{C}$-equivalence, which uses the ideal I_g in $\mathcal{E}_n$ generated by the components of the map germ g, the symmetric $\mathcal{C}$-tangent space consists of vector fields along g of the form $N(x)g(x)$, where $N(x)$ is even. We therefore use

$$I_g^- = \mathcal{E}_n^+ \left\{g_j \mid j = 1, \ldots, n\right\}.$$

The germs of symmetric vector fields on the source is denoted θ_n^- (as in Section 9.1); it is the module over $\mathcal{E}_n^+$ generated by the $x_i\partial_{x_j}$. The tangent space of the germ g is then the submodule of $\theta(g)^-$ given by

$$T\mathcal{K}^{\mathbb{Z}_2}{\cdot}g = \mathrm{t}g(\theta_n^-) + I_g^-\theta(g)^+.$$

For the finite determinacy theorem, we need to use vector fields whose linear part vanishes at the origin (cf. $\mathfrak{m}_n T\mathcal{K}{\cdot}g$ in Theorem 13.2). We therefore define

$$T\mathcal{K}_1^{\mathbb{Z}_2}{\cdot}g = \mathfrak{m}_n^+ T\mathcal{K}^{\mathbb{Z}_2}{\cdot}g. \tag{23.3}$$

The finite determinacy and versal unfolding theorems for this scenario then read (proofs are left to the reader, but note that the generators of $(\mathfrak{m}_n^+)^k\theta(g)^-$ have order $2k+1$),

Theorem 23.1. *(i). If $(\mathfrak{m}_n^+)^k\theta(g)^- \subset T\mathcal{K}_1^{\mathbb{Z}_2}{\cdot}g$ then g is $2k-1$ determined with respect to $\mathcal{K}^{\mathbb{Z}_2}$-equivalence.*

(ii). Let $G\colon \mathbb{R}^n \times \mathbb{R}^a \to \mathbb{R}^n$ be a symmetric unfolding of g. It is versal in the workd of symmetric map germs if and only if

$$T\mathcal{K}^{\mathbb{Z}_2}{\cdot}g + \dot{G} = \theta(g)^-.$$

For a single state variable ($n = 1$), any finitely determined germ is $\mathbb{Z}_2$-contact equivalent to x^k where k is odd (and indeed right equivalent to it). The first two cases are given in the following examples.

Example 23.2. Consider the A_3 core, namely $g_0 = x^3$. Then

$$T\mathcal{K}^{\mathbb{Z}_2}\cdot g_0 = \left\langle x^3 \right\rangle.$$

Since $\theta(g_0)^- = \mathcal{E}_1^+\{x\} \lhd \mathcal{E}_1^-$ (a submodule), a cobasis is just $\{x\}$ and we have the versal unfolding

$$F(x;u) = x^3 + ux,$$

which is the pitchfork bifurcation. Since this is versal, the pitchfork bifurcation is of codimension 0 in the world of $\mathbb{Z}_2$-symmetric bifurcations.

The discriminant is clearly just the origin $u = 0$, and the classification of a–parameter $\mathbb{Z}_2$-symmetric bifurcations with core x^3 reduces to the study of paths $h\colon (\mathbb{R}^a, 0) \to \mathbb{R}$ up to $\mathcal{K}$-equivalence.

For example, with 1 parameter, the codimension-0 bifurcation is the pitchfork arising from $h(\lambda) = \lambda$, or $g(x;\lambda) = x^3 + \lambda x = 0$.

The codimension-1 bifurcation arises from the path $h(\lambda) = \pm\lambda^2$, and is

$$g(x;\lambda) = x^3 \pm \lambda^2 x. \tag{23.4}$$

The versal unfolding (within the $\mathbb{Z}_2$-symmetric universe) is

$$G(x;\lambda;a) = x^3 \pm (\lambda^2 + a)x.$$

Figure 23.1(A,B) show these unfoldings. These singular bifurcation problems of the form $x^3 + \lambda^k x$ are called ***degenerate pitchfork bifurcations***. ✐

Example 23.3. Now consider the A_5 core, namely $g_0 = x^5$. Then

$$T\mathcal{K}^{\mathbb{Z}_2}\cdot g_0 = \left\langle x^5 \right\rangle.$$

Since $\theta(g_0)^- = \langle x \rangle \lhd \mathcal{E}_1^-$, a cobasis is just $\{x, x^3\}$ and we have the versal unfolding

$$F(x;u,v) = x^5 + ux^3 + vx,$$

which is a bifurcation we haven't considered before (it is a section through the butterfly catastrophe). Before classifying the symmetric bifurcations with this core, we need to find the discriminant, which we do in the usual fashion. Namely, we eliminate x from the equations $F = F_x = 0$; that is,

$$x^5 + ux^3 + vx = 5x^4 + 3ux^2 + v = 0.$$

These two equations define the singular set Σ_F. Note that unlike the general (unconstrained) theory of versal unfoldings, here Z_F is given by $x^5 + ux^3 + vx = 0$,

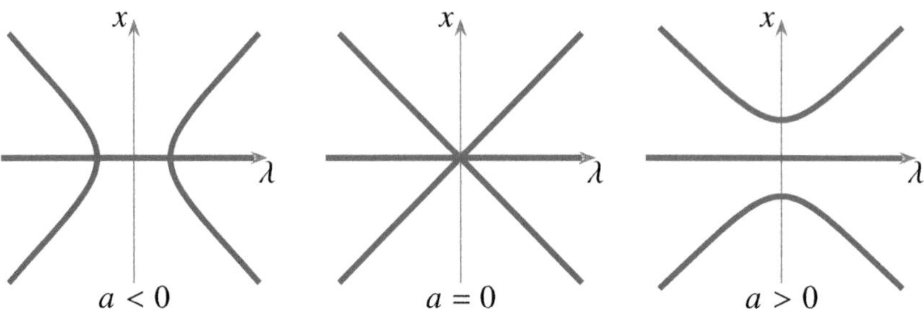

(A) The unfolding $x^3 - \lambda^2 x - ax = 0$: for $a < 0$ there are two symmetric pitchfork bifurcations, which coalesce as $a \to 0$, and disappear for $a > 0$.

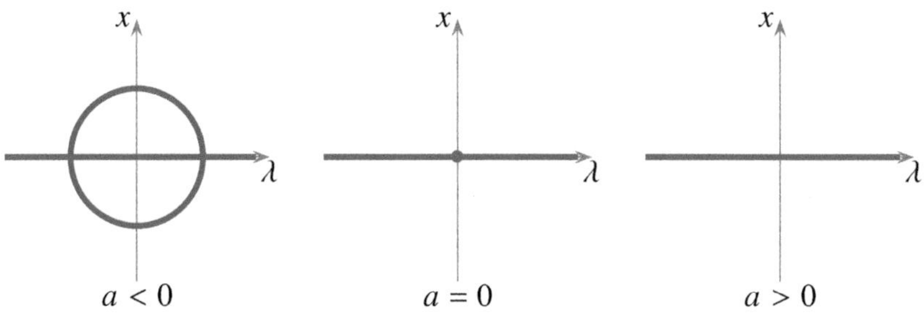

(B) The unfolding $x^3 + \lambda^2 x + ax = 0$: for $a < 0$ there are two symmetric pitchfork bifurcations, which collapse and disappear as a passes though 0.

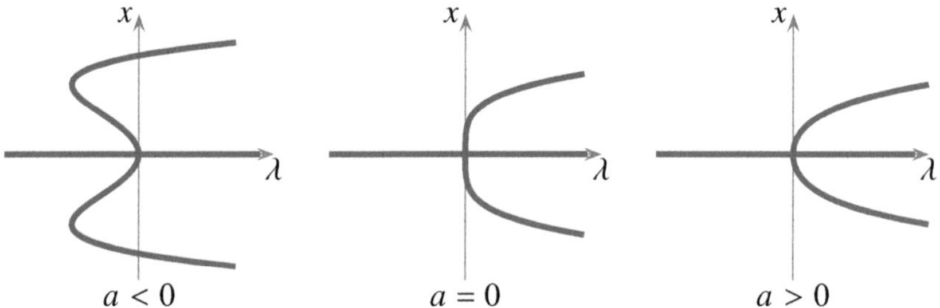

(C) The unfolding $x^5 - \lambda x + ax^3$: as a increases from negative to positive, so a pitchork and a symmetric pair of saddle-nodes merge into a single pitchfork.

FIGURE 23.1 Unfoldings of the three codimension-1 degenerate pitchforks within the class of symmetric bifurcations.

which is not a submanifold; indeed, it is a union of two submanifolds, $\{x = 0\}$ and $\{v = -x^4 - ux^2\}$. One finds similarly, Σ_F has two components,

$$\{x = 0,\ v = 0\} \text{ and } \{u = -2x^2,\ v = x^4\}.$$

Eliminating x from each of these components shows that the discriminant Δ also has two components:

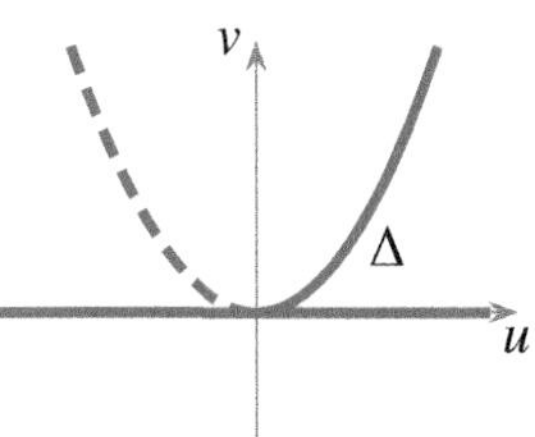

$$\Delta = \{v = 0\} \cup \{u^2 = 4v\}.$$

That is, Δ is given by $h = 0$, where $h = v(u^2 - 4v)$. See the figure on the right – the dashed line represents the real part of the complex discriminant (ie, with x imaginary, but u, v real); it is still a part of the algebraic set required for the path approach to bifurcations.

For 1–parameter symmetric bifurcations of this type, we consider paths $h\colon (\mathbb{R}, 0) \to (\mathbb{R}^2, 0)$, with $(u, v) = h(\lambda)$, up to $\mathcal{K}_\Delta$-equivalence. First one needs to find the module of vector fields tangent to Δ, which is

$$\theta_\Delta = \mathcal{E}_2 \left\{ \begin{pmatrix} u \\ 2v \end{pmatrix}, \begin{pmatrix} 2v \\ uv \end{pmatrix} \right\}. \tag{23.5}$$

Consider the path $h(\lambda) = (0, \lambda) = (u, v)$, with image along the v-axis. Then

$$T\mathcal{K}_\Delta{\cdot}h = \mathcal{E}_1 \left\{ \begin{pmatrix} 0 \\ 2\lambda \end{pmatrix}, \begin{pmatrix} 2\lambda \\ 0 \end{pmatrix} \right\} = \mathfrak{m}_1 \theta(h).$$

This shows h is 1-determined. The corresponding bifurcation is $g(x, \lambda) = x^5 + \lambda x = 0$. The extended tangent space is

$$T_e\mathcal{K}_\Delta{\cdot}h = \mathcal{E}_1 \left\{ \begin{pmatrix} 0 \\ 1 \end{pmatrix}, \begin{pmatrix} \lambda \\ 0 \end{pmatrix} \right\}.$$

This has codimension 1, and a versal unfolding is given by $H(\lambda; a) = (a, \lambda)$, and hence

$$G(x, \lambda; a) = x^5 + \lambda x + ax^3.$$

See Figure 23.1c, and Problem 23.3 for further information.

23.2 Bifurcations with fixed zero

For a second non-standard setting, we consider a class of bifurcation problems in one state variable for which $x = 0$ is always an equilibrium point. This could arise for example in a population model: if x represents the population then when $x = 0$ the population is extinct and can no longer evolve. Like the population model, other applications exist where there is a 1-sided constraint such as $x \geq 0$.

We consider the class of all smooth map germs (or bifurcation problems) $g\colon (\mathbb{R}^n \times \mathbb{R}^k, (0,0)) \to \mathbb{R}^n$ satisfying $g(0,\lambda) = 0$. Two such equations $g_1 = 0$ and $g_2 = 0$ are equivalent if (as usual) they are contact equivalent: there is a diffeomorphism

$$\Phi(x,\lambda) = (\phi(x,\lambda),\ \psi(\lambda)),$$

and an invertible-matrix-valued function $M(x,\lambda)$ such that

$$g_1 \circ \Phi = M g_2.$$

The one difference from the standard bifurcation theory is that, because $x = 0$ plays a special role, it must be preserved by ϕ; that is, $\phi(0,\lambda) = 0$ for all $\lambda \in \mathbb{R}^k$.

The core $g_0(x) = g(x,0)$ is therefore classified by $\mathcal{K}$-equivalence, and finite determinacy is the same. However, for unfoldings, the extended tangent space is the same as the ordinary $\mathcal{K}$ tangent space since the origin must be preserved.

Thus $G(x,u)$ is versal in this setting if and only if

$$T\mathcal{K}{\cdot}g_0 + \dot{G} = \mathfrak{m}_n\theta_n. \tag{23.6}$$

Note that $\dot{G} \subset \mathfrak{m}_n\theta_n$ since $G(0,u) = 0$ (for all u).

Example 23.4. Suppose the core is $g_0(x) = x^2$. Then $T\mathcal{K}{\cdot}g_0 = \langle x^2 \rangle$ and a versal unfolding is given by

$$G(x,u) = x^2 + ux.$$

This is a standard form of the transcritical bifurcation (see Section 18.2B). The discriminant is of course the origin $\{u = 0\}$, and the classification of k–parameter bifurcations with this A_1 core is given by the $\mathcal{K}$-classification of map germs $h\colon (\mathbb{R}^k, 0) \to (\mathbb{R}, 0)$, like the paths in Section 21.1.

With 1 parameter, this would just be $h(\lambda) = \lambda^k$ (or $\pm\lambda^k$, but the sign can be changed by changing the sign of x). The beginning of the classification reads (using the layout of the tables in Chapter 21),

codim 0:	$x^2 + \lambda x$	–
codim 1:	$x^2 + \lambda^2 x$	$+\,ax$ (see Figure 23.2)
codim 2:	$x^2 + \lambda^3 x$	$+\,(a\lambda + b)x$
etc.		

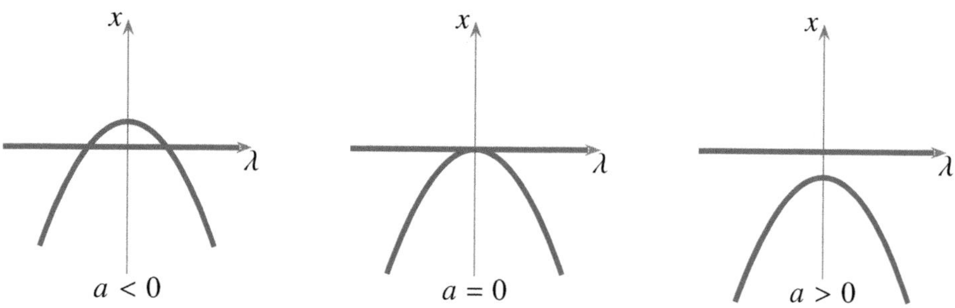

FIGURE 23.2 The versal unfolding $x^2 + \lambda^2 x + ax$ of the 1–parameter bifurcation of codimension 1, within the class of bifurcations with a fixed zero: the 'degenerate transcritical' at $a = 0$ deforms to two transcritical bifurcations on one side, and no bifurcations on the other. (As a general bifurcation, this usually has normal form $x^2 - \lambda^4$ and is of codimension 3; see Table 21.1.)

With two parameters, h would belong to the list of germs in Section 7.6 (there using $\mathcal{R}$-equivalence, but for low codimension the $\mathcal{R}$- and $\mathcal{K}$-equivalence classifications coincide). Thus, as functions of parameters λ, μ,

codim 0:	$x^2 + \lambda x$	–
codim 1:	$x^2 + (\lambda^2 \pm \mu^2)x$	$+\,ax$
codim 2:	$x^2 + (\lambda^3 + \mu^2)x$	$+\,(a\lambda + b)x$
etc.		

This can be continued for larger numbers of parameters. ✐

Example 23.5. Now suppose the core is $g_0(x) = x^3$. Then $T\mathcal{K}{\cdot}g_0 = \langle x^3 \rangle$ and a versal unfolding in this class of bifurcation problems is given by

$$G(x, u, v) = x^3 + ux^2 + vx,$$

using condition (23.6). (Note the difference with the versal unfolding in Section 18.1: here $x = 0$ must be a solution for all parameter values.) The bifurcation equation is degenerate if $G = G_x = 0$, which gives,

$$x(x^2 + ux + v) = 3x^2 + 2ux + v = 0.$$

Eliminating x (or projecting to the u-v plane) gives the bifurcation set Δ to be given

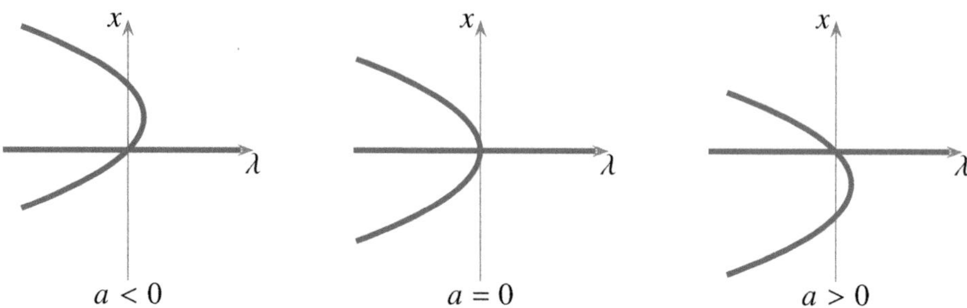

FIGURE 23.3 The versal unfolding $x^3 + \lambda x + ax^2$ of a pitchfork bifurcation, of codimension 1, within the class of bifurcations with a fixed zero (Example 23.5): the pitchfork at $a = 0$ deforms to a saddle-node plus a transcritical bifurcation as a becomes non-zero.

by

$$v(u^2 - 4v) = 0.$$

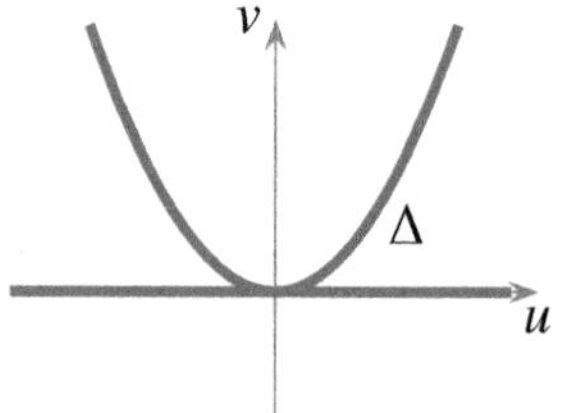

The component $v = 0$ corresponds to bifurcations at $x = 0$, while the other component corresponds to saddle-node bifurcations with $x \neq 0$, (cf. the discriminant in Example 23.3). The module of vector fields tangent to the discriminant is given in (23.5).

Now consider the 1–parameter bifurcation $g(x, \lambda) = x^3 + \lambda x$ (the pitchfork). It is induced by the path $h(\lambda) = (0, \lambda)$, for which

$$T_e\mathcal{K}_\Delta{\cdot}h = \left\{\begin{pmatrix}0\\1\end{pmatrix},\ \begin{pmatrix}\lambda\\0\end{pmatrix}\right\}.$$

The path h therefore has codimension 1 in this setting, and a versal unfolding is given by $H(\lambda; a) = (a, \lambda)$, which gives the bifurcation problem

$$x^3 + \lambda x + ax^2 = 0.$$

See Figure 23.3. ✐

Proposition 23.6. *Consider 1–parameter single variable bifurcations with a fixed zero. The codimension-0 and codimension-1 bifurcations are given by*

$$g(x, \lambda) + a\,u(x, \lambda) = 0,$$

where

codim 0:	$g = x^2 + \lambda x$	–	*(transcritical)*
codim 1:	$g = x^2 + \lambda^2 x$	$u = x$	
	$g = x^3 + \lambda x$	$u = x^2$	*(pitchfork).*

The proof is left to the reader (see the discussion of the pitchfork on p. 263).

We now proceed to consider two different 2–dimensional extensions of this idea of having a fixed zero: in the first, we require that the origin in $\mathbb{R}^2$ is a zero, while in the second we require that there are two 1-sided constraints, namely $x \geq 0,\ y \geq 0$ (as arising in population models for example).

Example 23.7. Consider the first corank-2 example with fixed zero (at the origin); in other words, the 'universe' we consider is map germs in $\mathfrak{m}_2\mathcal{E}_2^2$. The simplest corank-2 core would be of type $\mathrm{I}_{2,2}$, $g_0(x,y) = (x^2, y^2)$. With fixed zero at the origin, the versal unfolding is found from a cobasis for $T\mathcal{K}{\cdot}g_0$ in $\mathfrak{m}_2\theta(g_0)$. Thus,

$$F(x, y; u, v, s, t) = \begin{pmatrix} x^2 + ux + vy \\ y^2 + sx + ty \end{pmatrix}.$$

One finds (using for example MACAULAY2) that the module of vector fields tangent to the discriminant is given by

$$\theta_\Delta = \left\{ \begin{pmatrix} u \\ v \\ s \\ t \end{pmatrix}, \begin{pmatrix} u \\ 3v \\ -3s \\ -t \end{pmatrix}, \begin{pmatrix} vt \\ 0 \\ 2us + t^2 \\ 3vs - ut \end{pmatrix}, \begin{pmatrix} 3vs \\ u^2 \\ 7st \\ us + 4t^2 \end{pmatrix} \right\}.$$

For example the 1–parameter bifurcation problem in this context,

$$g(x, y, \lambda) = (x^2 + \lambda y, y^2 + \lambda x),$$

is induced by the path $h(\lambda) = (0, \lambda, \lambda, 0)$. This path has $\mathcal{K}_\Delta$-codimension equal to 5, with unfolding cobasis given by

$$(1, 0, 0, 0),\ (0, 1, -1, 0),\ (0, 0, 0, 1),\ (\lambda, 0, 0, 0),\ (0, 0, 0, \lambda).$$

✐

Example 23.8. Consider instead systems with two 1-sided constraints: $x \geq 0, y \geq 0$. A first-order differential equation would then be of the form

$$\dot{x} = x f_1(x, y), \quad \dot{y} = y f_2(x, y).$$

We therefore consider the $\mathcal{E}_2$-module M of map germs of the form

$$f\colon (x,y) \mapsto (xf_1(x,y),\ yf_2(x,y)),$$

with $f_1, f_2 \in \mathcal{E}_2$. Thus, we consider $f \in M$, where

$$M = \mathcal{E}_2 \left\{ \begin{pmatrix} x \\ 0 \end{pmatrix}, \begin{pmatrix} 0 \\ y \end{pmatrix} \right\}.$$

The zeros of such a map would correspond to equilibrium points of the system of ODES.

The changes of coordinates of the source would be generated by the vector fields, $x\partial_x$ and $y\partial_y$ (contained in $\mathfrak{m}_2\theta(2)$), and the relevant $\mathcal{C}$-equivalence would be the subset consisting of those preserving M, so generated by the matrices

$$\begin{pmatrix} 1 & 0 \\ 0 & 0 \end{pmatrix}, \begin{pmatrix} 0 & 0 \\ 0 & 1 \end{pmatrix}, \begin{pmatrix} 0 & x \\ 0 & 0 \end{pmatrix}, \begin{pmatrix} 0 & 0 \\ y & 0 \end{pmatrix}.$$

Write $\mathcal{K}_c$ for the resulting equivalence.

Consider first the core $g_0(x,y) = (x^2, y) \in M$. Then

$$T\mathcal{K}_c{\cdot}g_0 = \mathcal{E}_2 \left\{ \begin{pmatrix} x^2 \\ 0 \end{pmatrix}, \begin{pmatrix} 0 \\ y \end{pmatrix}, \begin{pmatrix} xy \\ 0 \end{pmatrix} \right\}.$$

This has codimension 1 in M, with cobasis $\{(x,0)\}$. The appropriate versal unfolding is therefore

$$F(x,y,u) = (x^2 + ux,\ y).$$

For non-zero u, this map has two zeros, one at the origin and the other on the x-axis. The bifurcation problems with this core are therefore classified by functions h with $u = h(\lambda)$, classified up to usual $\mathcal{K}$-equivalence.

More interesting is the core $g_0(x,y) = (x^2, y^2)$. Then

$$T\mathcal{K}_c{\cdot}g_0 = \mathcal{E}_2 \left\{ \begin{pmatrix} x^2 \\ 0 \end{pmatrix}, \begin{pmatrix} 0 \\ y^2 \end{pmatrix}, \begin{pmatrix} xy^2 \\ 0 \end{pmatrix}, \begin{pmatrix} 0 \\ x^2y \end{pmatrix} \right\}.$$

A cobasis in M is given by

$$\left\{ \begin{pmatrix} x \\ 0 \end{pmatrix}, \begin{pmatrix} xy \\ 0 \end{pmatrix}, \begin{pmatrix} 0 \\ y \end{pmatrix}, \begin{pmatrix} 0 \\ xy \end{pmatrix} \right\}.$$

One $\mathcal{K}_c$-versal unfolding is therefore

$$F(x,y;u,v,s,t) = (x^2 + uxy + vx,\ y^2 + sxy + ty),$$

and one finds that

$$\Theta_\Delta = \mathcal{E}_4\left\{\begin{pmatrix}u\\v\\-s\\0\end{pmatrix}, \begin{pmatrix}-u\\0\\s\\t\end{pmatrix}, \begin{pmatrix}v-ut\\0\\0\\0\end{pmatrix}, \begin{pmatrix}0\\0\\t-vs\\0\end{pmatrix}\right\}.$$

Consider the 1–parameter bifurcation problem in this context,

$$g(x,y,\lambda) = (x^2+\lambda x, y^2+\lambda y),$$

which is induced by the path $h(\lambda) = (0,\lambda,0,\lambda)$. Then

$$T_e\mathcal{K}_\Delta{\cdot}h = \mathcal{E}_1\left\{\begin{pmatrix}0\\1\\0\\1\end{pmatrix}\right\} + \mathfrak{m}_1\theta(h).$$

This is of codimension 3, with $\mathcal{K}_\Delta$-versal unfolding

$$H(\lambda;a,b,c) = (a,\lambda-c,b,\lambda+c),$$

which corresponds to the bifurcation problem

$$G(x,y,\lambda;a,b,c) = (x^2+axy+(\lambda-c)x,\ y^2+bxy+(\lambda+c)y).$$

An analysis of this constrained family is left to the reader. ✐

23.3 Bifurcation in variational problems

Given a family of functions $f\colon \mathbb{R}^n\times\mathbb{R}^k \to \mathbb{R}$, the corresponding variational problem is the equation $\nabla_x f(x,\lambda)=0$, where ∇_x is the gradient with respect to the x variables.

There are two possible approaches to using singularity theory for these problems: either right equivalence of the function or contact equivalence of the gradient. The two are not equivalent, and here we consider the former approach. However, there is a subtlety, namely that under right equivalence the number of critical *values* does not change: that is, if two functions defined on an open set are right equivalent, then they have the same number of critical values and that effects the notion of discriminant (see the definition below).

The alternative approach involving $\mathcal{K}$-equivalence of the gradient is developed (in a context with symmetry) in the book of Bridges and Furter [11]. Because the gradient maps do not 'see' the critical values, the bifurcation set is only the usual discriminant, and this has the effect of lowering the codimension, and there are bifurcation problems where right equivalence leads to infinite codimension, while the contact-gradient approach leads merely to finite codimension (see the end of Example 23.10).

Definition 23.9. Let $F\colon \mathbb{R}^n \times \mathbb{R}^a \rightarrowtail \mathbb{R}$ be a smooth family of functions. The ***full bifurcation set*** $\mathcal{B}$ of F is the set of parameter values where the number of critical values is not locally constant. ✯

In other words, a parameter value $\lambda_0 \notin \mathcal{B}$ if there is a neighbourhood U of λ_0 such that the number of critical values of f_λ is the same for every $\lambda \in U$. There are two ways this number can change: either through a degenerate critical point as before, or through two critical values becoming one, as for the function $f(x) = x^4 - x^2$. The locus of points in $\mathcal{B}$ where the function has at least two distinct critical points with the same value is called the ***Maxwell set*** (or Maxwell stratum). Adapting Saito's method for vector fields tangent to a discriminant (described in Section 19.2B), H. Terao [109] developed an algorithm giving generators for the module of vector fields tangent to the full bifurcation set.

Example 23.10. Consider the cusp family

$$F(x; u, v) = \tfrac{1}{4}x^4 + \tfrac{1}{2}ux^2 + vx.$$

The critical points are given by $x^3 + ux + v = 0$, and can be parametrized by $v = v(x, u)$. Now, the value of the critical point at $(x, u, v(x, u))$ is

$$F(x; u, v(x, u)) = -\tfrac{3}{4}x^4 - \tfrac{1}{2}ux^2.$$

To find where $f = F(\,\cdot\,; u, v)$ has two critical points with the same value, let $x_1 \neq x_2$ be the two points. Then we need to solve $f(x_1) = f(x_2)$, or

$$-\tfrac{3}{4}x_1^4 - \tfrac{1}{2}ux_1^2 = -\tfrac{3}{4}x_2^4 - \tfrac{1}{2}ux_2^2$$

together with $x_1^3 + ux_1 + v = x_2^3 + ux_2 + v = 0$. Solving these with $x_1 \neq x_2$ shows,

$$x_2 = -x_1, \;\; u = -x_1^2, \;\; v = 0.$$

In particular, the function has a double critical value when $v = 0$. The full bifurcation set $\mathcal{B}$ is therefore the union of the local bifurcation set which is the semicubical parabola, and the Maxwell set which is the u-axis. This union is given by $h = 0$ where $h = v(4u^3 + 27v^2)$ as shown below. Note that the positive u-axis corresponds to double critical values where the (distinct) critical points are not real.

For studying bifurcations using $\mathcal{K}_V$-equivalence, we need the module of vector fields tangent to the full bifurcation set $\mathcal{B}$. Since h is weighted homogeneous, the methods introduced in Chapter 20 apply here, giving, after some simplification,

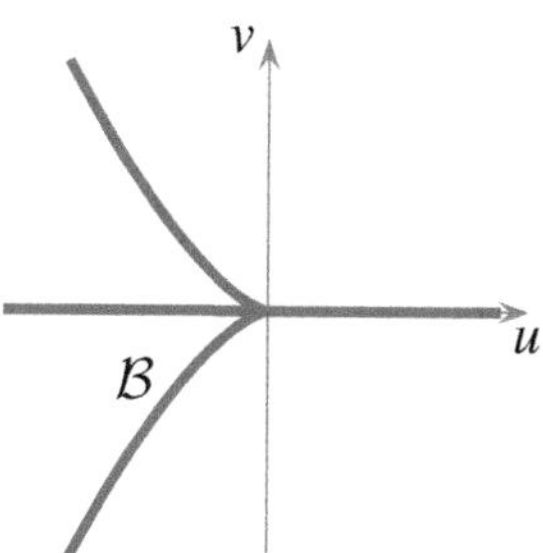

$$\theta_{\mathcal{B}} = \mathcal{E}_2 \left\{ \begin{pmatrix} 2u \\ 3v \end{pmatrix}, \begin{pmatrix} 9v^2 \\ -2u^2 v \end{pmatrix} \right\}.$$

Note that the determinant of the matrix formed from these two vector fields is equal to h (up to a constant multiple), a general property proved by Terao [109].

Consider the path $h\colon \lambda \mapsto (0, \lambda)$, which corresponds to the family $F(x, \lambda) = \frac{1}{4}x^4 + \lambda x$. Then

$$T\mathcal{K}_{\mathcal{B}} \cdot h = \mathcal{E}_1 \left\{ \begin{pmatrix} 0 \\ \lambda \end{pmatrix}, \begin{pmatrix} \lambda^2 \\ 0 \end{pmatrix} \right\},$$

whence, by Theorem 20.11, h is 1-determined, and

$$T_e\mathcal{K}_{\mathcal{B}} \cdot h = \mathcal{E}_1 \left\{ \begin{pmatrix} 0 \\ 1 \end{pmatrix}, \begin{pmatrix} \lambda^2 \\ 0 \end{pmatrix} \right\}.$$

Thus h is of codimension 2, with versal unfolding

$$H(\lambda; a, b) = (a + b\lambda, \lambda)$$

and corresponding bifurcation equation

$$\nabla \left(x^4 + (a + b\lambda)x^2 + \lambda x \right) = 0.$$

Note that $\nabla f = x^3 + \lambda = 0$ is the hysteresis bifurcation, which is merely of codimension 1.

More extreme is the path $\lambda \mapsto (\lambda, 0)$, which corresponds to the family $\frac{1}{4}x^4 + \lambda x^2$. This is of infinite codimension in the variational sense analyzed above, while its gradient is the pitchfork bifurcation which is of codimension 2. ✐

Problems

23.1 State and prove a Thom–Levine theorem for $\mathbb{Z}_2$ contact equivalence.

23.2 Let $g\colon (\mathbb{R}^n, 0) \to (\mathbb{R}^n, 0)$ be a symmetric germ. Show that

$$T\mathcal{K}^{\mathbb{Z}_2}\cdot g = T\mathcal{K}\cdot g \cap \theta(g)^-, \quad \text{and} \quad T\mathcal{K}_1^{\mathbb{Z}_2}\cdot g = T\mathcal{K}_1\cdot g \cap \theta(g)^-,$$

where $T\mathcal{K}_1\cdot g$ is defined in Problem 12.14.

23.3 Continue the calculation at the end of Example 23.3. Consider the path $h_r(\lambda) = (r\lambda, \lambda)$, with $r \in \mathbb{R}$. Show (i) h_r is 1-determined, and (ii) using the Thom–Levine theorem show that h_r is path equivalent to h_0.

23.4 Consider the codimension-1 bifurcation of Example 23.4, $G(x; \lambda; a) = x^2 + \lambda^2 x + ax = 0$, and show that as a varies this involves the merging and subsequent disappearing of two transcritical bifurcations.

23.5 Consider the symmetric germ $g_0(x, y) = (x^3, y^3)$. Show that g_0 is 3-determined in the world of symmetric (odd) map germs. Find a symmetric versal unfolding of g_0, and show g_0 is a *bimodal* singularity. (†)

23.6 Consider the fold family of functions from Chapter 2, $F(x; u) = \frac{1}{3}x^3 + ux$. In the variational context, find the codimension of the family $f(x, \lambda) = \frac{1}{3}x^3 + \lambda^2 x$, and a versal unfolding.

23.7 Referring to Example 23.10, find the codimension and a versal unfolding of the path $h(\lambda) = (c\lambda, d\lambda)$, for all values of the coefficients $c, d \in \mathbb{R}$.

23.8 In Example 23.2 we discussed the bifurcation problem

$$g(x; \lambda) = x^3 \pm \lambda^2 x = 0$$

as a symmetric bifurcation, and found it to be of codimension 1.

(i). Find its codimension and versal unfolding in the general, non-symmetric setting.

(ii). Find its codimension and versal unfolding in the setting of bifurcations with a fixed solution at $x = 0$;

(iii). Consider the cusp family $F(x; u, v) = \frac{1}{4}x^4 + \frac{1}{2}ux^2 + vx$. Find the codimension of the family $f(x, \lambda) = \frac{1}{4}x^4 + \frac{1}{2}\lambda x^2$ in the variational setting. (†)

Part IV

Appendices

A

Calculus of several variables

THIS FIRST APPENDIX CONSISTS OF two sections, both of which contain important background material: normal forms of linear maps (matrices) and differentiation of functions of several variables. The principal aim is to establish notation and very briefly describe some notions that are important to the main text. Complete proofs are not given here, but can be found in many texts on linear algebra and calculus of several variables respectively.

Throughout this text, we use the notation $f\colon \mathbb{R}^n \rightarrowtail \mathbb{R}^p$ to mean f is a smooth map whose domain is an open subset of $\mathbb{R}^n$. Whenever we have a map $f\colon \mathbb{R}^n \rightarrowtail \mathbb{R}^p$, we refer to $\mathbb{R}^n$ as the ***source*** of f, and $\mathbb{R}^p$ as its ***target***.

A.1 Linear maps and normal forms

One of the principal methods of singularity theory is to change coordinates to make the form of the map easier to compute with, or equations easier to solve. In this section we begin by showing how this works for linear maps, and then discuss briefly the sets of linear maps of fixed rank.

The space of linear maps $A\colon \mathbb{R}^n \to \mathbb{R}^p$ is denoted $L(\mathbb{R}^n, \mathbb{R}^p)$ or simply $L(n, p)$. Linear maps can be added together or multiplied by scalars, and so form a vector space.

The vector spaces $\mathbb{R}^n$ and $\mathbb{R}^p$ have *canonical* bases, namely

$$\left\{(1,0,\ldots,0)^T,\ (0,1,0,\ldots,0)^T,\ \ldots\ ,\ (0,0,\ldots,1)^T\right\},$$

(the $(\ldots)^T$ makes them into column vectors). The first vector is denoted $\mathbf{e}_1$ the second $\mathbf{e}_2$ etc. Using these canonical bases of $\mathbb{R}^n$ and $\mathbb{R}^p$ the map $A \in L(n,p)$ is represented by a $p \times n$ matrix $(a_{ij}) \in \mathsf{Mat}(p,n)$ and the jth column $(a_{1j}\ a_{2j}\ \ldots\ a_{pj})^T$ represents the image $A(\mathbf{e}_j)$ of the jth basis vector of $\mathbb{R}^n$. This representation provides an isomorphism of $L(n,p)$ with $\mathbb{R}^{np}$ (as vector spaces). Thus $\dim L(n,p) = np$.

In practice it is often useful to change to different bases in both source and target ($\mathbb{R}^n$ and $\mathbb{R}^p$), bases that reflect or are adapted to the map in question. Changing the bases affects the matrix as follows:

- changing basis in $\mathbb{R}^n$ corresponds to column operations, or multiplying by a change of basis matrix on the right, while
- changing basis in $\mathbb{R}^p$ corresponds to row operations, or multiplying by a change of basis matrix on the left.

For us the most important property of a matrix is its ***rank***. There are two definitions of the rank of a matrix: the *column rank* is the maximal number of linearly independent columns of the matrix, while the *row rank* is the maximal number of linearly independent rows. One of the important basic theorems of linear algebra is that these two ranks are equal, so just called the rank. We write the rank of a matrix (or linear map) A as $\operatorname{rk} A$. We write the subset of matrices in $L(n, p)$ of rank r by $L_r(n, p)$.

Example A.1. This example is an illustration of Proposition A.2 below. Let $n = 3,\ p = 4$ and consider the matrix

$$A = \begin{pmatrix} 1 & 1 & 0 \\ 0 & 1 & -1 \\ 2 & -1 & 3 \\ 3 & 0 & 3 \end{pmatrix}.$$

The rank of A is 2 because the first two columns are linearly independent, while col3 = col1 − col2. That is $A \in L_2(3, 4)$. To simplify the form of the matrix, we perform the column operation of adding col2 − col1 to col3, which is equivalent to multiplying A on the right by the change of basis matrix,

$$\Phi = \begin{pmatrix} 1 & 0 & -1 \\ 0 & 1 & 1 \\ 0 & 0 & 1 \end{pmatrix},$$

to get

$$A\Phi = \begin{pmatrix} 1 & 1 & 0 \\ 0 & 1 & 0 \\ 2 & -1 & 0 \\ 3 & 0 & 0 \end{pmatrix}.$$

Note that the third column of Φ is in the kernel of A, which explains why the third column of $A\Phi$ is zero. Now this new matrix $A\Phi$ can be simplified further by choosing the first two columns as part of a basis for the target $\mathbb{R}^4$. So we multiply on the left by the 4×4 matrix

$$\Psi = \begin{pmatrix} 1 & 1 & 0 & 0 \\ 0 & 1 & 0 & 0 \\ 2 & -1 & 1 & 0 \\ 3 & 0 & 0 & 1 \end{pmatrix}^{-1} = \begin{pmatrix} 1 & -1 & 0 & 0 \\ 0 & 1 & 0 & 0 \\ -2 & 3 & 1 & 0 \\ -3 & 3 & 0 & 1 \end{pmatrix}.$$

We now obtain

$$\Psi A \Phi = \begin{pmatrix} 1 & 0 & 0 \\ 0 & 1 & 0 \\ 0 & 0 & 0 \\ 0 & 0 & 0 \end{pmatrix},$$

which is as simple an expression as possible for a matrix of rank 2. ✐

As in the example above, if we apply a change of basis Φ of $\mathbb{R}^n$ and Ψ of $\mathbb{R}^p$, then the linear map is represented by a new matrix A' given by $A' = \Psi A \Phi$.

Proposition A.2. *Let $A \in L_r(n,p)$. Bases in $\mathbb{R}^n$ and $\mathbb{R}^p$ can be chosen so that the matrix of A is, in block form,*

$$A = \begin{bmatrix} I_r & 0 \\ 0 & 0 \end{bmatrix},$$

where I_r is the $r \times r$ identity matrix.

Such bases are often said to be ***adapted*** to the map A, and this form of the matrix is called its ***normal form***.

Remark A.3. This is simpler than Jordan canonical form because here $\mathbb{R}^n$ and $\mathbb{R}^p$ are considered to be different spaces (even if $n = p$) so the two choices of basis are independent, whereas in Jordan canonical form the spaces are the same, so there is only one choosing of a basis: $\Psi = \Phi^{-1}$. ❞

Proof: First we choose the basis in the source (guided by the example above). Since $\operatorname{rk} A = r$ it follows that $\dim\ker A = n - r$, so let $\{\mathbf{f}_{r+1}, \dots, \mathbf{f}_n\}$ be any basis for this kernel. Now choose $\mathbf{f}_1, \dots \mathbf{f}_r$ so that

$$\{\mathbf{f}_1, \dots, \mathbf{f}_r, \mathbf{f}_{r+1}, \dots, \mathbf{f}_n\}$$

is a basis of $\mathbb{R}^n$. With respect to such a basis, the last $n - r$ columns of the matrix of A will all be zero, so

$$A = \Big(\overbrace{\ A_1\ }^{r} \ \overbrace{\ 0\ }^{n-r} \Big),$$

where A_1 is a $p \times r$ matrix of rank r.

Now we choose a basis in the target. For $i = 1, \dots, r$ let $\mathbf{g}_i = A\mathbf{f}_i \in \mathbb{R}^p$. These r vectors are the columns of A_1 so are linearly independent (as $\operatorname{rk} A_1 = r$), and span the image of A. Now choose any $p - r$ vectors $\mathbf{g}_{r+1}, \dots \mathbf{g}_p$ so that $\{\mathbf{g}_1, \dots, \mathbf{g}_p\}$ forms a basis of $\mathbb{R}^p$. With respect to this basis, A will then have the desired form. ✔

A.1A Sets of linear maps of given rank

Let $L_r(n, p)$ be the set of linear maps of rank r, and $L_{\leq r}(n, p)$ be the set of those maps of rank at most r. Note of course that $L_{\leq r}(n, p) = L_{<r+1}(n, p)$. In practice, one usually calculates the rank of a given matrix by doing some intelligent row and column operations, so reducing it to echelon form. However, there is a more stupid method which consists in calculating the determinant of every $r \times r$ submatrix. Recall that given a $p \times n$ matrix A, an $r \times r$ submatrix is obtained from A by removing $p - r$ rows and $n - r$ columns, and the resulting determinant is called an $r \times r$ *minor* of A. There are $\binom{n}{r}\binom{p}{r}$ of these $r \times r$ minors of A.

Proposition A.4. *Let A be a $p \times n$ matrix. Then* $\operatorname{rk} A < r$ *if and only if every $r \times r$ minor of A is zero.*

Proof: First suppose $\operatorname{rk} A < r$, and let B be any $r \times r$ submatrix of A. The r rows used to define B are linearly dependent so that $\det B = 0$. This is true for all $r \times r$ submatrices.

Conversely, suppose $\operatorname{rk} A \geq r$. We wish to show that A has an $r \times r$ submatrix with nonzero determinant. Now, as A has rank at least r, it has r columns which are linearly independent. The $p \times r$ submatrix C formed of these columns has rank r and consequently C has r linearly independent rows. Choosing just these rows of C gives an $r \times r$ submatrix D of C (and hence of A) with rank r, and so $\det D \neq 0$, as required. ✔

Given a $p \times n$ matrix A, there are $\binom{p}{r}\binom{n}{r}$ $r \times r$ minors of A. Let $\mu_r(A)$ be the vector whose components are these minors, so

$$\mu_r : L(n, p) \longrightarrow \mathbb{R}^{\binom{p}{r}\binom{n}{r}}.$$

Proposition A.4 then says $\operatorname{rk} A < r$ if and only if $\mu_r(A) = 0$. By the definition of minors (determinants of $r \times r$ submatrices) each component of μ_r is a polynomial of degree r, and μ_r is therefore continuous.

Corollary A.5. *(i). For each r, $L_{<r}(n, p)$ is a closed subset of $L(n, p)$.*

(ii). If $r_0 = \min\{n, p\}$ then $L_{r_0}(n, p)$ is open in $L(n, p)$.

(iii). The closure $\overline{L_r(n, p)}$ of $L_r(n, p)$ is equal to $L_{\leq r}(n, p)$.

Recall that if X is a subset of a metric space (such as the vector space $L(n, p)$), then its *closure* $\overline{X}$ is the smallest closed set containing X. Points in $\overline{X}$ can be characterized as being the set of limits of sequences in X which converge in the ambient metric space.

PROOF: In this proof we suppress the (n, p), so for example $L = L(n, p)$ and $L_r = L_r(n, p)$.

(i) By Proposition A.4, $A \in L_{<r}$ if and only if $\mu_{r_0}(A) = 0$. It follows that $L_{<r_0}$ is a closed subset of L.

(ii) The complement of L_{r_0} is $L_{<r_0}$ which is closed. The complement of a closed set is open, so that indeed L_{r_0} is open.

(iii) Since by (i), $L_{\leq r}$ is closed, we have that $\overline{L_r} \subset L_{\leq r}$. For the reverse inclusion, we need to show that any $B \in L_{\leq r}$ there is a convergent sequence $(A_n) \to B$ with $A_n \in L_r$. If $\operatorname{rk} B = r$ then the constant sequence $A_n = B$ suffices. Otherwise let $k = \operatorname{rk} B < r$, and choose a basis such that

$$B = \begin{pmatrix} I_k & 0 \\ 0 & 0 \end{pmatrix}$$

(a basis adapted to B). Now for $n \in \mathbb{N}$ define A_n by

$$A_n = \begin{pmatrix} I_k & 0 & 0 \\ 0 & n^{-1} I_{r-k} & 0 \\ 0 & 0 & 0 \end{pmatrix}.$$

Then clearly $A_n \in L_r$ and $\lim_{n \to \infty} A_n = B$, as required. ✔

A.1B Bilinear and quadratic forms

A ***bilinear form*** on a vector space V (or on $\mathbb{R}^n$) is a map $B\colon V \times V \to \mathbb{R}$ which is linear in each argument, so

$$B(a\mathbf{u} + b\mathbf{v}, \mathbf{w}) = aB(\mathbf{u}, \mathbf{w}) + bB(\mathbf{v}, \mathbf{w})$$

and a similar statement for the second argument. B is said to be ***symmetric*** if $B(\mathbf{v}, \mathbf{u}) = B(\mathbf{u}, \mathbf{v})$, and ***skew-symmetric*** if $B(\mathbf{v}, \mathbf{u}) = -B(\mathbf{u}, \mathbf{v})$.

If V has a basis $\{\mathbf{e}_i\}$, then B can be represented by a square matrix $[B] = (b_{ij})$, by putting

$$B(\mathbf{u}, \mathbf{v}) = \mathbf{u}^T [B]\mathbf{v}$$

where on the right hand side $\mathbf{u}$ and $\mathbf{v}$ are written as column vectors with respect to the given basis: if $\mathbf{u} = \sum u_j \mathbf{e}_j$ and $\mathbf{v} = \sum v_j \mathbf{e}_j$ then

$$\mathbf{u}^T [B]\mathbf{v} = \sum b_{ij} u_i v_j.$$

The ***rank*** of the bilinear form is the rank of its matrix $[B] = (b_{ij})$, and a bilinear form is ***nondegenerate*** if $\det([B]) \neq 0$, or equivalently if one has

$$B(\mathbf{u}, \mathbf{v}) = 0 \ (\forall \mathbf{v} \in V) \Rightarrow \mathbf{u} = 0.$$

If one changes from a basis $\{\mathbf{e}_1, \ldots, \mathbf{e}_n\}$ to a different basis, $\{\mathbf{f}_1, \ldots, \mathbf{f}_n\}$ with change of basis matrix P, then

$$[B]_{\mathbf{f}} = P^T [B]_{\mathbf{e}} P,$$

where $[B]_{\mathbf{e}}$ means the matrix of B with respect to the basis $\{\mathbf{e}_j\}$, and similarly $[B]_{\mathbf{f}}$.

A ***quadratic form*** Q on V is a map $Q \colon V \to \mathbb{R}$ such that given any basis $\{\mathbf{e}_j\}$, there are real numbers q_{ij} such that

$$Q(\mathbf{u}) = \sum_{i,j} q_{ij} u_i u_j.$$

That is, Q is homogeneous of degree 2 in the components of $\mathbf{u}$.

Given a bilinear form B, one can associate a quadratic form Q by

$$Q(\mathbf{u}) = B(\mathbf{u}, \mathbf{u}).$$

This quadratic form Q is zero iff B is skew-symmetric (see Problem A.2).

On the other hand, given a quadratic form Q, one associates a bilinear form by the formula

$$B(\mathbf{u}, \mathbf{v}) := \tfrac{1}{2} \left(Q(\mathbf{u} + \mathbf{v}) - Q(\mathbf{u}) - Q(\mathbf{v}) \right), \tag{A.1}$$

the so-called ***polarization*** of the quadratic form. It is clearly a symmetric bilinear form. The matrix $[Q] = (q_{ij})$ of Q satisfies the same transformation rule under change of basis as the matrix of a bilinear form given above.

A standard theorem in linear algebra says that every real quadratic form can be diagonalized, with diagonal elements $0, \pm 1$, by a suitable choice of basis. The number of zeros is called the ***nullity*** of Q, the number of -1s the ***index*** and the difference between the number of $+1$s and -1s the ***signature***.

Extending the idea of bilinear map, a map $T\colon V \times V \times V \to \mathbb{R}$ which is linear in each argument is called a ***trilinear map***. A trilinear form is *symmetric* if $T(\mathbf{u},\mathbf{v},\mathbf{w}) = T(\mathbf{v},\mathbf{u},\mathbf{w}) = T(\mathbf{u},\mathbf{w},\mathbf{v})$ (and hence equal for all permutations of the arguments). And in a similar way as for bilinear maps, given a trilinear map, one defines a cubic form

$$C(\mathbf{u}) = T(\mathbf{u},\mathbf{u},\mathbf{u}).$$

We will usually write $B\mathbf{u}^2 = B(\mathbf{u},\ \mathbf{u})$ and $T\mathbf{u}^3 = T(\mathbf{u},\ \mathbf{u},\ \mathbf{u})$, and extending this, one writes $T\mathbf{u}^2\mathbf{v} = T(\mathbf{u},\mathbf{u},\mathbf{v})$ when T is symmetric.

A.1c Direct sum and cobasis

A final element of linear algebra we make use of is the following. Let V be a real vector space (possibly of infinite dimension), and U_1, U_2 be two subspaces of V. We write

$$V = U_1 + U_2 \tag{A.2}$$

(V is the sum of U_1 and U_2) to mean $\mathbf{v} \in V$ if and only if it can be written as a sum $\mathbf{v} = \mathbf{u}_1 + \mathbf{u}_2$ for some vectors $\mathbf{u}_1 \in U_1$ and $\mathbf{u}_2 \in U_2$. In general, for a given $\mathbf{v} \in V$ the vectors $\mathbf{u}_1, \mathbf{u}_2$ might not be unique. Indeed, suppose $\mathbf{v} = \mathbf{w}_1 + \mathbf{w}_2$ as well (with say $\mathbf{u}_1 \neq \mathbf{w}_1$), then subtracting shows that

$$\mathbf{u}_1 - \mathbf{w}_1 = \mathbf{w}_2 - \mathbf{u}_2.$$

The left-hand side is in U_1 the right hand side is in U_2, so in particular if the decomposition of $\mathbf{v}$ is not unique then $U_1 \cap U_2 \neq \{0\}$. If, on the other hand, (A.2) holds and $U_1 \cap U_2 = \{0\}$ then one writes

$$V = U_1 \oplus U_2, \tag{A.3}$$

and V is said to be the ***direct sum*** of U_1 and U_2.

Suppose now $V = U_1 \oplus U_2$, and that U_2 is finite–dimensional. Then we say U_1 has ***finite codimension*** in V. Moreover, if $\{\mathbf{e}_1, \dots, \mathbf{e}_n\}$ is a basis for U_2, then since every $\mathbf{v} \in V$ can be written uniquely as $\mathbf{v} = \mathbf{u}_1 + \mathbf{u}_2$, and $\mathbf{u}_2$ can be uniquely written as a sum $\mathbf{u}_2 = \sum_{j=1}^n \lambda_j \mathbf{e}_j$, we conclude that for given $\mathbf{v}$ there are unique real numbers λ_j, and a unique $\mathbf{u}_1 \in U_1$, such that

$$\mathbf{v} = \mathbf{u}_1 + \lambda_1\mathbf{e}_1 + \cdots + \lambda_n\mathbf{e}_n. \tag{A.4}$$

Such a basis for U_2 is called a ***cobasis*** for U_1 in V, and we call the number of elements in the cobasis the ***codimension*** of U_1 in V.

Thus, a ***cobasis*** for a subspace U of a vector space V is a set $\{\mathbf{e}_1, \ldots, \mathbf{e}_n\}$ of vectors in V such that any $\mathbf{v} \in V$ has a unique expression of the form (A.4), with $\lambda_j \in \mathbb{R}$ and $\mathbf{u}_1 \in U$. See also Problem 3.15.

A.2 Differentiation in several variables

Let $f\colon \mathbb{R}^n \rightarrowtail \mathbb{R}^p$ be a map. Written in coordinates, f has the form

$$f(x_1, \ldots, x_n) = (f_1(x_1, \ldots, x_n),\ f_2(x_1, \ldots, x_n),\ \ldots,\ f_p(x_1, \ldots, x_n)).$$

The functions $f_i(x_1, \ldots, x_n)$ are called the *components* of f. We assume throughout this book that f is ***smooth***, meaning that all the partial derivatives of all orders of all the components of f exist. In this case one says f is of class C^∞. In practice many of the results continue to hold if one only assumes all its partial derivatives exist up to some finite order.

A.2A The differential of a map

Let $f\colon \mathbb{R}^n \rightarrowtail \mathbb{R}^p$ be a smooth map. The ***differential*** $\mathsf{d}f_{x_0}$ of f at x_0 is the linear map defined as follows: for each $\mathbf{u} \in \mathbb{R}^n$ put

$$\mathsf{d}f_{x_0}(\mathbf{u}) = \lim_{t \to 0} \frac{f(x_0 + t\mathbf{u}) - f(x_0)}{t}. \tag{A.5}$$

This quantity is often called the *directional derivative* of f at x_0 in the direction $\mathbf{u}$, and it is a classical theorem of analysis that it depends linearly on $\mathbf{u} \in \mathbb{R}^n$. It is an important fact that the differential $\mathsf{d}f_{x_0}\colon \mathbb{R}^n \to \mathbb{R}^p$ is the *best linear approximation* to the map f at the point x_0.

Note that the definition does not require us to choose particular bases for $\mathbb{R}^n$ and $\mathbb{R}^p$, but since $\mathsf{d}f_{x_0}$ is a linear map, a choice of basis will allow us to represent it as a matrix. If $\{\mathbf{e}_1, \ldots, \mathbf{e}_n\}$ is a basis for $\mathbb{R}^n$ then equation (A.5) says that $\mathsf{d}f_{x_0}(\mathbf{e}_i) = \frac{\partial f}{\partial x_i}$ (which is a p-vector because f has p components). Thus, with respect to bases the differential becomes the $p \times n$ ***Jacobian matrix***,

$$\mathsf{d}f_{x_0} = \begin{pmatrix} \frac{\partial f_1}{\partial x_1}(x_0) & \cdots & \frac{\partial f_1}{\partial x_n}(x_0) \\ \vdots & & \vdots \\ \frac{\partial f_p}{\partial x_1}(x_0) & \cdots & \frac{\partial f_p}{\partial x_n}(x_0) \end{pmatrix}.$$

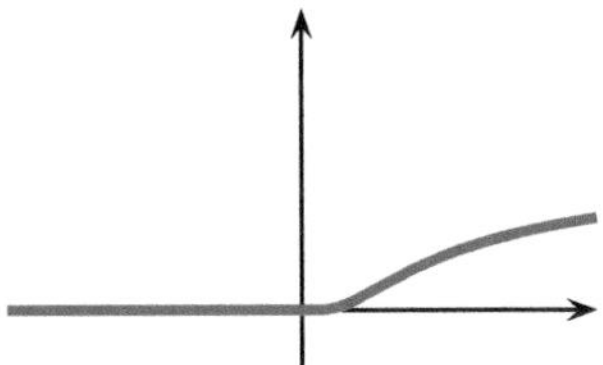

FIGURE A.1 The graph of the flat function h (see Example A.6(iv))

[*Note: in case it causes confusion,* x_0 *is a point in the source while* $x_1, \ldots, x_n$ *are coordinates.*]

Examples A.6. (i) The function $f\colon \mathbb{R} \to \mathbb{R}$, $f(x) = |x|$ is not differentiable at $x = 0$, though it is smooth everywhere else.
(ii) The function $f\colon \mathbb{R} \to \mathbb{R}$, $f(x) = |x|^{3/2}$ is differentiable everywhere, but not smooth at 0.
(iii) Any polynomial map $f\colon \mathbb{R}^n \to \mathbb{R}^p$ (ie, a map all of whose components are polynomials) is smooth.
(iv) The function $h\colon \mathbb{R} \to \mathbb{R}$ defined by

$$h(x) = \begin{cases} \mathrm{e}^{-1/x} & \text{if } x > 0, \\ 0 & \text{if } x \leq 0. \end{cases} \tag{A.6}$$

whose graph is shown in Figure A.1 is smooth but has zero Taylor series at the origin; such functions are said to be ***flat*** at the origin (see Problem A.1). ✐

Examples A.7. (i) If $f\colon \mathbb{R}^n \to \mathbb{R}^p$ is a linear map, then its best linear approximation is f itself, so $\mathsf{d}f_{x_0}(\mathbf{u}) = f(\mathbf{u})$ (as can easily be seen from (A.5)).
(ii) Let $f\colon \mathbb{R}^2 \to \mathbb{R}^3$ be given by $f(x, y) = (y + x^2,\ xy,\ x - y^3)$. Then

$$\mathsf{d}f_{(x,y)} = \begin{pmatrix} 2x & 1 \\ y & x \\ 1 & 3y^2 \end{pmatrix}.$$

In particular,

$$\mathsf{d}f_{(0,0)} = \begin{pmatrix} 0 & 1 \\ 0 & 0 \\ 1 & 0 \end{pmatrix}.$$

Thus $\mathsf{d}f_{(0,0)}(x, y) = (y, 0, x)^T$ which is clearly the linear part of f. ✐

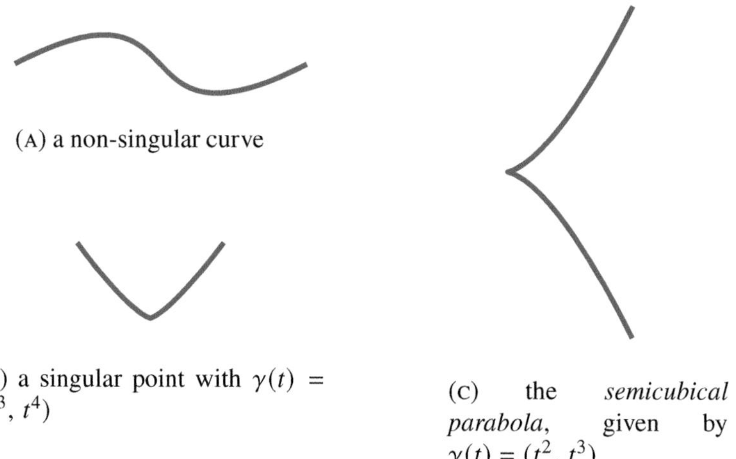

(A) a non-singular curve

(B) a singular point with $\gamma(t) = (t^3, t^4)$

(C) the *semicubical parabola*, given by $\gamma(t) = (t^2, t^3)$

FIGURE A.2 Three curves in the plane.

Composite of maps The differential of the composite of two maps is given by the chain rule: let $f\colon \mathbb{R}^n \to \mathbb{R}^p$ and $g\colon \mathbb{R}^p \to \mathbb{R}^q$, then $g \circ f\colon \mathbb{R}^n \to \mathbb{R}^q$ and

$$\mathrm{d}(g \circ f)_x = \mathrm{d}g_{f(x)}\mathrm{d}f_x.$$

The right hand side is the composite of linear maps; when bases are chosen and the linear maps are matrices this is just the product of the matrices.

Tangent vectors to a curve Let $\gamma\colon \mathbb{R} \rightarrowtail \mathbb{R}^n$ be a smooth map. The image of γ is then a curve in $\mathbb{R}^n$. The curve, or rather its parametrization, is *singular* at t_0 if $\frac{\mathrm{d}}{\mathrm{d}t}\gamma(t_0) = 0$, and if there is no such point it is a *non-singular curve*. See Figure A.2 for pictures of some singular curves in the plane.

To relate the usual notation $\frac{\mathrm{d}}{\mathrm{d}t}\gamma(t)$ to the notation introduced above, one has

$$\mathrm{d}\gamma_t(1) = \frac{\mathrm{d}}{\mathrm{d}t}\gamma(t).$$

Here 1 represents the unit tangent vector to $\mathbb{R}$ at the point t.

A.2B Higher derivatives

Higher derivatives of a map are successively more complicated, which is hardly surprising as there are more second order partial derivatives than first order, and more third order than second order, and so on. In this section we describe briefly the definition of higher–order differentials.

As we have seen, the differential at x of a smooth map $f\colon \mathbb{R}^n \rightarrowtail \mathbb{R}^p$ is a linear map $\mathsf{d}f_x \in L(\mathbb{R}^n, \mathbb{R}^p)$. Thus,

$$\begin{aligned} \mathsf{d}f\colon \mathbb{R}^n &\rightarrowtail L(\mathbb{R}^n, \mathbb{R}^p) \\ x &\longmapsto \mathsf{d}f_x. \end{aligned}$$

The second differential at x is the differential of this map. Thus

$$\mathsf{d}^2 f_x = \mathsf{d}(\mathsf{d}f)_x,$$

and $\mathsf{d}^2 f_x \in L(\mathbb{R}^n, L(\mathbb{R}^n, \mathbb{R}^p))$ with $\mathsf{d}^2 f_x(\mathbf{u}) \in L(\mathbb{R}^n, \mathbb{R}^p)$, and finally $\mathsf{d}^2 f_x(\mathbf{u})(\mathbf{v}) \in \mathbb{R}^p$. We will write $L(\mathbb{R}^n, L(\mathbb{R}^n, \mathbb{R}^p))$ more briefly as $L^{(2)}(\mathbb{R}^n, \mathbb{R}^p)$.

One usually writes $\mathsf{d}^2 f_x(\mathbf{u})(\mathbf{v}) = \mathsf{d}^2 f_x(\mathbf{u}, \mathbf{v})$; it is a *bilinear* map – meaning that it is linear in each of its two arguments (see Section A.1B). A basic theorem in calculus of several variables is that if f is smooth (or class C^2) then the bilinear map is symmetric: $\mathsf{d}^2 f_x(\mathbf{u}, \mathbf{v}) = \mathsf{d}^2 f_x(\mathbf{v}, \mathbf{u})$.

If $p = 1$, and if we are given a basis in $\mathbb{R}^n$, then $\mathsf{d}^2 f_x$ can be represented by a symmetric $n \times n$ matrix $H_f = (h_{ij})$ with

$$h_{ij} = \frac{\partial^2 f}{\partial x_i\, \partial x_j},$$

which means that

$$\mathsf{d}^2 f_x(\mathbf{u}, \mathbf{v}) = \mathbf{u}^T H_f\, \mathbf{v}.$$

The matrix B is called the ***Hessian*** matrix of f at x.

If $p > 1$ and we are given a basis of $\mathbb{R}^p$, then each component of f is a scalar-valued function so its second derivative is a matrix. Thus, if $p > 1$ the second differential is a 'vector' of symmetric matrices, one for each component of f.

Example A.8. Continuing the example from before (Example A.7(ii)), with $f\colon \mathbb{R}^2 \to \mathbb{R}^3$ given by $f(x, y) = (y + x^2, xy, x - y^3)$, one has

$$\mathsf{d}^2 f_{(x,y)} = \left(\begin{pmatrix} 2 & 0 \\ 0 & 0 \end{pmatrix}, \begin{pmatrix} 0 & 1 \\ 1 & 0 \end{pmatrix}, \begin{pmatrix} 0 & 0 \\ 0 & -3y \end{pmatrix} \right).$$

For example, the middle matrix is the Hessian matrix of the function xy. ✐

The third differential is defined from the second in the same way. So we have,

$$\begin{array}{rcl} \mathsf{d}^2 f\colon \mathbb{R}^n & \rightarrowtail & L^{(2)}(\mathbb{R}^n,\, \mathbb{R}^p) \\ x & \longmapsto & \mathsf{d}^2 f_x \end{array}$$

and the third differential is the differential of this: $\mathsf{d}^3 f_x = \mathsf{d}(\mathsf{d}^2 f)_x$ and so

$$\mathsf{d}^3 f_x \in L(\mathbb{R}^n,\, L^{(2)}(\mathbb{R}^n,\, \mathbb{R}^p)).$$

So given three vectors $\mathbf{u}, \mathbf{v}, \mathbf{w}$ one has $\mathsf{d}^3 f_x(\mathbf{u}, \mathbf{v}, \mathbf{w}) \in \mathbb{R}^p$ and this is linear in each argument, so making $\mathsf{d}^3 f_x$ a *trilinear* map. And by the theorem referred to earlier, it too is symmetric. And we write $L^{(3)}(\mathbb{R}^n,\, \mathbb{R}^p)$ for $L(\mathbb{R}^n,\, L^{(2)}(\mathbb{R}^n,\, \mathbb{R}^p))$, which is the same as $L(\mathbb{R}^n,\, L(\mathbb{R}^n,\, L(\mathbb{R}^n,\, \mathbb{R}^p)))$.

Higher differentials are defined in a similar way, each as the differential of the preceding one. So for each $k \in \mathbb{N}$,

$$\mathsf{d}^k f_x \in L^{(k)}(\mathbb{R}^n,\, \mathbb{R}^p).$$

And it is symmetric (assuming f is smooth).

To abbreviate the notation, we write $\mathsf{d}^k f_x(\mathbf{u}^k)$ instead of $\mathsf{d}^k f_x(\mathbf{u}, \mathbf{u}, \ldots, \mathbf{u})$, and for example $\mathsf{d}^3 f_x(\mathbf{u}, \mathbf{u}, \mathbf{v})$ will be written $\mathsf{d}^3 f_x(\mathbf{u}^2\mathbf{v})$ or just $\mathsf{d}^3 f_x\mathbf{u}^2\mathbf{v}$.

Theorem A.9 (Taylor's theorem). *Let $f\colon \mathbb{R}^n \rightarrowtail \mathbb{R}^p$ be a smooth map, $x \in \mathbb{R}^n$ and $k \in \mathbb{N}$. Then*

$$f(x+\mathbf{u}) = f(x) + \mathsf{d}f_x(\mathbf{u}) + \frac{1}{2}\mathsf{d}^2 f_x(\mathbf{u}^2) + \frac{1}{3!}\mathsf{d}^3 f_x(\mathbf{u}^3) + \cdots + \frac{1}{k!}\mathsf{d}^k f_x(\mathbf{u}^k) + o(\|\mathbf{u}\|^k).$$

Recall the *little-oh* notation: to say a function $h(\mathbf{u})$ is $o(\|\mathbf{u}\|^k)$, or just $o(k)$, at $\mathbf{u} = 0$, means that

$$\lim_{\mathbf{u}\to 0} \frac{h(\mathbf{u})}{\|\mathbf{u}\|^k} = 0.$$

It is perhaps worth reminding the reader that for analytic functions, the Taylor series converges to the function as $k \to \infty$, on some neighbourhood of the point x. On the other hand, for a function which is smooth but not analytic, the infinite Taylor series exists, but does not converge to the function on any neighbourhood. This possibility has been pointed out already in Example A.6 (iv).

Many authors denote the differentials at x by $\mathsf{d}f(x)$, $\mathsf{d}^2 f(x)$ etc., rather than $\mathsf{d}f_x$, $\mathsf{d}^2 f_x$. The advantage of using the subscript is to emphasize the difference between the argument x and the arguments $\mathbf{u}, \mathbf{v}$ of the multilinear maps: to the author, $\mathsf{d}^2 f_x(\mathbf{u}, \mathbf{v})$ seems marginally clearer than $\mathsf{d}^2 f(x)(\mathbf{u}, \mathbf{v})$.

Chain rule Let $g\colon \mathbb{R}^n \to \mathbb{R}^p$ and $f\colon \mathbb{R}^p \to \mathbb{R}^q$ then one defines the composite map $f \circ g\colon \mathbb{R}^n \to \mathbb{R}^q$, by $f \circ g(x) = f(g(x))$.

$$\mathbb{R}^n \xrightarrow{g} \mathbb{R}^p \xrightarrow{f} \mathbb{R}^q$$

Wring the second differential of $f \circ g$ in terms of differentials of g and f helps understand the structure of differentials. Recall the chain rule for first derivatives,

$$\mathsf{d}(f \circ g)_x(\mathbf{u}) = \mathsf{d}f_{g(x)}(\mathsf{d}g_x(\mathbf{u})).$$

Proposition A.10. *If g and f are as above, and $\mathbf{u}, \mathbf{v} \in \mathbb{R}^n$ then*

$$\mathsf{d}^2(f \circ g)_x(\mathbf{u}, \mathbf{v}) = \mathsf{d}^2 f_{g(x)}(\mathsf{d}g_x(\mathbf{u}), \mathsf{d}g_x(\mathbf{v})) + \mathsf{d}f_{g(x)}\left(\mathsf{d}^2 g_x(\mathbf{u}, \mathbf{v})\right).$$

The proof is left to the reader, and can be found in many books on calculus of several variables.

Similar expressions arise for higher derivatives, for example (with $\mathbf{u} = \mathbf{v} = \mathbf{w}$),

$$\mathsf{d}^3(f \circ g)_x(\mathbf{u}^3) = \mathsf{d}^3 f_{g(x)}\left(\mathsf{d}g_x(\mathbf{u})^3\right) + 3\mathsf{d}^2 f_{g(x)}(\mathsf{d}g_x(\mathbf{u}), \mathsf{d}g_x^2(\mathbf{u}^2)) + \mathsf{d}f_{g(x)}\left(\mathsf{d}^3 g_x(\mathbf{u}^3)\right).$$

The general case is the subject of Faà di Bruno's formula below.

A.2c Faà di Bruno's formula

For the record, though we will not be using it beyond the third derivative given above, there is a general formula for the nth derivative of a composition of two maps, known as Faà di Bruno's[1] formula.

For the first few cases, we introduce the notation, $A = \mathsf{d}f$, $B = \mathsf{d}^2 f$, $C = \mathsf{d}^3 f$ etc. (all derivatives taken at $g(x)$), and $\mathbf{a} = \mathsf{d}g(\mathbf{u})$, $\mathbf{b} = \mathsf{d}^2 g(\mathbf{u}^2)$, $\mathbf{c} = \mathsf{d}^3 g(\mathbf{u}^3)$ etc (all derivatives at x). Then,

$$\begin{aligned}
\mathsf{d}(f \circ g)_x(\mathbf{u}) &= A\mathbf{a}, \\
\mathsf{d}^2(f \circ g)_x(\mathbf{u}^2) &= A\mathbf{b} + B\mathbf{a}^2, \\
\mathsf{d}^3(f \circ g)_x(\mathbf{u}^3) &= A\mathbf{c} + 3B\mathbf{a}\mathbf{b} + C\mathbf{a}^3, \\
\mathsf{d}^4(f \circ g)_x(\mathbf{u}^4) &= A\mathbf{d} + 4B\mathbf{a}\mathbf{c} + 3B\mathbf{b}^2 + 6C\mathbf{a}^2\mathbf{b} + D\mathbf{a}^4, \\
\mathsf{d}^5(f \circ g)_x(\mathbf{u}^5) &= A\mathbf{e} + 5B\mathbf{a}\mathbf{d} + 10B\mathbf{b}\mathbf{c} + 10C\mathbf{a}^2\mathbf{c} + 15C\mathbf{a}\mathbf{b}^2 + 10D\mathbf{a}^3\mathbf{b} + E\mathbf{a}^5.
\end{aligned} \tag{A.7}$$

[1]Fr Francesco Faà di Bruno, 1825–1888, was an Italian priest and mathematician and studied mathematics under Cauchy in Paris. The formula was published in F. de Bruno, Note sur une nouvelle formule de calcul différentiel, *Quart. J. Math.* **1** (1856), 359–360.

Problem A.6 invites you to prove this. The general formula, due to Faà di Bruno, can be organised in several different ways, each involving some combinatorics. One is as follows. Write A_k for $\mathsf{d}^k f_{g(x)}$, and $\mathbf{a}_k$ for $\mathsf{d}^k g_x(\mathbf{u}^k)$. Then

$$\mathsf{d}^n (f \circ g)_x(\mathbf{u}^n) = \sum_{\pi \in \mathcal{P}'(n)} A_{|\pi|}(\mathbf{a}_{|S_1|}, \mathbf{a}_{|S_2|}, \ldots, \mathbf{a}_{|S_{|\pi|}|}) \tag{A.8}$$

The notation here is as follows:

- $\mathcal{P}'(n)$ is the set of partitions of the set $\bar{n} := \{1, 2, \ldots, n\}$,
- for each partition $\pi \in \mathcal{P}'(n)$, $|\pi|$ denotes the cardinality of that partition,
- for each $\pi \in \mathcal{P}'(n)$, the S_j ($j = 1, \ldots, |\pi|$) are the elements of π (which are subsets of $\bar{n}$).

For example, the 5 elements of $\mathcal{P}'(3)$ are:

$$\pi_1 = \{\{1\}, \{2\}, \{3\}\}, \quad \pi_2 = \{\{1\}, \{2, 3\}\}, \quad \pi_3 = \{\{2\}, \{1, 3\}\},$$

$$\pi_4 = \{\{3\}, \{1, 2\}\} \quad \text{and} \quad \pi_5 = \{\{1, 2, 3\}\}.$$

Here $|\pi_1| = 3$, while $|\pi_2| = |\pi_3| = |\pi_4| = 2$ and $|\pi_5| = 1$. So in the sum, π_1 contributes $A_3(\mathbf{a}_1, \mathbf{a}_1, \mathbf{a}_1)$ (which is written $C\mathbf{a}^3$ above), π_2, π_3 and π_4 each contributes $A_2(\mathbf{a}_1, \mathbf{a}_2)$ (which gives the $3B\mathbf{ab}$ term above), while π_5 gives $A_1(\mathbf{a}_3)$ (corresponding to the $A\mathbf{c}$ term above).

The formula above has many repetitions as there are many partitions with the same sizes of subsets (e.g. in $\mathcal{P}'(3)$, π_2, π_3 and π_4 all contribute to the same term), so there is a more compact form involving certain combinatorial coefficients:

$$\mathsf{d}^n (f \circ g)_x(\mathbf{u}^n) = \sum_{\mathbf{k} \in \mathcal{P}(n)} C(\mathbf{k})\, A_k(\mathbf{a}_{k_1}, \mathbf{a}_{k_2}, \ldots, \mathbf{a}_{k_r}) \tag{A.9}$$

Here $\mathcal{P}(n)$ is the set of all partitions of the integer n, so $\mathbf{k} \in \mathcal{P}(n)$ means $\mathbf{k} = (k_1, k_2, \ldots, k_r)$ with $k_1 + k_2 + \cdots + k_r = n$ and $k_1 \le k_2 \le \cdots \le k_r$. The coefficient $C(\mathbf{k})$ is the number of ways of partitioning the set $\bar{n}$ (see above) into r subsets of sizes $k_1, k_2, \ldots, k_r$. It is given by

$$C(\mathbf{k}) = \frac{n!}{k_1!\, k_2! \ldots k_r!\, m(\mathbf{k})}.$$

The quantity $m(\mathbf{k})$ here takes into account repetitions in $\mathbf{k}$: $m(\mathbf{k})$ is the product of $m!$ where m is the number of times each k_j is repeated. For example $\mathcal{P}(3) = \{(3), (1, 2), (1, 1, 1)\}$, and $C((1, 2)) = \frac{3!}{1!2!} = 3$, which is indeed the right value because there are 3 ways to partition $\{1, 2, 3\}$ into subsets of sizes 2 and 1 (namely π_2, π_3 and π_4 from above). And for $n = 6$ one has $C((3, 3)) = \frac{6!}{3!3!\,2!} = 10$ (and this would correspond to the term $10B\mathbf{c}^2$ in the expression for $\mathrm{d}^6(f \circ g)_x$), the final 2! occurring in the denominator because the e in $(3, 3)$ occurs twice. Similarly, for $n = 3$, $C((1, 1, 1)) = \frac{3!}{1!\,1!\,1!\,3!} = 1$, and for $n = 11$,

$$C((1, 2, 2, 3, 3)) = \frac{11!}{1!\,2!\,2!\,3!\,3!\,2!\,2!} = 69\,300.$$

For more information on this formula and its history, see [64].

A.2D Jets of smooth maps

Taylor's theorem above suggests a new definition, namely the kth order Taylor expansion of f about the point $x \in \mathbb{R}^n$. More succinctly this is called the ***k-jet of f at** x*, written $\mathrm{j}^k f_x$. Thus,

$$\mathrm{j}^k f_x(\mathbf{u}) = f(x) + \mathrm{d}f_x(\mathbf{u}) + \frac{1}{2}\mathrm{d}^2 f_x(\mathbf{u}^2) + \frac{1}{3!}\mathrm{d}^3 f_x(\mathbf{u}^3) + \cdots + \frac{1}{k!}\mathrm{d}^k f_x(\mathbf{u}^k).$$

For each x, $\mathrm{j}^k f_x$ is polynomial of degree k in $\mathbf{u}$. By Taylor's theorem, it is the best approximation to f at x among all degree k polynomials.

Examples A.11. (i) (One variable) Let $f(x) = (1 + x)^{-1}$. Then the familiar Taylor series at $x = 0$ of f shows that

$$\mathrm{j}^k f_0(u) = 1 - u + u^2 - u^3 + \cdots + (-u)^k.$$

On the other hand, at $x = 1$,

$$\mathrm{j}^k f_1(u) = \tfrac{1}{2} - \tfrac{1}{4}u + \tfrac{1}{8}u^2 - \tfrac{1}{16}u^3 + \cdots + \tfrac{1}{2^{k+1}}(-u)^k.$$

(ii) (Two variables) Let $f\colon \mathbb{R}^2 \to \mathbb{R}^2$ be given by $f(x, y) = (x^2, \mathrm{e}^y)$. Then for example,

$$\mathrm{j}^3 f_{(0,0)}(u, v) = (u^2,\ 1 + v + \tfrac{1}{2}v^2 + \tfrac{1}{6}v^3),$$

while

$$\mathrm{j}^3 f_{(1,2)}(u, v) = \left(1 + 2u + u^2,\ \mathrm{e}^2(1 + v + \tfrac{1}{2}v^2 + \tfrac{1}{6}v^3\right).$$

(iii) (Three variables) Let $f\colon \mathbb{R}^3 \to \mathbb{R}$ be given by $f(x, y, z) = \ln(x + yz)$. Then with $\mathbf{x} = (1, 2, 3)$ and $\mathbf{u} = (u, v, w)$,

$$\begin{aligned}
\mathrm{j}^0 f_{\mathbf{x}} &= \ln(7),\\
\mathrm{j}^1 f_{\mathbf{x}}(\mathbf{u}) &= \ln(7) + \tfrac{1}{7}(u + 3v + 2w),\\
\mathrm{j}^2 f_{\mathbf{x}}(\mathbf{u}) &= \ln(7) + \tfrac{1}{7}(u + 3v + 2w) + \tfrac{1}{98}(u^2 - 4uw - 6uv - 9v^2 + 2vw - 4w^2).
\end{aligned}$$

✐

A.3 Vector fields

A vector field on $U \subset \mathbb{R}^n$ is a vector $\mathbf{v}(x) \in \mathbb{R}^n$ at each point $x \in U$. We will always assume $\mathbf{v}$ depends smoothly on x. If we are given coordinates $x_1, \ldots, x_n$ on $\mathbb{R}^n$ (or on U), the vector field can be written

$$\mathbf{v}(x) = \begin{pmatrix} v_1(x) \\ v_2(x) \\ \vdots \\ v_n(x) \end{pmatrix}.$$

For example, $\mathbf{v}(x, y) = \frac{1}{2}\begin{pmatrix} y \\ -x \end{pmatrix}$ is a vector field on the plane, depicted here on the right. Note that in order to save space, we will often write vector fields as row vectors, so here $\mathbf{v}(x, y) = \frac{1}{2}(y, -x)$; though more correctly this should be written $\mathbf{v}(x, y) = \frac{1}{2}(y, -x)^T$.

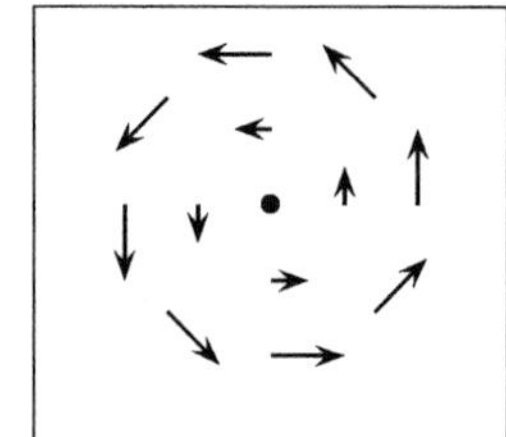

It is necessary also to consider *time-dependent* vector fields, written $\mathbf{v}(x, t)$, so the domain of definition is an open subset of $\mathbb{R}^n \times \mathbb{R}$ (i.e. it has $n + 1$ arguments, though it still has n components).

Vector fields as differential equations Vector fields are very closely related to first order ordinary differential equations (ODEs): we give a very brief overview here, but more details are given in Appendix C.

Let $\mathbf{v}(x)$ be a vector field. The associated differential equation is simply,

$$\dot{x} = \mathbf{v}(x),$$

or $\dot{x} = \mathbf{v}(x, t)$ if $\mathbf{v}$ is time-dependent. This is a first-order ODE. A solution is a differentiable curve $\gamma(t)$ for which $\dot{\gamma}(t) = \mathbf{v}(\gamma(t))$. Conversely, any first order ODE corresponds to a vector field, via the same equation.

The ***flow*** of a vector field or ODE is defined as follows. For each point x, let $\gamma(t)$ be the (unique) solution with $\gamma(0) = x$. Making the x-dependence explicit we can write $\gamma(t) = \phi(x, t)$ or $\phi_t(x)$. For each t, the map ϕ_t is a diffeomorphism, called the time-t flow (this is proved in Appendix C, Theorem C.6). Note that some care is needed in using this notion of flow, because many vector fields are not *complete*, meaning that ϕ_t may not exist for larger values of t. For $t = 0$, ϕ_0 is the identity so always exists, but for $|t| < \varepsilon$ say, the flow may only exist if the domain is restricted

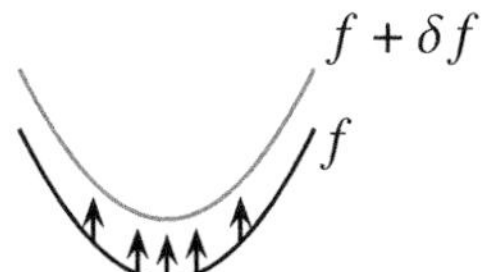

FIGURE A.3 Perturbing a map f and the resulting vector field along f.

to a suitably small set (depending on ε). Of course, for each x the local existence and uniqueness theorems guarantee there is an $\varepsilon > 0$ such that $\phi(x,t)$ exists for all $|t| < \varepsilon$. More details are given in Appendix C.

A.4 Vector fields along a map

It is useful in singularity theory to understand the notion of a vector field *along* a map. Let $f: \mathbb{R}^n \rightarrowtail \mathbb{R}^p$ be a smooth map. A ***vector field along*** f is simply another smooth map $\mathbf{v}: \mathbb{R}^n \rightarrowtail \mathbb{R}^p$ with the same domain as f. However, this second map $\mathbf{v}$ should be interpreted as assigning to each $x \in \mathrm{dom}(f)$ a vector $\mathbf{v}(x)$ which is based at $f(x)$; that is, $\mathbf{v}(x)$ belongs to the tangent space to $\mathbb{R}^p$ at $f(x)$, which can of course be identified with $\mathbb{R}^p$.

Such vector fields arise from perturbing the map f: let f_t be a one–parameter family of maps $f_t: \mathbb{R}^n \rightarrowtail \mathbb{R}^p$. Then for each $x \in \mathbb{R}^n$, let

$$\mathbf{v}(x) = \frac{\mathsf{d}f_t(x)}{\mathsf{d}t}\Big|_{t=0}.$$

This is a vector field along f_0. See Figure A.3.

Tangent map The differential $\mathsf{d}f$ of a map f sends tangent vectors to tangent vectors, while the tangent map $\mathsf{t}f$ maps vector fields on the domain of f to vector fields *along* f. Given a vector field $\mathbf{v}$ on $\mathbb{R}^n$ then $\mathsf{t}f(\mathbf{v})$ is the vector field along f defined by

$$\mathsf{t}f(\mathbf{v})(x) = \mathsf{d}f_x(\mathbf{v}(x)).$$

The special feature of $\mathsf{t}f(\mathbf{v})$ is that at each point $y = f(x)$, the vector $\mathsf{t}f(\mathbf{v})(x)$ is tangent at $f(x)$ to the image of f.

In terms of components, given a vector field $\mathbf{v}(x) = (v_1(x), \ldots, v_n(x))$ we have

$$\mathsf{t}f(\mathbf{v})(x) = \sum_{j=1}^{n} \frac{\partial f}{\partial x_j}(x)\, v_j(x).$$

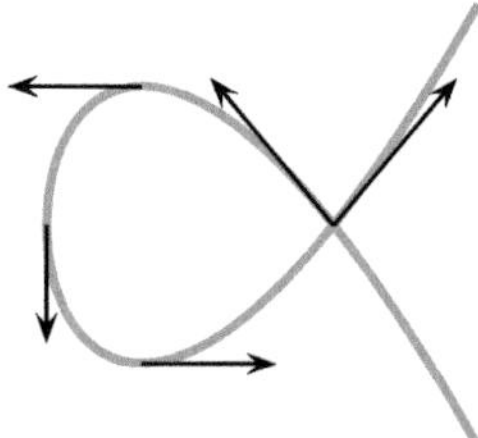

FIGURE A.4 A vector field along a curve in the image of $\mathrm{t}f$; see Example A.12.

And for each $x \in \mathbb{R}^n$ this is a vector in $\mathbb{R}^p$, since f has p components.

Example A.12. Figure A.4 shows the curve $\gamma(t) = (t^2, t^3 - t)$, with a depiction of the vector field along γ given by $\mathrm{t}\gamma(\mathbf{1})$ (the lengths of the vectors are not drawn to scale). Note firstly that this vector field along γ is only defined at points of the curve (the image of γ), and secondly that if a point in the image has more than one preimage, there can be a different vector for each preimage point. ✐

If $\mathbf{v}(x)$ is a vector field on $\mathbb{R}^n$ and ϕ_t the corresponding flow and if $f\colon \mathbb{R}^n \rightarrowtail \mathbb{R}^p$ is a smooth map, then

$$\frac{\mathrm{d}}{\mathrm{d}t}(f \circ \phi_t)\big|_{t=0} = \mathrm{t}f(\mathbf{v}).$$

Note that $f \circ \phi_t$ is a (right-)trivial family in the sense of Chapter 5. The equation above is a special case of Lemma A.14 below.

A third way that vector fields along f arise is described by the following example.

Example A.13. Let $f\colon \mathbb{R}^n \rightarrowtail \mathbb{R}^p$ be a smooth map, and $\mathbf{w}(y)$ a vector field on $\mathbb{R}^p$. Then $\mathbf{v}(x)$ defined by

$$\mathbf{v}(x) = \mathbf{w}(f(x))$$

(i.e., $\mathbf{v} = \mathbf{w} \circ f$) is a vector field along f, denoted by $\mathbf{v} = \omega f(\mathbf{w})$. Thus ωf is a linear map taking vector fields on $\mathbb{R}^p$ to vector fields along f. See Appendix D for further details. ✐

To end this chapter, we prove a lemma (essentially the chain rule) which is used frequently. Here f_s is a 1–parameter family of maps, with S an interval in $\mathbb{R}$, and $\dot{f}_s$ denotes the map $\frac{\mathrm{d}}{\mathrm{d}s} f_s$.

Lemma A.14. *Let $U \subset \mathbb{R}^n$ be an open set, and let $f_s : U \to \mathbb{R}^p$ be a smooth family of smooth maps ($s \in S \subset \mathbb{R}$ as above). If $\phi_s : U \to U$ is another smooth family of smooth maps then,*

$$\frac{\mathsf{d}}{\mathsf{d}s}(f_s \circ \phi_s(x)) = \dot{f}_s(y) + \mathsf{d}(f_s)_y\,(\mathbf{v}_s(y)),$$

where $y = \phi_s(x)$ and $\mathbf{v}_s(y) = \frac{\mathsf{d}}{\mathsf{d}s}\phi_s(x)$.

The right-hand side can be written more succinctly using the tangent map as $\dot{f}_s(y) + \mathsf{t}f_s(\mathbf{v}_s)(y)$, where $y = \phi_s(x)$.

PROOF: To differentiate $f_s \circ \phi_s(x)$ we work from first principles:

$$\begin{aligned}
\frac{\mathsf{d}}{\mathsf{d}s}(f_s \circ \phi_s(x)) &= \lim_{t\to 0} t^{-1}\left[f_{s+t}\circ\phi_{s+t}(x) - f_s\circ\phi_s(x)\right] \\
&= \lim_{t\to 0} t^{-1}\left[f_{s+t}\circ\phi_{s+t}(x) - f_s\circ\phi_{s+t}(x)\right] \\
&\qquad + \lim_{t\to 0} t^{-1}\left[f_s\circ\phi_{s+t}(x) - f_s\circ\phi_s(x)\right] \\
&= \dot{f}_s(\phi_s(x)) + \mathsf{d}(f_s)_{\phi_s(x)}\mathbf{v}_s(\phi_s(x)),
\end{aligned}$$

as required. ✔

One should think of $\mathbf{v}_s(\phi_s(x))$ as a vector field (for each $s \in S$): if we fix x, then we have a curve $\phi_s(x)$ parametrized by s, and its velocity vector (tangent vector) at $y = \phi_s(x)$ is therefore the derivative $\frac{\mathsf{d}}{\mathsf{d}s}\phi_s(x)$. Now, fixing s, $\mathbf{v}_s(\phi_s(x))$ can be written $\mathbf{v}_s(y)$, and since ϕ_s is a diffeomorphism, so $\mathbf{v}_s(y)$ is defined for each y. That is, $\mathbf{v}_s$ is a vector field defined on an open set in $\mathbb{R}^n$.

Problems

A.1 Show that the function h defined in (A.6) is smooth and has zero Taylor series at the origin. (†)

A.2 Let $\mathfrak{B}$ be the vector space of all bilinear forms on V, and $\mathfrak{Q}$ the vector space of all quadratic forms.

(i). Given $B \in \mathfrak{B}$, define B_s and B_a by

$$B_s(\mathbf{u},\mathbf{v}) = \tfrac{1}{2}\left(B(\mathbf{u},\mathbf{v}) + B(\mathbf{v},\mathbf{u})\right), \quad B_a(\mathbf{u},\mathbf{v}) := \tfrac{1}{2}\left(B(\mathbf{u},\mathbf{v}) - B(\mathbf{v},\mathbf{u})\right).$$

Show that B_s is symmetric and B_a skew-symmetric (anti-symmetric), and that

$$\mathfrak{B} = \mathfrak{B}_{\mathfrak{s}} \oplus \mathfrak{B}_{\mathfrak{a}}.$$

(ii). Let $q\colon \mathfrak{B} \to \mathfrak{Q}$ be the linear map associating to any bilinear map the corresponding quadratic form. Show that $\ker q = \mathfrak{B}_a$, and that the map $p\colon \mathfrak{Q} \to \mathfrak{B}$ giving the polarization of a quadratic form is the right inverse to q (ie, $q \circ p = \mathsf{Id}_{\mathfrak{Q}}$); see Equation (A.1).

(iii). Show that under a change of basis with matrix P, a quadratic form with matrix Q transforms to one with matrix $P^T QP$.

A.3 Let $f \in L(\mathbb{R}^n, \mathbb{R}^p)$ be a linear map (so represented by a matrix). Show that $\mathsf{d}f_x = f$ (meaning $\mathsf{d}f_x(\mathbf{u}) = f(\mathbf{u})$ for all $\mathbf{u} \in \mathbb{R}^n$ and all $x \in \mathbb{R}^n$), and that the higher derivatives all vanish: $\mathsf{d}^k f_x = 0$ for $k \geq 2$.

A.4 Let $f(x) = x^3$. Show that for each $a \in \mathbb{R}$,

$$j^2 f_a(u) = a^3 + 3a^2u + 3au^2.$$

Find the corresponding expression for $j^2 f(a)$ when $f(x) = \sin x$. (†)

A.5 Prove Proposition A.10.

A.6 Prove by hand Faà di Bruno's formulae appearing in (A.7).

A.7 Let $\phi_s : \mathbb{R}^2 \to \mathbb{R}^2$ be given by $\phi_s(x, y) = (x + sy,\ y - sx)$ and let $f_s(x, y) = x^2 + sxy + y^2$. Verify directly Lemma A.14 for this example.

B

Local geometry of regular maps

THROUGHOUT THIS TEXT, WE consider maps from $\mathbb{R}^n$ to $\mathbb{R}^p$, and the purpose of this chapter is twofold. Firstly, in order to understand singularities one should first understand non–singularities – that is points where the rank of the Jacobian matrix of the map takes its maximal possible value. The local geometry in the neighbourhood of such a point is governed by the inverse function theorem and its brethren such as the local immersion and local submersion theorems. Secondly the inverse function theorem is fundamental to singularity theory through the resulting changes of coordinates. The proof we give (in Chapter 5) of the inverse function theorem involves existence and uniqueness theorems for ordinary differential equations, and the resulting idea of flows, all of which is covered in Appendix C. The principal reason for giving the proof is that one of the main theorems of the subject (the finite determinacy theorem of Chapter 5) has an analogous, if slightly more complex, proof.

We use the notation $f\colon \mathbb{R}^n \rightarrowtail \mathbb{R}^p$ to mean that f is a map whose domain is an open subset of $\mathbb{R}^n$.

Definition B.1. Let $f\colon \mathbb{R}^n \rightarrowtail \mathbb{R}^p$ be a smooth map. The ***rank of the map*** at a point x is defined to be the rank of its Jacobian matrix $\mathsf{d}f_x$ at that point, and denoted $\mathrm{rk}_x(f)$. ✯

The Jacobian matrix $\mathsf{d}f_x$ is a $p \times n$ matrix so $\mathrm{rk}_x(f) \leq \min\{n, p\}$. This chapter is principally about the structure of maps f near points x where $\mathrm{rk}_x(f) = \min\{n, p\}$.

B.1 Changes of coordinates and diffeomorphisms

In linear algebra, a change of basis is implemented by an invertible linear map. If $\{\mathbf{e}_1, \ldots, \mathbf{e}_n\}$ is a basis for V, and $A \in \mathsf{GL}(n)$, then $\{\mathbf{e}'_1, \ldots, \mathbf{e}'_n\}$ with $\mathbf{e}'_i = A\mathbf{e}_i$ is another basis (because A is invertible). The coordinates of a point (vector) $\mathbf{v} \in V$ with respect to the basis $\{\mathbf{e}_i\}$ are the coefficients x_i appearing in the expression

$\mathbf{v} = \sum x_i \mathbf{e}_i$. In changing to the new basis the coordinates are also changed, to the coefficients x'_i in the expression $\mathbf{v} = \sum x'_i \mathbf{e}'_i$. And the two are related by $\mathbf{x} = A\mathbf{x}'$, that is $x_i = \sum_j a_{ij} x'_j$, so $x'_i = \sum_j b_{ij} x_j$, where $B = A^{-1}$.

On the other hand, when we are working with smooth maps and ignoring the linear structure on $\mathbb{R}^n$, we don't use bases so much as just the coordinates. This is because it is useful to consider nonlinear changes of coordinates which do not respect the nature of bases. Nonlinear changes of coordinates should be familiar from multiple integrals, for example changing from Cartesian coordinates to polar coordinates.

To define coordinates systems on $\mathbb{R}^n$, or on an open set $U \subset \mathbb{R}^n$, one can start with the usual Cartesian one on $\mathbb{R}^n$ and then say what constitutes a change of coordinates. The fundamental notion is the *diffeomorphism*.

Definition B.2. Let $U, V \subset \mathbb{R}^n$ be open sets. A smooth map $f\colon U \to V$ is a ***diffeomorphism*** if it has a smooth inverse; that is, if there is a smooth map $g\colon V \to U$ such that $f \circ g = \mathsf{Id}_V$ and $g \circ f = \mathsf{Id}_U$ (here Id_U is the identity map on U, and Id_V that on V). In this case one writes $g = f^{-1}$. ✯

Examples B.3.

(i). Let $f\colon \mathbb{R} \to \mathbb{R}$, $f(x) = x^2$. This is not invertible so is certainly not a diffeomorphism.

(ii). Restrict this map to $f\colon (0,\infty) \to (0,\infty)$, with again $f(x) = x^2$. Now f *is* invertible, with $f^{-1}(y) = \sqrt{y}$, and both f and f^{-1} are smooth (note that $0 \notin \mathrm{dom}(f^{-1})$).

(iii). Let $g\colon \mathbb{R} \to \mathbb{R}$ be defined by $g(x) = x^3$. This is invertible, with $g^{-1}(y) = y^{1/3}$, but g^{-1} is not differentiable at $y = 0$ so g is not a diffeomorphism. [This map $g(x) = x^3$ is an example of a *homeomorphism* which is not a diffeomorphism: it is a continuous and invertible map and its inverse is also continuous; however, its inverse is not differentiable.]

(iv). Let $U \subset \mathbb{R}^+ \times (-\pi, \pi)$ be any open set, where $\mathbb{R}^+$ is the set of strictly positive real numbers, and define

$$\phi(r,\theta) = (r\cos\theta,\ r\sin\theta).$$

Then ϕ is a diffeomorphism of U with $\phi(U)$ – the familiar change of coordinates between polar and Cartesian. Note that if one includes $r = 0$ in U then the map is defined but fails to be a diffeomorphism; indeed it fails to have an inverse as $\phi(0,\theta) = (0,0)$ for all θ.

(v). Any invertible linear map on $\mathbb{R}^n$ is a diffeomorphism, because both the map and its inverse are smooth. If we choose bases so the linear map is represented by a matrix, then the inverse map is represented by the inverse matrix. ✐

Returning to coordinate systems, suppose we start out with a given coordinate system on an open set $U \subset \mathbb{R}^n$ (possibly the usual Cartesian one), then a new coordinate system $\{x'_1, \ldots, x'_n\}$ is related to the original one by $x'_i = \phi_i(x_1, \ldots, x_n)$, for for some functions ϕ_i with $i = 1, \ldots, n$. These components define a map $\phi\colon U \to \mathbb{R}^n$ by

$$\phi(x_1, \ldots, x_n) = (\phi_1(x_1, \ldots, x_n), \ \ldots, \ \phi_n(x_1, \ldots, x_n)).$$

What properties should this map ϕ have to be a change of coordinates? Firstly, it must be invertible, otherwise we cannot change back to the original coordinates. Moreover ϕ and ϕ^{-1} should both be differentiable (or better, smooth), otherwise a function which is differentiable (or smooth) in one coordinate system might not be in the other system (for an example, see Problem B.3). In short, ϕ must be a diffeomorphism. In Example B.3(iv) the diffeomorphism ϕ is the familiar change of coordinates from polar to Cartesian.

Thus a change of coordinates *is* a diffeomorphism, and vice versa: the new coordinates are the components of the diffeomorphism. That is, if we call the new coordinates $(y_1, \ldots, y_n)$ then

$$y_i = \phi_i(x_1, \ldots, x_n).$$

We will use the expression ***coordinates about a point*** q to mean coordinates defined on a neighbourhood of q, whose components are all equal to 0 at the point q.

One should think about coordinate systems and diffeomorphisms as follows, and this becomes particularly useful when we define submanifolds below.

Let $U \subset \mathbb{R}^n$ be an open set and $q \in U$. Now consider another copy of $\mathbb{R}^n$ with its usual Cartesian coordinates $(x_1, \ldots, x_n)$. A system of coordinates on U is then a map $\phi\colon U \to \mathbb{R}^n$ which is a diffeomorphism with its image $\phi(U)$, the coordinates being the components of ϕ; see Figure B.1. Thus a single coordinate, ϕ_1 say, is just a function on U, $\phi_1\colon U \to \mathbb{R}$, and is often written $x_1 \circ \phi$ since x_1 is a coordinate function on $\mathbb{R}^n$. Furthermore, ϕ defines a coordinate system *about* q if $\phi(q) = 0$.

In the next section we see how to determine whether a given smooth map is a diffeomorphism, at least in some neighbourhood of a given point. Before doing that, we give one final definition.

Definition B.4. Two subsets S_1, S_2 of $\mathbb{R}^n$ are ***diffeomorphic*** if there are neighbourhoods U_1 of S_1 and U_2 of S_2 and a diffeomorphism $\phi\colon U_1 \to U_2$ which maps S_1 to S_2. ★

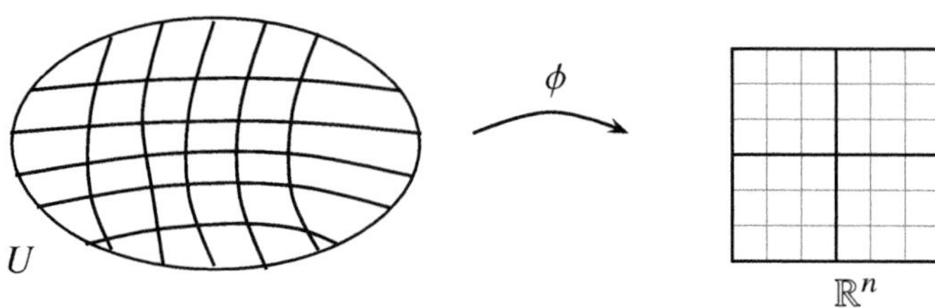

FIGURE B.1 A coordinate system on U

Maps and changes of coordinates Let $f\colon \mathbb{R}^n \rightarrowtail \mathbb{R}^p$ be a (smooth) map. The expression for the map will depend on the coordinates used on its domain. For example the function $f\colon \mathbb{R}^2 \to \mathbb{R}$ given in Cartesian coordinates by $f(x,y) = x^2 - y^2$, becomes in polar coordinates (see Example B.3(iv)) $f \circ \phi(r,\theta) = r^2 \sin 2\theta$. Here ϕ maps *from* $\mathbb{R}^+ \times (-\pi/\pi)$ *to* the domain of f.

We would often express this in reverse: if $U \subset \mathrm{dom}(f)$, then new coordinates on U would be defined by a diffeomorphism $\phi\colon U \to V$. Then the new expression for f would then be $f(x) = f \circ \phi^{-1}(y)$.

B.2 Inverse function theorem

Many of the methods of singularity theory involve calculations using infinitesimal data (differentials and vector fields) and making deductions about maps. The inverse function theorem is the archetype of such results, and not only is it a result of central importance, but the proof we give (in Chapter 5) provides a model for the proofs of several important theorems in this text and so is worthwhile understanding in detail.

Before stating the theorem, we show its converse: *if U and V are open sets in $\mathbb{R}^n$ and $f\colon U \to V$ is a diffeomorphism, then the differential $\mathsf{d}f_x$ is invertible for all $x \in U$.*

This is not surprising: if there is any justice in mathematics (and there is!), then the best linear approximation to an invertible map ought to be invertible. Indeed, since f is invertible, let $g = f^{-1}\colon V \to U$ be its inverse, so that $g \circ f\colon U \to U$ is the identity map on U. Applying the chain rule gives

$$\mathsf{d}g_y\, \mathsf{d}f_x = \mathsf{Id}_n,$$

where $y = f(x)$ and Id_n is the $n \times n$ identity matrix, so $[\mathsf{d}f_x]^{-1} = \mathsf{d}g_y$.

The inverse function is the converse of this observation. It says that if the best linear approximation (the differential) to a smooth map at a particular point is invertible, then so is the map itself – at least in some neighbourhood of that point. More formally:

Theorem B.5 (Inverse function theorem). *Let $f\colon \mathbb{R}^n \rightarrowtail \mathbb{R}^n$ be a smooth map with $x_0 \in \mathrm{dom}(f)$, and suppose f has rank n at x_0. Then there is a neighbourhood U of x_0 such that the restriction $f_{|U}\colon U \to f(U)$ is a diffeomorphism.*

The statement that f has rank n at x_0 is equivalent to saying that the Jacboian matrix $\mathsf{d}f_{x_0}$ is invertible. In most texts, this theorem is proved using the contraction mapping principle. We prove it using the 'homotopy method' in Chapter 5 (p. 72).

Remark B.6. Since diffeomorphisms are equivalent to smooth changes of coordinates, the statement of the theorem can also conclude that there is a change of coordinates in, say, the target, such that the map takes the form

$$f(x_1, \ldots, x_n) = (x_1, \ldots, x_n).$$

Namely, the change of coordinates is f^{-1}, for of course $f^{-1} \circ f = \mathsf{Id}$. Likewise, if one fixed the coordinates in the target, there is a change of coordinates in the source such that f takes the same form, since in this case $f \circ f^{-1} = \mathsf{Id}$.

Such choices of coordinates are a particular case of ***linearly adapted coordinates***; we will see more of this idea below. ❞

Example B.7. Let $f\colon \mathbb{R}^2 \to \mathbb{R}^2$ be the smooth map

$$(u_1, u_2) = f(x_1, x_2) = (x_1 + x_2^2,\ x_2 + x_1^2).$$

Then at the origin $\mathsf{d}f_{(0,0)} = \mathsf{Id}$ which is invertible. Consequently there is a neighbouhood U of the origin in the source such that $f\big|_U\colon U \to f(U)$ is a diffeomorphism.

Moreover, if we let $y_1 = x_1 + x_2^2$ and $y_2 = x_2 + x_1^2$ then y_1, y_2 defines a coordinate system in a neighbourhood of $(0,0)$ in the source, and with these coordinates $f(y_1, y_2) = (y_1, y_2)$.

Likewise, if we let v_1, v_2 be such that $u_1 = v_1 + v_2^2$ and $u_2 = v_2 + v_1^2$ then v_1, v_2 defines a coordinate system in a neighbourhood of $(0,0)$ in the target, and with these coordinates $f(x_1, x_2) = (x_1, x_2)$.

Note that to find v_1, v_2 in terms of u_1, u_2, involves solving the equations which in general is not possible. ✐

B.3 Immersions & submersions

The inverse function theorem describes the local structure of a map $f\colon \mathbb{R}^n \rightarrowtail \mathbb{R}^p$ of maximal rank when $n = p$. There are similar descriptions for maps of maximal rank when $n \neq p$, and these are derived from the inverse function theorem. We turn to these now: first we consider $n < p$ (immersions) and then below $n > p$ (submersions).

B.3A Immersions

Definition B.8. A smooth map $f\colon \mathbb{R}^n \rightarrowtail \mathbb{R}^p$ is an ***immersion at*** x if f has rank n at that point. The map is an ***immersion*** if it is an immersion at every point in $\mathrm{dom}(f)$. ✯

Of course, this is only possible if $n \leq p$, and if $n = p$ then the property is precisely that of being a local diffeomorphism. See Figure B.2 for an example: notice that f is not 1–1 as there are two points mapping to the same point of intersection.

Example B.9. The graph of any smooth map $f\colon \mathbb{R}^n \rightarrowtail \mathbb{R}^p$ defines an immersion

$$g\colon \mathrm{dom}(f) \hookrightarrow \mathbb{R}^n \times \mathbb{R}^p,$$

given by $g(x) = (x, f(x))$. The proof is an exercise (see Problem B.13). ✐

In fact the graph is more than an immersion, it is an example of an ***embedding***: an immersion $f\colon N \to P$ is an embedding if (i) it is 1–1, and (ii) for every open set U in N there is an open set $V \subset P$ for which $U = f^{-1}(V)$. Another way to say this second condition is (ii)′ if (x_j) is a sequence in N such that $f(x_j) \to f(q)$ for some $q \in N$ then $x_n \to q$, but we will not pursue this further here. The map depicted in Figure B.2 is an immersion but not an embedding, as it violates condition (i) (see Problem B.21 for an example of a 1-1 immersion which violates condition (ii)). As was mentioned earlier, being an immersion is a local property. On the other hand, being an embedding is definitely not local.

Theorem B.10 (Local immersion theorem)**.** *Let $f\colon \mathbb{R}^n \rightarrowtail \mathbb{R}^p$ be a smooth map with $n < p$. Suppose that f is an immersion at $x \in \mathrm{dom}(f)$. Then there is a neighbourhood U of x and local coordinates in the target about $f(x)$ such that*

$$f(x_1, \ldots, x_n) = (x_1, \ldots, x_n, 0, \ldots, 0).$$

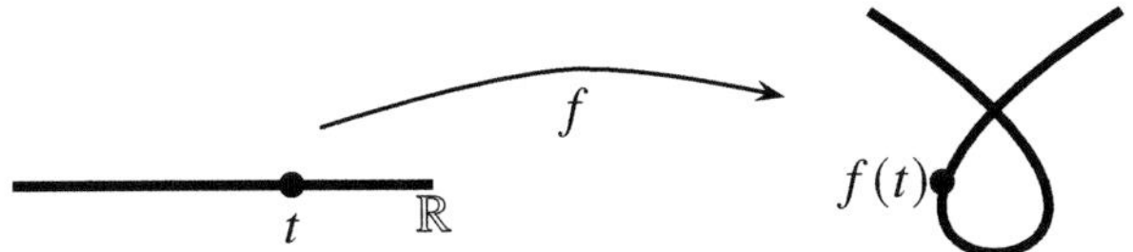

FIGURE B.2 An immersion: $f(t) = (t^3 - t,\ t^2)$.

PROOF: The proof is as follows, but the details are left to the reader. First permute the basis vectors of $\mathbb{R}^p$ so that

$$\mathrm{d}f_x = \begin{bmatrix} A \\ B \end{bmatrix},$$

where A is an invertible $n \times n$ matrix (this is done by permuting the rows of $\mathrm{d}f_x$ so that the first n rows are linearly independent). Next define $F\colon \mathbb{R}^n \rightarrowtail \mathbb{R}^n$ by

$$F(x_1, \ldots, x_n, y_1, \ldots, y_{p-n}) = f(x_1, \ldots, x_n) + (0, \ldots, 0, y_1, \ldots, y_{p-n}).$$

It is easy to check that F satisfies the hypotheses of the inverse function theorem. Finally, one changes coordinates on $\mathbb{R}^p$ using F^{-1}. ✔

B.3B Submersions

The other major theorem on regular behaviour of maps is the submersion theorem, and the essentially equivalent implicit function theorem. We deduce this theorem from the inverse function theorem, although it can also be proved directly using the homotopy method.

Definition B.11. A smooth map $f\colon \mathbb{R}^n \rightarrowtail \mathbb{R}^p$ is a ***submersion at*** x if f has rank p at that point. It is a ***submersion*** if it is a submersion at every point in its domain $\mathrm{dom}(f)$. ✩

Example B.12. The projection $f\colon \mathbb{R}^p \times \mathbb{R}^k \to \mathbb{R}^p$ given by $f(x, y) = x$ is a submersion. ✎

In fact this example is the archetype of a submersion, at least locally, as the next theorem shows.

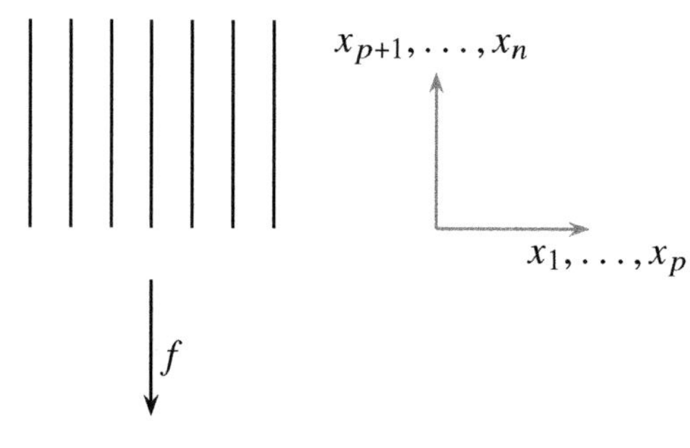

FIGURE B.3 A submission

Theorem B.13 (Local submersion theorem). *Let $f \colon \mathbb{R}^n \rightarrowtail \mathbb{R}^p$ be a smooth map with $n > p$, and suppose f is a submersion at $x_0 \in \mathrm{dom}(f)$, and $y_0 = f(x_0)$. Given any system of coordinates about y_0 in $\mathbb{R}^p$, there is a neighbourhood U of x_0 and coordinates about x_0 for which f takes the form*

$$f(x_1, \dots, x_p, x_{p+1}, \dots, x_n) = (x_1, \dots, x_p).$$

That is, a submersion is locally a projection; see Figure B.3. Notice moreover that the image of any submersion is open. This theorem, like the next, follows from the extended implicit function theorem – the proof is given below.

In terms of applications, the implicit function theorem is the most important theorem of this appendix. A great deal of information about the theorem, its history and its applications can be found in [66].

Theorem B.14 (Implicit function theorem). *Let $f \colon \mathbb{R}^p \times \mathbb{R}^k \rightarrowtail \mathbb{R}^p$ be a smooth map with $f(x_0, y_0) = z_0$, and suppose $\mathsf{d}f_{(x_0,y_0)}$ has the form*

$$\mathsf{d}f_{(x_0,y_0)} = [A \;\; B], \tag{B.1}$$

where A is an invertible $p \times p$ matrix, and B is any $p \times k$ matrix. Then there are neighbourhoods U of x_0 and V of y_0 in $\mathbb{R}^p$ and $\mathbb{R}^k$ respectively, and a smooth map $h \colon V \to U$, with $h(y_0) = x_0$, such that, for $(x, y) \in U \times V$,

$$f(x, y) = z_0 \iff x = h(y).$$

In other words, the equation $f(x, y) = z_0$ can (in principle) be solved for x as a function of y. This function h is the eponymous 'implicit function'. It is unusual to be able to find an explicit form for this implicit function, although its Taylor series to any order can be computed using implicit differentiation.

Example B.15. As a very simple example, consider the equation $x^3 - xy - y^4 = 5$. One sees that the point $(x, y) = (2, 1)$ satisfies the equation. Can the set of solutions near to this point be written as x being a function of y? Answer: yes! Because $\frac{\partial f}{\partial x}(2, 1) \neq 0$. And because $\frac{\partial f}{\partial y}(2, 1) \neq 0$, it can also be solved (locally) for y as a function of x. See Problem B.9. ✐

Both the submersion theorem and the implicit function theorem follow from the inverse function theorem, via the following result, which has a pleasing symmetry about its conclusion.

Theorem B.16 (Extended implicit function theorem). *Assume f satisfies the hypothesis of the implicit function theorem above. Then there are neighbourhoods U, V, W of x_0, y_0 and z_0 in $\mathbb{R}^p$, $\mathbb{R}^k$ and $\mathbb{R}^p$ respectively and a smooth map $g\colon V \times W \to U$, satisfying for $(x, y, z) \in U \times V \times W$,*

$$f(x, y) = z \iff g(z, y) = x.$$

Proof: Define a map $F\colon \mathbb{R}^p \times \mathbb{R}^k \rightarrowtail \mathbb{R}^p \times \mathbb{R}^k$ by

$$F(x, y) = (f(x, y), y).$$

The Jacobian matrix of F at (x_0, y_0) is

$$\mathsf{d}F_{(x_0,y_0)} = \begin{bmatrix} A & B \\ 0 & I_k \end{bmatrix}.$$

This is invertible, with inverse $\begin{bmatrix} A^{-1} & -A^{-1}B \\ 0 & I_k \end{bmatrix}$ (as is readily checked), so by the inverse function theorem there is a neighbourhood U_1 of (x_0, y_0) (which we can take to be of the form $U \times V$) and a map $G\colon W \times V \to U$, where $W = f(U \times V)$ which is a neighbourhood of z_0, satisfying $F \circ G = \mathsf{Id}_V$. Moreover G has the form

$$G(z, y) = (g(z, y), y)$$

for some smooth map $g\colon V \to \mathbb{R}^p$ (exercise: prove this). Since $G = F^{-1}$ it follows that

$$F(x, y) = (z, y) \iff (x, y) = G(z, y),$$

and this is equivalent to $f(x, y) = z \iff x = g(z, y)$ as required. ✔

This extended implicit function theorem can be viewed as a parametrized version of the inverse function theorem: here y is the parameter, and for each value of y the maps f_y and g_y are mutually inverse. From this theorem we can derive the local submersion and the (ordinary) implicit function theorems as follows.

PROOF OF THEOREM B.13: As a first step, permute the columns of $\mathsf{d}f_{x_0}$ so that the first p columns are linearly independent; this amounts to permuting the basis vectors in $\mathbb{R}^n$. Then $\mathsf{d}f_{x_0}$ has the form (B.1), so we can apply the extended implicit function theorem, and write $x_0 = (u_0, y_0) \in \mathbb{R}^p \times \mathbb{R}^k$ where $k = n - p$.

The map $F(u, y) = (f(u, y), y)$ is therefore a diffeomorphism in a neighbourhood of x_0 (cf. the proof of Theorem B.16). Consider the change of coordinates F^{-1}. Then $f \circ F^{-1}$ has the required form. ✔

PROOF OF THEOREM B.14: Here one just defines $h(y) = g(z_0, y)$. ✔

Definition B.17. The k–dimensional ***suspension*** of a map $f : \mathbb{R}^n \to \mathbb{R}^p$ is the map

$$\begin{array}{ccc} \mathbb{R}^n \times \mathbb{R}^k & \longrightarrow & \mathbb{R}^p \times \mathbb{R}^k \\ (x, y) & \longmapsto & (f(x), y). \end{array}$$

☆

It is easy to see that if f is an immersion, submersion or diffeomorphism, then so, correspondingly, is any suspension of f.

B.4 Submanifolds and local straightening

The idea of submanifolds is a crucial concept in singularity theory, as well as in many other branches of mathematics such as topology, geometry, differential equations, classical mechanics and many more besides.

Definition B.18. A subset $M \subset \mathbb{R}^n$ is a ***submanifold*** if there is an integer d (called the ***dimension*** of M) such that for each point $q \in M$ there is a neighbourhood U of q in $\mathbb{R}^n$ and a *diffeomorphism* $\Phi : U \to V$, where $V \subset \mathbb{R}^n$ is an open set, such that

$$\Phi(M \cap U) = (\mathbb{R}^d \times \{0\}) \cap V \subset \mathbb{R}^d \times \mathbb{R}^{n-d}.$$

The diffeomorphism Φ is called a ***local straightening map*** for M at q (or in a neighbourhood of q). One also says that M is of ***codimension*** $(n - d)$ in $\mathbb{R}^n$. ☆

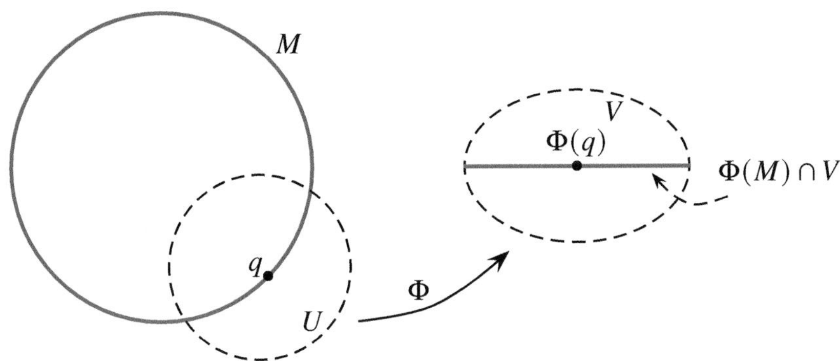

FIGURE B.4 A local straightening.

It is worth emphasizing that the straightening map Φ must be a diffeomorphism, and U and V are both open sets in $\mathbb{R}^n$. Such a submanifold is often referred to as an *embedded* submanifold, as distinct from an immersed submanifold, which we'll encounter later in this appendix.

Examples B.19 (of submanifolds).

(i). Let $M = \{(x, y) \in \mathbb{R}^2 \mid y = x^2\}$ (the parabola in the plane). In this case we can define a single *global* straightening map $\Phi\colon \mathbb{R}^2 \to \mathbb{R}^2$ by

$$\Phi(x, y) = (x, y - x^2).$$

Then $\Phi(M) = \mathbb{R} \times \{0\}$, showing that the parabola is a submanifold of dimension 1 (and codimension 1).

(ii). Consider the unit circle in the plane, centre the origin, usually denoted $S^1 \subset \mathbb{R}^2$. This needs more than one straightening map: for $q = (q_1, q_2) \in S^1$, define

- if $q_2 > 0$ put $\Phi(x, y) = (x, y - \sqrt{1 - x^2})$, which is a diffeomorphism for $U = \{(x, y) \in S^1 \mid y > 0\}$;
- if $q_2 < 0$ put $\Phi(x, y) = (x, y + \sqrt{1 - x^2})$, which is a diffeomorphism for $U = \{(x, y) \in S^1 \mid y < 0\}$;
- if $q_1 > 0$ put $\Phi(x, y) = (x - \sqrt{1 - y^2}, y)$, which is a diffeomorphism for $U = \{(x, y) \in S^1 \mid x > 0\}$;
- if $q_1 < 0$ put $\Phi(x, y) = (x + \sqrt{1 - y^2}, y)$, which is a diffeomorphism for $U = \{(x, y) \in S^1 \mid x < 0\}$.

FIGURE B.5 None of these sets are submanifolds: the figure–8, the semicubical parabola (cusp) and the cone; each has one singular point.

These four straightening maps together cover the entire circle, which is enough to show it is a submanifold (of dimension 1). ✐

The following result is often used to show a subset is a submanifold. Recall that the graph of a map $f\colon X \to Y$ is the set

$$\Gamma_f = \{(x, y) \in X \times Y \mid y = f(x)\}.$$

Proposition B.20. *Let $f\colon \mathbb{R}^n \to \mathbb{R}^p$ be a smooth map. The graph Γ_f of f is a submanifold of $\mathbb{R}^n \times \mathbb{R}^p$ of dimension n.*

PROOF: Γ_f has a *global* straightening map, $\Phi\colon \mathbb{R}^n \times \mathbb{R}^p \to \mathbb{R}^n \times \mathbb{R}^p$ defined simply by $\Phi(x, y) = (x, y - f(x))$. This is a diffeomorphism, as its inverse is

$$\Phi^{-1}(x, z) = (x, z + f(x)),$$

as can be readily checked, and both Φ and Φ^{-1} are smooth. Moreover,

$$\Phi(x, f(x)) = (x, f(x) - f(x)) = (x, 0),$$

so $\Phi(\Gamma_f) = \mathbb{R}^n \times \{0\}$, as required. ✔

Clearly in this setting, the codimension of the graph is $\operatorname{codim}(\Gamma_f) = p$.

Local parametrizations One of the main features of manifolds is they have (local) parametrizations, or coordinates. Let $M \subset \mathbb{R}^n$ be a submanifold of dimension d and let $p \in M$, and let $\Phi\colon U \to \mathbb{R}^n$ be a local straightening map defined in a neighbourhood of p. Write $\Phi(x) = (\Psi(x), \chi(x))$, so $\Psi(x) \in \mathbb{R}^d$, and $\chi(x) = 0$ if and only if $x \in M \cap U$.

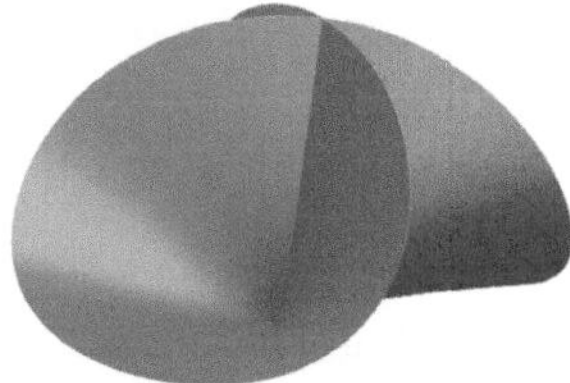

FIGURE B.6 Two views of the Cayley cross–cap – not a submanifold: it has a half–line of singular points. It is the image of the map $(x, y) \mapsto (x, xy, y^2)$ which fails to be an immersion only at $(0, 0)$.

If we restrict to $x \in M$, then the restriction $\Psi\colon M \cap U \to \mathbb{R}^d$ defines a smooth map whose inverse is $\Psi^{-1}(y) = \Phi^{-1}(y, 0)$. Thus $\Psi^{-1}\colon U' \to M$ is a smooth map of rank d whose image is a neighbourhood of x in M; here $U' = \Psi(M \cap U) \subset \mathbb{R}^d$. If we use coordinates $(x_1, \ldots, x_d)$ on U', then $\Psi^{-1}(x_1, \ldots, x_d)$ gives a smooth parametrization of a neighbourhood of x in M. Although (at this point) we have very little explicit information on what M can be like, this local parametrization allows us to define what is meant for a function on M to be *smooth*, namely $f\colon M \to \mathbb{R}$ is ***smooth*** if and only if $f \circ \Psi^{-1}\colon U' \to \mathbb{R}$ is smooth (see Problem B.16).

Example B.21. Continuing part (ii) of the previous example, with $M = S^1$ and $q = (0, 1)$, we have $\Psi(x, y) = x$, and since on M we have $x = \sqrt{1 - y^2}$, it follows that $\Psi^{-1}(x) = (x, \sqrt{1 - x^2})$, which is a parametrization of the upper half of the circle S^1 (i.e., of that part with $y > 0$). Moreover, the function $f(x, y) = \sin(y)$ is smooth on the portion of the circle parametrized by Ψ^{-1} because $f \circ \Psi^{-1}(x) = \sin(\sqrt{1 - x^2})$ which is smooth provided $|x| < 1$. ✐

Remark B.22. Readers familiar with the abstract definition of manifold will recognize the local parametrizations as the basis of that definition; indeed the map $\Psi|_M$ defines a local coordinate chart. The compatibility between different coordinate charts follows automatically here, since the composition of two diffeomorphisms is a diffeomorphism. ❞

Immersions and submersions If $f\colon \mathbb{R}^k \rightarrowtail \mathbb{R}^n$ is an immersion, the local immersion theorem tells us that locally (in $\mathbb{R}^k$) the image of f is a submanifold of $\mathbb{R}^n$ of dimension k. If on the other hand $f\colon \mathbb{R}^n \twoheadrightarrow \mathbb{R}^p$ is a submersion, then the local submersion theorem implies that all the level sets of f (the subsets of the form $f^{-1}(y)$) are submanifolds of dimension $n - p$; indeed, the reader can check that the change of coordinates in the definition provides the local straightening maps. It is useful to formalize this as the *Regular Value Theorem* – a regular value of a map f is a point y in the image for which f is a submersion at every $x \in f^{-1}(y)$.

Theorem B.23 (Regular value theorem). *Let $f\colon \mathbb{R}^n \rightarrowtail \mathbb{R}^p$, with $n > p$, be a smooth map, and $y \in \mathbb{R}^p$ a regular value of f. Then $f^{-1}(y)$ is a submanifold of $\mathbb{R}^n$ of codimension p, and hence dimension $n - p$.*

The proof follows from the local submersion theorem, and details are left to the reader.

Tangent spaces Let $X \subset \mathbb{R}^n$ be a submanifold with $x \in X$. Let $u\colon \mathbb{R} \rightarrowtail X$ be a smooth parametrized curve in X with $u(0) = x$. Then the vector $\dot{u}(0) = \frac{\mathsf{d}}{\mathsf{d}t}u(0)$ is called a ***tangent vector*** to X at x. The set of all such vectors is the ***tangent space*** of X at x written,

$$T_xX = \{\dot{u}(0) \mid u\colon \mathbb{R} \rightarrowtail X \text{ with } u(0) = x\}.$$

Proposition B.24. *Let X be a submanifold of $\mathbb{R}^n$ of dimension k and let $x \in X$.*

(i). If $g\colon \mathbb{R}^k \rightarrowtail \mathbb{R}^n$ is an immersion with image X then the tangent space

$$T_xX = \operatorname{im} \mathsf{d}g_q,$$

where $x = g(q)$.

(ii). If $f\colon \mathbb{R}^n \rightarrowtail \mathbb{R}^p$ is a submersion with $X = f^{-1}(0)$ (where $p = n - k$) then

$$T_xX = \ker \mathsf{d}f_x.$$

PROOF: (i) Let $v\colon \mathbb{R} \rightarrowtail \mathbb{R}^k$ be a path with $v(0) = q$. Then $u = g \circ v$ is a path in X, and its velocity vector is $\dot{u} = \mathsf{d}g_q\dot{v}$ so that $\dot{u} \in \operatorname{im} \mathsf{d}g_q$. Furthermore, any curve in X through x can be obtained in this way, so that indeed $T_xX = \operatorname{im} \mathsf{d}g_q$.
(ii) Let u be a path in X. Then $f \circ u \equiv 0$. Differentiating with respect to t gives $\mathsf{d}f_x\dot{u}(0) = 0$. Thus $T_xX \subset \ker \mathsf{d}f_x$. However, these two spaces have the same dimension, namely $k = n - p$, so they must be equal. ✔

B.5 Example: the set of matrices of given rank

Consider the vector space $\mathsf{Mat}(p, n)$ of $p \times n$ matrices (representing linear maps $\mathbb{R}^n \to \mathbb{R}^p$). Let $\Delta_r = \Delta_r(p, n) \subset \mathsf{Mat}(p, n)$ be the subset consisting of those matrices of rank equal to r.

Theorem B.25. *The subset* $\Delta_r(p,n) \subset \mathsf{Mat}(p,n)$ *is a submanifold of dimension* $r(n+p-r)$.

PROOF: Let $A \in \Delta_r$, and choose a basis so that A takes the form given in Proposition A.2. Now any matrix $A' \in \mathsf{Mat}(p,n)$ can be written in block form as

$$A' = \begin{bmatrix} I_r + B & C \\ D & E \end{bmatrix},$$

where I_r is the $r \times r$ identity matrix, $B \in \mathsf{Mat}(r,r)$, $C \in \mathsf{Mat}(r,n-r)$, $D \in \mathsf{Mat}(p-r,r)$ and $E \in \mathsf{Mat}(p-r,n-r)$. Choose a neighbourhood U of A by assuming that $B \in U$ is sufficiently small that $I_r + B$ is invertible. We now apply row operations on A', to show

$$\mathsf{A}' \sim \begin{bmatrix} I_r & (I_r+B)^{-1}C \\ D & E \end{bmatrix} \sim \begin{bmatrix} I_r & (I_r+B)^{-1}C \\ 0 & E - D(I_r+B)^{-1}C \end{bmatrix}.$$

This last matrix is of rank r if and only if $E = D(I_r+B)^{-1}C$. That is, in the neighbourhood U, the set Δ_r can be expressed as a graph of E as a function of B, C, D, and so (by Proposition B.20) Δ_r is indeed a submanifold of dimension $\dim \mathsf{Mat}(r,r) + \dim \mathsf{Mat}(r,n-r) + \dim \mathsf{Mat}(p-r,r)$, which is equal to $r(n+p-r)$. ✔

Notice in the proof that the rank r condition is written as a condition on E, which can be interpreted as saying that Δ_r has codimension $(n-r)(p-r)$ (equal to the number of entries in E). See Problems B.19 and B.20 for information about the tangent space to Δ_r.

B.6 Maps of constant rank

Let $f\colon \mathbb{R}^n \rightarrowtail \mathbb{R}^p$ be a smooth map. Since the entries of the matrix $\mathsf{d}f_x$ depend continuously on the point $x \in \mathrm{dom}(f)$, it follows that for each $r \in \mathbb{N}$ the set

$$S_r(f) := \{x \in \mathrm{dom}(f) \mid \mathrm{rk}_x(f) \le r\}$$

is a *closed* subset of $\mathrm{dom}(f)$. Equivalently, the rank function $x \mapsto \mathrm{rk}_x(f)$ is *upper semicontinuous*, meaning that if x_j is a sequence of points in $\mathrm{dom}(f)$ converging to x_0, then

$$\lim_{j\to\infty} \mathrm{rk}_{x_j}(f) \ge \mathrm{rk}_{x_0}(f),$$

if the limit exists.

In all the theorems above, the rank of the map is maximal, and it follows therefore that it is constant in a neighbourhood of the point in question. There is a more general theorem which extends all the previous ones.

Theorem B.26 (Constant rank theorem)**.** *Let $f\colon \mathbb{R}^n \rightarrowtail \mathbb{R}^p$ be a smooth map of constant rank k, and let $x_0 \in \operatorname{dom}(f)$. Then there is a neighbourhood of x_0 and coordinates $x_1, \ldots, x_n$ about x_0 and $y_1, \ldots, y_p$ around $f(x_0)$ such that in these coordinates f is given by*

$$f(x_1, \ldots, x_n) = (x_1, \ldots, x_k, 0, \ldots, 0).$$

Notice that one consequence is that image(f) is a k–dimensional immersed submanifold of $\mathbb{R}^p$ and each level set $f^{-1}(y)$ is a submanifold of $\mathbb{R}^n$ of dimension $n-k$.

This theorem contains the previous three as special cases (although its proof uses the implicit function theorem, which in turn relies on the inverse function theorem). We will prove this using 'linearly adapted coordinates' later in this chapter (see p. 354).

It is worth emphasizing one distinction between the four local theorems. In the inverse function theorem one can specify coordinates in the source or in the target, and deduce coordinates in the other so that the map has the special form (namely, the identity). In the local immersion theorem, one needs to choose (or change) the coordinates in the target, in the regular value theorem one needs to choose coordinates in the source, while in the constant rank theorem, one needs to choose coordinates in both in order to write f in the form given by the theorem.

B.7 Transversality

Two submanifolds X and Y of $\mathbb{R}^n$ are ***transverse*** at $x \in \mathbb{R}^n$, written $X \pitchfork_x Y$, if either $x \notin X \cap Y$ or their tangent spaces satisfy

$$T_xX + T_xY = \mathbb{R}^n.$$

A map $f\colon \mathbb{R}^k \rightarrowtail \mathbb{R}^n$ is ***transverse*** to a submanifold $X \subset \mathbb{R}^n$ at $x \in X$, written $f \pitchfork_x X$ if

$$\operatorname{im} \mathrm{d}f_q + T_xX = \mathbb{R}^n,$$

for all $q \in f^{-1}(x)$. Two maps $f\colon \mathbb{R}^k \rightarrowtail \mathbb{R}^n$ and $g\colon \mathbb{R}^\ell \rightarrowtail \mathbb{R}^n$ are ***transverse*** at $x \in \mathbb{R}^n$ (written $f \pitchfork_x g$) if

$$\operatorname{im} \mathrm{d}f_q + \operatorname{im} \mathrm{d}g_p = \mathbb{R}^n,$$

for all q and p satisfying $f(q) = g(p) = x$. Finally, two such objects are simply ***transverse*** if they are transverse at x for all $x \in \mathbb{R}^n$.

The principal use of transversality is the following.

Theorem B.27. *Suppose X, Y are two submanifolds of $\mathbb{R}^n$.*

(i). If X and Y are transverse then their intersection $X \cap Y$ is also a submanifold. Moreover

$$\operatorname{codim}(X \cap Y) = \operatorname{codim}(X) + \operatorname{codim}(Y).$$

(ii). Let $f\colon \mathbb{R}^k \to \mathbb{R}^n$ be transverse to X. Then $f^{-1}(X)$ is a submanifold of $\mathbb{R}^k$, with

$$\operatorname{codim}(f^{-1}(X)) = \operatorname{codim}(X),$$

where the first codimension is in $\mathbb{R}^k$ and the second in $\mathbb{R}^n$.

PROOF: We will prove (ii) and leave (i) to the reader. Suppose X is of dimension m. Let $q \in f^{-1}(X)$ and let U be a neighbourhood of $f(q)$ on which there is a straightening map for X: that is a diffeomorphism $\Phi\colon U \to \mathbb{R}^n$ with $\Phi(U \cap X)$ an open subset of $\mathbb{R}^m \times \{0\}$. We now proceed with the 'straightened version' of X; that is, we write

$$\mathbb{R}^n = \mathbb{R}^m \times \mathbb{R}^{n-m}$$

with $X \cap U \subset \mathbb{R}^m \times \{0\}$. Accordingly, write f as $f(x) = (f_1(x), f_2(x))$ with $f_1\colon \mathbb{R}^k \rightarrowtail \mathbb{R}^m$ and $f_2\colon \mathbb{R}^k \rightarrowtail \mathbb{R}^{n-m}$.

Now $T_{f(q)}X = \mathbb{R}^m \times \{0\}$, and hence f is transverse to X if and only if $\mathsf{d}f_2$ has rank $n - m$ at p; that is, f_2 is a submersion there. Moreover, for $x \in f^{-1}(U)$, $f(x) \in X$ if and only if $f_2(x) = 0$. Equivalently, $f^{-1}(U \cap X) = f_2^{-1}(0)$. The local result (in a neighbourhood of q) then follows from the local submersion theorem (Theorem B.13), including the statement about the codimension of $f^{-1}(U \cap X)$. Since this holds for all $q \in f^{-1}(X)$, the result follows. ✔

B.8 Linearly adapted coordinates

Recall that the rank of a map f at x is the rank of the Jacobian matrix $\mathsf{d}f_x$. Both words 'singular' and 'regular' have different meanings in mathematics depending on context. We use the following definition.

Definition B.28. A ***singular point*** or ***singularity*** of a smooth map $f\colon \mathbb{R}^n \rightarrowtail \mathbb{R}^p$ is a point $x_0 \in \operatorname{dom}(f)$ where the rank $\operatorname{rk}_x(f) < p$. If $\operatorname{rk}_x(f) = p$ then x is a ***regular point*** of f. ★

Note that if $p > n$ then every point in the source is singular. However, if $p > n$ and $\operatorname{rk}_(f) = n$ then one talks of f being a ***regular parametrization*** of its image (which by the local immersion theorem is a submanifold of $\mathbb{R}^p$, at least in a neighbourhood of x).

Note that if $p = 1$ (scalar-valued functions), a singular point is a point where all partial derivatives vanish, so a singular point is the same as a critical point.

A first step in the study of singularities is to introduce coordinates adapted to the situation. For non-singular points, the appropriate coordinates are the ones of the inverse, implicit, local immersion or constant rank theorems described above. Extending these to the case where the map is singular is the subject of the following theorem.

Theorem B.29 (Linearly adapted coordinates). *Let $f\colon \mathbb{R}^n \rightarrowtail \mathbb{R}^p$ be a smooth map with $f(0) = 0$, and $\operatorname{rk}_0(f) = k$. Then there is a neighbourhood U of 0 in $\mathbb{R}^n$ and coordinates $u_1, \dots, u_k, x_1, \dots, x_{n-k}$ on U and $(v_1, \dots, v_k, y_1, \dots, y_{p-k})$ in a neighbourhood of 0 in $\mathbb{R}^p$, and a smooth map $g\colon U \to \mathbb{R}^{p-k}$ with $g(u,0) = 0$ and $\mathsf{d}g_0 = 0$, such that f takes the form,*

$$f(u,x) = (u, g(u,x)).$$

That is, in a neighbourhood of 0, f is given by

$$\begin{cases} v_i = u_i & (i = 1, \dots, k), \\ y_j = g_j(u,x) & (j = 1, \dots, p-k). \end{cases}$$

Proof: Since $\operatorname{rk}(\mathsf{d}f_0) = k$ we can choose bases in $\mathbb{R}^n$ and $\mathbb{R}^p$ so that

$$\mathsf{d}f_0 = \begin{pmatrix} I_k & 0 \\ 0 & 0 \end{pmatrix}. \tag{B.2}$$

Writing the target accordingly as $\mathbb{R}^p = \mathbb{R}^k \times \mathbb{R}^{p-k}$ we write

$$f(u, x) = (f_1(u, x), f_2(u, x)),$$

where $f_1\colon \mathbb{R}^n \to \mathbb{R}^k$. Now f_1 has rank k so it follows from the local submersion theorem that there are coordinates for which $f_1(u,x) = u$. In these coordinates, we have

$$f(u,x) = (u, f_2(u,x)).$$

Now make a further change of coordinates in the target by

$$(v,w) \longmapsto (v, w - f_2(v,0)).$$

Then in these coordinates, f becomes

$$f(u,x) = (u, f_2(u,x) - f_2(u,0)),$$

so putting $g(u,x) = f_2(u,x) - f_2(u,0)$ we get the desired property that $g(u,0) = 0$.
✔

Definition B.30. Let $f\colon \mathbb{R}^n \rightarrowtail \mathbb{R}^p$ be a smooth map of rank k at the origin. Using linearly adapted coordinates we write $f(u,x) = (u,\ g(u,x))$ as in the theorem above. The map $g_0\colon \mathbb{R}^{n-k} \rightarrowtail \mathbb{R}^{p-k}$ defined by $g_0(x) = g(0,x)$ (which has rank 0 at the origin) is called the ***core*** of f (at the origin). ✯

The core of a map is important in the study of bifurcations: this is because the set of solutions of $f = 0$ is essentially the same as the set of solutions of $g_0 = 0$. As an expression in its coordinates, the core is not uniquely defined because linearly adapted coordinates are not unique. However, any two cores of a given map are 'contact equivalent' (in the sense of Chapter 11).

Example B.31. Consider the map $f\colon \mathbb{R}^3 \to \mathbb{R}^2$ given by $f(x,y,z) = (x+y, x^2 + y^2 - z^2)$. The differential of f at the origin is

$$\mathsf{d}f_0 = \begin{pmatrix} 1 & 1 & 0 \\ 0 & 0 & 0 \end{pmatrix}.$$

This has rank 1, so we change basis so that $\mathsf{d}f$ is in the required form (B.2). So let $\mathbf{e}_1 = \frac{1}{2}(1,1,0)^T$ and $\mathbf{e}_2 = (1,-1,0)^T$ and $\mathbf{e}_3 = (0,0,1)^T$ in the source, and leave the basis in the target as given. Note that $\{\mathbf{e}_2, \mathbf{e}_3\}$ spans $\ker \mathsf{d}f_0$. Then we have coordinates (u,v,w), with $(x,y,z) = u\mathbf{e}_1 + v\mathbf{e}_2 + w\mathbf{e}_3$ so that $x = \frac{1}{2}u + v$ and $y = \frac{1}{2}u - v$ and $z = w$. In these new coordinates, f takes the form

$$f(u,v,w) = \left(u,\ \tfrac{1}{2}u^2 + 2v^2 - w^2\right).$$

This is the expression of f in linearly adapted coordinates, with $g(u,v,w) = \frac{1}{2}u^2 + v^2 - w^2$, and the core is $g_0(v,w) = g(0,v,w) = v^2 - 2w^2$. Notice that $f(u,v,w) = 0$ if and only if $u = 0$ and $g_0(v,w) = 0$. ✐

The calculation in this example and the proof above, is formalized by a procedure called *Lyapunov–Schmidt reduction*; see the section below. But first we give a proof delayed from earlier.

PROOF OF THE CONSTANT RANK THEOREM B.26: Let $q \in \mathrm{dom}(f)$, then $\mathrm{rk}\,\mathsf{d}f_q = k$. We can use linearly adapted coordinates in a neighbourhood U of q with $f(u,x) = (u, g(u,x))$. Then at any point $(u,x) \in U \subset \mathrm{dom}(f)$,

$$\mathsf{d}f_{(u,x)} = \begin{pmatrix} I_k & 0 \\ \mathsf{d}_u g & \mathsf{d}_x g \end{pmatrix},$$

where $\mathsf{d}_x g$ is the Jacobian matrix of the map $x \mapsto g(u,x)$ (consisting of partial derivatives with respect to the x variables), and $\mathsf{d}_u g$ analogously. This matrix has rank k if and only if $\mathsf{d}_x g(u,x) = 0$, and if this block of partial derivatives is identically zero, then $g(u,x)$ must be independent of x. That is,

$$f(u,x) = (u,\ g(u)).$$

The image of f is thus the graph of g, and we can change coordinates again on $\mathbb{R}^p$ by, $\psi(u,y) = (u, y - g(u))$. Then

$$\psi \circ f(u,x) = (u,0)$$

as required. ✔

The following 'reduction' procedure for cores is often useful.

Proposition B.32. *Let $f\colon \mathbb{R}^n \rightarrowtail \mathbb{R}^a \times \mathbb{R}^b$ with $f(x) = (f_1(x), f_2(x))$ accordingly. Suppose $f(0) = (0,0)$ and that $f_1\colon \mathbb{R}^n \to \mathbb{R}^a$ is a submersion at 0. Then f and the restriction $f_2|_V$ have the same core at the origin, where $V = f_1^{-1}(0)$.*

Note that saying cores are 'the same', one means that we can choose coordinates so that they agree. One cannot expect more than that, given that – as mentioned above – the core is not unique.

PROOF: Since f_1 is a submersion there is a neighbourhood of the origin in $\mathbb{R}^n$ on which there are coordinates for which $f_1(x,y) = x$, with $x \in \mathbb{R}^a$, $y \in \mathbb{R}^{n-a}$. In these coordinates, $V = \{(x,y) \mid x = 0\} = \mathbb{R}^{n-a}$. These are linearly adapted coordinates for f, since $f(x,y) = (x,\ f_2(x,y))$. The core of f is the same as the core of $f_2(0,y)$, and clearly $y \mapsto f_2(0,y)$ is the same as $f_2|_V$. ✔

B.9 Lyapunov–Schmidt reduction

Lyapunov–Schmidt reduction is a practical procedure for finding the core of a map, or linearly adapted coordinates, and is a straightforward application of the implicit function theorem. It is included here as it is often used as a first step in applications of bifurcation theory. This procedure is also valid in an infinite–dimensional context for bifurcation problems on Banach spaces (possibly arising from PDEs), although the required functional analysis would take us too far afield (see for example [49, 50] for details). See also [47] for further properties of this procedure.

Let $f\colon \mathbb{R}^n\times\mathbb{R}^k \rightarrowtail \mathbb{R}^p$ be a smooth map. We are interested in studying solutions of the 'bifurcation equation' $f(x;\lambda) = 0$ (here we think of $\lambda \in \mathbb{R}^k$ as parameters, and $x \in \mathbb{R}^n$ as state variables) in a neighbourhood of a point (x_0, λ_0) which we take to be the origin.

There are two steps: (i) solve the largest non-degenerate part of f for the appropriate variables (using the implicit function theorem), and (ii) substitute those values into the remainder of f.

We write $f_0(x) = f(x, 0)$. Then $\mathsf{d}(f_0)_0$ is the Jacobian matrix of f with respect to the x-variables (a $p\times n$ matrix) at the origin. Let $K = \ker \mathsf{d}f_0 \subset \mathbb{R}^n$ be its kernel, $R \subset \mathbb{R}^p$ its image (range) and $Q = \mathbb{R}^p/R$ its cokernel. In practice one identifies Q with a subspace of $\mathbb{R}^p$ by writing $\mathbb{R}^p = R \oplus Q$. Let $\Pi\colon \mathbb{R}^p \to Q$ be the Cartesian projection with kernel R. Then $(I - \Pi)$ is the projection to R.

Write

$$f_1 = (I - \Pi) \circ f, \quad \text{and} \quad f_2 = \Pi \circ f. \tag{B.3}$$

Then $f(x, \lambda) = 0$ if and only if $f_1(x, \lambda) = 0$ and $f_2(x, \lambda) = 0$.

Step (i): consider $f_1\colon \mathbb{R}^n \times \mathbb{R}^k \to R$. This is a submersion (in a neighbourhood of 0) and with $\mathbb{R}^n = Y \oplus K$ (for any choice of complementary subspace Y) one has $\mathsf{d}_x(f_1)_0 = [A \;\; 0]$ with A invertible. Then by the implicit function theorem there is a neighbourhood of the origin and a map $h\colon K\times\mathbb{R}^k \rightarrowtail Y$ such that $f_1(x, \lambda) = 0$ if and only if $x \in \Gamma_h$ (the graph of h). In other words, writing $(v, \lambda) \in K\times\mathbb{R}^k$, $f_1(x, \lambda) = 0$ if and only if there is a $v \in K$ for which $(x, \lambda) = (h(v, \lambda), v, \lambda) \in Y \times K \times \mathbb{R}^k$ (all restricted to appropriate neighbourhoods of the origin).

Step (ii): define $g\colon K \times \mathbb{R}^k \rightarrowtail Q$ by

$$g(v, \lambda) = f_2(h(v, \lambda), v, \lambda).$$

Clearly, $f(x, \lambda) = 0$ if and only if there exists $v \in K$ for which $(x, \lambda) = (h(v, \lambda), v, \lambda)$ and $g(v, \lambda) = 0$. That is, the set of zeros of f forms a graph over the set of zeros of g. The map g is the result of Lyapunov–Schmidt reduction.

If $\mathsf{d}(f_0)_0$ has rank r, then $K \simeq \mathbb{R}^{n-r}$ and $Q \simeq \mathbb{R}^{p-r}$, so the reduction reduces the dimension by r in both source and target. Note also that $\mathsf{d}(g_0)_0 = 0$, so no further reduction is possible, and moreover that g_0 is the core of f_0.

See Problem B.18 for an example.

It should be pointed out that the resulting reduced map g depends on the choices of Q and Y (the splittings of $\mathbb{R}^n$ and $\mathbb{R}^p$). However, different choices lead to 'equivalent' maps (in particular, contact equivalent, or with the parameter, $\mathcal{K}_{\text{un}}$-equivalent, as described in Chaps 11 and 14).

Problems

B.1 At which points is $f\colon \mathbb{R} \to \mathbb{R}$, $f(x) = \cos x$ a local diffeomorphism? (†)

B.2 Let ϕ, ψ be local diffeomorphisms, ϕ defined in a neighbourhood of 0, and ψ in a neighbourhood of $\phi(0)$. Show that $\psi \circ \phi$ is a local diffeomorphism in a neighbourhood of 0.

B.3 Let x be the usual coordinate on $\mathbb{R}$, and let $x' = \phi(x) = x^3$. Consider the smooth function $f(x) = x^2$. Express f in terms of x', and show it is not differentiable at $x' = 0$. [*Note: this ϕ is a homeomorphism (i.e. ϕ is invertible and both ϕ and ϕ^{-1} are continuous). This example demonstrates why we need to consider changes of coordinates to be diffeomorphisms rather than just homeomorphisms.*]

B.4 Let $a \in \mathbb{R}$ and consider the map $f_a\colon \mathbb{R} \to \mathbb{R}$ defined by $f_a(x) = x + ax^2$. Use the inverse function theorem to show that f_a is a diffeomorphism on some neighbourhood of the origin. Now by explicitly solving $y = f_a(x)$ find the largest open interval U containing 0 such that $f\colon U \to f_a(U)$ is a diffeomorphism (the answer will depend on a). (†)

B.5 Consider the so-called ***folded handkerchief*** map $f\colon \mathbb{R}^2 \to \mathbb{R}^2$, $f(x, y) = (x^2, y^2)$. Find the singular points of f. Find also the image of f and the image of the singular points. (†)

B.6 Show that $f\colon \mathbb{R}^2 \to \mathbb{R}^2$ defined by $f(x, y) = (\mathsf{e}^x \cos y, \mathsf{e}^x \sin y)$ is a local diffeomorphism at each point $(x, y) \in \mathbb{R}^2$, but that it is not a (global) diffeomorphism.

B.7 Define $f\colon \mathbb{R}^3 \to \mathbb{R}$ by $f(x, y, z) = x^2 + y^2 - z^2$.

(i). Show that 0 is the only singular value of f,

(ii). Show that if a and b have the same sign, then $f^{-1}(a)$ and $f^{-1}(b)$ are diffeomorphic. [*Hint*: *consider scalar multiplication by a suitable constant on* $\mathbb{R}^3$.]

(iii). If a and b are of opposite signs, show that $f^{-1}(a)$ and $f^{-1}(b)$ are not diffeomorphic. [*Hint*: *show for example that one is connected and the other not.*]

B.8 Let $f\colon \mathbb{R}^n \rightarrowtail \mathbb{R}^p$ be a submersion. Show it is an open map: that is, if $U \subset \mathrm{dom}(f)$ is open then $f(U)$ is open in $\mathbb{R}^p$. (†)

B.9 Consider the relation $x^2 + 1 - y^4 - y = 0$. Show that this can be solved for y as a function of x in a neighbourhood of the point $(1, 1)$. By using implicit differentiation, find the Taylor series of this function to order 3 about $x = 1$.

B.10 Consider the map $f\colon \mathbb{R}^3 \to \mathbb{R}^2$, $f(x, y, z) = (x + y^2 - xz, y - z^2)$. Define a new map $F\colon \mathbb{R}^3 \to \mathbb{R}^3$ whose first two components are the same as f and whose third component is z (this exercise illustrates the proof of the submersion theorem).

(i). Show that f is a submersion at the origin in $\mathbb{R}^3$.

(ii). Show that F defines a local diffeomorphism at 0.

(iii). Show that $f \circ F^{-1}(X, Y, Z) = (X, Y)$ (as in the conclusion to the submersion theorem).

B.11 Find linearly adapted coordinates for the map $f(x, y) = (x, xy, y + x^2)$. (†)

B.12 Prove the Lagrange multiplier theorem: *Suppose* $f\colon \mathbb{R}^n \rightarrowtail \mathbb{R}$ *and* $g\colon \mathbb{R}^n \rightarrowtail \mathbb{R}^k$ *are smooth, and let* $p \in \mathrm{dom}(f) \cap \mathrm{dom}(g)$. *Suppose* $c = g(p)$ *is a regular value of* g *and let* $M = g^{-1}(c)$. *Then* p *is a critical point of the restriction* $f|_M$ *of* f *to* M *if and only if there are* $\lambda_1, \ldots, \lambda_k \in \mathbb{R}$ *such that*

$$\mathsf{d}f_p = \sum_{i=1}^{k} \lambda_i\, \mathsf{d}(g_i)_p.$$

[*Hint*: *start by applying the regular value theorem to* g.]

B.13 Prove the statement in Example B.9. Show moreover that the map defining the graph is an embedding.

B.14 Show that $S^n = \{x \in \mathbb{R}^{n+1} \mid \sum x_i^2 = 1\}$ is a submanifold of $\mathbb{R}^{n+1}$ of dimension n. It is called the n-sphere.

B.15 Let $X = \{(\mathbf{u}, \mathbf{v}) \in \mathbb{R}^3 \times \mathbb{R}^3 \mid \|\mathbf{u}\|^2 = \|\mathbf{v}\|^2 = 1,\ \mathbf{u} \cdot \mathbf{v} = 0\}$. Show that X is a submanifold of $\mathbb{R}^6$ and determine its dimension. (†)

B.16 Let X be a subset of $\mathbb{R}^n$, and $f\colon X \to \mathbb{R}$ a function. One says f is ***smooth*** if for each $p \in X$ there is a neighbourhood U of p in $\mathbb{R}^n$ and a smooth function $\bar{f}\colon U \to \mathbb{R}$ such that $f = \bar{f}|_X$. Show that if X is a submanifold, this definition is equivalent to the one via straightening maps given on p. 347.

B.17 Consider the 3–dimensional vector space Sym(2) consisting of 2×2 symmetric matrices. Show that the subset consisting of symmetric matrices of rank r is a submanifold of Sym(2), for $r = 0, 1, 2$. What are their dimensions?

B.18 Apply Lyapunov–Schmidt reduction at the origin, to the bifurcation problem $f\colon \mathbb{R}^3 \times \mathbb{R} \to \mathbb{R}^3$ given by

$$f(x, y, z, \lambda) = (z + x^2 - \lambda y,\ y + z - 2\lambda x, z - \lambda). \qquad (\dagger)$$

(This is a sufficiently simple example that everything can be done explicitly.)

B.19 Let $A \in \Delta_r$, the subset of $\mathsf{Mat}(p, n)$ consisting of those matrices of rank r (see Section B.5). Let $K \in \mathsf{Mat}(n, n)$ be any matrix of rank $n - r$ such that $AK = 0$, and let $Q \in \mathsf{Mat}(p, p)$ be any matrix of rank $p - r$ such that $QA = 0$. Show that the tangent space to Δ_r at A is given by

$$T_A\Delta_r = \{V \in \mathsf{Mat}(p, n) \mid QVK = 0\}.$$

[*Hint*: *choose bases so that A takes the normal form in Proposition A.2.*]

B.20 Let $A \in \Delta_r \subset \mathsf{Mat}(p, n)$. Use the previous problem to show that $PA \in T_A\Delta_r$ for all $P \in \mathsf{Mat}(p, p)$ and $AN \in T_A\Delta_r$ for all $N \in \mathsf{Mat}(n, n)$.

B.21 Show that the following 'figure-8 without double point' is an immersion but not an embedding. The map is $f\colon (0,\ 2\pi) \to \mathbb{R}^2$

$$f(t) = (\sin(t),\ \sin(2t)).$$

The diagram shows the figure-8 which is the image of f. Note that $f(\pi) = (0, 0)$ so that the origin is in the image and there is no gap, and if either endpoint of the interval $[0,\ 2\pi]$ were included in the domain the map would no longer be injective. The arrows are there to emphasize the behaviour of f as $t \to 0$ and $t \to 2\pi$. [*Hint*: *to prove that f is not an embedding, consider the open set $U = (\pi - \varepsilon,\ \pi + \varepsilon) \subset (0, 2\pi)$. It is enough to show there is no neighbourhood V of $(0, 0)$ in the target space $\mathbb{R}^2$ such that $f^{-1}(V) = U$.*]

C

Differential equations and flows

ERE WE PROVIDE A BRIEF REVIEW of the basic theorems from the theory of ordinary differential equations needed in this text, in particular in the proofs using the so-called homotopy method.

The theorems we need are the basic existence and uniqueness theorems for solutions of ODEs, and the fact that near a zero of a vector field, there are solutions for any finite time, and simple properties of the resulting diffeomorphisms. We are not concerned with analysing exactly what degrees of differentiability are needed, so we will assume everything in sight is smooth (C^∞).

There are many books on ODEs, all of which will prove at least some of the theorems stated below (and often only assuming finite differentiability). One book I would particularly recommend is the text of V. I. Arnold [1], it also contains proofs of the theorems in Section C.1. His point of view starting with vector fields is similar to the one we present in this appendix.

C.1 Existence and uniqueness

Vector fields were defined in Appendix A (see p. 330). Let $U \subset \mathbb{R}^n$ be an open set and $\mathbf{v}$ a smooth vector field on U. Associated to $\mathbf{v}$ is a first order ordinary differential equation (ODE),

$$\dot{x} = \mathbf{v}(x, t), \tag{C.1}$$

or $\dot{x} = \mathbf{v}(x)$ if $\mathbf{v}$ is time-independent. A solution of this ODE is a smooth curve $\gamma\colon \mathbb{R} \rightarrowtail \mathbb{R}^n$ for which $\dot{\gamma}(t) = \mathbf{v}(\gamma(t), t)$ (or $\dot{\gamma}(t) = \mathbf{v}(\gamma(t))$ in the time-independent case).

Such a first order ODE describes how a point moves in time, and a point can keep moving until it reaches the boundary of U where the vector field may no longer be defined; if $U = \mathbb{R}^n$ then the point could go to infinity in a finite time, as occurs in Example C.2 below. Such considerations imply that in general one can only expect to have *local* solutions; that is, solutions valid for a finite interval of time.

In our applications of ODEs, we have vector fields that depend on time as well as on x, so-called *non-autonomous* ODEs. There is a simple trick which reduces this to the case of autonomous ODEs, just by including t as one of the variables. Let $\mathbf{v}(x,t)$ be the time-dependent vector field, and define for $(x,s) \in U \times \mathbb{R}$

$$\hat{\mathbf{v}}(x,s) := (\mathbf{v}(x,s), 1)$$

where the last component corresponds to the s-axis. This is now an autonomous vector field on $U \times \mathbb{R}$ (or an open subset of $U \times \mathbb{R}$), and $\gamma(t)$ is a solution to the non-autonomous equation if and only if $\hat{\gamma}(t) := (\gamma(t), t)$ is a solution to the autonomous one (using $\hat{\mathbf{v}}$), as is readily checked. This simple trick shows that existence and uniqueness theorems about autonomous ODEs apply also to non-autonomous ones.

So now we consider ODEs of the form

$$\dot{x} = \mathbf{v}_t(x), \tag{C.2}$$

where we have written $\mathbf{v}_t(x)$ in place of $\mathbf{v}(x,t)$. If we specify an initial condition x_0, then (C.2) becomes an *initial value problem* (or IVP),

$$\begin{cases} \dot{x} &= \mathbf{v}_t(x) \\ x(0) &= x_0. \end{cases} \tag{C.3}$$

A local solution to (C.3) is a path $\gamma\colon I \to U$, where I is some interval in $\mathbb{R}$ containing 0, satisfying

$$\begin{cases} \dot{\gamma}(t) &= \mathbf{v}_t(\gamma(t)) \quad \text{for } t \in I \\ \gamma(0) &= x_0. \end{cases} \tag{C.4}$$

See Figure C.1.

Theorem C.1 (Local existence and uniqueness). *Let $\mathbf{v}_t$ be a smooth time-dependent vector field defined on an open subset U of $\mathbb{R}^n$. Then for each $x_0 \in U$ there is an interval I containing 0, and a unique smooth map $\gamma\colon I \to U$ satisfying (C.4).*

The proof of this theorem involves an application of the contraction mapping theorem to show that successive approximations (determined by Picard's method) converge to a unique solution. Details of this and proofs of the theorems below can be found in many texts, for example in [1].

Let us now include the initial value in the notation for the solution, and write $\phi(x_0,t)$ for the unique solution $\gamma(t)$ given above with $\gamma(0) = x_0$. Note that by definition, $\phi(x,0) = x$, for all x in the domain.

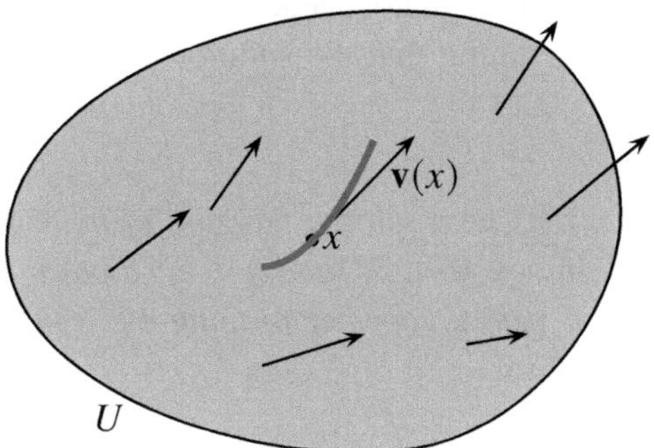

FIGURE C.1 A local solution through x.

Example C.2. Consider the ODE on $U = \mathbb{R}$ given by $\dot{x} = x^2$. Solutions can be found by the method of separation of variables, obtaining

$$\phi(x,t) = \frac{x}{1-tx}.$$

Notice two things: (i) for $x \neq 0$, $\phi(x,t)$ is not defined for all time (it 'blows up' as $t \to 1/x$), and (ii) if we write $y = x/(1-tx)$ then this can be solved for x giving

$$x = \frac{y}{1+ty}.$$

This means that for each t, the map $x \mapsto \phi(x,t)$ is a local diffeomorphism wherever it is defined.

Theorem C.3 (Smooth dependence on initial conditions). *For each $x_0 \in U$ there is a neighbourhood V of x_0 and $\varepsilon > 0$ such that the map $\phi\colon V \times (-\varepsilon, \varepsilon) \to U$ is defined and smooth. Explicitly ϕ is defined by the properties*

$$\begin{cases} \frac{\partial}{\partial t}\phi(x,t) &= \mathbf{v}_t(\phi(x,t)) \\ \phi(x,0) &= x. \end{cases} \tag{C.5}$$

If we define the maps ϕ_t by $\phi_t(x) = \phi(x,t)$ then the last condition says that ϕ_0 is the identity map.

Remark C.4. If $\mathbf{v}(x)$ is a vector field, then the flow ϕ_t of the autonomous differential equation $\dot{x} = \mathbf{v}(x)$ satisfies the ***local group property***,

$$\phi_s \circ \phi_t = \phi_{s+t}$$

whenever both sides are defined (hence the 'local' in the name).

Parametrized version Although not needed in this text, this is a natural extension of the previous theorems, stating that the solutions also depend smoothly on any parameters, as follows.

Theorem C.5. *Let $\mathbf{v}(x,t,\lambda)$ be a smooth function of its arguments, and a family of vector fields defined for $x \in U \subset \mathbb{R}^n$ and $\lambda \in \mathbb{R}^k$ a parameter. Then for each λ there is a solution map as before, together forming $\phi(x,t,\lambda)$. This map ϕ depends smoothly on all its arguments.*

C.2 Vector fields and flows

Let $U \subset \mathbb{R}^n$ be an open set and $\mathbf{v}_t$ a time-dependent smooth vector field on U. Associated to $\mathbf{v}$ we have the first order ODE, $\dot{x} = \mathbf{v}_t(x)$.

With the notation of Theorem C.3, for each t, we have a map

$$\phi_t : V \to U$$

defined by $\phi_t(x) := \phi(x,t)$. The defining property of ϕ_t is that, for each $x \in U$,

$$\frac{\mathrm{d}}{\mathrm{d}t}\phi_t(x) = \mathbf{v}_t(\phi_t(x)). \tag{C.6}$$

The collection of maps ϕ_t is called the ***flow*** of the vector field $\mathbf{v}$, or of the ODE. In Example C.2, for each t,

$$\phi_t(x) = \frac{x}{1-tx}.$$

Notice that indeed, $\phi_0 = \mathrm{Id}$.

This example illustrates that some care is needed in using this notion of flow, because many vector fields are not *complete*, meaning that ϕ_t may not exist for larger values of t. For $t = 0$, ϕ_0 is the identity so always exists, but for $|t| < \varepsilon$ say, the flow may only exist if the domain is restricted to a suitably small set (depending on ε).

Theorem C.6 (Flow is a diffeomorphism). *Let $\mathbf{v}_t$ and U be as above and let $x_0 \in U$. With ε, V as in Theorem C.3, for each $t \in (-\varepsilon, \varepsilon)$ the map ϕ_t is a diffeomorphism of V onto $\phi_t(V)$.*

PROOF: We know from Theorem C.3 that for each $s \in (-\varepsilon,\varepsilon)$ the map $\phi_s : V \to \phi_s(V)$ is smooth; we need to show it has an inverse which is also smooth. Define the 'reverse' vector field by $\mathbf{w}_t(x) := -\mathbf{v}_{s-t}(x)$, and let ψ_t be the flow for the associated ODE,

$$\begin{cases} \dot{y} &= \mathbf{w}_t(y) \\ y(0) &= y_0 \end{cases} \tag{C.7}$$

Let $y_0 = \phi_s(x_0)$. Then by the theorems above there is a neighbourhood V' of y_0 on which the flow ψ_t is defined, for t in some neighbourhood of 0. We claim that $\phi_s \circ \psi_s = \mathsf{Id}_{V'}$ and $\psi_s \circ \phi_s = \mathsf{Id}_V$. It then follows that V' can be taken to be $\phi_s(V)$, and so ϕ_s has a smooth inverse and is therefore a diffeomorphism.

To see that $\psi_s \circ \phi_s = \mathsf{Id}_V$ note that if $\gamma(t)$ is a local solution to (C.3), as in (C.4) then $t \mapsto \gamma(s-t)$ is a solution to (C.7). But this is true for arbitrary x_0 and $y_0 = \phi_s(x_0)$, so the claim holds. ✔

C.3 The flow near a singular point

In our applications, we will need to let t range over all of the interval $[0, 1]$ which in general is not possible unless it is known that the vector field is complete. However if $\mathbf{v}(x_0) = \mathbf{0}$ then it can be done, after choosing the neighbourhood V appropriately.

Theorem C.7. *With the same notation as the previous theorems, suppose* $\mathbf{v}_t(x_0) = \mathbf{0}$ *for all* $t \in [0, 1]$. *Then there is a neighbourhood* $V \subset U$ *of* x_0 *and a flow* $\Phi\colon V \times [0, 1] \to U$ *solving the* IVP *in (C.3).*

For example, in Example C.2 where $U = \mathbb{R}$ and $\mathbf{v}(x) = x^2$ which satisfies $\mathbf{v}(0) = 0$, one can take the neighbourhood V of 0 to be $(-1, 1)$.

PROOF: For an autonomous system, the flow satisfies the 'local group property', see Remark C.4, namely $\phi_t \circ \phi_s = \phi_{s+t}$. However, this needs adapting for non-autonomous ODEs. Given the vector field $\mathbf{v}(x, t)$, define the 'time-displaced' vector field $\mathbf{v}^s(x, t) := \mathbf{v}(x, s + t)$, and denote the resulting flow by $\phi^s(x, t) = \phi^s_t(x)$. Then one can show the local group property should be replaced by

$$\phi^s_t \circ \phi_s = \phi_{s+t}. \tag{C.8}$$

(For the proof, see Problem C.3.)

The proof of the theorem relies on the compactness of $[0, 1]$. For each $s \in [0, 1]$ let ε_s and V_s be the value of ε and the neighbourhood V of x_0 whose existence are guaranteed by Theorem C.3. In particular, if $x \in V_s$ and $t \in (0, \varepsilon_s)$ then $\phi^s_t(x) \in U$.

Consider the cover $(-\varepsilon_0, \varepsilon_0) \cup \bigcup_{s\in[0,1]}(s, s+\varepsilon_s)$ of $[0,1]$. Because $[0,1]$ is compact, there is a finite sub-cover, say

$$[0,1] \subset (-\varepsilon_0, \varepsilon_0) \cup \bigcup_{j=1}^{N}(s_j, s_j + \varepsilon_{s_j})$$

of $[0,1]$, ordered so that $s_j < s_{j+1}$ and write $s_0 = 0$ and $s_N = 1$. In particular then, for each $j = 0, \ldots, N$, the ε_{s_j} satisfy $s_{j+1} < s_j + \varepsilon_{s_j}$ (otherwise it is not a cover of $[0,1]$).

Now, we know that if $x \in V_0$ then $\phi_{s_1}(x) \in U$, since $s_1 < \varepsilon_1$. To proceed further, we require $\phi_{s_1}(x) \in V_{s_1}$, so let $W_1 = \phi_{s_1}^{-1}(V_{s_1})$. This is a neighbourhood of x_0 since $\phi_s(x_0) = x_0$ for all s. Thus, for $x \in W_1$ we deduce that $\phi_{s_1}(x) \in V_{s_1}$ and hence $\phi^{s_1}_{s_2-s_1}(\phi_{s_1}(x)) \in U$. But by Equation (C.8) $\phi^{s_1}_{s_2-s_1}(\phi_{s_1}(x)) = \phi_{s_2}(x)$, and hence

$$x \in W_1 \;\Rightarrow\; \phi_{s_2}(x) \in U.$$

Continuing in this manner, let $W_2 = \phi_{s_2}^{-1}(V_{s_2}) \cap W_1$, and we deduce that

$$x \in W_2 \;\Rightarrow\; \phi_{s_3}(x) \in U.$$

After a finite number of steps, we will have constructed a neighbourhood W_N of x_0 such that

$$x \in W_N \;\Rightarrow\; \phi_t(x) \in U, \quad (\forall t \in [0,1])$$

as required. ✔

In many of our applications, not only is the value $\mathbf{v}_t(x_0) = \mathbf{0}$ but $\mathbf{v}_t$ may vanish to some higher–order at x_0.

Theorem C.8. *Let $\mathbf{v}_t, \phi_t$ be as in Theorem C.7, and suppose in addition that the k-jet $\mathrm{j}^k(\mathbf{v}_t)_{x_0} = 0$. Then $\mathrm{j}^k(\phi_t)_{x_0} = \mathrm{Id}$ for all $t \in [0,1]$, where Id is the identity map.*

Proof: We prove this for $k = 1$, the general idea is the same. We have that,

$$\frac{\mathrm{d}}{\mathrm{d}t}\phi_t(x) = v_t(\phi_t(x)).$$

Differentiating with respect to x and commuting the partial derivatives gives

$$\frac{\mathrm{d}}{\mathrm{d}t}\mathrm{d}(\phi_t)_0 = \mathrm{d}(v_t)_0,$$

since $\phi_t(0) = 0$ for all $t \in [0,1]$. But the right hand side is zero by hypothesis, so that $\mathrm{d}(\phi_t)_0$ is constant (independent of t). When $t = 0$, $\phi_0 = \mathrm{Id}$ and therefore $\mathrm{d}(\phi_t)_0 = \mathrm{Id}$ for all $t \in [0,1]$. ✔

C.4 Linear differential equations

Linear vector fields are ones of the form $\mathbf{v}(x) = Lx$, where L is a linear map. If L is constant (independent of t), then the solution of $\dot{x} = Lx$ with initial condition x_0 is

$$x(t) = \exp(tL)\, x_0,$$

where,

$$\exp(A) = I + A + \tfrac{1}{2}A^2 + \tfrac{1}{3!}A^3 + \cdots .$$

This shows that the solutions of a linear autonomous differential equation exist for all time; that is linear vector fields on $\mathbb{R}^n$ are *complete*. However, if L is not constant then there is no closed formula for the solution. On the other hand, the following is not hard to prove.

Theorem C.9. *Let $L(t)$ be a time dependent $p \times p$ matrix, and consider the* ODE *in* $\mathsf{Mat}(p,n)$*,*

$$\dot{A} = L(t)A.$$

Then there is a time dependent invertible matrix $X(t) \in \mathsf{GL}_p(\mathbb{R})$, defined for all $t \in \mathbb{R}$, such that the solution to the IVP *with $A(0) = A_0$ is $A(t) = X(t)A_0$.*

The matrix solution $X(t)$ is called the ***fundamental solution*** of the ODE. A discussion of linear non-autonomous systems can be found in many texts; see for example [1, § 27].

By taking the transpose of the ODE, it follows that the analogous theorem applies to ODEs of the form $\dot{A} = A\,L(t)$.

Problems

C.1 Let $\mathbf{v}(x)$ be a vector field on $\mathbb{R}^n$ and ϕ_t the corresponding flow and let $f\colon \mathbb{R}^n \rightarrowtail \mathbb{R}^p$ be a smooth map. Show that

$$\frac{\mathsf{d}}{\mathsf{d}t}(f \circ \phi_t) = \mathsf{t}f(\mathbf{v}) \circ \phi_t. \qquad (\dagger)$$

C.2 Let $\mathbf{v}$ be a smooth vector field on $\mathbb{R}^n$, and suppose that $\mathbf{v}(x_0) \neq 0$. Show that the integral curve through x_0 is a 1–dimensional immersed submanifold in some neighbourhood of x_0. [The *integral curve* through x_0 is the solution $x(t)$ with initial condition $x(0) = x_0$.]

C.3 Prove the 'non-autonomous local group property' of the flow of a non-autonomous vector field $\mathbf{v}(x,t)$, as detailed in the proof of Theorem C.7, and in particular Equation (C.8). [*Hint*: *translate the local group property of Remark* C.4 *to the given statement for non-autonomous* ODEs *using the non-autonomous to autonomous trick described on p.* 362.]

C.4 Let $\mathbf{v}$ be a vector field defined on a neighbourhood U of the origin in $\mathbb{R}^n$ with $\mathbf{v}(0) = 0$, and consider the ODE on U, $\dot{x} = \mathbf{v}(x)$. Let ϕ_t be the integral flow associated to $\mathbf{v}$. (a) Adapt the proof of Theorem C.7 to show that for any compact interval containing 0 there is a neighbourhood V for which the flow exists. (b) Deduce that the germ at 0 of ϕ_t is defined for all $t \in \mathbb{R}$.

D

Rings, ideals and modules

MANY OF THE CALCULATIONS INVOLVED in singularity theory are algebraic, and in this appendix we present/recall the required algebraic background. The earlier chapters (on right equivalence of functions) just require rings and ideals, while later material (on maps) relies on some knowledge of modules.

D.1 Rings

A ring is a set together with the two operations of addition and multiplication. The archetypal example is the set $\mathbb{Z}$ of all integers.

Definition D.1. A ***commutative ring*** is a set R endowed with two binary operations called addition and multiplication, denoted $+$ and $\cdot$, satisfying the following axioms. For all $r, s, t \in R$

R1 $r + s = s + r$

R2 $r + (s + t) = (r + s) + t$

R3 there exists an element $0 \in R$ (called zero) such that $0 + r = r$

R4 for each $r \in R$ there exists an element $r' \in R$ such that $r + r' = 0$.

R5 $r \cdot (s + t) = r \cdot s + r \cdot t$

R6 there exists an element $1 \in R$ (called the identity) such that $1 \cdot r = r$.

R7 $rs = sr$ (ring is commutative)

R8 $r \cdot (s \cdot t) = (r \cdot s) \cdot t$

One usually writes the product as rs rather than $r \cdot s$. ☆

Henceforth we will just talk of a ring, assuming it to be commutative.

In any (commutative) ring R, an element $a \in R$ is called a ***unit*** if it has a multiplicative inverse: that is, if there exists $b \in R$ such that $a \cdot b = 1$; in this case b can be written a^{-1}. A ***field*** is a ring in which every element except zero is a unit.

Remarks D.2. (i) The element r' in [R4] is usually written $(-r)$, so that $r+(-r) = 0$. This allows one to define subtraction by $r - s := r + (-s)$.
(ii) In some texts, axiom [R6] is omitted, and in such texts our definition would be called a 'commutative ring with unit'.

Examples D.3. (i). The set of integers $\mathbb{Z}$ forms a ring; the only units are ± 1.

(ii). The set $\mathbb{Z}_n$ of integers modulo n is a ring. The units are the integers coprime to n, and it is a field if and only if n is a prime number.

(iii). The sets $\mathbb{Q}, \mathbb{R}, \mathbb{C}$ (rational numbers, real numbers, complex numbers) are fields.

(iv). $\mathbb{R}[x_1, \ldots, x_n]$, the set of polynomials in n variables with real coefficients forms a ring;

(v). more generally, if R is any ring, then $R[x_1, \ldots, x_n]$ is the ring of polynomials in n variables with coefficients in R;

(vi). if R is any ring, then $R[[x_1, \ldots, x_n]]$ is the ring of formal power series in n variables with coefficients in R (see below);

(vii). $C^\infty(U)$ the set of smooth real-valued functions on any open subset $U \subset \mathbb{R}^n$; the units are the functions which are nowhere zero (so it is not a field).

In Example D.3(vi), a ***formal power series*** is similar to a polynomial, except that the series may be infinite. For example

$$1 + x + x^2 + x^3 + \cdots + x^n + \cdots = \sum_{j=0}^{\infty} x^j$$

is a formal power series. This example converges in a neighbourhood of the origin (to $(1 - x)^{-1}$), but in general a formal power series may or may not converge. An example of a power series in $\mathbb{Z}[[x, y]]$ is

$$\sum_{j=0}^{\infty} \sum_{k=0}^{\infty} (j + k)x^j y^k.$$

An example of a formal power series which does not converge on any neighbourhood of 0 is $\sum_{n=0}^{\infty} n!\, x^n$.

D.2 Ideals

Definition D.4. A subset $J \subset R$ of a ring R is an ***ideal*** if it has the following 2 properties: for all $a \in R$ and all $g, h \in J$,

(i). $g + h \in J$;

(ii). $ah \in J$.

We write $J \lhd R$ to mean that J is an ideal in R. ☆

It follows for example that $0 \in J$ and that $g \in J \Rightarrow -g \in J$. It is useful to notice that if an ideal J contains a unit then J is equal to the whole ring: this follows immediately from the definition and is left as an exercise.

Let R be any ring, and let $h_1, \ldots, h_r \in R$. The ***ideal generated by*** $h_1, \ldots, h_r$, denoted $\langle h_1, \ldots, h_r \rangle$, is the set of all linear combinations of the elements $h_1, \ldots, h_r$, with coefficients taken from the ring; that is,

$$\langle h_1, \ldots, h_r \rangle := \left\{ a_1 h_1 + \cdots + a_r h_r \mid a_j \in R \right\}.$$

It is not hard to see that this is the smallest ideal containing the given elements. An ideal $J \lhd R$ is said to be ***finitely generated*** if there is a finite set of elements $h_1, \ldots, h_r$ such that $J = \langle h_1, \ldots, h_r \rangle$.

Examples D.5. The following are examples of ideals, as is readily checked (proofs are left to the reader):

(i). Let $n \in \mathbb{Z}$. The set $n\mathbb{Z}$ of all multiples of n is an ideal in $\mathbb{Z}$; it has a single generator, namely n, so $n\mathbb{Z} = \langle n \rangle$.

(ii). $J_0 = \{p \in \mathbb{R}[x_1, \ldots, x_n] \mid p(0, \ldots, 0) = 0\}$.

(iii). R is itself an ideal in R, and so is $\{0\}$, all other ideals are called ***proper ideals***.

(iv). If I and J are ideals in R then so is their intersection $I \cap J$.

(v). The product of two ideals is defined to be the set $I \cdot J = \{i \cdot j \mid i \in I,\ j \in J\}$; it is an ideal. Powers of an ideal are defined recursively: $J^1 = J$ and $J^n = J \cdot J^{n-1}$ for $n > 1$.

(vi). If I is a finitely generated ideal in R and S is a ring containing R, then we write SI or $S.I$ for the ideal in S generated by the generators of I.

Given an ideal J in a ring R, define the equivalence relation of elements modulo J by $a \equiv b \bmod J$ if and only if $a - b \in J$. The set of equivalence classes is denoted R/J and called the ***quotient*** of R by J. Given $a \in R$ the equivalence class containing a is then the subset $a + J \subset R$.

Proposition D.6. *Let R be a commutative ring and J an ideal. The quotient R/J has a natural ring structure inherited from that of R.*

PROOF: Addition and multiplication are defined 'in the obvious way':
- $(a + J) + (b + J) = (a + b) + J$,
- $(a + J) \cdot (b + J) = (a \cdot b) + J$.

The reader should check that these binary operations on R/J are well defined. Together with defining zero to be $0 + J = J$ and the identity to be $1 + J$, it is easy to show that the axioms of a ring are satisfied. ✔

Definition D.7. A map $\phi \colon R \to S$ between two rings is a ***homomorphism*** if it respects the operations of addition and multiplication:

$$\phi(r + s) = \phi(r) + \phi(s), \quad \text{and} \quad \phi(r \cdot s) = \phi(r) \cdot \phi(s).$$

It is an ***isomorphism*** if it is invertible; that is, there is a ring homomorphism $\psi \colon S \to R$ such that $\phi \circ \psi = \mathsf{Id}_S$ and $\psi \circ \phi = \mathsf{Id}_R$. ☆

Definition D.8. An ideal J in R is ***maximal*** if there are no ideals in R containing J except for J and R themselves. A ring is said to be a ***local ring*** if it has a unique maximal ideal. ☆

Example D.9. The ideal J_0 in Examples D.5(ii) is a maximal ideal. To see this, suppose I is an ideal strictly containing J_0, and let $p \in I \setminus J_0$. Then $p(0) \neq 0$. Define $p_0 := p - p(0)$. Then $p_0 \in J_0$ by the definition of J_0. Consequently, by axiom (1) of an ideal, $p(0) = p - p_0 \in I$. But $p(0)$ is a non-zero constant so is a unit in the ring of polynomials. Thus I contains a unit so is equal to the entire ring.

Moreover, given any point $x_0 \in \mathbb{R}^n$ the ideal J_{x_0} of polynomials which are zero at x_0 also form a maximal ideal, by the same reasoning as for J_0. The polynomial ring is therefore *not* a local ring.

Theorem D.10 (First fundamental theorem for rings). *Let* $\phi\colon R \to S$ *be a ring homomorphism, then* $\ker\phi$ *is an ideal and* ϕ *provides an isomorphism*

$$\operatorname{im}\phi \simeq {}^{R}\!\big/_{\ker\phi}.$$

In fact every ideal is the kernel of a ring homomorphism: if I is the ideal then the natural homomorphism $R \to {}^{R}\!/_{I}$ has kernel I.

D.3 Ideals and varieties

Given any integer $n \geq 1$, the set of polynomials in n variables $x_1, \ldots, x_n$, with real coefficients is denoted $\mathbb{R}[x_1, x_2, \ldots, x_n]$. With the usual definitions of addition and multiplication of polynomials this forms a ring.

Similarly, one defines $\mathbb{C}[x_1, x_2, \ldots, x_n]$ to be the polynomials in $x_1, \ldots, x_n$ with complex coefficients.

Let I be an ideal in $\mathbb{R}[x_1, \ldots, x_n]$. Then the associated subset of $\mathbb{R}^n$ is

$$V(I) = \{\mathbf{x} \in \mathbb{R}^n \mid p(\mathbf{x}) = 0,\ \forall p \in I\}.$$

In other words, $V(I)$ is defined by a collection of polynomial equations, and in particular one for each generator of I (see Problem D.5). A subset defined in this way is called an ***algebraic variety***. An analogous definition holds for ideals in $\mathbb{C}[x_1, \ldots, x_n]$.

Conversely, given an algebraic variety V in $\mathbb{R}^n$ one defines the associated ideal to be

$$I(V) = \left\{ f \in \mathbb{R}[x_1, \ldots, x_n] \mid f\big|_V = 0 \right\}.$$

And analogously for $V \subset \mathbb{C}^n$. For example, the maximal ideal J_0 is the ideal corresponding to the set $\{0\}$.

Many different ideals may give rise to the same variety, for example J_0 and J_0^2 both have associated variety $V = \{0\}$. On the other hand, two different varieties will always have different ideals.

The discussion above can also be applied to the ring of analytic functions (or germs) on $\mathbb{R}^n$, giving rise to (germs of) *analytic varieties*.

D.4 Derivations

Extending the product rule $(uv)' = u'v + uv'$ for differentiation of functions with respect to a single variable, one defines a ***derivation*** on a ring as follows.

Definition D.11. A ***derivation*** on a ring R is a map $\mathcal{D}\colon R \to R$ satisfying the ***Leibniz rule*** (or product rule):

$$\mathcal{D}(rs) = (\mathcal{D}r)s + r(\mathcal{D}s).$$

☆

The case of interest to us is when R is a ring of infinitely differentiable functions (such as analytic or smooth functions or polynomials). In that case, by the product rule for differentiation, any first order partial differential defines a derivation:

$$\frac{\partial}{\partial x_j}\colon f \mapsto \frac{\partial f}{\partial x_j}$$

More generally, given functions $v_j \in R$, the map $\mathcal{D}\colon R \to R$ defined by

$$\mathcal{D}\colon f \mapsto \sum v_j \frac{\partial f}{\partial x_j}$$

is a derivation, again by the product rule. (Higher derivatives are *not* derivations.)

Let $U \subset \mathbb{R}^n$ be an open set, and $C^\infty(U)$ the ring of smooth functions on U. Let $\mathbf{v}$ be a vector field on U. Then $\mathbf{v}$ defines a derivation on $C^\infty(U)$ by differentiating in the $\mathbf{v}$ direction (as in directional derivatives):

$$(\mathcal{D}f)(x) = \mathsf{d}f_x(\mathbf{v}(x)). \tag{D.1}$$

That is, in coordinates, $(\mathcal{D}f)(x) = \sum_j v_j(x)\frac{\partial f}{\partial x_j}$, as above. Because of this, one often writes the vector field $\mathbf{v}$ as

$$\mathbf{v} = \sum_j v_j \frac{\partial}{\partial x_j}. \tag{D.2}$$

Conversely, given any derivation $\mathcal{D}$ of $C^\infty(U)$, there is a unique vector field $\mathbf{v}$ on U satisfying (D.1): using any system of coordinates $\{x_1, \ldots, x_n\}$, let

$$v_j(x) = \mathcal{D}x_j,$$

then the vector field $\mathbf{v}$ is simply the one given by (D.2).

D.5 Modules

Modules are to rings what vector spaces are to fields: one can take linear combinations of elements of a module, where the coefficients lie in the given ring. The formal definition is as follows.

Definition D.12. Let R be a commutative ring. A ***module over*** R (or an R-module) is a set M in which elements can be added and subtracted, and can be multiplied by elements of R. The operations have to be compatible in that they satisfy the following axioms: for all $u, v, w \in M$ and all $r, s \in R$,

M1 $u + v = v + u$,

M2 $u + (v + w) = (u + v) + w$,

M3 $r \cdot (u + v) = r \cdot u + r \cdot v$,

M4 $(r + s) \cdot u = r \cdot u + s \cdot u$,

M5 $r \cdot (s \cdot u) = (rs) \cdot u$,

M6 there exists an element $0 \in M$ (called zero) such that $0 + u = u$,

M7 for each $u \in M$ there is an element $(-u) \in M$ such that $u + (-u) = 0$,

M8 for each $u \in M$, one has $1.u = u$, where 1 is the identity element of R.

★

As with rings, one usually just writes the scalar multiplication as ru rather than $r \cdot u$.

Example D.13. These examples are given without proof but are all straightforward to check.

(i). If R is a field, then M is an R-module if and only if it is a vector space over R.

(ii). The set $\mathbb{Z}_n$ of integers modulo n is a $\mathbb{Z}$-module (as well as being a ring in its own right).

(iii). Let $U \subset \mathbb{R}^n$ be an open set, and let $\Theta(U)$ be the set of all smooth vector fields on U, then $\Theta(U)$ is a $C^\infty(U)$-module (any smooth function times a smooth vector field is a smooth vector field).

(iv). The same is true for germs: the set θ_n of germs at 0 of vector fields on $\mathbb{R}^n$ is a module over $\mathcal{E}_n$.

(v). If R is a ring and I an ideal, then I and R/I are also R-modules.

✐

A module is ***finitely generated*** if there are finitely many elements $u_1, \dots, u_r$ of M which generate M. That is for any element $u \in M$ there are elements $a_1, \dots, a_r \in R$ such that

$$u = \sum_{i=1}^{r} a_i u_i.$$

We write $M = \langle u_1, \dots, u_r \rangle$, or $M = R\{u_1, \dots, u_r\}$ for such a module.

In the examples above, if R is a field then M is finitely generated iff it is of finite dimension as a vector space; the $\mathbb{Z}$-module $\mathbb{Z}_n$ is generated by the class $1 \in \mathbb{Z}_n$; $\theta(U)$ is generated over $C^\infty(U)$ by the basic vector fields $\mathbf{e}_1, \dots, \mathbf{e}_n$ (given by the basis vectors of $\mathbb{R}^n$ considered as constant vector fields).

Definition D.14. If R and S are rings with $R \subset S$, and M is a finitely generated R-module, then we write SM or $S.M$ for the module over S generated by the generators of M (cf. Example D.5(vi)): explicitly, if $M = R\{u_1, \dots, u_r\}$ then

$$S.M = S\{u_1, \dots, u_r\} = \left\{ \sum_j s_j u_j \mid s_j \in S \right\}.$$

☆

Definition D.15. Let M be a module over R. A subset $N \subset M$ is a ***submodule*** if it is a module in its own right, using the operations of M. In other words, if it is closed under the operations of addition and scalar multiplication. We write $N \lhd M$ to mean N is a submodule of M. ☆

D.5A Vector fields and modules

In this section we look briefly at some algebraic properties of vector fields associated to a smooth map $f\colon \mathbb{R}^n \rightarrowtail \mathbb{R}^p$, with $U = \mathrm{dom}(f)$. Let us write $\theta(U)$ for the $C^\infty(U)$-module of all smooth vector fields on U, and $\theta_f(U)$ the $C^\infty(U)$-module of all smooth vector fields along f (see Section A.4).

The set $\theta_f(U)$ can be seen as a module in two different ways. Firstly there is the more obvious structure of being a module over the ring $C^\infty(U)$: $(g\mathbf{v})(x) = g(x)\mathbf{v}(x)$. But secondly it is also a module over the ring $C^\infty(\mathbb{R}^p)$, as follows. If $h \in C^\infty(\mathbb{R}^p)$ then we define

$$(h.\mathbf{v})(x) = h(f(x))\mathbf{v}(x). \tag{D.3}$$

We leave it to the reader to check that this does indeed define a module structure. (See Problem D.9.)

We define two maps into $\theta_f(U)$,

$$\mathrm{t}f\colon \theta(U) \to \theta_f(U) \quad \text{and} \quad \omega f\colon \theta(\mathbb{R}^p) \to \theta_f(U)$$

by

$$\mathrm{t}f(\mathbf{v}) = \sum_i v_i(x)\frac{\partial f}{\partial x_i}(x), \quad \text{and} \quad \omega f(\mathbf{w})(x) = \mathbf{w}(f(x)).$$

We claim that both of these are homomorphisms. Firstly, it is straightforward to see that $\mathrm{t}f$ is a homomorphism of $C^\infty(U)$-modules (Problem D.10). On the other hand, ωf is a homomorphism of $C^\infty(\mathbb{R}^p)$-modules, because with $\mathbf{w} \in \theta(\mathbb{R}^p)$, and $h \in C^\infty(\mathbb{R}^p)$,

$$\omega f(h\mathbf{w})(x) = (h\mathbf{w})(f(x)) = h(f(x))\,\mathbf{w}(f(x)) = h(f(x))\,\omega f(\mathbf{w})(x).$$

That is,

$$\omega f(h\mathbf{w}) = h.\omega f(\mathbf{w}).$$

We use these homomorphisms $\mathrm{t}f$ and ωf many times in the main text, but at the level of germs.

D.6 Noetherian rings and modules

The Noetherian property (named after Emmy Noether) is about objects being finitely generated:

- A ring is ***Noetherian*** if every ideal is finitely generated.
- A module is ***Noetherian*** if every submodule is finitely generated.

If we consider a ring as a module over itself, then the submodules are precisely the ideals, so the first definition is a special case of the second.

The simplest example is the ring $\mathbb{Z}$ of integers which is Noetherian because every ideal is of the form $n\mathbb{Z}$ for some $n \in \mathbb{Z}$; that is every ideal in $\mathbb{Z}$ is generated by just one element ($n\mathbb{Z}$ is generated by the integer n).

Theorem D.16. *The following rings and modules are Noetherian:*

(i). $\Bbbk[x_1, \dots, x_n]$, *the ring of all polynomials in n variables, where* $\Bbbk = \mathbb{R}$ *or* $\mathbb{C}$, *corresponding to real or complex polynomials respectively,*

(ii). $\mathcal{O}_n$, *the ring of germs at a point of holomorphic functions on* $\mathbb{C}^n$ *(Rückert's basis theorem),*

(iii). any finitely generated module over a Noetherian ring.

On the other hand, if $U \subset \mathbb{R}^n$ is open, the ring $C^\infty(U)$ of smooth functions on U is not Noetherian, and neither is the ring $\mathcal{E}_n$ of germs of smooth functions at the origin in $\mathbb{R}^n$. For example, neither the ideal of functions in $C^\infty(U)$ whose Taylor series vanishes at a given point $x_0 \in U$, nor the ideal of flat functions in $\mathcal{E}_n$ are finitely generated: see Problem D.8.

Proofs of parts (i) and (iii) of the theorem can be found in many books on commutative algebra, for example Atiyah & MacDonald [7]. A proof of part (ii) can be found in the book by Ebeling [34].

D.7 Nakayama's lemma

In this section we prove a very useful result from commutative algebra known as Nakayama's lemma. The proof requires us to understand what it means for a matrix with entries in a ring to be invertible, and the simple necessary and sufficient condition for invertibility.

Let Λ be an $n \times n$ matrix whose entries are elements of a ring R; write $M_n(R)$ for the set (in fact a non-commutative ring) of all such matrices. We say $\Lambda \in M_n(R)$ is ***invertible over*** R if there is a matrix written $\Lambda^{-1} \in M_n(R)$ such that $\Lambda\Lambda^{-1} = \Lambda^{-1}\Lambda = \mathrm{Id}$. The set (indeed, group) of such invertible matrices is denoted $\mathsf{GL}_n(R)$.

Example D.17. Let $R = \mathbb{Z}$. Then $\Lambda = \begin{pmatrix} 3 & 2 \\ 1 & 1 \end{pmatrix} \in M_2(\mathbb{Z})$ is invertible, with $\Lambda^{-1} = \begin{pmatrix} 1 & -2 \\ -1 & 3 \end{pmatrix}$. On the other hand $K = \begin{pmatrix} 2 & 1 \\ 0 & 1 \end{pmatrix}$ is not invertible over $\mathbb{Z}$. Of course K is invertible over $\mathbb{R}$ (or even over $\mathbb{Q}$), with inverse $K^{-1} = \begin{pmatrix} 1/2 & -1/2 \\ 0 & 1 \end{pmatrix}$, but this inverse is clearly not in $M_2(\mathbb{Z})$. ✐

Given a square matrix $\Lambda \in M_n(R)$, it would be useful to have a condition for it to be invertible over R. In fact a simple (and familiar) criterion does exist: recall the classical formula for the inverse of a matrix Λ over $\mathbb{R}$ (or $\mathbb{C}$):

$$\Lambda^{-1} = \frac{1}{\det \Lambda}\,\mathrm{adj}(\Lambda).$$

This formula is purely formal: it does not require that the entries of the matrix be real numbers, only that they commute (and be in a ring, so they can be multiplied and added). The *adjugate matrix* $\mathrm{adj}(\Lambda) \in M_n(R)$, which is the transpose of the matrix of cofactors, is obtained by calculating the minors of all elements of the matrix, and involves only the ring operations. In fact, for any matrix Λ over any (commutative) ring R one has

$$\Lambda\,\mathrm{adj}(\Lambda) = (\det\Lambda)\mathrm{Id}. \tag{D.4}$$

It follows that Λ^{-1} exists if and only if $\det\Lambda$ is invertible in R. Thus we have the following useful result.

Lemma D.18. *Let R be a commutative ring. A square matrix over R is invertible over R if and only if its determinant is a unit in R.*

PROOF: The 'if' part is proved above. The 'only if' part is straightforward: if Λ is invertible, then

$$1 = \det(\mathrm{Id}) = \det(\Lambda^{-1}\Lambda) = \det(\Lambda^{-1})\ \det(\Lambda),$$

showing that $\det(\Lambda)$ is invertible (with $\det(\Lambda)^{-1} = \det(\Lambda^{-1})$). ✔

For $R = \mathbb{R}$ or $\mathbb{C}$, the only non-unit is 0, so Λ is invertible if and only if $\det\Lambda \neq 0$, as should be familiar. Indeed, the same is true for any field. For $R = \mathbb{Z}$ the only units are ± 1, and indeed the matrix Λ in Example D.17 has determinant equal to 1, but K has determinant equal to 2 (not a unit in $\mathbb{Z}$).

Example D.19. Let $R = \mathbb{R}[x]$ (the ring of polynomials in x with real coefficients); the only units in R are ± 1. Let $\Lambda = \begin{pmatrix} 1+x & x \\ -x & 1-x \end{pmatrix}$, then $\det\Lambda = 1$ so Λ belongs to $\mathsf{GL}_2(R)$. Exercise: find its inverse. ✎

Theorem D.20 (Nakayama's lemma). *Let R be a commutative ring, and $\mathcal{M}$ an ideal with the property that $x \in \mathcal{M} \Rightarrow 1 + x$ is a unit in R. Let A, B be submodules of an R-module M with A finitely generated. Then*

$$A \subset B + \mathcal{M}A \ \Rightarrow\ A \subset B.$$

It follows that if $A = B + \mathcal{M}A$ then $A = B$, for if $B + \mathcal{M}A \subset A$ then a fortiori $B \subset A$. Although this theorem is stated for submodules A, B, it is often applied when A, B are ideals in the ring R (recall that ideals in R are the same as submodules of R).

PROOF: Let $a_1, a_2, \ldots, a_r$ be the generators of A. Since $A \subset B + \mathcal{M}A$ we can find for each a_i an element $b_i \in B$ and $\lambda_{ij} \in \mathcal{M}$ for which

$$a_i = b_i + \sum_j \lambda_{ij} a_j.$$

Let Λ denote the $r \times r$ matrix (λ_{ij}), then these equations can be written,

$$\mathbf{a} = \mathbf{b} + \Lambda\mathbf{a}$$

where $\mathbf{a}$ is the column vector $(a_1, a_2, \ldots, a_r)^T$ and $\mathbf{b} = (b_1, b_2, \ldots, b_r)^T$. Rearranging gives

$$(I - \Lambda)\mathbf{a} = \mathbf{b}.$$

To solve for $\mathbf{a}$ we need to invert the matrix $I - \Lambda$. But this can be done because $\det(I - \Lambda) = 1 + \lambda$ for some $\lambda \in \mathcal{M}$, and by hypothesis all such elements are units, so the invertibility of $I - \Lambda$ follows from Lemma D.18. Thus,

$$\mathbf{a} = (I - \Lambda)^{-1}\mathbf{b},$$

and $(I - \Lambda)^{-1} \in M_r(R)$, so that each $a_i \in B$ and $A \subset B$ as required. ✔

Example D.21. Let $R = \mathcal{E}_n$ and $\mathcal{M} = \mathfrak{m}_n$. Consider the two ideals $I = \mathfrak{m}_n$ and

$$J = \langle x_1 + h_1,\ x_2 + h_2,\ \ldots, x_n + h_n \rangle$$

where each $h_i \in \mathfrak{m}_n^2$. We want to show $J = I$. Clearly $J \subset I$, but to show $I \subset J$ we use Nakayama's lemma after first showing that $I \subset J + \mathfrak{m}_n I$. Note that $I = \mathfrak{m}_n$ is finitely generated (Hadamard's lemma).

To show $I \subset J + \mathfrak{m}_n I$, consider $x_i \in I$ (a generator of I). Now $x_i = (x_i + h_i) - h_i$ and $h_i \in \mathfrak{m}_n^2 = \mathfrak{m}_n I$ so that indeed $x_i \in J + \mathfrak{m}_n I$. This is valid for each x_i so that $I \subset J + \mathfrak{m}_n I$ as required. Therefore, $I = J$. ✐

Remark D.22. The hypothesis that A is finitely generated is essential. Indeed, Nakayama's lemma is sometimes used to show that a module is not finitely generated. ❞

D.7A An alternative version

The following, seemingly simpler, theorem is in fact equivalent to Nakayama's lemma as stated above (see Problem D.7).

Theorem D.23 (Nakayama's lemma, v2). *Suppose $R, \mathcal{M}$ are as in the theorem above, and let A be a finitely generated module (or ideal). Then*

$$A = \mathcal{M}A \implies A = 0.$$

Proof: The proof is by contradiction. Suppose A is non-zero, and let $\{a_1, \dots, a_r\}$ be a minimal set of generators of A. Now consider a_1. By hypothesis $a_1 \in \mathcal{M}A$, which implies

$$a_1 = \sum_j g_j a_j, \quad (g_i \in \mathcal{M}).$$

In particular

$$(1 - g_1)a_1 = \sum_{j=1}^{r} g_j a_j.$$

Since $(1-g_1)$ is a unit, it follows that a_1 belongs to the module (or ideal) generated by $a_2, \dots, a_n$, so implying that the original set of generators was not minimal. ✔

D.8 Weighted homogeneous functions

A (real or complex) polynomial $f(x_1, \dots, x_n) \in \Bbbk[x_1, \dots, x_n]$ (where $\Bbbk = \mathbb{R}$ or $\mathbb{C}$) is said to be ***homogeneous of degree*** d if

$$f(\lambda x_1, \lambda x_2, \dots, \lambda x_n) = \lambda^d f(x_1, x_2, \dots, x_n), \quad \forall \lambda \in \Bbbk.$$

For example, $f(x, y) = x^3 + 2xy^2 - y^3$ is homogeneous of degree 3.

This concept has a natural generalization as follows. Assign a a positive integer w_j to each variable x_j, called its *weight*. A function is ***weighted homogeneous*** of degree d with respect to these weights if

$$f(\lambda^{w_1} x_1, \lambda^{w_2} x_2, \dots, \lambda^{w_n} x_n) = \lambda^d f(x_1, x_2, \dots, x_n), \quad \forall \lambda \in \Bbbk. \tag{D.5}$$

For example, with two variables, if $w_1 = 2, w_2 = 3$, then the function $f(x, y) = x^3 + y^2$ is weighted homogeneous of degree 6, because

$$f(\lambda^2 x, \lambda^3 y) = (\lambda^2 x)^3 + (\lambda^3 y)^2 = \lambda^6 (x^3 + y^2).$$

One usually writes the weights as a vector: $\mathbf{w} = (w_1, w_2, \dots, w_n) \in \mathbb{Z}^n$, and then a monomial $\mathbf{x}^{\mathbf{a}} = x_1^{a_1} x_2^{a_2} \dots x_n^{a_n}$ is weighted homogeneous of degree d if and only if $\sum w_i a_i = d$, which we can also write as the scalar product $\mathbf{w} \cdot \mathbf{a} = d$.

Clearly, a polynomial is homogeneous if and only if it is weighted homogeneous with weight vector $\mathbf{w} = (1, 1, \ldots, 1)$.

Recall that the powers of the maximal ideal $\mathfrak{m}_n \subset \mathcal{E}_n$ satisfy

$$\mathfrak{m}_n^k \mathfrak{m}_n^\ell = \mathfrak{m}_n^{k+\ell},$$

and $\mathfrak{m}_n^r$ is generated by the monomials of degree r. For weighted homogeneous functions a similar property holds, but it is not so strong: fix the weight vector $\mathbf{w}$ and let J_r be the ideal generated by

$$J_r = \langle x^{\mathbf{a}} \mid \mathbf{w} \cdot \mathbf{a} \geq r \rangle . \tag{D.6}$$

The product of two such ideals satisfies $J_k J_\ell \subset J_{k+\ell}$, but in general one does not have equality (see Problem D.11).

Definition D.24. The vector field (or derivation)

$$\mathfrak{e} = \sum_{j=1}^{n} w_j x_j \frac{\partial}{\partial x_j}$$

is called the ***Euler vector field*** for the weights $w_1, \ldots, w_n$. It is a linear vector field, whose matrix is

$$L = \begin{pmatrix} w_1 & 0 & \cdots & 0 \\ 0 & w_2 & \cdots & 0 \\ \vdots & \vdots & \ddots & \vdots \\ 0 & 0 & \cdots & w_n \end{pmatrix}.$$

☆

For example, for homogeneous functions where all weights are equal to 1, the Euler vector field is

$$\mathfrak{e} = x_1 \frac{\partial}{\partial x_1} + x_2 \frac{\partial}{\partial x_2} + \cdots + x_n \frac{\partial}{\partial x_n}.$$

The matrix for this is the identity matrix; geometrically on $\mathbb{R}^n$ it is a vector field everywhere pointing radially away from the origin. If f is homogeneous of degree r, then the conclusion of Problem D.12 becomes

$$\mathfrak{e}(f) = \sum_{j=1}^{n} x_j \frac{\partial f}{\partial x_j}(x) = r\, f(x) .$$

See Problem D.12 for the weighted homogeneous version.

D.9 Elimination

There is an extensive theory of elimination of variables between simultaneous polynomial equations (extending Gaussian elimination for linear equations). Here we briefly outline the basic step, eliminating one variable between two equations using the so-called resultant. Proofs can be found in [113]. The general setting, while based on this case, uses Gröbner bases and a thorough account can be found in the books [25, 26].

Consider two polynomial equations P and Q, respectively of degrees n and m in x, whose coefficients a_k, b_k are polynomials in other variables (say $y_1, \dots, y_r$),

$$\begin{aligned} P: \quad a_n x^n + a_{n-1}x^{n-1} + \cdots + a_1 x + a_0 &= 0, \\ Q: \quad b_m x^m + b_{m-1}x^{m-1} + \cdots + b_1 x + b_0 &= 0. \end{aligned} \tag{D.7}$$

We assume a_n, b_m are both non-zero polynomials in the y_j.

Now form the ***resultant***, which is the following $(m+n) \times (m+n)$ determinant,

$$R(P,Q) = \begin{vmatrix} a_n & a_{n-1} & \dots & a_1 & a_0 & 0 & 0 & \cdots & 0 \\ 0 & a_n & \dots & a_2 & a_1 & a_0 & 0 & \dots & 0 \\ \vdots & & & & \vdots & & & \vdots & \\ 0 & 0 & \dots & a_n & a_{n-1} & a_{n-2} & \dots & a_1 & a_0 \\ b_m & b_{m-1} & \dots & b_1 & b_0 & 0 & 0 & \cdots & 0 \\ 0 & b_m & \dots & b_2 & b_1 & b_0 & 0 & \dots & 0 \\ \vdots & & & & \vdots & & & \vdots & \\ 0 & 0 & \dots & 0 & b_m & b_{m-1} & \dots & b_1 & b_0 \end{vmatrix}.$$

(there are m rows with the a_j coefficients and n rows with the b_j coefficients). The resultant determinant R is then a polynomial in $y_1, \dots, y_r$. One finds that if the two original polynomials are (weighted) homogeneous of degrees d_1 and d_2 respectively (in $x, y_1, \dots, y_r$), then the resultant is (weighted) homogeneous of degree $d_1 d_2$ (in $y_1, \dots, y_r$).

Theorem D.25. *Given values for $y_1, \dots, y_r \in \mathbb{C}$, these satisfy the resultant equation $R = 0$ if and only if there is an $x \in \mathbb{C}$ such that equations (D.7) are satisfied.*

Note that if there are no supplementary variables y_j and P, Q are simply polynomials in x, the resultant R is a constant which vanishes if and only if the two polynomials have a common root.

Example D.26. Consider the two polynomials

$$P = x^3 + y_1 x + y_2, \quad Q = 3x^2 + y_1.$$

The resultant determinant is

$$R(y_1, y_2) = \begin{vmatrix} 1 & 0 & y_1 & y_2 & 0 \\ 0 & 1 & 0 & y_1 & y_2 \\ 3 & 0 & y_1 & 0 & 0 \\ 0 & 3 & 0 & y_1 & 0 \\ 0 & 0 & 3 & 0 & y_1 \end{vmatrix} = 4y_1^3 + 27y_2^2,$$

an expression we see several times in the main text. Note that in this example, P and Q are weighted homogeneous of degrees 3 and 2 respectively, and with the required weights R is homogeneous of degree 6. ✐

D.10 Nullstellensatz

Some of the chapters contain a section detailing a geometric criterion for finiteness (finite codimension in particular). These are all based on the following theorem.

For this fundamental theorem, relating algebra and geometry, one needs to assume the ground field is algebraically closed. Let R be either of the two rings,

(i). $R = \mathbb{C}[x_1, \ldots, x_n]$, the ring of polynomials in n complex variables, or

(ii). $R = \mathcal{O}_n = \mathbb{C}\{x_1, \ldots, x_n\}$, the ring of convergent power series in n complex variables (or equivalently, the ring of germs at 0 of complex analytic functions).

Theorem D.27 (Nullstellensatz). *Let $I \triangleleft R$ be an ideal and let V be the locus of common zeros of all elements of I, as described in Section D.3. Let $I(V)$ be the ideal in R of all elements vanishing on V. Then $I(V)$ is equal to the radical of I: $I(V) = \sqrt{I}$.*

Here the ***radical*** of an ideal I is the ideal

$$\sqrt{I} = \{a \in R \mid \exists n > 0,\ a^n \in I\}.$$

For the polynomial (algebraic) case this is called Hilbert's Nullstellensatz. A proof can be found in many texts on commutative algebra, for example [7, 36]. The analytic Nullstellensatz is due to Rückert, and a proof can be found in [108].

In this text we only make use of the following particular case:

Corollary D.28. *Let I be an ideal of R (R as above) with the property that $V(I) = \{0\}$. Then I is of finite codimension in R.*

PROOF: If $V(I) = \{0\}$ then it follows from the Nullstellensatz that $\sqrt{I} = \mathfrak{m}_n$ (the maximal ideal in R of functions vanishing at 0). Thus for each $j = 1, \ldots, n$ there is an n_j such that $x_j^{n_j} \in I$. Let $k = \sum_j n_j$. Then $\mathfrak{m}_n^k \subset I$ so that I is indeed of finite codimension. ✔

Problems

D.1 Show that an ideal $J \lhd R$ is maximal if and only if the quotient ring R/J is a field.

D.2 Let $\phi\colon R \to S$ be a ring homomorphism, and $I \lhd R$, and $J \lhd S$ with $\phi(I) \subset J$.
(1) Show that the map $r + I \mapsto \phi(r) + J$ is a well-defined homomorphism $R/I \to S/J$; denote this by $\bar{\phi}$. It is called the homomorphism *induced* by ϕ.
(2) Show that if ϕ is surjective and $\phi(I) = J$ then $\bar{\phi}$ is an isomorphism.

D.3 Let I, J be two ideals in R. Define

$$I + J = \{r + s \mid r \in I, s \in J\}.$$

Show that $I + J$ is an ideal in R, and is the smallest ideal containing both I and J.

D.4 Let R be a ring and I a finitely generated ideal in R. Write $I = \langle f_1, \ldots, f_r \rangle \lhd R$. Now let J be any ideal of R. Show that $I \subset J$ if and only if $f_i \in J$ for each $i = 1, \ldots, r$. (†)

D.5 Let $I = \langle p_1, p_2, \ldots, p_k \rangle \lhd \mathbb{R}[x_1, \ldots, x_n]$. Show that a point $x \in V(I)$ (see p. 373) if and only if

$$p_1(x) = p_2(x) = \cdots = p_k(x) = 0.$$

(That is, it suffices to check the values of the generators at x.)

D.6 Let R, S be rings and $\phi\colon R \to S$ a homomorphism. Show that,
(i) $\ker \phi \lhd R$,
(ii) and more generally, if $J \lhd S$ then $\phi^{-1}(J) := \{r \in R \mid \phi(r) \in J\}$ is an ideal in R,
(iii) if ϕ is surjective, then $I \lhd R \Rightarrow \phi(I) \lhd S$.

D.7 Show that the two versions of Nakayama's lemma (Theorems D.20 and D.23) are equivalent. [*Hint*: *One direction is straightforward, for the other consider the module* $A/(A \cap B)$.]

D.8 Let Φ be the ideal in $\mathcal{E}_1$ of all germs of *flat* functions of 1 variable: those with zero Taylor series (i.e., all derivatives at the origin are zero). Use Nakayama's lemma to show Φ is not finitely generated. (†)

D.9 Let $\phi\colon R \to S$ be a homomorphism ot rings, and let M be an S-module. Show that M becomes an R-module by defining $r.m := \phi(r)m$ for $r \in R$ and $m \in M$. Use this to show that (D.3) indeed makes $\theta_f(U)$ into a $C^\infty(\mathbb{R}^p)$-module. [*Hint*: *put* $\phi = f^*$.]

D.10 Check that indeed $\mathrm{t}f\colon \theta(n) \to \theta(f)$ is a homomorphism of $\mathcal{E}_n$-modules, as claimed in Section D.5A.

D.11 For a given weight vector $\mathbf{w} \in \mathbb{Z}^n$, let J_k be the ideal defined in Equation (D.6). Show $J_k J_\ell \subset J_{k+\ell}$. Now consider the weight vector $\mathbf{w} = (2,3) \in \mathbb{Z}^2$. Write down generators for the ideals J_r for $r \leq 6$. Show that $J_6 \neq J_3^2$. (†)

D.12 Suppose $f \in \mathcal{E}_n$ is weighted homogeneous of degree r with weight vector $\mathbf{w} \in \mathbb{Z}^n$. By differentiating (D.5) with respect to λ, show that

$$\sum_{j=1}^{n} w_j x_j \frac{\partial f}{\partial x_j} = r f .$$

Deduce that $f \in Jf$ (the Jacobian ideal of f).

D.13 Let V be the vector space of (real or complex) homogeneous polynomials of degree at most d in n variables. Let $f \in V$ and let $\mathcal{O}_f$ be the set of all polynomials in V obtained from f by a linear change of coordinates (or change of basis of $\mathbb{R}^n$ or $\mathbb{C}^n$). Show that the tangent space $T_f\mathcal{O}_f$ at f to this set is equal to the subspace spanned by the n^2 polynomials $x_i \frac{\partial f}{\partial x_j}$. [*Note*: *this is the homogeneous part of the ideal* $\mathfrak{m}_n Jf$.] Can this be extended to weighted homogeneous polynomials?

D.14 Find the resultant with respect to x of $P(x,y,z) = x^2 - y^2$ and $Q = x^2 - z^2$ to see that the resultant does not necessarily give the reduced equation after eliminating x.

D.15 Let $I = \langle x^2 + y^2 \rangle \lhd \mathbb{R}[x,y]$. Find $V \subset \mathbb{R}^2$, the real variety of I. Show that $I(V) \neq \sqrt{I}$, and therefore that the Nullstellensatz does not hold for the ring $\mathbb{R}[x,y]$.

E

Solutions to selected problems

As the title suggests, in this appendix we provide solutions to a selection of problems (those marked with a dagger (†) in the problem sections). Some of the solutions are given in greater detail than others, but I hope there is enough in each case for the reader to be able to provide any remaining argument.

Chapter 1

Problem 1.3 The two bifurcation diagrams are shown below, where we employ the convention used in bifurcation of dynamical systems that dashed lines represent unstable equilibrium points, solid lines stable ones.

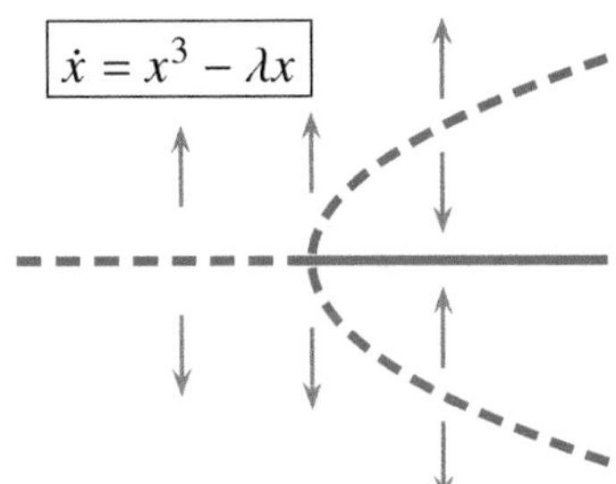

The *subcritical* pitchfork bifurcation

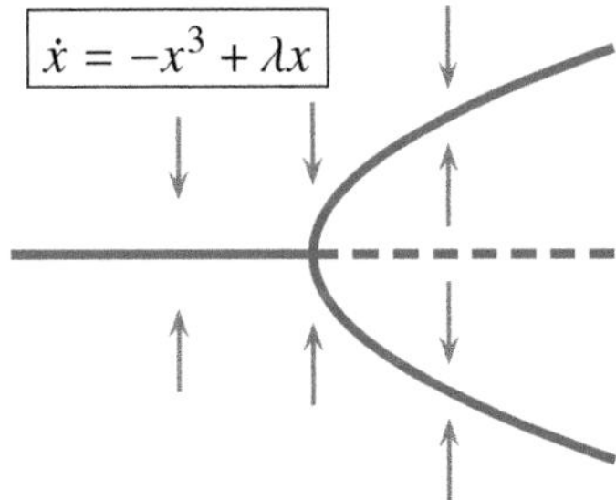

The *supercritical* pitchfork bifurcation

The direction of the arrows show the sign of $\dot{x}$ in each region, from which one can easily see which equilibria are attracting and which are repelling. (The λ–axis points to the right, x upwards.)

Problem 1.5 The equation $h_u(x) = c$ is a cubic equation in x so has at most three real roots. Sketching the curve should convince you that if $u \leq 0$ then the function h_u is monotonically increasing and for each c there is only one solution. On the other hand, if $u > 0$ there is a range of values of c in which there are three solutions and two values of c for which there are precisely two solutions. The question is to determine how this range depends on u.

To rephrase in our new language, the number of roots can only change when there is a double root, which is to say these bifurcations occur where $h'_u(x) = 0$ (as usual). That is, $3x^2 - 3u = 0$, so $x = \pm\sqrt{u}$ (if $u \geq 0$). For such a value of x,

$$c = h_u(x) = (\pm\sqrt{u})^3 - 3u(\pm\sqrt{u}) = \mp 2u\sqrt{u}.$$

Squaring both sides shows $c^2 = 4u^3$, which is the equation of a *semicubical parabola*, or cusp curve, and this is the discriminant, shown in red below. Finally, for $c^2 < 4u^3$ the level sets have three points while for $c^2 > 4u^3$ they only have one point.

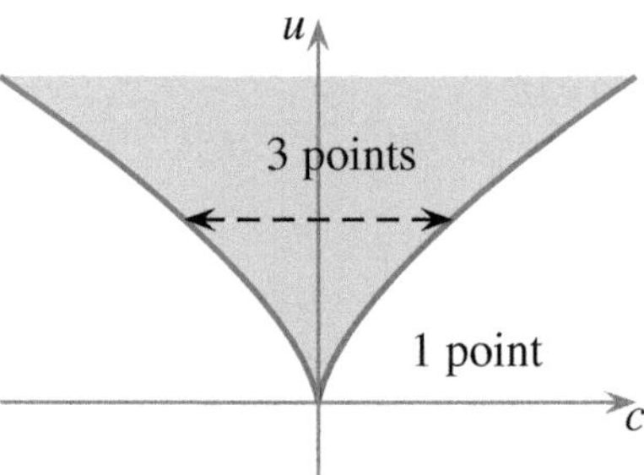

The discriminant for this problem is a semicubical parabola.

The dashed line shows, for a fixed value of $u > 0$, the range of values of c for which the level set contains three points.

Problem 1.6 First note that, as described, λ can be solved for x in the equation $g(x, \lambda) = 0$: it is simply $\lambda = -x^2$, and this function of λ has a local maximum at $x = \lambda = 0$. We solve this problem in several steps, with repeated applications of the implicit function theorem (IFT).

(i) First consider $G(x, \lambda, u) = 0$. Since $\frac{\partial G}{\partial \lambda}(0, 0, 0) = 1 \neq 0$ the IFT guarantees this equation can be solved for λ as a function of (x, u) – say, $\lambda = h(x, u)$ (of course, $h(x, 0) = -x^2$).

(ii) Now we want to show that, as a function of x (i.e. with u constant) this function has a nondegenerate maximum. This requires simply that $\frac{\partial h}{\partial x} = 0$ for some x (possibly depending on u) and that $\frac{\partial^2 h}{\partial x^2} < 0$ at that point.

(iii) We wish therefore to solve $\frac{\partial h}{\partial x}(x, u) = 0$ for x as a function of u. The IFT tells us this is possible (in a neighbourhood of $(0,0)$) provided $h_{xx} \neq 0$. Since $h_{xx}(0,0) = -2$, this holds.

(iv) The nondegeneracy of the critical point of λ as a function of x (for each u) follows by continuity: since $h_{xx}(0,0) = -2 < 0$ it follows by continuity that $h_x x < 0$ on a neighbourhood of $(0,0)$.

Problem 1.8 For $c = 0$ the surface $x^2 + y^2 - z^2 = c = 0$ is a circular cone, with axis along the z-axis. For $c > 0$, it is a one-sheeted hyperboloid, while for $c < 0$ it is a two-sheeted hyperboloid.

Problem 1.10 Starting from the outside region, there are two critical points. Crossing over the evolute (discriminant) once, increases the number to four. And in the single 'central' region there are six.

If a discriminant is just a smooth curve, it is not possible to tell just from the picture which side has more critical points and which side fewer. On the other hand, and as will be seen in later chapters, on the 'inside' of a cusp there are always two more critical points than on the outside – see the illustration of this in the solution to Problem 1.5.

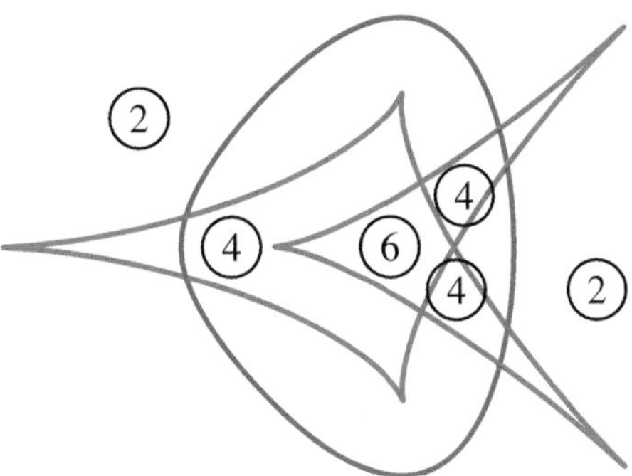

Chapter 2

Problem 2.1 We have $\mathrm{d}f = (4x(x^2 - 1), 2y)$. Critical points are points where this is zero, so at points satisfying $y = 0$ and $x(x^2 - 1) = 0$. There are therefore three critical points: $A(0,0)$, $B(1,0)$, $C(-1,0)$. The Hessian matrix is $H_f = \mathrm{d}^2 f = \begin{pmatrix} 12x^2 - 4 & 0 \\ 0 & 2 \end{pmatrix}$. The index at the three points is therefore: 1 at A and 0 at each of B and C (recall: index = number of negative eigenvalues).

Problem 2.3 Consider the Taylor series of each. For example

$$f(x) = x^2(x - \tfrac{1}{3!}x^3 + \cdots)^5 = x^2(x^5 - \cdots) = x^7 + \cdots$$

It follows that f has an A_6 critical point at the origin. In a similar manner, one finds g has a critical point of type A_1.

Problem 2.4

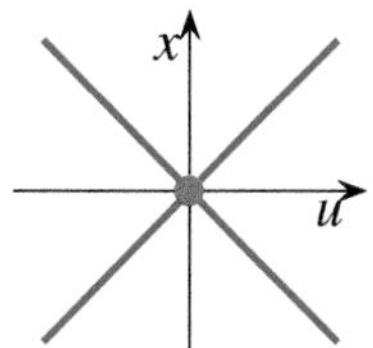

We have $C_F = \{(x,u) \mid 3x^2 - 3u^2 = 0\}$. The Hessian is just $6x$ which is degenerate (in this 1 variable case that means zero) when $x = 0$, so Σ_F is the origin.

Chapter 3

Problem 3.3 Let $p = (u, v)$. First, if both u and v are non-zero then there is a neighbourhood U of p such that $A \cap U = B \cap U = \emptyset$, so they are germ equivalent at p. Now suppose p is such that $u \neq 0, v = 0$ (so p lies on the 'x-axis'). Then there is a neighbourhood U of p (disjoint from the 'y-axis') such that $A \cap U = B \cap U = \{x\text{-axis}\} \cap U$, so A and B are germ equivalent at such p as well. Finally, suppose $p = (0, v)$ (for any v). Then for any neighbourhood U of p, $A \cap U$ contains points of the y-axis, while $B \cap U$ does not, so these intersections are always different, however small U is chosen.

Problem 3.5 Suppose f_1, f_2 are germ equivalent at 0, and similarly g_1, g_2. Then there are neighbourhoods U, V of 0 such that $f_1\big|_U = f_2\big|_U$ and $g_1\big|_V = g_2\big|_V$. Let $W = U \cap V$. Then W is a neighbourhood of 0 and it follows that for all $x \in W$,

$$f_1(x) + g_1(x) = f_2(x) + g_2(x), \quad \text{and} \quad f_1(x)g_1(x) = f_2(x)g_2(x).$$

Consequently, $f_1 + g_1$ is germ equivalent at 0 to $f_2 + g_2$. And $f_1 g_1$ is germ equivalent at 0 to $f_2 g_2$.

Problem 3.6 Let f, g be two functions defined on neighbourhoods V_1 and V_2 of A. We say they are germ equivalent along A if there is a neighbourhood U of A with $U \subset V_1 \cap V_2$ such that $f\big|_U = g\big|_U$. The ring structure is left to the reader (see previous solution).

Problem 3.10 (i) Suppose f, g are smooth germ equivalent functions defined in neighbourhoods of 0 in $\mathbb{R}^p$, and suppose they coincide on the neighbourhood U of

0 in $\mathbb{R}^p$. Let $V = \phi^{-1}(U) \subset \mathbb{R}^n$. Since ϕ is continuous, V is an open set containing 0. Now, for $x \in V$, $\phi(x) \in U$ and hence $f(\phi(x)) = g(\phi(x))$, and therefore $\phi^* f$ and $\phi^* g$ are germ equivalent.
(ii) Now let $f, g \in \mathcal{E}_p$. We need to show that $\phi^*(f+g) = \phi^*(f) + \phi^*(g)$, and $\phi^*(fg) = \phi^*(f)\,\phi^*(g)$ (definition of ring homomorphism: it preserves the ring structure; see Definition D.7). Now $\phi^*(f+g) = (f+g)\circ\phi = f\circ\phi + g\circ\phi = \phi^* f + \phi^* g$. The middle equality is just the definition of addition of functions: $(f+g)(x) = f(x) + g(x)$.
The argument for multiplication is similar.
(iii) The inverse to ϕ^* is readily shown to be $(\phi^{-1})^*$.

<u>*Problem 3.14*</u> $y^5 = y\alpha + (y^2 - x)\beta$, where $\alpha = x^2$ and $\beta = xy + y^3$ are the two generators of the ideal. This is found by asking, how can any polynomial involving y^5 be found from the two generators? The only possibility is $y^2\beta$, and then proceed from there:

$$\begin{aligned} y^5 &= y^2\beta - xy^3 \\ &= y^2\beta - x(\beta - xy) \\ &= y^2\beta - x\beta + y\alpha \\ &= y\alpha + (y^2 - x)\beta. \end{aligned}$$

<u>*Problem 3.22*</u> We give two solutions, both using induction on n. First let $P(n)$ be the statement that, for all r, there are $\binom{n+r-1}{r}$ monomials of degree r in n variables. Certainly $P(1)$ is true (ie, for each r there is only 1 monomial of degree r in 1 variable). Suppose now $P(n)$ is true. Now consider $n+1$ variables, $x_1, x_2, \ldots, x_n, z$. Any monomial of degree r is of the form $z^{n-k}M$, where M is a monomial of degree k in the n variables $x_1, \ldots, x_n$, for $0 \le k \le r$. It follows that the total number of such monomials is equal to

$$\binom{n+0-1}{0} + \binom{n+1-1}{1} + \cdots + \binom{n+r-1}{r}. \qquad (*)$$

That this is equal to $\binom{n+r}{r}$ is one of the standard properties of the $\binom{a}{b}$ function (and can be proved by induction). This establishes $P(n+1)$.

The second, more elegant, solution to Problem 3.22 doesn't assume we know the formula beforehand. It also illustrates the general method of generating functions.

Let $C(n, r)$ be the number of monomials of degree r in n variables. For each n, define the so-called generating function

$$F_n(t) := \sum_{r=0}^{\infty} C(n, r) t^r.$$

We claim that $F_n(t) = \frac{1}{(1-t)^n}$ (for $|t| < 1$). Call this statement $P(n)$. We know that $C(1, r) = 1$ so that

$$F_1(t) = \sum_{r \geq 0} t^r = 1/(1-t),$$

thus proving $P(1)$. Now, by the same observation as in the first solution (using the variables x_i and z), we have

$$C(n+1, r) = \sum_{k=0}^{r} C(n, k).$$

Then

$$\begin{aligned} F_{n+1}(t) &= \textstyle\sum_{r \geq 0} \sum_{k=0}^{r} C(n, k) t^r &= \textstyle\sum_{k=0}^{r} \sum_{r \geq k} C(n, k) t \\ &= \textstyle\sum_{k=0}^{r} C(n, k) \sum_{r \geq k}^{r} t^r &= \textstyle\sum_{k=0}^{r} C(n, k) \frac{t^k}{1-t} \\ &= \frac{1}{1-t} F_n(t). \end{aligned}$$

Since $F_1(t) = (1-t)^{-1}$, we deduce $F_2(t) = (1-t)^{-2}$, $F_3(t) = (1-t)^{-3}$, etc, so finding inductively that $F_n(t) = (1-t)^{-n}$ as claimed.

There now remains to show that the coefficient $C(n, r)$ of t^r in $F_n(t)$ is equal to $\binom{n+r-1}{r}$. Using the Taylor series of F_n, we have

$$C(n, r) = \frac{1}{r!} \frac{\mathrm{d}^r}{\mathrm{d}t^r} \left(\frac{1}{(1-t)^n} \right) |_{t=0}.$$

It is a standard exercise to deduce that indeed,

$$C(n, r) = \binom{n+r-1}{r}.$$

Chapter 4

Problem 4.6 We have $Jf = \langle x^2, y^2 \rangle$ and $Jg = \langle 3x^2 + 2xy^2, 3y^2 + 2x^2y \rangle$. That $Jg \subset Jf$ is easy to see; for example $3x^2 + 2xy^2 = 3(x^2) + 2x(y^2) \in Jf$. For the other inclusion, we have $\mathfrak{m}Jf = \langle x^3, x^2y, xy^2, y^3 \rangle = \mathfrak{m}_2^3$. Consequently $Jf \subset Jg + \mathfrak{m}Jf$ so that (by Nakayama's lemma) $Jf \subset Jg$ as required.

Problem 4.8 (i) $Jf = \langle x^2, y^2 \rangle$ so a cobasis for Jf is $\{x, y, xy\}$, and $\operatorname{codim}(f) = 3$. (ii) $Jf = \langle x^2, y^2, z \rangle$ and this has the same cobasis for Jf as in (i), so again

$\operatorname{codim}(f) = 3$.
(iii) This time $Jf = \langle x^2, y^2, z^2 \rangle$. It is not so easy to draw Newton's diagram for 3 variables, but at least one can make a list of all the monomials in each degree, and cross off the ones in the ideal. This gives:

$$\begin{array}{c} x \quad y \quad z \\ \cancel{x^2} \quad \cancel{y^2} \quad \cancel{z^2} \quad xy \quad yz \quad xz \\ \cancel{x^3} \quad \cancel{y^3} \quad \cancel{z^3} \quad \cancel{x^2y} \quad \cancel{xy^2} \quad \cancel{y^2z} \quad \cancel{yz^2} \quad \cancel{x^2z} \quad \cancel{xz^2} \quad xyz \end{array}$$

At the next level (degree 4), all monomials contain a square, so everything is crossed out; that is, $\mathfrak{m}_3^4 \subset Jf$. Therefore Jf has as cobasis

$$\{x,\ y,\ z,\ xy,\ yz,\ zx,\ xyz\}$$

and $\operatorname{codim}(f) = 7$.
(iv) $Jf = \langle x^a, y^b \rangle$. The natural cobasis (ie, using monomials) for Jf is

$$\{x^r y^s \mid 0 \le r < a,\ 0 \le s < b,\ (r,s) \ne (0,0)\}\,.$$

This has $ab - 1$ elements, so that $\operatorname{codim}(f) = ab - 1$.
(v) This germ has infinite codimension, as $Jf = \langle xy^2, x^2y \rangle$, which contains no monomials of the form x^a or y^b.
(vi) This is a little tricky, but we claim the codimension is $\operatorname{codim}(f) = 3(a-1)$. The argument is as follows. We have $Jf = \langle 3x^2 + y^a, xy^{a-1} \rangle$, and three monomials in Jf are, x^3, xy^{a-1} and y^{2a-1}. The monomials missing are therefore,

$$\begin{cases} y^r, & 0 < r \le 2a-2 \\ xy^s & 0 \le s \le a-2 \\ x^2y^t & 0 \le t \le a-2. \end{cases}$$

This gives a total of $4a - 4$ monomials. However, the monomials are not unrelated, as $3x^2 + y^a \in Jf$ so in $\mathfrak{m}/_{Jf}$ we have $y^{a+k} \equiv -3x^2y^k$, for $k \ge 0$. This removes, say, powers y^r for $r \ge a$ from our list above, so removing $(a-1)$ elements. This leaves $(4a-4) - (a-1) = 3(a-1)$ elements for our basis, as claimed. And a cobasis is:

$$\left\{y^r,\ xy^{r-1},\ x^2y^{r-1} \mid 1 \le r \le a-1\right\}.$$

<u>*Problem 4.9*</u> We have $g = f \circ \phi$. Then by the chain rule, putting $x = \phi(y)$,

$$\frac{\partial g}{\partial y_i}(y) = \sum_j \frac{\partial f}{\partial x_j}(x)\,\frac{\partial \phi_j}{\partial y_i}(y), \quad \text{and} \quad \frac{\partial f}{\partial x_j}(x) = \sum_i \frac{\partial g}{\partial y_i}(y)\,\frac{\partial \bar\phi_i}{\partial x_j}(x),$$

where $\bar{\phi} = \phi^{-1}$. The first equality expresses the generator $\frac{\partial g}{\partial y_i}$ of Jg as a combination of $\frac{\partial f}{\partial x_j} \circ \phi$ with coefficients $\frac{\partial \phi_j}{\partial x_i} \in \mathcal{E}_n$, which shows $\frac{\partial g}{\partial y_i} \in \phi^* Jf$. The second equality shows similarly that $Jf \subset \bar{\phi}^* Jg$. Since ϕ^* and $\bar{\phi}^*$ are mutual inverses, it follows that indeed $Jg = \phi^* Jf$.

(ii) We are given that $\mathfrak{m}_n = Jf + \mathbb{R}\{h_1, \dots, h_r\}$ (where $r = \operatorname{codim}(f)$), and we wish to show that

$$\mathfrak{m}_n = Jg + \mathbb{R}\{h_1 \circ \phi, \dots, h_r \circ \phi\}. \tag{E.1}$$

Write $V = \mathbb{R}\{h_1, \dots, h_r\}$ which is a vector subspace of $\mathfrak{m}_n$, and $\mathfrak{m}_n = Jf \oplus V$. Now ϕ^* also provides an isomorphism $\phi^* \colon \mathfrak{m}_n \to \mathfrak{m}_n$. Therefore $\mathfrak{m}_n = \phi^*(Jf) \oplus \phi^*(V) = Jg \oplus \phi^*(V)$, and hence any basis for $\phi^* V$ is a cobasis for Jg in $\mathfrak{m}_n$. But $\{h \circ \phi, \dots, h_r \circ \phi\}$ is clearly one such cobasis.

(iii) Suppose now f and g are R^+-equivalent. Let $f_1 = f - f(0) + g(0)$; then f_1 and g are $\mathcal{R}$-equivalent. Firstly, $Jf = Jf_1$, and secondly by part (i), $Jg = \phi^* Jf_1 = \phi^* Jf$. If f is of finite codimension, then by (ii) so is g and the cobases have the same number of elements, whence $\operatorname{codim}(f) = \operatorname{codim}(g)$. On the other hand, if f is of infinite codimension, then so must g be.

Problem 4.12 (a) The Hessian matrix at 0 has rank 2, so corank = 3 − 2 = 1. We have $f(x, y, 0) = xy$ which has a nondegenerate critical point at the origin (indeed, a saddle point). Now, $f_x = y$ and $f_y = x + 2yz$, and these are both zero if and only if $x = y = 0$. Thus $h(z) = f(0, 0, z) = z^3$.
(b) This time $f_x = y - z^2$ and $f_y = x + 2yz$. These both vanish when $y = z^2$ and $x = -2yz = -2z^3$. Then

$$h(z) = f(-2z^3,\, z^2, z) = -2z^5 + z^5 + 2z^5 + z^3 = z^3 + z^5.$$

Problem 4.13 (i) The Hessian matrix of f at the origin is,

$$H_f = \begin{pmatrix} 2 & 0 & 0 \\ 0 & 0 & 0 \\ 0 & 0 & 0 \end{pmatrix},$$

which has rank 1, so the corank of f is 3 − 1 = 2.

The restriction of f to $\mathbb{R} \times \{0\}$ is $f(x, 0, 0) = x^2$ which indeed has a nondegenerate critical point at the origin. To find the remainder term $h(y, z)$, we differentiate f with respect to x, and then solve $f_x = 0$ for x in terms of y, z. Thus $f_x = 2(x + y^2)$ and this vanishes for $x = -y^2$.

Consequently, $h(y,z) = f(-y^2, y, z) = -y^4 + yz^2$.

(ii) We put $x^2 + 2xy^2 + yz^2 = X^2 - y^4 + yz^2$. Thus,

$$X^2 = x^2 + 2xy^2 + y^4 = (x + y^2)^2.$$

So $X = x + y^2$ gives the change of coordinate.

Problem 4.14 We apply the splitting lemma to the the function $F(t,x) = f_t(x)$ of $n+1$ variables. Since, for $t = 0$, f_0 has a nondegenerate critical point at 0, there is a neighbourhood U of the origin $(0,0)$ in $\mathbb{R}^{n+1}$ on which we can change coordinates via $x = x(X,t)$ so that

$$F(x(X,t),t) = Q(X) + h(t),$$

where Q is a nondegenerate quadratic form. This has a nondegenerate critical point at $X = 0$, for all t for which $(0,t)$ lies in the given neighbourhood. The map ψ is $\psi(t) = x(0,t)$. (Furthermore, the critical point of f_t has the same index as that of f_0.)

Chapter 5

Problem 5.2 Now $\frac{\mathrm{d}}{\mathrm{d}s} f_s(x) = 3x^2y$, so we wish to find a smooth vector field $\mathbf{v}_s(x)$ satisfying $\mathrm{t}f_s(v_s)(x) = x^2y$. This is a simple computation. Since we require $v_s(0) = 0$, we can write

$$\mathbf{v}_s(x) = \begin{pmatrix} xa_1 + ya_2 \\ xb_1 + yb_2 \end{pmatrix}.$$

Here $a_1, \ldots b_2$ are functions of s, x, y. The condition $\mathrm{t}f_s(\mathbf{v}_s) = \frac{\mathrm{d}}{\mathrm{d}s} f_s$ becomes

$$\begin{aligned} x^2y &= (xa_1 + ya_2)(x^2 + 2sxy) + (xb_1 + yb_2)(sx^2 + y^2) \\ &= x^3(a_1 + sb_1) + x^2y(2sa_1 + a_2 + sb_2) + xy^2(2sa_2 + b_1) + y^3(b_2). \end{aligned}$$

Equating coefficients shows that one solution is

$$a_1 = \frac{2s^2}{4s^3 + 1}, \quad a_2 = \frac{1}{4s^3 + 1}, \quad b_1 = -\frac{2s}{4s^3 + 1}, \quad b_2 = 0$$

(indeed this is the unique *linear* vector field satisfying the equation – on the other hand there will be many nonlinear ones where a_1 etc. have higher–order terms).

Since $4s^3 + 1 \neq 0$ provided s is small enough, It follows that the required vector field exists and f_s is a trivial family. [*Remark*: If s satisfies $4s^3 = -1$ then f_s has a repeated factor so cannot be equivalent to f_0.]

Problem 5.4 Now, $\dot{f}_s = x^2 + 2sy^2$ and

$$Jf_s = \langle 2(1+s)x,\ 2s^2y\rangle = \begin{cases} \langle x,\ y\rangle & \text{if } s \neq 0 \\ \langle x\rangle & \text{if } s = 0, \end{cases}$$

and the first part is clear. On the other hand, when $s = 0$, $f_0 = x^2$ which is a degenerate critical point, while for $s \neq 0$ the critical point is nondegenerate so this cannot be a trivial family. Solving the equation for a vector field, one finds s in the denominator, so the vector field is not smooth in s when $s = 0$ and the Thom–Levine principle does not apply.

Problem 5.6 (i) $Jf = \langle x^2 + y^2,\ xy\rangle$ so $\mathfrak{m}_2 Jf = \langle x^3 + xy^2,\ x^2y + y^3,\ x^2y,\ xy^2\rangle = \mathfrak{m}_2^3$. Then by the finite determinacy theorem f is 3-determined.
(ii) Here $Jf = \langle x(3x+2y),\ x^2\rangle$. Every element of the ideal has x as a factor, so for all k, $y^k \notin Jf$. Thus f is not finitely determined.
(iii) $Jf = \langle x,\ y^{k-1}\rangle$, so that $\mathfrak{m}_2^{k-1} \subset Jf$. It follows that f is k-determined.
(iv) $Jf = \langle xy, x^2 + ky^{k-1}\rangle$. Then

$$\mathfrak{m}\, Jf = \langle x^2y, xy^2, x^3 + xy^{k-1}, x^2y + ky^k\rangle = \langle x^2y, xy^2, x^3, y^k\rangle$$

so $\mathfrak{m}^k \subset \mathfrak{m}Jf$ showing f is k-determined.
(v) One shows that $\mathfrak{m}_2^4 \subset \mathfrak{m}_2 Jf + \mathfrak{m}_2^5$. For example,

$$x^4 = x^2(f_x) - x^2y^3 \in \mathfrak{m}_2 Jf + \mathfrak{m}_2^5.$$

(vi) $Jf = \langle x + x^2 - y^2,\ xy\rangle = \langle \alpha,\ \beta\rangle$. Now $x\alpha + y\beta = x^2 + x^3 = x^2(1+x)$ so that

$$x^2 = (1+x)^{-1}(x\alpha + y\beta) \in \mathfrak{m}_2 Jf.$$

(Since $(1+x)$ is not zero at the $x = 0$, it follows that $(1+x)^{-1}$ is a smooth function near $x = 0$.) Also

$$y^3 = -y\alpha + \beta + x\beta \in Jf$$

so $\mathfrak{m}_2^4 \subset \mathfrak{m}_2 Jf$ and f is 4-determined.
(vii) $\mathfrak{m}_2^9 \subset \mathfrak{m}_2^2 Jf$ so f is 8-determined.

Problem 5.8 (i) is straightforward. (ii) The functions can be chosen to be $k_1 = x^3$ and $k_2 = x^2y + y^4$. (iii) Let $f_s = f + sh$. Then $\dot{f_s} = h \in T\mathcal{R}{\cdot}f_s$ smoothly in s; this is because both h and the generators of $T\mathcal{R}{\cdot}f_s$ are independent of s, so writing $s = \mathrm{t}f_s(\mathbf{v}_s)$ can be done independently of s. That f_0 and $f = f_1$ are $\mathcal{R}$-equivalent then follows from Theorem 5.4.

Problem 5.12 Last part: because the vector field isn't in Jf_s for all $s \in [0, 1]$ (in particular for $s = 1$).

Chapter 6

Problem 6.3 The Hessian matrix is zero, so f is of corank 2. Let α, β be the two generators of Jf. Then $\alpha - x\beta = xy(1 - 3y)$. Since $(1 - 3y)$ is a unit in $\mathcal{E}_2$ it follows that $xy \in Jf$. The codimension is then 3, so the critical point is of type D_4. Moreover, it is 3-determined, since $\mathfrak{m}_2^3 \subset \mathfrak{m}_2 Jf$ so that the x^4 term can be removed, and we obtain

$$f \sim_{\mathcal{R}} 2(x^2y + y^3) \sim_{\mathcal{R}} x^2y + y^3$$

(the last by rescaling x, y), so we finally see that it's a D_4^+ critical point.

Problem 6.4 For (ii): for example, $\alpha = Y(cX + dY)$ and $\beta = X(aX + bY)$ (other choices are possible).

Problem 6.6 A calculation: for example let $\mathrm{wt}(x) = p^{-1}$, $\mathrm{wt}(y) = q^{-1}$ and $\mathrm{wt}(z) = r^{-1}$. Then f has degree 1 for the parabolic case. On the other hand, whatever weights are given to x, y, z this is not homogeneous for the hyperbolic case, as $\frac{1}{p} + \frac{1}{q} + \frac{1}{r} \neq 1$, so the degree of xyz cannot be the same as the degree of the other terms.

Chapter 7

Problem 7.3 We have $G(x, v) = F(x, \phi(v))$, and so $g_v = f_{\phi(v)}$. Suppose $\phi(v) \in \Delta_F$. Then $f_{\phi(v)}$ has a degenerate critical point and therefore so does g_v which means $v \in \Delta_G$. And conversely.

Problem 7.6 (This family is trivial because each f_u is equivalent to f_0. We have $C_F = \{(x,u) \mid x = u\}$ and since $(f_u)''(x) = 6(x-u)$ we have $\Sigma_F = C_F$ (ie, all critical points are degenerate). In this case $\pi_F(x) = x$ (if we parametrize C_F by x), hence π_F is a diffeomorphism. If the converse to Proposition 7.1 were true then Σ_F would be empty, which it isn't. Thus the converse does not hold (it does hold if the unfolding is *versal* – see Proposition 7.14).

Problem 7.7 We have $g_u(x,y) = f_u(x) + \sum_i \pm y_i^2$. Then

$$\mathsf{d}(g_u)(x,y) = (\mathsf{d}(f_u)(x),\ \pm 2y_1, \ldots, \pm 2y_k),$$

so $(x,y) \in C_G$ iff $x \in C_F$ and $y_1 = \cdots = y_\ell = 0$. That is,

$$C_G = C_F \times \{0\} \subset \mathbb{R}^n \times \mathbb{R}^a \times \mathbb{R}^k.$$

Moreover, for each u,

$$H_{g_u} = \begin{pmatrix} H_{f_u} & 0 \\ 0 & D \end{pmatrix}$$

where D is a diagonal matrix with ± 2s down the diagonal. Thus $H_{g_u}(x,0)$ is degenerate iff $H_{f_u}(x)$ is degenerate. So $u \in \Delta_G$ iff $u \in \Delta_F$, which means $\Delta_G = \Delta_F$.

Problem 7.12 (i) This is just induced by $u = \phi(\lambda) = -\lambda$. The constant term is $C(\lambda) = -\lambda^2$. Then

$$F(x,-\lambda) = G(x,\lambda) + C(\lambda).$$

(ii) If we put $u = \phi(\lambda) - \lambda$, the equivalence is essentially swapping x and y over, but involves also moving the critical point as we will see. It is helpful to use different letters for the variables in F and in G, so we write

$$\begin{aligned} G(X,Y,\lambda) &= X^3 + \lambda X + Y^2 + \mu Y \\ &= X^3 + \lambda X + (Y + \frac{1}{2}\mu)^2 - \frac{1}{4}\mu^2 \end{aligned}$$

If we put $x = Y + \frac{1}{2}\mu$ and $y = X$ then

$$F(x,y,-\lambda) = G(X,Y,\lambda) + \frac{1}{4}\mu^2.$$

This is an equivalence, with $\psi(X,Y,\lambda,\mu) = (Y + \frac{1}{2}\mu),\ X)$ and $C(\lambda,\mu) = \frac{1}{4}\mu^2$.

(iii) Here one must 'complete the cube' in G, to eliminate the y^2 term:

$$\begin{aligned} G(y,\lambda) &= (y-\tfrac{1}{3}\lambda)^3 - \tfrac{1}{3}\lambda^2 y + \tfrac{1}{27}\lambda^3 \\ &= (y-\tfrac{1}{3}\lambda)^3 - \tfrac{1}{3}\lambda^2 (y-\tfrac{1}{3}\lambda) - \tfrac{2}{27}\lambda^3 \\ &= x^3 - \tfrac{1}{3}\lambda^2 x - \tfrac{2}{27}\lambda^3, \end{aligned}$$

where $x = \lambda - \frac{1}{3}\lambda$. Comparing this with F one sees that,

$$F(\psi(y,\lambda),\ \phi(\lambda)) = G(y,\lambda) + C(\lambda),$$

where $\phi(\lambda) = -\frac{1}{3}\lambda^2$ and $\psi(y,\lambda) = y - \frac{1}{3}\lambda$, and of course $C(\lambda) = -\frac{2}{27}\lambda^3$.

(iv) This one is at first sight more difficult than the previous ones, and it might be thought that λ should correspond in some way to u (as both are coefficients of the quadratic part in x, y). However this is not the case – and one can check that for $\lambda \neq 0$ the function g_λ has 4 critical points, whereas if $v = w = 0$ the function $f_{(u,0,0)}$ has at most two so they cannot be equivalent.

Looking closely one can see a similarity between this and the previous problem (iii), but now in each of the two variables. So to solve this, one should 'complete the cube' in each of the two variables in G:

$$\begin{aligned} G(X,Y;\lambda) &= X^3 + \lambda X^2 + Y^3 - \lambda Y^2 \\ &= (X+\tfrac{1}{3}\lambda)^3 - \tfrac{1}{3}\lambda^2(X+\tfrac{1}{3}\lambda) + (Y-\tfrac{1}{3}\lambda)^3 - \tfrac{1}{3}\lambda^2(Y-\tfrac{1}{3}\lambda) \\ &= x^3 + y^3 - \frac{1}{3}\lambda^2 x + \frac{1}{3}\lambda^2 y, \end{aligned}$$

where $x = X + \frac{1}{3}\lambda$ and $y = Y - \frac{1}{3}\lambda$. Thus we have

$$F(\psi(X,Y,\lambda),\ \phi(\lambda)) = G(X,Y,\lambda) + C(\lambda),$$

where $\phi(\lambda) = (0, -\frac{1}{3}\lambda^2, \frac{1}{3}\lambda^2)$ and $\psi(X,Y,\lambda) = (X + \frac{1}{3}\lambda,\ Y - \frac{1}{3}\lambda)$, and this time $C(\lambda) = 0$.

Problem 7.14 [The swallowtail] (i) We have $Jf = \langle x^4 \rangle \lhd \mathcal{E}_1$, and $\dot{F} = \mathbb{R}\{x^3, x^2, x\}$. It follows that $\mathcal{E}_1 = \mathbb{R} + \dot{F} + Jf$ so that the unfolding is indeed versal.

Secondly, $C_F = \{(x,a,b,c) \in \mathbb{R}^4 \mid x^4 + ax^2 + bx + c = 0\}$. To see this is a submanifold we can write it is as a graph of the map $(x,a,b) \mapsto -(x^4 + ax^2 + bx)$. Or even more directly, we can apply the straightening map

$$\Phi(x,a,b,c) = (x,\ a,\ b,\ c + (x^4 + ax^2 + bx)).$$

Then $(x, a, b, c) \in C_F$ if and only if $\Phi(x, a, b, c) \in \mathbb{R}^3 \times \{0\}$ so that C_F is a submanifold of dimension 3.

(ii) We have $G(x, u) = \frac{1}{5}x^5 + \frac{2}{3}ux^3 + ux$. Then $g_u'(x) = x^4 + 2ux^2 + u = (x^2 + u)^2$. Thus g_u has a critical point iff $x^2 = -u$, so iff $x = \pm\sqrt{-u}$ (which are real points if $u < 0$ and imaginary if $u > 0$). Furthermore, $g_u''(x) = 4x(x^2 + u)$ which is also zero at these critical points, so the critical points are both degenerate. [At these points, $g_u'''(x) \neq 0$, so the critical points are of type A_2.]

(iii) We have $H(x, a) = \frac{1}{5}x^5 + \frac{1}{3}x^3$. Then $h_a'(x) = x^4 + ax^2 = x^2(x^2 + a)$, and $h_a''(x) = 4x^3 + 2ax = 2x(2x^2 + a)$. So for each $a \neq 0$, h_a has three critical points, at $x = \pm\sqrt{-a}$ and $x = 0$. The first two are nondegenerate while the third ($x = 0$) is degenerate of type A_2 as $h_a'''(0) \neq 0$.

(iv) We have $h_x(x) = \frac{1}{5}x^5 - \frac{1}{3}x^3 + cx$. Then $h_c'(x) = x^4 - x^2 + c$, so critical points of h_c satisfy $x^2 = \frac{1}{2}(1 \pm \sqrt{1 - 4c})$.

- For $c < 0$, one of these values of x^2 is positive and one negative, so that h_c has two real critical points and two imaginary (they are all nondegenerate). (These are also nondegenerate.)
- For $c = 0$ this is the $H(x, -a)$ we had above, so the function has two nondegenerate critical points and one degenerate one.
- For $c \in (0, \frac{1}{4})$, both values of x^2 are real and positive, and h_c has four distinct real critical points.
- For $c = \frac{1}{4}$ this is the $G(x, \frac{1}{2})$ from above which has two degenerate critical points.
- For $c > \frac{1}{4}$, both values of x^2 are imaginary, so all four critical points of h_c are complex.

(v) In region A the function has no critical points, in region B it has two and in region C (the 'pocket') it has four.

(vi) A critical point is of type A_3 if $f'(x) = f''(x) = f'''(x) = 0$ but $f^{(iv)}(x) \neq 0$. Apply this to $f_{(a,b,c)}$:

$$\begin{aligned} x^4 + ax^2 + bx + c &= 0 \\ 4x^3 + 2ax + b &= 0 \\ 12x^2 + 2a &= 0. \end{aligned}$$

These can be solved for a, b, c in terms of x, giving

$$a = -6x^2, \quad b = 8x^3, \quad c = -3x^4.$$

This gives the locus in parametric form. One can also eliminate x to obtain the locus in Cartesian form, $27b^2 = -8a^3$ and $12c = -a^2$. (One could also write down $27b^4 = -4096c^3$, but this follows from the other two.) This curve is the cuspidal edge in Figure 7.1 (and we already know that the discriminant of an A_3 singularity contains a cusp).

Problem 7.15

Here we just give the answers.

(i) This is just a computation.

(ii) C_F has equation $x^2 + y^2 = a^2$ so for $a \neq 0$ represents a circle of radius $|a|$. Overall, this is the equation of a circular cone in $\mathbb{R}^3$.

(iii) $f_{(a,0,0)}$ has four critical points. The indices are ± 2 (at the origin, and according to the sign of a), and index 1 at the other three points. That is, it has one local minimum or maximum and three saddle points.

(iv) If we parametrize the circle from part (ii) by $x = a\cos\theta$ and $y = a\sin\theta$, we find that any non-zero kernel vector of the Hessian at (x, y) is tangent to the circle precisely for $\theta = 0, \pm 2\pi/3$.

Problem 7.17 Write f_u as a Taylor series in x:

$$f_u(x) = a(u) + b(u)x + \tfrac{1}{2}c(u)x^2 + d(x, u)x^3.$$

The hypotheses are equivalent to $b(u) \equiv 0$, and $c(0) = 0$, and, for u non-zero, $c(u)$ has the same sign as u. Critical points of f_u are then given by

$$x[c(u) + xe(x, u)] = 0,$$

for some smooth function e. Now choose $u_1 > 0$ and $u_2 < 0$ for which $c(u_1) > 0$ and $c(u_2) < 0$. By hypothesis, and the continuity of e, there is a neighbourhood U of 0 such that for all $x \in U$, $c(u_1) + xe(x, u_1) > 0$ and $c(u_2) + xe(x, u_2) < 0$. Now instead fix $x \in U$, and consider the function $u \mapsto c(u) + xe(x, u)$. Since it is positive for $u = u_1$ and negative for $u = u_2$, there is an intermediate point $u = u_x$ for which $c(u) + xe(x, u) = 0$, and x is therefore a critical point of f_{u_x} as required.

Chapter 8

Problem 8.1 We consider the plane curves $x^2 + y^4 + ay^2 + by + c = 0$, and refer to the top diagram in Figure 8.4 (on p. 115). For $(a, b, c) = 0$ the 'curve' reduces to a point, and if c is increased (with $a = b = 0$) the curve is empty. Thus, above the swallowtail surface (region A in the figure) the curve is empty. Inside the 'pocket' of the swallowtail (region C) the curve consists of two disjoint ovals, and as the parameter tends to the double point D, the two ovals shrink to two distinct isolated points. Along the bottom edge of the pocket (marked H) the curve is a figure-8. At other points of the boundary of the pocket, one oval remains and the other shrinks to a point. In region B there is just a single oval.

Chapter 9

Problem 9.2 The even Jacobian ideal is

$$J^+f = \left(\mathfrak{m}_3 \left\langle x^3, y^3, z\right\rangle\right) \cap \mathcal{E}_n^+ = \left\langle x^4, x^3y, xy^3, y^4, xz, yz, z^2\right\rangle \lhd \mathcal{E}_n^+$$

A cobasis for this ideal in $\mathfrak{m}_3^+$ is given by $\{x^2, xy, y^2, x^2y^2\}$, and hence $\operatorname{codim}^+(f) = 4$, and an $\mathcal{R}^+$-versal unfolding is given by

$$F(x, y, z; r, s, t, u) = x^4 + y^4 + z^2 + rx^2 + sxy + ty^2 + ux^2y^2.$$

Chapter 10

Problem 10.2 Firstly, $Z_G = \{(x, y, u, v) \in \mathbb{R}^4 \mid u = x^2 + y^2,\ v = xy\}$. This can be parametrized by (x, y).

Next $dg_{(u,v)} = \begin{pmatrix} 2x & 2y \\ y & x \end{pmatrix}$. Thus $\det(dg_{(u,v)}) = 2(x^2 - y^2)$, so that

$$\Sigma_G = \{(x, y, u, v) \in Z_G \mid x^2 = y^2\}.$$

This condition for being an element of Σ_G implies $x = \pm y$, so that

$$\Sigma_G = \{(x, y, u, v) \in Z_G \mid x = y\} \cup \{(x, y, u, v) \in Z_G \mid x = -y\}.$$

When $x = y$ we have $u = 2x^2$ and $v = x^2$, while when $x = -y$ we have $u = 2x^2, v = -x^2$. Thus Δ_G is the union of two half-lines (only half-lines because $x^2 \not< 0$):

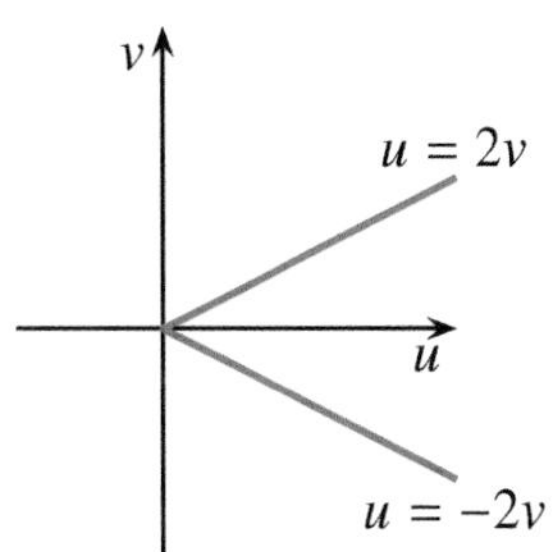

$$\Delta_G = \{(u,v) \in \mathbb{R}^2 \mid u = 2v,\ v \geq 0\} \bigcup$$

$$\{(u,v) \in \mathbb{R}^2 \mid u = -2v,\ v \leq 0\}.$$

For (u,v) inside the discriminant (so $u > 2|v|$) the equation $g_{(u,v)} = (0,0)$ has four solutions (try for example $u = 1, v = 0$). Outside the discriminant, one has no solutions (eg, any $u < 0$). On the discriminant there are precisely two solutions.

A nice geometric way to analyze these is as follows: The equations $g_{(u,v)}(x,y) = (0,0)$ has two components. The first one is $x^2+y^2-u = 0$, which is satisfied by the points on the circle centre $(0,0)$ radius $\sqrt{u}$ (assuming $u > 0$). The second equation is $xy = v$ which is satisfied by the points on a hyperbola. Now, if we fix v we have a given hyperbola, and it is clear that if u is large enough the circle will intersect the hyperbola in 4 points. Reducing the radius of the circle at some point the circle will be tangent to the hyperbola (in two places simultaneously) and if u is reduced further the circle and hyperbola will not meet at all. The four, two or zero points of intersection are the solutions of the equation $g_{(u,v)} = 0$.

<u>*Problem 10.3*</u> Equilibria occur where $\dot{x} = \dot{y} = 0$. Thus $x(x^2 - \lambda) = 0$ and $y = x^2$. If $\lambda \leq 0$ there is a unique equilibrium, at $(x,y) = (0,0)$, while if $\lambda > 0$ there are 3 equilibria, at $(x,y) = (0,0)$ and $(x,y) = (\pm\sqrt{\lambda}, \lambda)$. (This is the pitchfork bifurcation seen in the previous problem.) The phase diagrams are,

$\lambda \leq 0$ ← ⊖ → x

$\lambda > 0$ ← ⊖ → ⊖ ← ⊖ → x

<u>*Problem 10.6*</u> The set Z_F is the graph of f which can naturally be identified with $\mathbb{R}^n$ via $x \mapsto (x, f(x))$. Under this identification, $\Sigma_F = \Sigma_f$. Moreover, the discriminants of the two coincide, as subsets of $\mathbb{R}^p$.

Chapter 11

Problem 11.1 Write the maps as column vectors, so in (a) $f(t) = \begin{pmatrix} t^2 \\ 0 \end{pmatrix}$.

(i) Use ϕ is the identity, and $M(t) = \begin{pmatrix} 1 & 0 \\ t^{a-2} & 1 \end{pmatrix}$. Then $g(t) = M(t)f(t)$.

(ii) $g = Mf$ for $M = \begin{pmatrix} 1 & 1 \\ 1 & -1 \end{pmatrix}$

(iii) Let $x = u + v$, $y = u - v$. Then $g(x, y) = g \circ \phi(u, v) = (2u^2 + 2v^2, u^2 - v^2)$. So $g \circ \phi(u, v) = Mf(u, v)$ for $M = \begin{pmatrix} 2 & 2 \\ 1 & -1 \end{pmatrix}$.

(iv) $f(x, y) = M(x, y)g(x, y)$ if $M = \begin{pmatrix} 1 & 0 \\ y & 1 \end{pmatrix}$ (and ϕ is the identity again).

Problem 11.6 $m_A(f) = \dim Q(f)$. Here $I_f = \langle x,\ y^3 + xy \rangle = \langle x,\ y^3 \rangle$, so $Q(f) = \mathcal{E}_2/I_f \simeq \mathbb{R}\left\{1, y, y^2\right\}$ which has dimension three.

Problem 11.8 **(i)** Write $g(x, u) = x$ (linearly adapted coordinates, $x \in \mathbb{R}^{n-k}$, $u \in \mathbb{R}^k$). Then $M = g^{-1}(0) = \{0\} \times \mathbb{R}^k$ (i.e. M is given by $x_1 = x_2 = \cdots = x_{n-k} = 0$, and is parametrized by $u_1, u_2, \dots, u_k$). Now consider $f \in \mathcal{E}_n$. Then the restriction $f_{|M}$ is a function of u, and $f_{|M}(u) = f(0, u)$. In these coordinates the Jacobian matrix of (f, g) at the origin is

$$\mathsf{d}(f, g) = \begin{pmatrix} \mathsf{d}_x f & \mathsf{d}_u f \\ \mathsf{d}_x g & \mathsf{d}_u g \end{pmatrix} = \begin{pmatrix} \mathsf{d}_x f & \mathsf{d}_u f \\ I_{n-k} & 0 \end{pmatrix}.$$

Note that the first row is just a single row. This matrix has rank at least $n - k$, and at most $n - k + 1$ (as the bottom $n - k$ rows are linearly independent, and it has only $n - k + 1$ rows in all). The matrix is row equivalent to

$$\left(\begin{array}{c|c} 0 \ \cdots \ 0 & \frac{\partial f}{\partial u_1} \ \cdots \ \frac{\partial f}{\partial u_k} \\ I_{n-k} & 0 \end{array}\right)$$

which has rank $n-k$ if only if $\frac{\partial f}{\partial u_1} = \cdots = \frac{\partial f}{\partial u_k} = 0$, which is to say that $\mathsf{d}(f_{|M}) = 0$. Thus (f, g) is singular (is not a submersion) if and only if $f_{|M}$ has a critical point at M.

(ii) Write $I = \langle f, g \rangle$ (ideal in $\mathcal{E}_n$), and $J = \langle f \circ j \rangle$ (ideal in $\mathcal{E}_k$).

From j^* define $\rho\colon \mathcal{E}_n \twoheadrightarrow \mathcal{E}_k/J$, by $\rho(h) = j^*(h) + J = h \circ j + J$. Since j^* is surjective, so is ρ.

Now, $\ker\rho = \{h \in \mathcal{E}_n \mid h \circ j \in J\}$. I claim that $\ker\rho = I$.

To see this, first suppose $h \in I$. Then $h(x) = a(x)f(x) + \sum_j b_j(x)g_j(x)$ (by definition of I). Then $j^*(h) = h \circ j = (a \circ j)(f \circ j) + 0$ (as $g_j \circ j = 0$). Consequently, $j^*(h) \in J$ so $h \in \ker\rho$.

Conversely, suppose $\rho(h) = 0$, which means $j^*(h) \in J$. So we can write $h \circ j = a(f \circ j)$ for some $a \in \mathcal{E}_k$. Let us write $u = (x_1, \ldots, x_k)$ and $y = (x_{k+1}, \ldots, x_n)$. Now $a \in \mathcal{E}_k$ can be considered as an element $\bar{a} \in \mathcal{E}_n$ by putting $\bar{a}(u, y) = a(y)$. Then

$$0 = (h \circ j)(u) - a(u)(f \circ j)(u) = h(u, 0) - \bar{a}(u, 0)f(u, 0)$$

It follows that $(h - \bar{a}\, f) \in \langle x_{k+1}, \ldots, x_n\rangle$ (by Hadamard's lemma). But $\langle x_{k+1}, \ldots, x_n\rangle = \langle g_1, \ldots, g_{n-k}\rangle$, so $(h - \bar{a}\, f) \in \langle g\rangle$, or equivalently, $f \in \langle f, g\rangle = I$ as required.

Thus we have shown that $\ker\rho = I$. The required result now follows from the first isomorphism theorem for rings:
if $\rho\colon R \twoheadrightarrow S$ is a ring homomorphism, then the image of ρ is isomorphic to $R/\ker\rho$.
In our setting, as has been pointed out already, ρ is surjective, so $\mathrm{image}(\rho) = \mathcal{E}_k/J$, and as we have shown $\ker\rho = I$, so that indeed

$$\mathcal{E}_n/I \simeq \mathcal{E}_k/J.$$

Chapter 12

<u>*Problem 12.1*</u> The diffeomorphisms of $\mathcal{R}_1$-equivalence are of the form $\phi(y) = y + O(y^2)$. A 1–parameter family $\phi_t(y)$ of such diffeomorphisms only varies in the part that is of degree 2 and higher, so differentiating with respect to t gives a vector field which lies in $\mathfrak{m}_n^2\theta(n)$. It follows by the same argument as for $T\mathcal{R} \cdot f$ that $T\mathcal{R}_1 \cdot f = \mathrm{t}f(\mathfrak{m}_n^2\theta(n))$.

<u>*Problem 12.6*</u> Basic idea of proof: Differentiating $f \circ \phi_s(x)$ gives (by definition of $\mathrm{t}f$ and the relation between ϕ_s and $\mathbf{v}$),

$$\frac{\mathrm{d}}{\mathrm{d}s}(f(\phi_s(x))) = \mathrm{t}f(\mathbf{v})(\phi_s(x)) = 0.$$

It follows that, for each x, $f(\phi_s(x))$ is constant (independent of s), so equal to $f(\phi_0(x)) = f(x)$. That is, for all s, $f \circ \phi_s = f$.

However, we need to be more careful as f is a germ, not a function defined on some open set, and that is where the final part of the question arises.

So let us work with a representative f_1 of the germ f; that is, f_1 is a function defined on an open set U_1. Similarly let $\mathbf{v}_1$ be a representative of the germ $\mathbf{v}$, defined on some open set U_2. The fact that (as germs) $\mathrm{t}f(\mathbf{v}) = 0$ means that on some (possibly smaller) neighbourhood $U_3 \subset U_1 \cap U_2$ we have $\mathrm{t}f_1(\mathbf{v}_1) = 0$.

Now let $x \in U_3$, and apply the 'basic idea' described above. However, this is only valid provided $\phi_s(x) \in U_3$ as well. For a general vector field $\mathbf{v}$ this would only be true for suitably small values of s (as s increases, the point $\phi_s(x)$ may leave U_3). However, if $\mathbf{v} \in \mathfrak{m}_n\theta(n)$ (so $\mathbf{v}(0) = 0$), then for each $s > 0$ we can chose a sufficiently small nbhd U_4 such that $x \in U_4 \Rightarrow \phi_t(x) \in U_3$ $(\forall |t| < s)$. Thus for each s, the germ of $f \circ \phi_s$ at 0 is equal to f.

On the other hand, if $\mathbf{v}(0) \neq 0$ then $\phi_s(0) \neq 0$ for $s \neq 0$ and so the expression $f \circ \phi_s$ is meaningless, as the germ f isn't defined at $\phi_s(0)$. (And a representative of the germ might not be defined there while different representatives could have different values.)

Problem 12.10 One has, with $I = \mathbb{R}$,

$$T_{\mathrm{rel}}\,\mathcal{R}{\cdot}f_s = \langle y + 2sx,\ x - 2sy\rangle \lhd \mathcal{E}_{n,\mathbb{R}}.$$

Now,

$$\begin{pmatrix} x \\ y \end{pmatrix} = \frac{1}{(1+4s^2)}\begin{pmatrix} 2s & 1 \\ 1 & -2s \end{pmatrix}\begin{pmatrix} y+2sx \\ x-2sy \end{pmatrix}$$

Since $1 + 4s^2$ is non-zero forall $s \in \mathbb{R}$, it follows that

$$T_{\mathrm{rel}}\,\mathcal{R}{\cdot}f_s = \langle x,\ y\rangle \lhd \mathcal{E}_{n,\mathbb{R}}$$

which is constant. Moreover, $h = x^2 - y^2 \in T\mathcal{R}{\cdot} \cdot f_0$, and it follows from Theorem 12.21 that this family is $\mathcal{R}$-trivial.

Problem 12.13 If $f \in \mathcal{E}_x^p$ with $f^*\colon \mathcal{E}_y \to \mathcal{E}_x$ then one finds

$$\begin{aligned} T_e\mathcal{L}_{\mathrm{un}} \cdot F &= \omega F(\theta_{y/u}) + \mathrm{t}_u F(\theta_u), \\ T_e\mathcal{A}_{\mathrm{un}} \cdot F &= \mathrm{t}_x F(\theta_{x/u}) + \omega F(\theta_{y/u}) + \mathrm{t}_u F(\theta_u). \end{aligned}$$

Chapter 13

Problem 13.1 We need to calculate $T\mathcal{K}\cdot f$. We have

$$\begin{aligned} \mathrm{t}f(\mathfrak{m}_2\theta(2)) &= \mathfrak{m}_2\,\mathrm{t}f(\theta(2)) = \mathfrak{m}_2\left\{\begin{pmatrix}2x\\ y\end{pmatrix}, \begin{pmatrix}2y\\ x\end{pmatrix}\right\} \\ &= \mathcal{E}_2\left\{\begin{pmatrix}2x^2\\ xy\end{pmatrix}, \begin{pmatrix}2xy\\ y^2\end{pmatrix}, \begin{pmatrix}2xy\\ x^2\end{pmatrix}, \begin{pmatrix}2y^2\\ xy\end{pmatrix}\right\}. \end{aligned}$$

Moreover

$$I_f\theta(2) = \mathcal{E}_2\left\{\begin{pmatrix}x^2+y^2\\ 0\end{pmatrix}, \begin{pmatrix}0\\ x^2+y^2\end{pmatrix}, \begin{pmatrix}xy\\ 0\end{pmatrix}, \begin{pmatrix}0\\ xy\end{pmatrix}\right\}.$$

Combining these and simplifying (removing redundant generators) gives

$$T\mathcal{K}\cdot f = \mathcal{E}_2\left\{\begin{pmatrix}x^2\\ 0\end{pmatrix}, \begin{pmatrix}xy\\ 0\end{pmatrix}, \begin{pmatrix}y^2\\ 0\end{pmatrix}, \begin{pmatrix}0\\ x^2\end{pmatrix}, \begin{pmatrix}0\\ xy\end{pmatrix}, \begin{pmatrix}0\\ y^2\end{pmatrix}\right\} = \mathfrak{m}_2^2\theta(f).$$

It follows that f is 2-determined.

Problem 13.3 The codimensions are 4 and $r-1$, respectively.

Problem 13.6 One might expect this to be 3-determined, so let's see. We have

$$T\mathcal{K}\cdot g = \mathcal{E}_2\left\{\begin{pmatrix}x^3\\ 0\end{pmatrix}, \begin{pmatrix}x^2y\\ 0\end{pmatrix}, \begin{pmatrix}y^3\\ 0\end{pmatrix}, \begin{pmatrix}0\\ x^3\end{pmatrix}, \begin{pmatrix}0\\ xy^2\end{pmatrix}, \begin{pmatrix}0\\ y^3\end{pmatrix}\right\}.$$

Clearly then $\mathfrak{m}_2^3\theta(f) \not\subset T\mathcal{K}\cdot f$. However, after multiplying both sides by $\mathfrak{m}_2$ one gets $\mathfrak{m}_2^4\theta(f) \subset \mathfrak{m}_2 T\mathcal{K}\cdot f$, whence f is indeed 3-determined. A few calculations with Newton diagrams shows that f has $\mathcal{K}$-codimension equal to 12.

Problem 13.8 We begin by writing $f(x,\mathbf{y}) = (g(x),\mathbf{y})$, where $g\colon (\mathbb{R},0) \to (\mathbb{R},0)$ and $\mathbf{y}\in\mathbb{R}^{n-1}$ (linearly adapted coordinates). There remains to classify g, and it is easy to check that f and g have the same $\mathcal{K}$-codimension. Now write the Taylor series of g. For some $\ell > 0$ we have

$$g(x) = ax^\ell + h(x)$$

with $a \neq 0$ and $h \in \mathfrak{m}_1^{\ell+1}$ (the remainder in the Taylor series). Then g is ℓ-determined by Theorem 13.2. Thus g is $\mathcal{K}$-equivalent to x^ℓ. Now, the function x^ℓ has $\mathcal{K}$-codimension $\ell-1$, and since we assume $d(f,\mathcal{K}) = k$ we have $\ell - 1 = k$, so f is $\mathcal{K}$-equivalent to $(x,y) \mapsto (x^{k+1}, y)$.

Chapter 14

Problem 14.1 Differentiation of f gives,

$$\mathrm{t}f(\theta(2)) = \left\langle \begin{pmatrix} 2x \\ y \\ 0 \end{pmatrix}, \begin{pmatrix} 0 \\ x \\ 2y \end{pmatrix} \right\rangle; \qquad I_f\theta(f) = \mathfrak{m}_2^2\theta(f).$$

A cobasis for $T\mathcal{K}_e \cdot f$ in $\theta(f)$ is (for example)

$$\left\{ \begin{pmatrix} 1 \\ 0 \\ 0 \end{pmatrix}, \begin{pmatrix} 0 \\ 1 \\ 0 \end{pmatrix}, \begin{pmatrix} 0 \\ 0 \\ 1 \end{pmatrix}, \begin{pmatrix} y \\ 0 \\ 0 \end{pmatrix}, \begin{pmatrix} 0 \\ 0 \\ x \end{pmatrix}, \begin{pmatrix} 0 \\ x \\ 0 \end{pmatrix}, \begin{pmatrix} 0 \\ y \\ 0 \end{pmatrix} \right\},$$

and $\operatorname{codim}(f, \mathcal{K}) = 7$. The versal deformation using this cobasis is,

$$F(x, y; r, s, t, a, b, c, d) = (x^2 + ay + r,\ xy + cx + dy + s,\ y^2 + bx + t).$$

The zero-set Z_F can be parametrized by x, y, a, b, c, d, and is of dimension 6. The projection $Z_F \to \mathbb{R}^7$ is

$$\pi_F(x, y, a, b, c, d) = (-x^2 - ay,\ -xy - cx - dy,\ -y^2 - bx, a, b, c, d),$$

and in this case Δ_F is the image of π_F, which is given by a lengthy polynomial of weighted degree 8 (obtained by elimination of x, y from the equations $F = 0$).

Problem 14.6 From equation (14.1) (on p. 93), we have

$$F(x;\ a, b, c, d) = (xy + a,\ x^2 + y^2 + bx + cy + d).$$

The zero-set Z_F is therefore

$$Z_F = \left\{(x, y, a, b, c, d) \in \mathbb{R}^2 \times \mathbb{R}^4 \mid a = -xy,\ d = -(x^2 + y^2 + bx + cy)\right\}.$$

This is a submanifold of dimension 4 since it can be written as a graph (with a, d as functions of (x, y, b, c)). By definition, Σ_F is the set of points in Z_F where f_u has rank less than 2. With $u = (a, b, c, d)$,

$$\mathrm{d}f_u(x, y) = \begin{pmatrix} y & x \\ 2x + b & 2y + c \end{pmatrix}.$$

Thus f_u is singular if the determinant is zero, so $y(2y + c) - x(2x + b) = 0$, or $2y^2 - 2x^2 + cy - bx = 0$ which is the equation of a hyperbola (for fixed b, c).

I'll leave the sketches to the reader, but note that when $b = \pm c$, the equation factorizes and the hyperbola degenerates to a pair of (orthogonal) lines; for example if $b = c$ then $2y^2 - 2x^2 + cy - bx = (y - x)(2y + 2x + c)$.

Problem 14.7 Very similar to previous question, but the curves are now ellipses in general.

Problem 14.11 Suppose $(x_0, u_0) \in Z_F \setminus \Sigma_F$, so $\mathsf{d}_x F(x, u)$ is invertible. It then follows from the implicit function theorem that the equation $F(x, u) = 0$ can be solved (in a neighbourhood V of (x_0, u_0)) for x as a function of u; call this function g. Then (close to (x_0, u_0))

$$(x, u) \in Z_F \iff x = g(u).$$

It follows that Z_F can be locally parametrized by u (and it is automatically a submanifold of dimension a), and in this parametrization, the projection π_F becomes simply $\pi_F(u) = u$, which is clearly a local diffeomorphism.

Problem 14.12 Choose linearly adapted coordinates for f, that is $f(x,y) = (\bar{f}(x,y), y)$, and this is contact equivalent to $(x,y) \mapsto (f_0(x), y)$ where $f_0(x) = \bar{f}(x,0)$ is the core of f. Thus f is $\mathcal{K}$-equivalent to a suspension of f_0.

Similarly, we can write $F(x,y,u) = (F_0(x,u), y)$. Since F is a regular unfolding, so is F_0, and we can use the implicit function theorem to choose the parameters so that $u = (v, s)$, with $F_0(x, v, s) = F_1(x, v) - s$. Then $Z_F = \{(x, v, s, y) \mid s = F_1(x, v),\ y = 0\}$, and then, using (x, v) as parameters for Z_F, $\pi_F(x, v) = (F_1(x, v), v)$. And this is $\mathcal{K}$-equivalent to $(x, v) \mapsto (F_1(x, 0), v) = (f_0(x), v)$ which is a suspension of f_0.

Problem 14.13 With $\varepsilon \neq 0$, change coordinates near the origin by $(x, y) \mapsto (X, y)$ with $X = \varepsilon x + x^2 + y^2$. Then f_ε becomes $(X, y) \longmapsto (X, h(X, y))$, for some smooth germ $h \in \mathcal{E}_2$ (depending on ε). This map is $\mathcal{C}$-equivalent to $(X, y) \mapsto (X, h(y, 0))$, and a local Taylor series calculation shows that $h(y, 0) = -\frac{1}{\varepsilon^3} y^6 + O(8)$, which means that for $\varepsilon \neq 0$, this has an A_5 singularity at the origin, as required.

Problem 14.18 Write $f_t = (g, h)$. Since the point $\mathbf{x}_t \neq (0, 0)$, there is a nighbourhood of $\mathbf{x}_t$ on which we can solve $g = 0$ for y as a function of x and then substitute for y into $h(x, y)$. One obtains

$$h(x,\ y(x)) \;=\; \frac{1}{x^b}(x + t^b)^{a+b}$$

In the neighborhhod in question, x^b is a unit, and therefore the map has a singularity of type A_{a+b-1} at $x = -t^b$ (and hence $y = -t^b$) as required. [*Note: the nonexistence of similar unfoldings for the* $\mathrm{II}_{a,b}$ *singularities is shown in* [68]. *On the other hand, it follows from this problem and Problem* 14.14 *that* $\mathrm{II}_{a,b}$ *contains* A_{a+b-2} *singularities in its versal unfolding.*]

Chapter 15

Problem 15.3 Let us consider the more general case of the ellipse

$$\gamma(s) = (\alpha \cos s,\ \beta \sin s),$$

with $\alpha \neq \beta$ and both positive (cf. Example 1.1). The normal N_s through the point $\gamma(s)$ is given by

$$(\mathbf{x} - \gamma(s)) \cdot \begin{pmatrix} -\alpha \sin s \\ \beta \cos s \end{pmatrix} = 0,$$

where the second vector is the tangent vector to the curve. Explicitly,

$$G(s,x,y) := -\alpha x \sin s + \beta y \cos s + \tfrac{1}{2}(\alpha^2 - \beta^2) \sin 2s = 0.$$

In the notation above, we have $a(s) = -\alpha \sin s$ and $b(s) = \beta \cos s$. The bifurcation points occur where $g(s) = g'(s) = 0$, which is satisfied for

$$x = \frac{(\alpha^2 - \beta^2)}{\alpha} \cos^3 s, \quad y = \frac{(\beta^2 - \alpha^2)}{\beta} \sin^3 s.$$

(This is a linear transformation of the astroid.) Finally, $g = g'' = 0$ implies $\sin 2s = 0$, so the cusps occur for s any integer multiple of $\pi/2$. And at these points (in fact at all points) $L_s \neq L_s'$ so they are genuine cusps (cf. Example 15.6), as depicted in Figure 1.6.

Problem 15.6 Differentiating this map (using an overdot for derivates with respect to t and a prime for those with respect to s, thus $\dot{\gamma}$ is tangent to the curve), we obtain,

$$\mathsf{d}F = (\gamma',\ \dot{\gamma}).$$

This 2×2 matrix is singular if and only if γ' and $\dot{\gamma}$ are linearly dependent.

On the other hand, differentiating the identity (15.2) one has,

$$\mathbf{n} \cdot \dot{\gamma} = 0, \quad \text{and} \quad \mathbf{n}' \cdot \gamma + \mathbf{n} \cdot \gamma' + c' = 0.$$

The first of these equations is equivalent to $\mathbf{n}$ being normal to the curve. The second tells us that $\frac{\partial G}{\partial s} = 0$ if and only if $\mathbf{n} \cdot \gamma' = 0$, which in turn is equivalent to γ' being parallel to $\dot{\gamma}$, as required.

Chapter 16

Problem 16.4 (i) Here use $\phi\colon \mathbb{R}^2 \to \mathbb{R}^3$ given by $\phi(x, y) = (x^2, xy, y^2)$. We leave the details to the reader (they are similar to (ii)).
(ii) Here we use $\phi(x, y) = (x^2, y^2)$. Then $\mathcal{E}_2/I_\phi \simeq \mathbb{R}\{1, x, y, xy\}$ and the required expression for f follows immediately from the preparation theorem.
(iii) If $f(x, y) = f(-x, -y)$ then, firstly, this implies

$$xh_1(x^2, xy, y^2) + yh_2(x^2, xy, y^2) = 0,$$

whence $f(x, y) = h_0(x^2, xy, y^2)$.

Similarly, $f(x, y) = f(-x, -y)$ also implies $xg_1(x^2, y^2) + yg_2(x^2, y^2) = 0$, whence $f(x, y) = g_0(x^2, y^2) + xyg_3(x^2, y^2)$, as required.

If $f(-x, -y) = -f(x, y)$ then a similar analysis shows,

$$\begin{aligned} f(x, y) &= xh_1(x^2, xy, y^2) + yh_2(x^2, xy, y^2), \\ f(x, y) &= xg_1(x^2, y^2) + yg_2(x^2, y^2). \end{aligned}$$

This shows that the h_j can be chosen independent of their second argument (if f is odd). [*Note*: *in (i), the three quadratic functions generate the ring of invariants of* $\mathbb{Z}_2$ *acting on* $\mathbb{R}^2$ *by* $\mathbb{Z}_2 = \{\pm I_2\}$. *On the other hand the polynomials in (ii) generate the ring of invariants for the natural action of* $\mathbb{Z}_2 \times \mathbb{Z}_2$ *on* $\mathbb{R}^2 = \mathbb{R} \times \mathbb{R}$ *(one copy of* $\mathbb{Z}_2$ *acting on each copy of* $\mathbb{R}$*).*]

Problem 16.7 First write the Taylor series of f to order 5:

$$f(t) = a_0 + a_1 t + \cdots + a_5 t^5 + t^6 f_1(t).$$

To apply the preparation theorem (to f_1) consider the map germ $\phi(t) = (t^3, t^4)$. Then $\{1, t, t^2, t^5\}$ is a cobasis for I_ϕ in $\mathcal{E}_1$. It follows that there are smooth functions $h_0, h_1, h_2, h_5 \in \mathcal{E}_2$ such that

$$f_1(t) = h_0(t^3, t^4) + th_1(t^3, t^4) + t^2 h_2(t^3, t^4) + t^5 h_5(t^3, t^4).$$

Then

$$t^6 f_1(t) = (t^3)^2 h_0(t^3,t^4) + (t^3)(t^4)h_1(t^3,t^4) + (t^4)^2 h_2(t^3,t^4) + (t^3)(t^4)^2 h_5(t^3,t^4).$$

which is a smooth function of t^3 and t^4, so let $h(t^3,t^4) = a_0 + a_3t^3 + a_4t^4 + t^6 f_1(t)$ (which is also a smooth function of t^3, t^4). Then $f(t)$ has the desired form.

Chapter 17

Problem 17.3 Since f and g are $\mathcal{K}$-equivalent, their versal unfoldings $F(x,u) = f(x) - u$ and $G(x,u) = g(x) - u$ are $\mathcal{K}_{\text{un}}$-equivalent. Consequently (by Martinet's first theorem), π_F and π_G are $\mathcal{A}$-equivalent. But $\pi_F = f$ and $\pi_G = g$.

Problem 17.4 Similarly to Example 17.5, there are three types: (i) the submersion $(x_1,\ldots,x_n) \mapsto (x_1,x_2)$, (ii) the fold $(x_1,\ldots,x_n) \mapsto (x_1, \sum_{j>1} \pm x_j^2)$, and the cusp $(x_1,\ldots,x_n) \mapsto (x_1, x_2^3 + x_1x_2 + \sum_{j>2} \pm x_j^2)$. It is simple to check in the third one that the singular set is a smooth curve and the discriminant is indeed a cusp.

Problem 17.6 Use $f\colon (\mathbb{R},0) \to (\mathbb{R}^n,0)$ given by $f(t) = (t^2,0,\ldots,0)$.

Chapter 18

Problem 18.2 Put $x = y + \beta/3$, $\lambda = -\mu - \beta^2/3$, $a = -\beta/3$ and $b = \alpha - \beta^3/27$.

Chapter 19

Problem 19.1 The projection to the x axis is the complement of the origin $\{c \in \mathbb{R} \mid x \neq 0\}$. To express this as a semialgebraic set (Definition 19.1), we write this set as $(\mathbb{R} \setminus S_x) \cup (\mathbb{R} \setminus S_{-x})$.

Problem 19.3 The module is generated by, $\theta_V = \mathcal{E}_n \left\{u_j \frac{\partial}{\partial u_j}\right\}_{j=1,\ldots,n}$.

Problem 19.5 The vector fields are tangent to the spheres $h(x, y, z) = r^2$ (where $r > 0$ is the radius).

Problem 19.8 To show $f \notin I_W$, proceed by contradiction: suppose there is a smooth germ g such that $f = gh$, and let U be a neighbourhood of the origin on which g has a representative. Restrict the equation to $z = a$ with $a < 0$, a constant for which $(0, 0, a) \in U$, and then restrict the equation $f = gh$ to the plane $z = a$. Put $b = \mathrm{e}^{1/a} > 0$. One obtains $f(x, y, a) = bx$ while $h(x, y, a) = ax^2 - y^2$, and therefore $g(x, y, a) = \frac{bx}{ax^2-y^2}$ which is evidently not smooth at $(x, y) = (0, 0)$.

Chapter 20

Problem 20.5 Just follow the definition.

Problem 20.7 The list is, $f(\lambda) = (\lambda, \lambda)$, (λ, λ^2), (λ, λ^3), and (λ^2, λ^2).

Problem 20.8 One sees that $T\mathcal{K}_V \cdot f \supset T\mathcal{R} \cdot f = \theta(f)$.

Chapter 21

Problem 21.2 One versal unfolding is

$$G(x; \lambda, \mu; a, b, c, d) = x^3 + \mu x + \lambda^3 + a\lambda\mu + b\lambda + c\mu + d = 0.$$

Problem 21.3 $T_{\mathrm{rel}}\mathcal{K}_\Delta \cdot h_s$ contains $\mathcal{E}_{1,I}\,\mu\theta(h)$, for $I = \mathbb{R}$, and $\dot{h}_s \in \mu\theta(h)$.

Chapter 22

Problem 22.1 Let G be the versal deformation of g_0:

$$G(t, u, r, s) = (t^2 - r, ut - s).$$

The discriminant of this unfolding is the Cayley cross-cap (see Problem 14.8). The module of vector fields tangent to this discriminant is given in Section 19.3.

Consider the map $\gamma\colon \mathbb{R}^3, 0 \to \mathbb{R}^3, 0$ given by

$$(u, r, s) = \gamma(\lambda, \mu, \nu) = (\lambda^2 \pm \mu,\ \nu,\ \mu).$$

The rest is left to the reader . . .

Chapter 23

Problem 23.8 (i) The path is $h(\lambda) = (\lambda^2, 0)$ which is of $\mathcal{K}_\Delta$-codimension 5 (which is too high for our classification in previous chapters, so it doesn't appear there). A versal unfolding of the path is,

$$H(\lambda; \alpha, \beta_0, \dots, \beta_3) = (\lambda^2 + \alpha,\ \beta_3\lambda^3 + \cdots + \beta_0).$$

The corresponding bifurcation problem is then

$$G(x, \lambda; \alpha, \beta_0, \dots, \beta_3) = x^3 + (\lambda^2 + \alpha)x + \beta_3\lambda^3 + \cdots + \beta_0 = 0.$$

(ii) As a problem with constraint, we refer to Example 23.5. The versal unfolding of the core is

$$G(x; u, v) = x^3 + ux^2 + vx,$$

and the discriminant Δ has equation $v(u^2 - 4v) = 0$. The path we are considering is therefore $h(\lambda) = (0, \lambda^2)$. Using the vector fields tangent to Δ given in (23.5), one finds

$$T\mathcal{K}_\Delta{\cdot}h = \mathcal{E}_2 \left\{ \begin{pmatrix} 0 \\ \lambda \end{pmatrix}, \begin{pmatrix} \lambda^2 \\ 0 \end{pmatrix} \right\}.$$

This has codimension 3, with unfolding

$$H(\lambda; \alpha, \beta, \gamma) = (\alpha\lambda + \beta,\ \lambda^2 + \gamma),$$

with resulting bifurcation

$$G(x; \lambda; \alpha, \beta, \gamma) = x^3 + (\alpha\lambda + \beta)x^2 + (\lambda^2 + \gamma)x = 0.$$

(iii) The cusp family is discussed in Example 23.10. For the $\mathcal{R}$-versal unfolding $F(x; u, v) = \frac{1}{4}x^4 + \frac{1}{2}ux^2 + vx$. The full bifurcation set $\mathcal{B}$ has equation $v(4u^3 + 27v^2) = 0$ with vector fields given on p. 311. The path in this case is $h(\lambda) = (\lambda, 0)$, which is

of infinite codimension. This is explained by the fact that the given family always has two critical points with equal critical values.
(iv) [Not asked in the question] If we modify (iii) by looking at the codimension of the gradient problem, we have $\nabla G = x^3 + ux + v$, and $\nabla g = x^3 + \lambda^2 x$, which therefore reduces to part (i) of this problem. (Here ∇ is just $\mathrm{d}/\mathrm{d}x$.)

Problem 23.5 One finds,

$$T\mathcal{K}_1^{\mathbb{Z}_2}\cdot g_0 = \mathcal{E}_2^+\left\{\begin{pmatrix}x^3\\0\end{pmatrix}, \begin{pmatrix}x^2y\\0\end{pmatrix}, \begin{pmatrix}xy^4\\0\end{pmatrix}, \begin{pmatrix}y^5\\0\end{pmatrix}, \begin{pmatrix}0\\x^5\end{pmatrix}, \begin{pmatrix}0\\x^4y\end{pmatrix}, \begin{pmatrix}0\\xy^2\end{pmatrix}, \begin{pmatrix}0\\y^3\end{pmatrix}\right\}$$

It follows that $(\mathfrak{m}_2^+)^2\theta(g)^- \subset T\mathcal{K}_1^{\mathbb{Z}_2}\cdot g_0$ and hence by Theorem 23.1 that g_0 is 3-determined. A cobasis for $T\mathcal{K}^{\mathbb{Z}_2}\cdot g_0$ in $\theta(g)^-$ consists of

$$\left\{\begin{pmatrix}x\\0\end{pmatrix}, \begin{pmatrix}y\\0\end{pmatrix}, \begin{pmatrix}xy^2\\0\end{pmatrix}, \begin{pmatrix}0\\x\end{pmatrix}, \begin{pmatrix}0\\y\end{pmatrix}, \begin{pmatrix}0\\x^2y\end{pmatrix}\right\}$$

so the $\mathcal{K}^{\mathbb{Z}_2}$-codimension is 6, and one can easily write down the versal unfolding. In the versal unfolding, the parameters that are coefficients of $(xy^2, 0)^T$ and $(0, x^2y)^T$ are moduli (or modal parameters). This is because the codimension of

$$(x^3 + axy^2,\ y^3 + bx^2y)$$

is independent of a, b; the symmetric singularity g_0 is therefore *bimodal*.

Appendix A

Problem A.1 Away from 0 the smoothness is clear; at 0, show by induction that for each k there is a polynomial p_k of degree $2k$ such that for $x > 0$, $f^{(k)}(x) = p_k(x^{-1})f(x)$, and then use the fact that for any polynomial $p(t)$ one has $\lim_{t\to\infty} p(t)e^{-t} = 0$.

Problem A.4 For any $a \in \mathbb{R}$ we have $f(a+u) = (a+u)^3$ so

$$f(a+u) = a^3 + 3a^2u + 3au^2 + u^3.$$

For j^2 we only use the terms up to u^2, so we ignore the final u^3 term, giving the answer required. For the second part, $f(a+u) = \sin(a+u) = \sin a \cos u + \cos a \sin u$. Expanding these to terms in u^2 gives

$$j^2 f_a(u) = \sin a + (\cos a)u - \tfrac{1}{2}(\sin a)u^2.$$

Appendix B

Problem B.1 Use the inverse function theorem: we need $f'(x) \neq 0$. Here, $f'(x) = -\sin x$ and this is non-zero provided $x \neq n\pi$ for any $n \in \mathbb{Z}$. [Draw a picture: these 'bad' points are precisely the points where $\cos(x)$ has a critical point.]

Problem B.2 Let U be the neighbourhood of 0 on which ϕ is a diffeomorphism, and V a neighbourhood of $\phi(0)$ on which ψ is a diffeomorphism (with $\psi(V_1)$). Now let $V_1 = V \cap \phi(U)$. Then ψ restricted to V_1 is also a diffeomorphism. Let $U_1 = \phi^{-1}(V_1) \subset U$. This is an open set, so $\phi\colon U_1 \to V_1$ is a diffeo, and $\psi\colon V_1 \to W_1 := \psi(V_1)$ is a diffeo. Therefore $\psi \circ \phi\colon U_1 \to W_1$ is a diffeo because (a) it is smooth (the composite of smooth maps is smooth, and (b) its inverse is defined $(\psi \circ \phi)^{-1} = \phi^{-1} \circ \psi^{-1}\colon W_1 \to U_1$.

Problem B.4 Clearly, $\mathsf{d}(f_a)(0) = 1$ so the inverse function theorem applies. How large is U? This depends on a. Indeed, writing $y = f_a(x)$ and solving for x (it's a quadratic equation) gives U to be the interval $U = \left(-\frac{1}{2a}, \infty\right)$, and so $V = f(U) = \left(-\frac{1}{4a}, \infty\right)$. If one makes U any larger, the function will not be a diffeomorphism (it won't be one-to-one).

Problem B.5 $\mathsf{d}f_{(x,y)} = \begin{pmatrix} 2x & 0 \\ 0 & 2y \end{pmatrix}$, and this fails to be invertible iff $xy = 0$, so the singular set is the union of the two axes. The image of the singular set is therefore the union of the two *positive* axes in $\mathbb{R}^2$, and the image of f is the positive quadrant.

Problem B.8 Let $v \in f(U)$, so $v = f(u)$ for some $u \in U$. Since the map is a submersion, there is a nbhd U' of u in U and coordinates on U' about u such that $f(x_1, \ldots, x_n) = (x_1, \ldots, x_p)$. Now consider a small ε-ball $B_\varepsilon(u)$ about u (in these coordinates). Then $f(B_\varepsilon(u) = B_\varepsilon(v)$ (why?), so that $f(U)$ contains this nbhd of v. This is true for every $v \in f(U)$ so that $f(U)$ is indeed open.

Problem B.11 The linear part (differential at 0) of f is $(x, y) \mapsto (x, 0, y)$. The first step is to change coordinates in $\mathbb{R}^3$ simply by swapping the last two components: write $\psi(X, Y, Z) = (X, Z, Y)$. Then

$$\psi \circ f(x, y) = (x, y + x^2, xy).$$

Now we change coordinates in the source $\mathbb{R}^2$ in order that the second component of the map (i.e. $y+x^2$) becomes simply y. So let $\bar{y} = y+x^2$ (equivalently, $y = \bar{y}-x^2$). Then $x = x$, $y+x^2 = \bar{y}$ and $xy = x(\bar{y}-x^2)$. If we write $(x,y) = \phi(x,\bar{y}) = (x,\bar{y}-x^2)$ then f becomes

$$\psi \circ f \circ \phi(x,\bar{y}) = (x,\bar{y},x(\bar{y}-x^2)).$$

That is, in these new coordinates, f takes the form $(x,v) \mapsto (x,v,xv-x^3)$.

Problem B.15 Consider the map $f\colon \mathbb{R}^3 \times \mathbb{R}^3 \to \mathbb{R}^3$ given by

$$f(\mathbf{u},\mathbf{v}) = \left(\|\mathbf{u}\|^2,\ \|\mathbf{v}\|^2,\ \mathbf{u}\cdot\mathbf{v}\right).$$

Let $y = (1,1,0) \in \mathbb{R}^3$. We are interested in the set $X = f^{-1}(y)$. Now the differential $\mathsf{d}f$ at $(\mathbf{u},\mathbf{v})$ is the 3×6 matrix,

$$\mathsf{d}f = \begin{pmatrix} 2\mathbf{u}^T & 0 \\ 0 & 2\mathbf{v}^T \\ \mathbf{v}^T & \mathbf{u}^T \end{pmatrix}.$$

This matrix is of rank 3 at $(\mathbf{u},\mathbf{v})$ if and only if $\mathbf{u}$ and $\mathbf{v}$ are linearly independent. Now, at every point $(\mathbf{u},\mathbf{v}) \in X$ the vectors $\mathbf{u}$ and $\mathbf{v}$ are indeed linearly independent, whereby the result follows from the regular value theorem (Theorem B.23).

Problem B.18 We have $f(x,y,z,\lambda) = (z+x^2-\lambda y,\ y+z-2\lambda x,\ z-\lambda)$. Then, with $\lambda = 0$

$$\mathsf{d}(f_0)_0 = \begin{pmatrix} 0 & 0 & 1 \\ 0 & 1 & 1 \\ 0 & 0 & 1 \end{pmatrix}.$$

Then the kernel is $K = \mathbb{R}\{e_1\}$, range is $R = \mathbb{R}\{e_1+e_3,\ e_2\}$. We can identify the cokernel Q with any complement to R and Y with any complement to K; we take $Q = \mathbb{R}\{e_1\}$ and $Y = \mathbb{R}\{e_2,e_3\}$. Then $\Pi\colon \mathbb{R}^3 \to Q$ is the projection $\Pi(u,v,w) = (u-w,0,0)$ (which indeed has kernel R), whence $(I-\Pi)(u,v,w) = (w,v,w)$. Then defining f_1 and f_2 by (B.3) gives

$$f_1(x,y,z,\lambda) = (z-\lambda, y+z-2\lambda, z-\lambda), \quad f_2(x,y,z,\lambda) = (x^2-\lambda y+\lambda,0,0)$$

Step (i): Solving $f_1 = 0$ can be done explicitly, to get $z = \lambda$, $y = -\lambda + 2\lambda x$.
Step (ii): substitute these expressions for y and z into f_1 to get

$$g(x,\lambda) = x^2 - 2\lambda^2 x + \lambda^2 + \lambda.$$

In general, step (i) can only be done implicitly, and using Taylor series one can obtain solutions to any order in (x,λ).

Appendix C

Problem C.1 This follows from the chain rule:

$$\frac{\mathrm{d}}{\mathrm{d}t}(f \circ \phi_t(x)) = \mathrm{d}f_{\phi_t(x)} \frac{\mathrm{d}}{\mathrm{d}t}(\phi_t(x))$$

But $\frac{\mathrm{d}}{\mathrm{d}t}(\phi_t(x)) = \mathbf{v}_t(\phi_t(x))$ by definition, and consequently,

$$\frac{\mathrm{d}}{\mathrm{d}t}(f \circ \phi_t(x)) = \mathrm{d}f_{\phi_t(x)} \mathbf{v}_t(\phi_t(x)) = \mathrm{t}f(\mathbf{v})(\phi_t(x)).$$

Appendix D

Problem D.4 Clearly if $I \subset J$ then each element $f_i \in J$. Conversely, suppose all $f_i \in J$ (for $i = 1, \dots, r$). To show $I \subset J$, let $g \in I$. Then g can be written as $g = \sum_{i=1}^r f_i h_i$ for some $h_i \in R$. But since the $f_i \in J$ it follows from the definition of ideal that $g \in J$. Thus $I \subset J$.

Problem D.8 Let $f \in \Phi$. Since $f(0) = 0$ we can write $f(x) = xg(x)$ (by Hadamard's lemma), for some smooth germ g. It is clear that the Taylor series of g also vanishes. It follows that $\Phi \subset \mathfrak{m}_1\Phi$, and if Φ were finitely generated then we would deduce from Nakayama's lemma (in its alternative version from Problem D.7) that $\Phi = 0$. However $f(x) = \exp(-1/x^2)$ is in Φ so it is non-zero. Hence Φ is not finitely generated.

Problem D.11 One finds, $J_1 = \mathfrak{m}_2$ and

$$\begin{aligned} J_2 &= \langle x, y \rangle & J_3 &= \langle x^2, y \rangle \\ J_4 &= \langle x^2, xy, y^2 \rangle & J_5 &= \langle x^3, xy, y^2 \rangle \\ J_6 &= \langle x^3, x^2y, y^2 \rangle \end{aligned}$$

Then $J_3^2 = \langle x^4, x^2y, y^2 \rangle \neq J_6$.

Problem D.15 Clearly V consists of just the origin. Thus $I(V) = \langle x, y \rangle$. However, $\sqrt{I} = I$ (since over $\mathbb{R}$, $x^2 + y^2$ does not have any non-trivial divisors), and so $I(V) \neq \sqrt{I}$.

References

[1] V.I. Arnold, *Ordinary Differential Equations*. MIT Press, 1978.

[2] V.I. Arnold, *Catastrophe Theory*. Springer. 1998.

[3] V.I. Arnold, Critical points of smooth functions and their normal forms. *Russ. Math. Surv.* **30**:5 (1975), 1–75.

[4] V.I. Arnold, Critical points of functions on a manifold with boundary, the simple Lie groups B_k, C_k, and F_4 and singularities of evolutes. *Russ. Math. Surv.* **33**:5 (1978), 99–116.

[5] V.I. Arnold, V. Goryunov, O.V. Lyashko & V.A. Vasil'ev, *Singularity Theory*, Vol. I. Springer. 1993.

[6] V.I. Arnold, S. Gusein-Zade & A. Varchenko, *Singularities of Differentiable Maps*, Vol. 1. Birkauser, 1985.

[7] M.F. Atiyah & I.G. MacDonald, *Introduction to Commutative Algebra*. Addison-Wesley, 1969.

[8] T. Banchoff, T. Gaffney & C. McCrory. *Cusps of Gauss Mappings*. Pitman, 1982. Available online at https://www.emis.de/monographs/CGM

[9] M.V. Berry, Waves and Thom's theorem. *Adv. Phys.* **25** (1976), 1–26.

[10] E. Bierstone, Local properties of smooth maps equivariant with respect to finite group actions. *J. Diff. Geom.* **10** (1975), 523–540.

[11] T. Bridges & J.E. Furter, *Singularity Theory and Equivariant Symplectic Maps*. Lecture Notes Math. **1558**. Springer, 1993.

[12] Th. Bröcker, *Differentiable Germs and Catastrophes*. Lond. Math. Soc. Lecture Note Series **17**. Cambridge University Press, 1975.

[13] J.W. Bruce, Functions on discriminants. *J. Lond. Math. Soc.* **30** (1984), 551–567.

[14] J.W. Bruce, Vector fields on discriminants and bifurcation varieties. *Bull. Lond. Math. Soc.* **17** (1985), 257–262.

[15] J.W. Bruce, Classifications in singularity theory and their applications. In *New Developments in Singularity Theory*. NATO Sci. Ser. II Math. Phys. Chem., 21, Kluwer, 2001, 3–33.

[16] J.W. Bruce & T.J. Gaffney, Simple singularities of mappings $\mathbb{C}, 0$ to $\mathbb{C}^2, 0$. *J. Lond. Math. Soc.* **26** (1982) 465–474.

[17] J.W. Bruce, T. Gaffney, & A.A du Plessis, On left equivalence of map germs. *Bull. Lond. Math. Soc.* **16** (1984), 303–306.

[18] J.W. Bruce & P.J. Giblin, *Curves and Singularities*, 2nd ed. Cambridge University Press, 1992.

[19] J.W. Bruce, P.J. Giblin & F. Tari, Families of surfaces: focal sets, ridges and umbilics, *Math. Proc. Camb. Phil. Soc.* **125** (1999), 243–268.

[20] J.W. Bruce, N.P. Kirk & A.A. du Plessis, Complete transversals and the classification of singularities. *Nonlinearity* **10** (1997), 253–275.

[21] J.W. Bruce, A.A du Plessis, & C.T.C. Wall, Determinacy and unipotency. *Invent. Math.* **88** (1987), 521–554.

[22] J.W. Bruce & M. Roberts, Critical points of functions on analytic varieties. *Topology* **27** (1988), 57–90.

[23] M. do Carmo, *Differential Geometry of Curves and Surfaces*. Prentice–Hall, 1976.

[24] T. Cooper, D. Mond & R. Wik Atique, Vanishing topology of codimension 1 multi-germs over $\mathbb{R}$ and $\mathbb{C}$. *Compositio Mathematica* **131** (2002), 121–160.

[25] D. Cox, J. Little & D. O'Shea, *Using Algebraic Geometry*, 2nd ed. Springer, 2005.

[26] D. Cox, J. Little & D. O'Shea, *Ideals, Varieties, and Algorithms*, 3rd ed. Springer, 2007.

[27] J. Damon, The unfolding and determinacy theorems for subgroups of $\mathcal{A}$ and $\mathcal{K}$. *Mem. Am. Math. Soc.* **50** (1984), 306.

[28] J. Damon, Deformations of sections of singularities and Gorenstein surface singularities. *Am. J. Math.* **109** (1987), 695–721.

[29] J. Damon, $\mathcal{A}$-equivalence and the equivalence of sections of images and discriminants. In *Singularity Theory and Its Applications, Part I*, Springer Lecture Notes **1462**, 1991, 93–121.

[30] A.A. Davydov, G. Ishikawa, S. Izumiya & W.-Z. Sun, Generic singularities of implicit systems of first order differential equations on the plane. *Japan. J. Math.* **3** (2008), 93–119.

[31] A. Dimca, *Topics on Real and Complex Singularities*. Springer, 1987.

[32] A. Dimca, *Singularities and Topology of Hypersurfaces*. Springer, 1992.

[33] P.G. Drazin, *Nonlinear Systems*. Cambridge University Press, 1992.

[34] W. Ebeling, *Functions of Several Complex Variables and Their Singularities*, Graduate Stud. Math. **83**, AMS, 2007.

[35] J. Eggers & N. Suramlishvili, Singularity theory of plane curves and its applications. *Eur. J. Mech. B/Fluids* **65** (2017), 107–131.

[36] D. Eisenbud, *Commutative Algebra, with a View toward Algebraic Geometry*. Springer, 2002.

[37] J.-E. Furter & A.M. Sitta, Nondegenerate umbilics, the path formulation and gradient bifurcation problems. *Int. J. Bifur. Chaos* **19** (2009), 2965–77.

[38] J.-E. Furter, A.M. Sitta & I.N. Stewart, Algebraic path formulation for equivariant bifurcation problems. *Math. Proc. Camb. Phil. Soc.* **124** (1998), 275–304.

[39] J.-E. Furter, A.M. Sitta & I.N. Stewart, Singularity theory and equivariant bifurcation problems with parameter symmetry. *Math. Proc. Camb. Phil. Soc.* **120** (1998), 547–578.

[40] T. Gaffney, A note on the order of determination of a finitely determined germ. *Invent. Math.* **52** (1979), 127–135.

[41] C.G. Gibson, *Singular Points of Smooth Mappings*. Pitman, 1979.

[42] C.G. Gibson, K. Wirthmuller, A.A. du Plessis & E.J.N. Looijenga, *Topological Stability of Smooth Mappings*. Lecture Notes Math., **552**. Springer, 1976.

[43] C.G. Gibson & C.A.Hobbs Simple singularities of space curves. *Math. Proc. Camb. Phil. Soc.* **113** (1993), 297–310.

[44] R. Gilmore, *Catastrophe Theory for Scientists and Engineers*. Dover, 1993.

[45] M. Golubitsky & V. Guillemin, *Stable Mappings and Their Singularities*. Springer, 1974.

[46] M. Golubitsky & W.F. Langford, Classification and unfoldings of degenerate Hopf bifurcations. *J. Diff. Eqns.* **41** (1981), 375–415.

[47] M. Golubitsky, J. Marsden, I. Stewart, & M. Dellnitz, The constrained Lyapunov–Schmidt procedure and periodic orbits, *Fields Inst. Comm.*, **4** (1995), 81–127.

[48] M. Golubitsky & D. Schaeffer, A theory for imperfect bifurcation theory via singularity theory. *Comm. Pure Appl. Math.* **32** (1979), 21–98,
Imperfect bifurcation in the presence of symmetry. *Comm. Math. Phys.* **67** (1979), 205–232.

[49] M. Golubitsky & D. Schaeffer, *Singularities and Groups in Bifurcation Theory*, Vol 1. Springer, 1985.

[50] M. Golubitsky, I. Stewart & D. Schaeffer, *Singularities and Groups in Bifurcation Theory*, Vol 2. Springer, 1988.

[51] M. Golubitsky & I. Stewart, *The Symmetry Perspective*. Birkhauser, 2002.

[52] D.R. Grayson & M.E. Stillman, *Macaulay2, a software system for research in algebraic geometry.* Available online at https://faculty.math.illinois.edu/Macaulay2/

[53] G.-M. Greuel, C. Lossen & E. Shustin, *Introduction to Singularities and Deformations*. Springer, 2007.

[54] J. Guckenheimer, Review of [111]. *Bull. Am. Math. Soc.* **79** (1973), 878–890.

[55] J. Guckenheimer & P. Holmes, *Nonlinear Oscillations, Dynamical Systems and Bifurcations of Vector Fields*. Springer, 1983.

[56] V. Guillemin & A. Pollack, *Differential Topology*. Prentice Hall, 1974.

[57] M. Hirsch, *Differential Topology*. Springer, 1976.

[58] L. Hörmander, *The Analysis of Linear Partial Differential Operators I. Distribution Theory and Fourier Analysis*. Springer, 1983.

[59] K. Houston, On the classification of complex multi-germs of corank one and codimension one. *Math. Scand.* **96** (2005), 203–222.

[60] K. Ikeda & K. Murota, *Imperfect Bifurcation in Structures and Materials*, 2nd ed. Appl. Math. Sci. **149**, Springer (2010).

[61] S. Izumiya, M.C. Romero Fuster, M. Ruas & F. Tari, *Differential Geometry from a Singularity Theory Viewpoint*. World Scientific, 2016.

[62] S. Janeczko & M. Roberts, Classification of symmetric caustics II: caustic equivalence. *J. Lond. Math. Soc.* **48** (1993), 178–192.

[63] K. Jänich, Caustics and catastrophes. *Math. Ann.* **209** (1974), 161–180.

[64] W.P. Johnson, The curious history of Faà di Bruno's formula. *Am. Math. Monthly* **109** (2002), 217–234.

[65] T. de Jong & G. Pfister, *Local Analytic Geometry*. Vieweg, 2000.

[66] S.G. Krantz & H.R. Parks, *The Implicit Function Theorem, History, Theory, and Applications*. Birkhäuser, 2003.

[67] Y.A. Kuznetsov, *Elements of Applied Bifurcation Theory*, 2nd ed. Springer, 1998.

[68] L. Lander, The structure of the Thom–Boardman singularities of stable classes with type $\Sigma^{2,0}$. *Proc. Lond. Math. Soc.* **33** (1976), 113–137.

[69] A. Lari-Lavassani & Y-C. Lu, The stability theorems for subgroups of $\mathcal{A}$ and $\mathcal{K}$. *Can. J. Math.* **46** (1994), 995–1006.

[70] H. Levine, Singularities of differentiable maps. (Notes on lectures by R. Thom, originally published as mimeographed notes by the University of Bonn, 1960). Reprinted in [114], 1–89.

[71] E. Looijenga, *Isolated Singular Points on Complete Intersections*. Lond. Math. Soc. Lecture Notes **77**. Cambridge University Press, 1984.

[72] B. Malgrange, *Ideals of Differentiable Functions*. Oxford University Press, 1967.

[73] J. Martinet, Déploiements versels des applications différentiables et classification des applications stables. In: Burlet O., Ronga F. (eds) *Singularités d'Applications Différentiables*. Lecture Notes Math. **535**. pp. 1–44. Springer 1976.

[74] J. Martinet, Déploiements stables des germes de type fini, et détermination finie des applications différentiables. *Bol. Soc. Bras. Mat.* **7** (1976), 89–109.

[75] J. Martinet, *Singularities of Smooth Functions and Maps*. Lond. Math. Soc. Lecture Note Series **58**. Cambridge University Press, 1982.

[76] J.N. Mather, Stability of C^∞ mappings: **I**. The division theorem. *Ann. Math.* **87**:2 (1968), 89–104; **II**. Infinitesimal stability implies stability. *Ann. Math.* **89**:2 (1969)

254–291; **III**. Finitely determined map germs. *Publ. Math. Inst. Hautes Etud. Sci.* **35** (1969), 127–156; **IV**. Stability of C^∞ mappings IV: Classification of stable germs by R-algebras. *Publ. Math. Inst. Hautes Etud. Sci.* **37** (1969), 223–248; **V**. Transversality. *Adv. Math.* **4** (1970), 301–336; **VI**. The nice dimensions. In *Proceedings of Liverpool Singularities Symposium, I*. Lect. Notes Math., **192**, Springer, 1971, 207–253.

[77] J. Milnor, *Morse Theory*, Princeton University Press, 1963.

[78] J. Milnor, *Singular Points of Complex Hypersurfaces*, Princeton University Press, 1968.

[79] D. Mond, On the classification of germs of maps from $\mathbb{R}^2$ to $\mathbb{R}^3$. *Proc. Lond. Math. Soc.* **50** (1985), 333–369.

[80] D. Mond & J. Montaldi, Deformations of maps on complete intersections, Damon's $\mathcal{K}_V$-equivalence and bifurcations. In *Singularities*, Lond. Math. Soc. Lecture Note Ser. **201**, Cambridge University Press, 1994, 263–284.

[81] D. Mond & J.J. Nuño-Ballesteros, *Singularities of Mappings*. Springer, 2020.

[82] J. Montaldi, *Contact, with applications to submanifolds of* $\mathbb{R}^n$. PhD thesis, University of Liverpool, 1983. Available online at https://livrepository.liverpool.ac.uk/3052146/

[83] J. Montaldi, On contact between submanifolds. *Mich. Math. J.* **33** (1986), 195–199.

[84] J. Montaldi, Caustics in time reversible Hamiltonian systems. In *Singularity Theory and its Applications, Part II*. Lecture Notes Math. **1463**, Springer, 1991, 266–277.

[85] J. Montaldi, The path formulation of bifurcation theory. In *Dynamics, Bifurcation and Symmetry*. NATO Adv. Sci. Inst. Ser. C Math. Phys. Sci. **437**, Kluwer, 1994, 259–278.

[86] J. Montaldi, Multiplicities of critical points of invariant functions. *Mat. Contemp.* **5** (1993), 93–135.

[87] J. Montaldi, Bifurcations of relative equilibria near zero momentum in Hamiltonian systems with spherical symmetry. *J. Geom. Mech* **6** (2014), 237–260.

[88] J. Montaldi & D. van Straten, Quotient spaces and critical points of invariant functions for $\mathbb{C}^*$-actions. *J. Reine Angew. Math.* **437** (1993), 55–99.

[89] M.D. Neusel & L. Smith, *Invariant Theory of Finite Groups*, Math. Surv. Monogr. **94**, AMS, 2002.

[90] V.P. Palamodov, Multiplicity of holomorphic mappings. *Funct. Anal. Appl.* **1** (1967), 218–226.

[91] M. Peters, Classification of two–parameter bifurcations. In *Singularity Theory and Its Applications, Part II*, Lecture Notes Math. **1463**, Springer, 1991, 294–300..

[92] M. Peters, *Classification of two–parameter bifurcations*. PhD thesis, University of Warwick, 1991. Available online at https://webcat.warwick.ac.uk/record=b3226436~S15

[93] A.A. du Plessis, T. Gaffney & L.C. Wilson, Map germs determined by their discriminants. In *Stratifications, Singularities and Differential Equations, I* Travaux en Cours **54**, Hermann, 1997, 1–40.

[94] A.A. du Plessis & C.T.C. Wall, *The Geometry of Topological Stability*. Clarendon Press, 1995.

[95] I.R. Porteous, Probing singularities. In *Singularities*. Proc. Symposia in Pure Math. **40** Part II. AMS, 1983, 395–406.

[96] I.R. Porteous, *Topological Geometry*. 2nd ed. Cambridge University Press, 1981.

[97] I.R. Porteous, *Geometric Differentiation: For the Intelligence of Curves and Surfaces*, 2nd ed. Cambridge University Press, 2001.

[98] T. Poston & I. Stewart, *Catastrophe Theory and Applications*. Dover, 1996. (Original edition Prentice Hall, 1978.)

[99] R.M. Roberts, Equivariant Milnor numbers and invariant Morse approximations. *J. Lond. Math. Soc.* **31**:2 (1985), 487–500.

[100] K. Saito, Quasihomogene isolierte Singularitäten von Hyperflächen. *Invent. Math.* **14** (1971), 123–142.

[101] K. Saito, Theory of logarithmic differential forms and logarithmic vector fields. *J. Fac. Sci. Univ. Tokyo*, Section 1A (Mathematics), **27** (1980), 265–281.

[102] P.T. Saunders, *An Introduction to Catastrophe Theory*. Cambridge University Press, 1980.

[103] G.W. Schwarz, Smooth functions invariant under the action of a compact Lie group. *Topology* **14** (1975), 63–68.

[104] J.-P. Serre, Géométrie algébrique et géométrie analytique. *Ann. l'Inst. Fourier* **6** (1956), 1–42.

[105] I.N. Stewart, Applications of catastrophe theory to the physical sciences. *Phys. D* **2** (1981), 245–305.

[106] I.N. Stewart, Beyond elementary catastrophe theory. *Appl. Math. Comp.* **14** (1984), 25–31.

[107] Surfer. Software for the visualization of algebraic surfaces. Available online at https://imaginary.org/program/surfer

[108] J.L. Taylor, *Several Complex Variables with Connections to Algebraic Geometry and Lie Groups*. Graduate Stud. Math. **46**, AMS, 2002.

[109] H. Terao, The bifurcation set and logarithmic vector fields. *Math. Ann.* **263** (1983), 313–321.

[110] R. Thom, Les singularités des applications différentiables. *Ann. Inst. Fourier* **6** (1956), 43–87.

[111] R. Thom, *Structural Stability and Morphogenesis*. W. A. Benjamin, 1975. (Translation of *Stabilité Structurelle et Morphogenèse*, (same publishers) 1972.)

[112] F. TREVES, *Introduction to Pseudodifferential and Fourier Integral Operators*, Vol. 2. Plenum Press, 1980.

[113] B.L. VAN DER WAERDEN, *Algebra*, Vol. 1. Springer, 1991.

[114] C.T.C. WALL (ed.), *Proceedings of Liverpool Singularities Symposium, I (1969/70)*. Lecture Notes Math. **192**, Springer, 1971.

[115] C.T.C. WALL, Finite determinacy of smooth map germs. *Bull. Lond. Math. Soc.* **13** (1981), 481–539.

[116] H. WHITNEY, The general type of singularity of a set of $2n-1$ smooth functions of n variables. *Duke Math. J.* **10** (1943), 161–172.

[117] H. WHITNEY, The singularities of a smooth n-manifold in $(2n-1)$-space. *Ann. Math.* **45** (1944), 247–293.

[118] H. WHITNEY, On singularities of mappings of Euclidean spaces. I. Mappings of the plane into the plane. *Ann. Math.* **62** (1955), 374–410.

[119] R. WIK ATIQUE, On the classification of multi-germs of maps from $\mathbb{C}^2$ to $\mathbb{C}^3$ under $\mathcal{A}$-equivalence. In *Real and Complex Singularities*, Chapman & Hall/CRC, 2000, 119–133.

[120] E.C. ZEEMAN, Catastrophe theory. *Sci. Am.* **234** (1976), 65–83.

[121] E.C. ZEEMAN, *Catastrophe Theory. Selected Papers, 1972–1977*. Addison-Wesley, 1977.

Index

Figures are in **bold**. Individual singularity or bifurcation types are listed under 'singularity type' or 'bifurcation type' respectively